GCSE

Suitable for all exam boards

Maths in a year

Dave Capewell
Formerly Westfield School, Sheffield
Peter Mullarkey
Netherhall School, Maryport, Cumbria
Katherine Pate
Maths Publishing Consultant

OXFORD
UNIVERSITY PRESS

Great Clarendon Street, Oxford OX2 6DP

Oxford University Press is a department of the University of Oxford.
It furthers the University's objective of excellence in research, scholarship,
and education by publishing worldwide in

Oxford New York

Auckland Cape Town Dar es Salaam Hong Kong Karachi
Kuala Lumpur Madrid Melbourne Mexico City Nairobi
New Delhi Shanghai Taipei Toronto

With offices in

Argentina Austria Brazil Chile Czech Republic France Greece
Guatemala Hungary Italy Japan South Korea Poland Portugal
Singapore Switzerland Thailand Turkey Ukraine Vietnam

First published 2007

British Library Cataloguing in Publication Data

Data available

ISBN 978 019 915156 1

10 9 8 7 6 5 4 3

Printed in China by Printplus

Paper used in the production of this book is a natural,
recyclable product made from wood grown in sustainable forests.
The manufacturing process conforms to the environmental
regulations of the country of origin.

Acknowledgements

The Publisher would like to thank Nicola Fleet and Brian Jefferson for their
authoritative guidance in preparing this book.

The image on the cover is reproduced courtesy of ImageDJ.

The Publisher would like to thank the following for permission to
reproduce photographs:
P11 Oxford University Press; **p17** Paul Doyle/Alamy; **p81** Science Photo
Library; **p86** Brendan Regan/Corbis; **p127** TopFoto UK Ltd; **p135** Oxford
University Press; **p149** Hemera/Oxford University Press; **p150** Photodisc/
Oxford University Press; **p155** Oxford University Press; **p176** NASA/Oxford
University Press; **p182** Oxford University Press; **p195** Oxford University
Press; **p197** The Photolibrary Wales/Alamy; **p243** John La Gette/Alamy;
p285 India Images/Dinodia Images/Alamy; **p287** Stockfolio/Alamy; **p304**
World Pictures Ltd/Alamy; **p308** Oxford University Press; **p315** Andrew
Syred/Science Photo Library; **p316l** Bob Croxford/Atmosphere Picture
Library/Alamy; **p316r** Ordnance Survey.

Contents

About this book

This book has been specifically produced to help you get your best possible grade in your GCSE Mathematics examinations. It is designed for students who are on a one year course and want to achieve a grade C.

The authors are experienced teachers and consultants who have an excellent understanding of the GCSE specifications and so are well qualified to help you successfully meet your objectives.

The book is made up of units that are based on the specifications of all the UK exam boards and provide coverage of the National Curriculum strands at Key Stage 4.

The units are:

Each unit contains double page spreads for each lesson. These are shown on the full contents list.

Problem solving is integrated throughout the material as suggested in the National Curriculum.

How to use this book

This book is made up of units of work that are colour-coded: Number (orange), Algebra (green), Shape, space and measures (pink), and Data (blue).

Each unit starts with an overview of the content, so that you know exactly what you are expected to learn.

This unit will show you how to

- Understand place value and order numbers
- Multiply and divide numbers by powers of ten
- Represent decimal numbers as positions on a number line
- Read scales, dials and timetables
- Order negative numbers using a number line

The first page of a unit also provides prior knowledge questions to help you revise before you start – then you can apply your knowledge later in the unit:

Before you start ...

You should be able to answer these questions.

1 Put these numbers in order from lowest to highest.
0.37 0.4 0.312 0.35

2 What number is the arrow pointing to?

Inside each unit, the content develops in double page spreads that all follow the same structure.

Each spread starts with a list of the learning outcomes and a summary of the keywords:

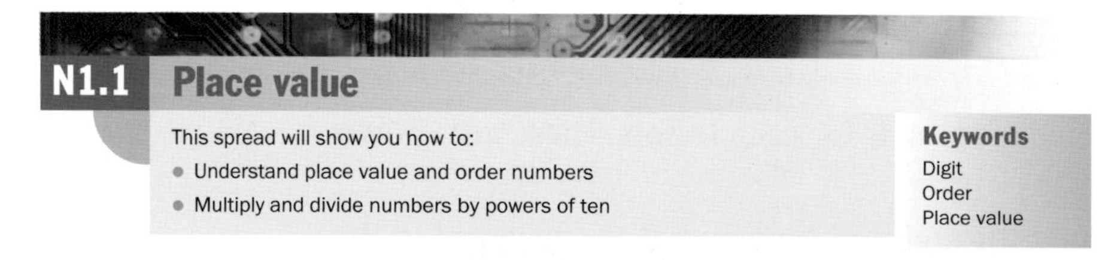

Key points are highlighted in the text so you can see the facts you need to learn:

- Area of triangle = $\frac{1}{2}\times$ base $\times$ height

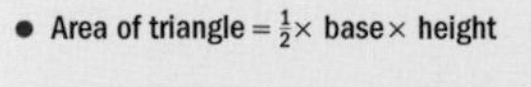

The **height** must be **perpendicular** to the **base**.

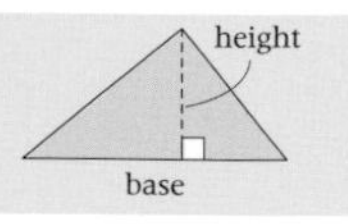

Perpendicular means at right angles.

Examples showing the key skills and techniques you need to develop are shown in boxes. Also, margin notes show tips and reminders you may find useful:

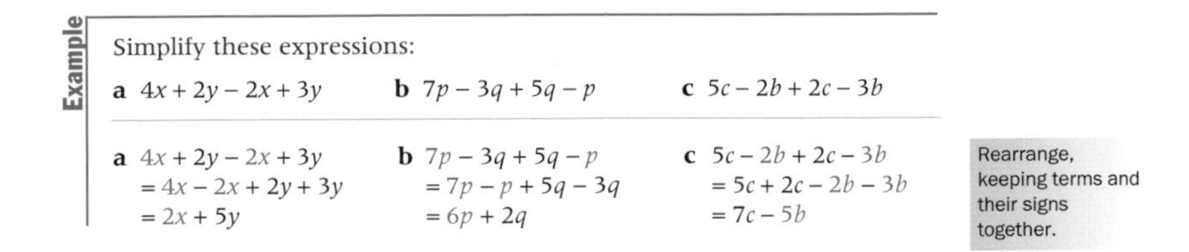
Example

Simplify these expressions:

a $4x + 2y - 2x + 3y$ **b** $7p - 3q + 5q - p$ **c** $5c - 2b + 2c - 3b$

a $4x + 2y - 2x + 3y$
$= 4x - 2x + 2y + 3y$
$= 2x + 5y$

b $7p - 3q + 5q - p$
$= 7p - p + 5q - 3q$
$= 6p + 2q$

c $5c - 2b + 2c - 3b$
$= 5c + 2c - 2b - 3b$
$= 7c - 5b$

Rearrange, keeping terms and their signs together.

Each exercise is carefully graded, set at three levels of difficulty:

- The first few questions are mainly repetitive to give you confidence, and simplify the content of the spread
- The questions in the middle of the exercise consolidate the topic, focusing on the main techniques of the spread
- Later questions extend the content of the spread – some of these questions may be problem-solving in nature, and may involve different approaches.

At the end of the unit is an exam review page so that you can revise the learning of the unit before moving on. The key objectives are identified:

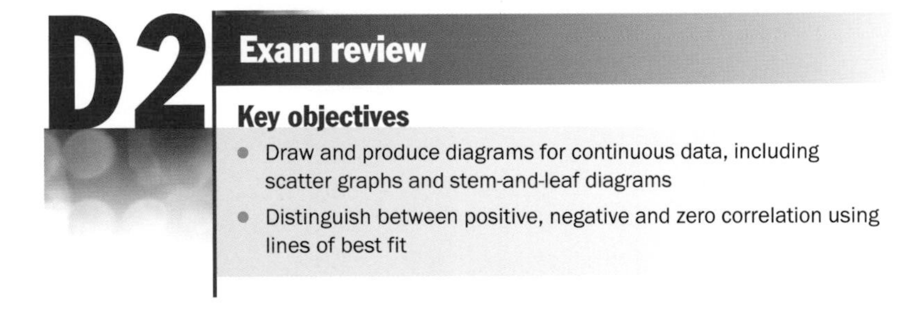
D2 Exam review

Key objectives

- Draw and produce diagrams for continuous data, including scatter graphs and stem-and-leaf diagrams
- Distinguish between positive, negative and zero correlation using lines of best fit

Summary questions, including a past exam questions from all the UK exam boards, are provided to help you check your understanding of the key concepts covered and your ability to apply the key techniques.

Foundation and Higher formulae pages are provided near the end of the book so you can see what information you will be given in the exam.

Foundation and Higher Practice Exam Papers are also provided for revision purposes.

You will find the answers to all exercises at the back of the book so that you can check your own progress and identify any areas that need work.

Contents

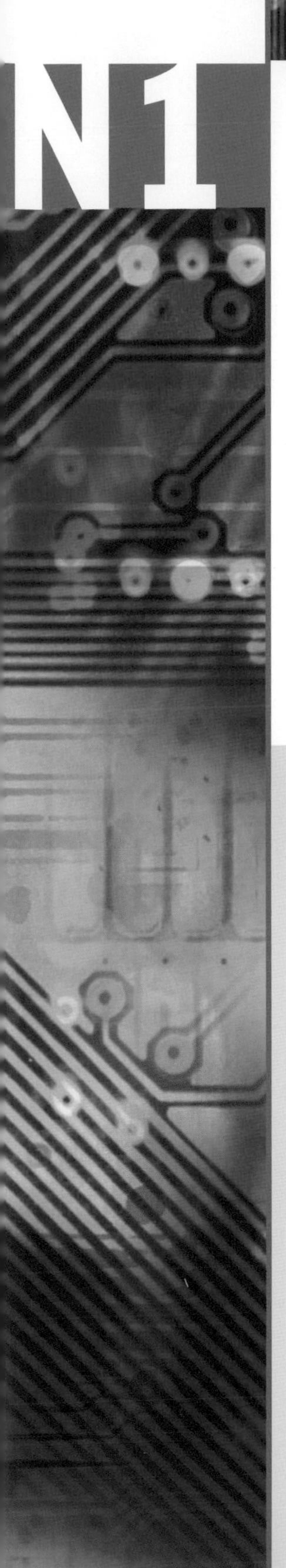

Integers and decimals

This unit will show you how to

- Understand place value and order numbers
- Multiply and divide numbers by powers of ten
- Represent decimal numbers as positions on a number line
- Read scales, dials and timetables
- Order negative numbers using a number line
- Add, subtract, multiply and divide with negative numbers
- Express a whole number as a product of its factors
- Understand and use simple divisibility tests
- Find the highest common factor and least common multiple of two numbers
- Recognise prime factors and express a number as a product of its prime factors
- Find the HCF and LCM of two numbers

Before you start ...

You should be able to answer these questions.

1 Put these numbers in order from lowest to highest.

0.37 0.4 0.312 0.35

2 What number is the arrow pointing to?

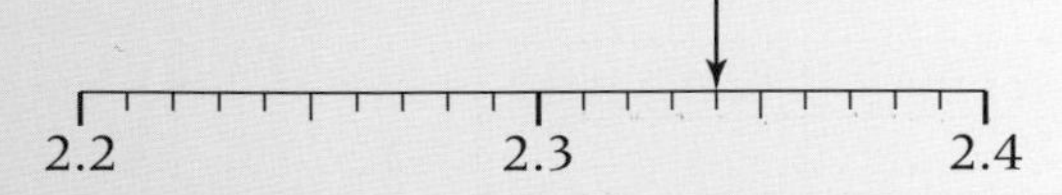

3 Put these numbers in order starting with the smallest.

−8, −1, 2, −5, −3, 4

4 Calculate.

a $10 - 3$ **b** $-6 - 4$

c $3 \times (-4)$ **d** $-6 \div 2$

5 Write all the factors of 48.

N1.1 Place value

This spread will show you how to:

- Understand place value and order numbers
- Multiply and divide numbers by powers of ten

Keywords
Digit
Order
Place value

- **The value of each digit in a number depends upon its place value.**

You can use place value to compare or **order** two or more numbers.

Example

Put these numbers in order from lowest to highest.

0.47 0.5 0.512 0.55 0.52

Look at each number to see the place value of the first non-zero digit.

0.47 0.5 0.512 0.55 0.52

You can see that 0.47 is the smallest number. The other four numbers all have a 5 in the first decimal place, so now look at the second digit.

0.50 0.512 0.55 0.52

You can now order the numbers: 0.47 0.5 0.512 0.52 0.55

The digit 4 stands for 4 tenths and the digit 5 stands for 5 tenths.

You can use a place value table to multiply and divide.

- **To multiply a number by 100, move all the digits two places to the left.**

6.7×100

Hundreds	Tens	Units	•	tenths	
		6	•	7	× 100
6	7	0	•		

$6.7 \times 100 = 670$

The 0 holds the digits in place.

- **To divide a number by 10, move all the digits one place to the right.**

$73.2 \div 10$

Tens	Units	•	tenths	hundredths	
7	3	•	2		÷ 10
	7	•	3	2	

$73.2 \div 10 = 7.32$

Example

Jake is told that $23 \times 48 = 1104$. Use this information to help Jake calculate:

a 2.3×48 **b** 0.23×4.8

a $2.3 \times 48 = \frac{23 \times 48}{10} = \frac{1104}{10} = 110.4$

b $0.23 \times 4.8 = \frac{23}{100} \times \frac{48}{10} = \frac{1104}{1000} = 1.104$

$0.23 = \frac{23}{100}$

$4.8 = \frac{48}{100}$

Exercise N1.1

1 Write each of these numbers in words.

a 456 **b** 13 200 **c** 115 020 **d** 460 340 **e** 4 325 400

f 55 670 345 **g** 45.8 **h** 367.03 **i** 4503.34 **j** 2700.02

2 Write each of these numbers in figures.

a five hundred and thirty-eight

b fifteen thousand, six hundred and three

c four hundred and seventeen point three

d five hundred and thirty-seven point four zero three

3 What number lies exactly halfway between

a 25 and 26 **b** 1.8 and 1.9 **c** 30 and 70

d 4.9 and 5 **e** 1.25 and 1.5 **f** 0.7 and 0.71?

4 Put these lists of numbers in order, starting with the smallest.

a	5.103	5.099	5.2	5.12	5.007
b	0.545	0.55	0.525	0.5	0.509
c	7.302	7.403	7.35	7.387	7.058
d	0.4	4.2	0.42	42	2.4
e	27.6	26.9	27.06	26.97	27.1
f	13.3	14.15	13.43	13.19	14.03

5 Calculate each of these without using a calculator.

a 3.2×100 **b** 0.4×10 **c** $152 \div 100$ **d** $14.6 \div 100$

e 2.37×10 **f** 24.3×100 **g** $1.23 \div 100$ **h** $45.9 \div 10$

i 3.4×1000 **j** 13.56×10 **k** $0.236 \div 10$ **l** 1.745×10

m 0.0392×10 **n** $72.8 \div 100$ **o** $12.4 \div 1000$ **p** 0.0814×100

6 Use the information given to work out each of these calculations without using a calculator.

a $23 \times 42 = 966$ What is 2.3×42?

b $91 \times 103 = 9373$ What is 9.1×103?

c $39 \times 57 = 2223$ What is 0.39×57?

d $34 \times 71 = 2414$ What is 340×71?

7 Marie knows that $32 \times 57 = 1824$. Use this information to help Marie calculate

a 3.2×57 **b** 0.32×5.7

N1.2 Reading scales

This spread will show you how to:

- Represent decimal numbers as positions on a number line
- Read scales, dials and timetables

Keywords
Estimate
Number line
Scale
Timetable

- **You can represent decimal numbers as a position on a number line.**

The arrow is pointing between 4 and 5.

3 4 5

There are 10 spaces between 4 and 5.
10 spaces represent 1 unit.
1 space represents $1 \div 10 = \frac{1}{10} = 0.1$ unit.
The arrow is pointing to the number 4.3.

The arrow is pointing between 2.2 and 2.3.

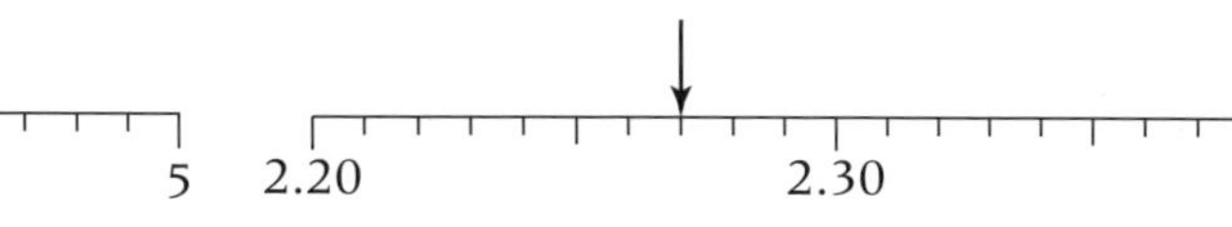

There are 10 spaces between 2.2 and 2.3.
10 spaces represent 0.1 unit.
1 space represents $0.1 \div 10 = 0.01$ unit.
The arrow is pointing to the number 2.27.

You can write 2.2 as 2.20, and so on.

You can **estimate** a measurement from a **scale**.

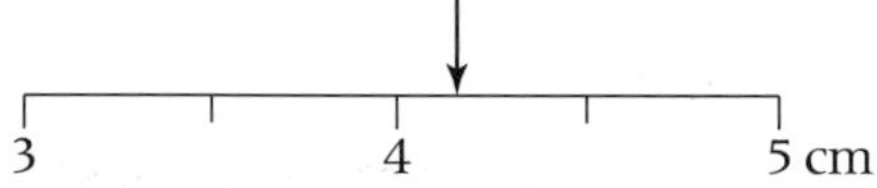

Most of the scales you read are number lines.

The reading is between 4 and 5 cm.
There are only two spaces between 4 and 5 cm.
The pointer is a little under a quarter of the way between 4 and 5.
A good estimated reading is 4.2 cm.

You should be able to read **timetables** for buses and trains.

Example

How long does it take for the 07:40 train from Clitheroe to get to Colne?

Station	Time of leaving	Time of leaving	Time of leaving
Clitheroe	07:10	07:40	08:10
Blackburn	07:28	07:58	08:28
Nelson	08:23	08:53	09:23
Colne	08:41	09:11	09:41
Bradford	09:52	10:22	10:52

The train leaves Clitheroe at 07:40.
It arrives at Colne at 09:11.

From 07:40 to 08:00 = 20 minutes
From 08:00 to 09:00 = 60 minutes
From 09:00 to 09:11 = 11 minutes
Total journey time = 91 minutes = 1 hour 31 minutes

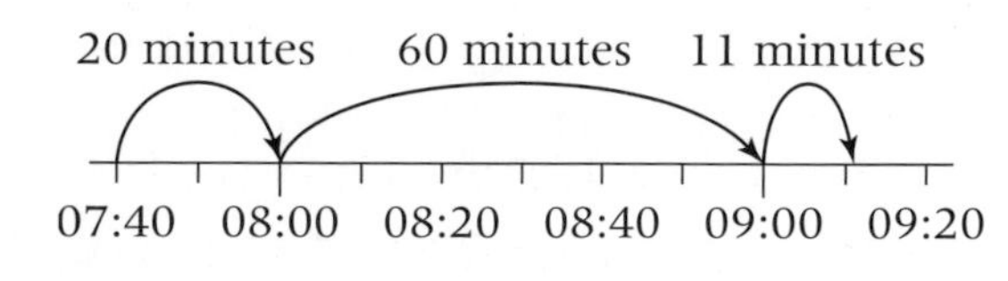

Exercise N1.2

1 Write the number each of the arrows is pointing to.

a

3 4 5

b

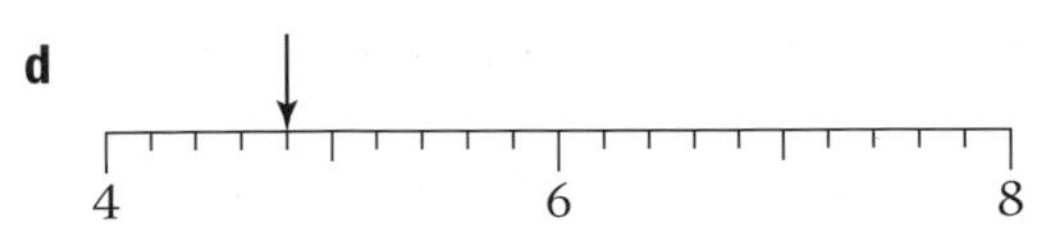

c

200 400 600

d

4 6 8

2 Write the reading shown on each scale.

a

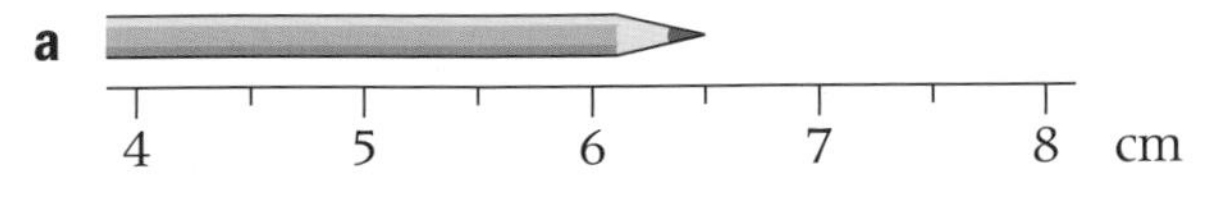

b

4 4.5 5 5.5 6 6.5 kg

c

1150 1160 1170 °C

d

3.2 3.25 3.3 tonnes

3 Use the scales to write a good estimated reading for each question.

a

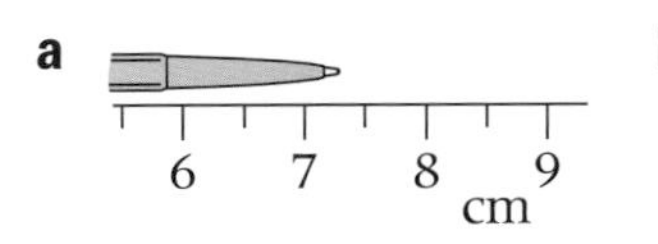

b

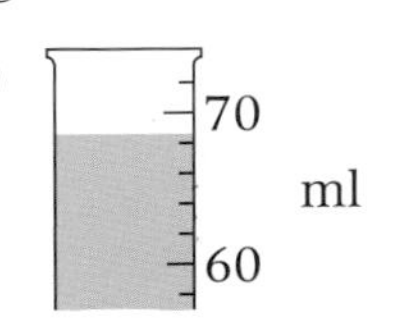

c

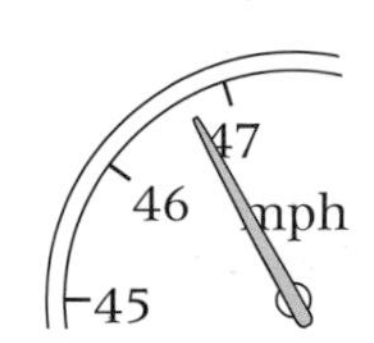

d

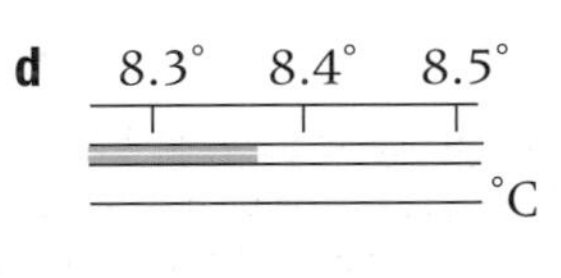

e

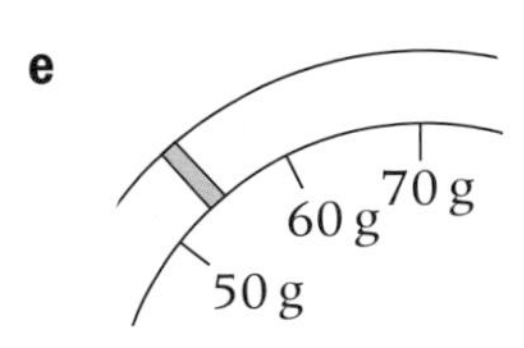

f

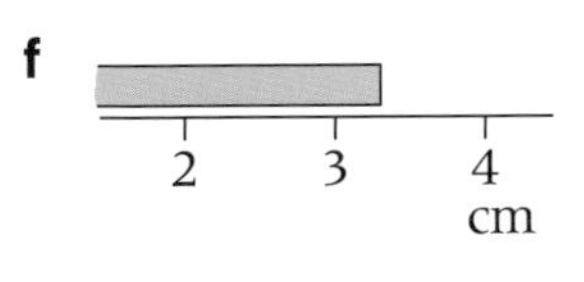

g

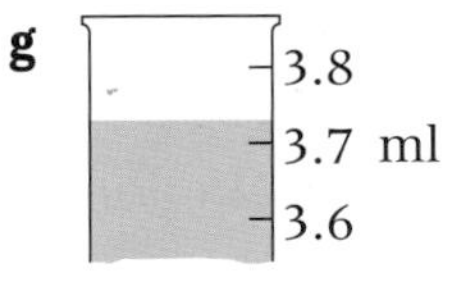

h

3000°C 2000°C 4000°C

4

Carlisle–Hexham bus timetable			
Carlisle	09:25	11:50	15:00
Crosby on Eden	09:41	12:11	15:21
Lanercost Priory	09:58	12:28	15:38
Birdoswald Fort	10:10	12:40	15:50
Chesters Fort	11:18	13:48	16:58
Hexham	11:32	13:59	17:09

a What time does the 11:50 bus from Carlisle arrive at Hexham?

b What time does the 15:50 bus from Birdoswald Fort leave Crosby on Eden?

c How long does it take the 09:25 bus from Carlisle to travel to Hexham?

d Harry catches the 12:28 bus at Lanercost Priory. He gets off at Chesters Fort. How long is his journey?

N1.3 Adding and subtracting negative numbers

This spread will show you how to:

- Order negative numbers using a number line
- Add and subtract with negative numbers

Keywords
Add
Negative numbers
Order
Subtract

- **Negative numbers** are numbers below zero.

The temperature in the fridge is −5°C or 5 degrees below freezing.

You can order negative numbers using a number line.

Example

Place these numbers in **order**, starting with the smallest.

−13, −14, 2, −5, −3, 4

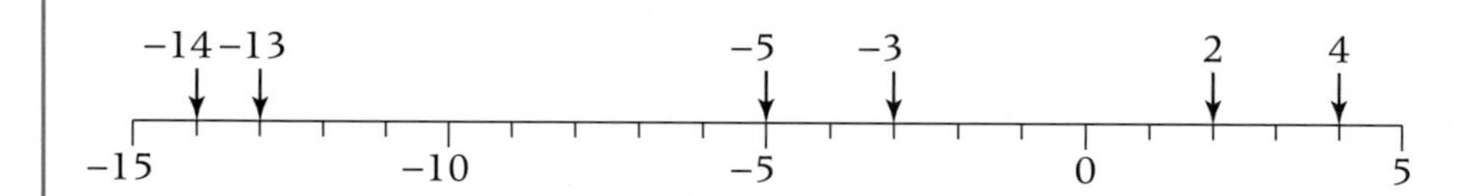

−14 is further away from zero than −13, so it is smaller.

The correct order is −14, −13, −5, −3, 2, 4.

You can use a number line to help you **add** to or **subtract** from a negative number.

To add, you move to the right.

To subtract, you move to the left.

Example

Calculate **a** 5−12 **b** −3 + 8 **c** −5 − 4

a Start at 5 and subtract 12 (move to the left).

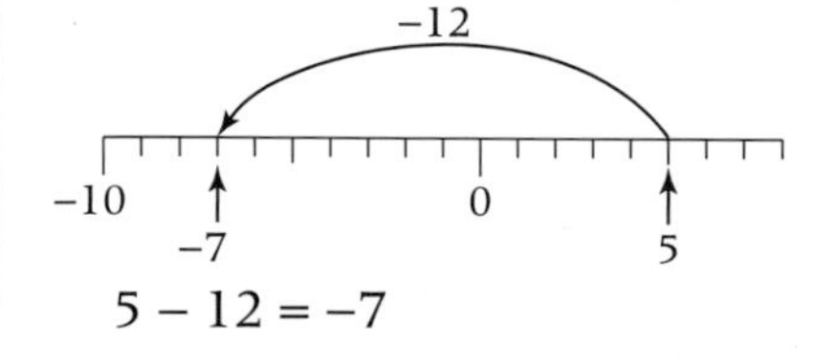

5 − 12 = −7

b Start at −3 and add 8 (move to the right).

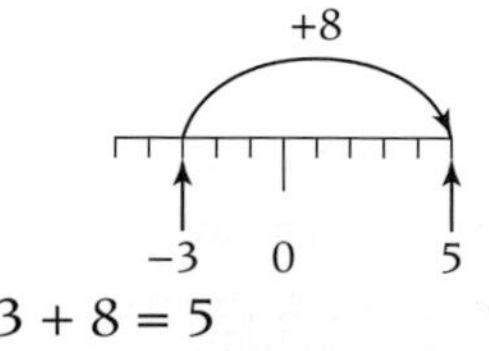

−3 + 8 = 5

c Start at −5 and subtract 4.

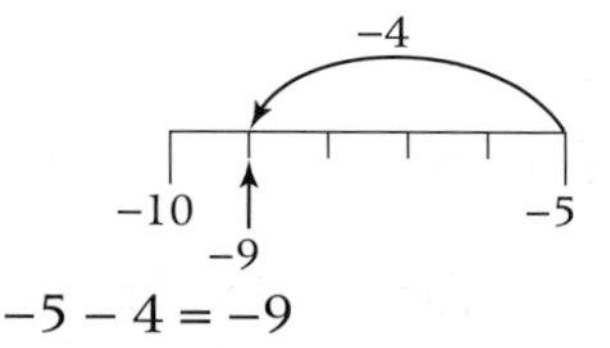

−5 − 4 = −9

There are two rules for adding and subtracting negative numbers.

- **Adding a negative number is the same as subtracting a positive number.**

18 + −3 = 18 − 3
= 15

- **Subtracting a negative number is the same as adding a positive number.**

18 − −3 = 18 + 3
= 21

Exercise N1.3

1 Put these lists of numbers in order from lowest to highest.

a	−13	−6	0	17	−12	15
b	0	−5	−6	−8	−3	−7
c	2	1	−2	4	3	−5
d	−1.5	3	9	−3	2	−8
e	−3	2	−5	−4.5	3	−2
f	3	8	6	−9	−1	2
g	−1	−3	0	−4.5	5.5	−2.5
h	−5	−5.1	−6	−5.8	−5.7	−5.4

2 Calculate.

a $4 + 12$ **b** $5 - 12$ **c** $7 - 3$ **d** $14 + 23$
e $34 - 17$ **f** $8 - 15$ **g** $-3 + 12$ **h** $-23 + 12$
i $-15 + 7$ **j** $-13 + 34$ **k** $-5 - 3$ **l** $-5 + 3$
m $-12 - 6$ **n** $21 - 17$ **o** $-8 + 3$ **p** $-4 + 8 - 2$
q $-12 - 3 - 5$ **r** $13 - 8 + 5$ **s** $-5 + 4 - 7$ **t** $-12 - 4 - 12$

3 Find the number that lies exactly halfway between each of these pairs of numbers.

a 28 and 34 **b** −5 and −17 **c** −6 and 14 **d** −18 and 4
e 3 and 8 **f** −4 and 9 **g** −20 and 35 **h** −3.5 and 2.5

4 Calculate.

a $13 + -5$ **b** $6 + -8$ **c** $12 + -3$ **d** $4 + -4$
e $-5 + -8$ **f** $-3 + -11$ **g** $-11 + -3$ **h** $15 - -5$
i $4 - -8$ **j** $-2 - -5$ **k** $-12 - -7$ **l** $-14 - -8$
m $-16 - -20$ **n** $-13 + -12$ **o** $-13 - -12$ **p** $13 + -12$
q $-12 + 7 -4$ **r** $-12 + -7 - 4$ **s** $-12 - -7 - 4$ **t** $-12 - 7 + -4$

5 Find the value of each expression when $a = -2$, $b = 3$, and $c = -4$.

a $a + b$ **b** $a + a$ **c** $a + c - b$
d $b - a$ **e** $a - c$ **f** $10 - a - b$

6 Find the final temperatures in these experiments.

a Starting temperature 36°C. It goes up 28°, then down 50°.
b Starting temperature 12°C. It goes up 17°, then down 39°.
c Starting temperature −8°C. It goes down 13°, then up 25°.

N1.4 Multiplying and dividing negative numbers

This spread will show you how to:

- Multiply and divide with negative numbers

Keywords
Divide
Multiply
Negative number

You can use a number line to help you multiply or divide **negative numbers**.

- -2×4 can be represented on a number line as four lots of -2.

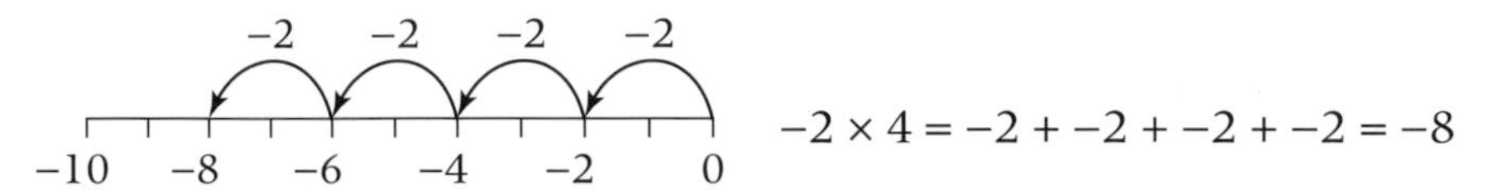

- **Negative number × positive number = negative number.**

- $-16 \div -4$ can be represented on a number line as 'how many lots of -4 are needed to make -16?'

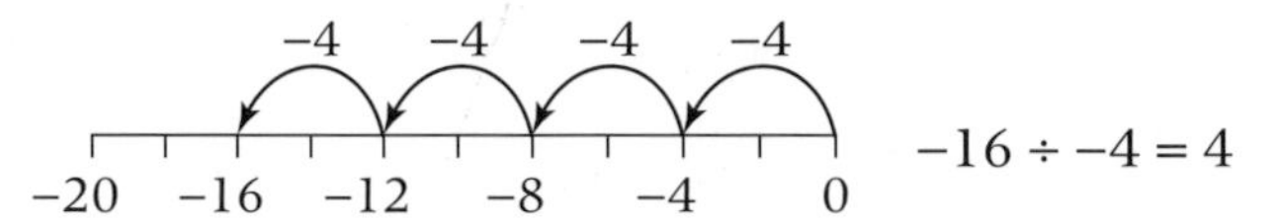

- **Negative number ÷ negative number = positive number.**

To **multiply** a negative number by a negative number, look for patterns in multiplication tables.

-5×2	=	-10	-5×0	=	0
-5×1	=	-5	-5×-1	=	5

- **Negative number × negative number = positive number.**

Remember these rules:

- Negative × positive = negative
  $-2 \times 4 = -8$
- Negative × negative = positive
  $-2 \times -4 = 8$
- Positive ÷ negative = negative
  $-8 \div -2 = -4$
- Negative ÷ positive = negative
  $-8 \div 2 = -4$
- Negative ÷ negative = positive
  $-8 \div -2 = 4$

×/÷	+	−
+	+	−
−	−	+

If the signs are different the answer will be negative.
If the signs are the same the answer will be positive.

Exercise N1.4

1 Choose a number card to make each of these calculations correct.

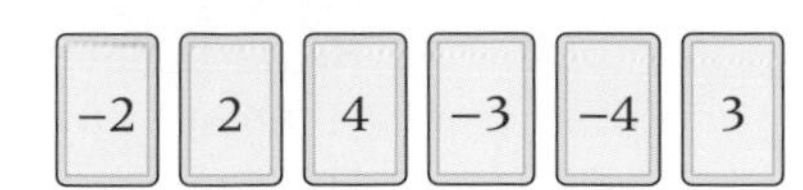

a $7 \times \square = -14$ **b** $-30 \div \square = -15$

c $-5 \times \square = 15$ **d** $-27 \div \square = -9$

e $\square \times -5 = 20$ **f** $-6 \times \square = -24$

2 Calculate.

a 3×-2 **b** 6×5 **c** 3×-7 **d** -4×-2 **e** -5×4

f -6×-4 **g** 8×-3 **h** -6×-7 **i** -8×-2 **j** -5×-10

k $-10 \div -2$ **l** $-40 \div 5$ **m** $-30 \div -6$ **n** $-45 \div -9$ **o** $54 \div -6$

p -9×7 **q** -7×7 **r** -8×-9 **s** $72 \div -8$ **t** $-42 \div 7$

u $-12 \div 3$ **v** $100 \div -10$ **w** $-81 \div -9$ **x** -7×8 **y** $-130 \div 10$

3 Here are a set of multiplication and division questions that have been marked by the teacher.
Explain why each of the answers that has been marked wrong is incorrect and write the correct answer.

a $4 \times -5 = 20$ ✗	b $3 \times -2 = -6$ ✓
c $-5 \times -6 = -30$ ✗	d $-14 \times -2 = -28$ ✗
e $7 \times -3 = -21$ ✓	f $8 \times 5 = 20$ ✗
g $7 \times -5 = 35$ ✗	h $-40 \div -5 = -8$ ✗
i $30 \div -5 = -6$ ✓	j $-14 \times -5 = -70$ ✗

4 Calculate these, using either a mental or a written method.
Remember to check the sign of your answer.

a -8×15 **b** -12×-11 **c** 25×-9 **d** 21×-7 **e** -9×13

f -19×7 **g** -23×18 **h** $-240 \div 6$ **i** $221 \div -17$ **j** -21×3.2

5 Find the value of each expression when $p = -2$, $q = 3$ and $r = -5$.

a $2p + q$ **b** $3(p - q)$ **c** $p^2 + q$ **d** $4p - 2q$

e $2q + 3r$ **f** $5p - 3r$ **g** $4(p + q + r)$

N1.5 Factors and multiples

This spread will show you how to:

- Express a whole number as a product of its factors
- Understand and use simple divisibility tests
- Find the highest common factor and least common multiple of two numbers

Keywords
Factor
Highest common factor
Least common multiple
Multiple
Product

- **Any whole number can be written as the product of two factors.**

$48 = 4 \times 12$ so 4 and 12 are **factors** of 48.

You can use simple divisibility tests to find the factors of a number.

Here are some simple tests:

Factor	Test	120
2	the number ends in a 0, 2, 4, 6 or 8	120
3	the sum of the digits is divisible by 3	$1+2+0=3$
4	the last two digits of the number are divisible by 4	$20 \div 4 = 5$
5	the number ends in 0 or 5	120
6	the number is divisible by 2 *and* by 3	60, 40
7	there is no simple check for divisibility by 7	
8	half of the number is divisible by 4	15
9	the sum of the digits is divisible by 9	–
10	the number ends in 0	120

A factor divides into a number exactly, with no remainder.

Factors of 120:
1×120
2×60
3×40
4×30
5×24
6×20
8×15
10×12

The multiples of 15 are 15, 30, 45, 60, ...

- You can find the **highest common factor** (HCF) of two numbers by listing all the factors of both numbers.
- You can find the **least common multiple** (LCM) of two numbers by listing the first few multiples of each number.

The **multiples** of a number can be divided exactly by the number, leaving no remainder.

Example

Find the HCF and LCM of 10 and 15.

The factors of 10 are: 1 2 5 10
The factors of 15 are: 1 3 5 15

1 and 5 are **common factors** of 10 and 15.
5 is the highest common factor of 10 and 15.

The first six multiples of 10 are: 10 20 30 40 50 60
The first six multiples of 15 are: 15 30 45 60 75 90

30 and 60 are **common multiples** of 10 and 15.
30 is the least common multiple of 10 and 15.

Exercise N1.5

1 Look at this list of numbers.

2	3	4	5	6	8	10
12	15	16	17	18	19	20

a Write all the numbers that are factors of 20.
b Write all the numbers that are factors of 192.
c Write all the numbers that are multiples of 5.
d Write all the numbers that are prime numbers.

A **prime number** has exactly two factors 1 and itself. 23 has factors 1 and 23, so it is a prime number.

2 Write all the factor pairs of each of these numbers.

a 24 **b** 45 **c** 66 **d** 100 **e** 120
f 132 **g** 160 **h** 180 **i** 360 **j** 324
k 224 **l** 264 **m** 312 **n** 325 **o** 432

3 Write the first three multiples of each of these numbers.

a 17 **b** 29 **c** 42 **d** 25 **e** 47
f 35 **g** 90 **h** 120 **i** 95 **j** 208

4 Find a number between 300 and 400 that has exactly 15 factors.

5 Find the highest common factor of

a 6 and 4 **b** 25 and 40 **c** 18 and 30 **d** 24 and 56
e 30 and 75 **f** 36 and 54 **g** 50 and 125 **h** 24, 36 and 72
i 30, 75 and 105

6 Find the least common multiple of

a 6 and 4 **b** 5 and 8 **c** 12 and 18 **d** 15 and 25
e 14 and 21 **f** 30 and 75

7 **a** Two hands move around a dial. The faster hand moves around in 24 seconds, and the slower hand in 30 seconds. If the two hands start together at the top of the dial, how many seconds does it take before they are next together at the top?

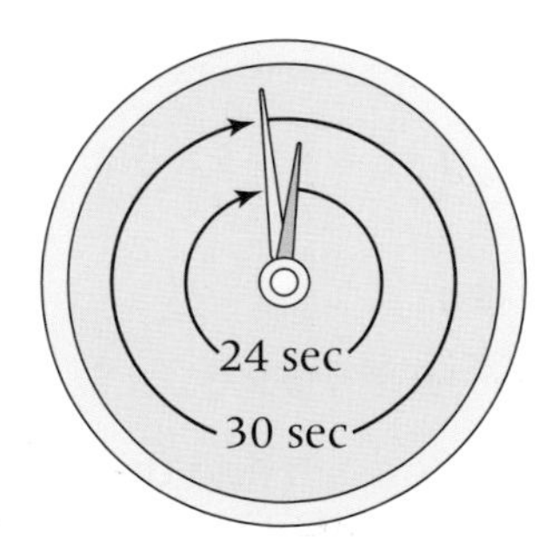

b A wall measures 234 cm by 432 cm. What is the largest size of square tile that can be used to cover the wall, without needing to cut any of the tiles?

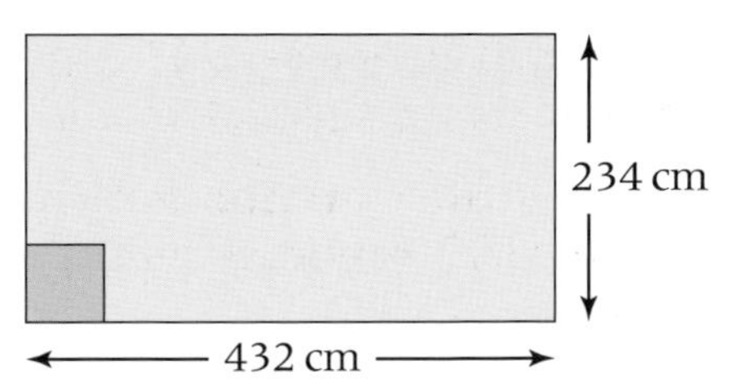

N1.6 Prime factor decomposition

This spread will show you how to:

- Recognise prime factors and express a number as a product of its prime factors
- Find the HCF and LCM of two numbers

Keywords
Factor
HCF
Index
LCM
Prime number

- A **prime factor** is a factor of a number which is also prime.

Factors of 28 are {1, 2, 4, 7, 14, 28}. Prime factors of 28 are {2, 7}

Look back to page 10 to remind yourself about factors.

- Every whole number can be written as the product of its **prime factors.**

$18 = 2 \times 3 \times 3$

Here are two common methods to find prime factors.

Factor trees

Split the number into a **factor** pair. Continue splitting until you reach a prime factor.

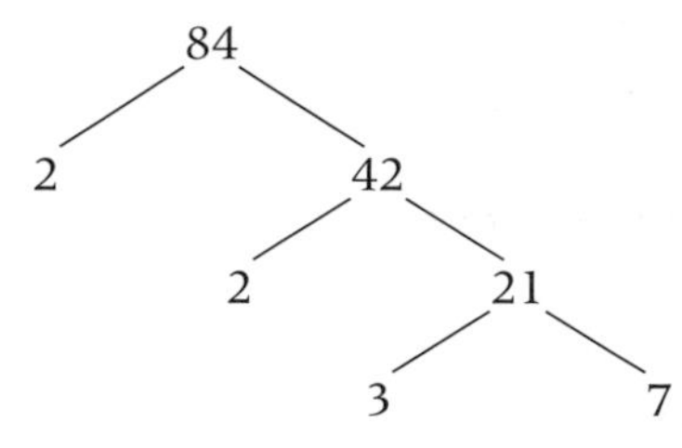

$84 = 2 \times 2 \times 3 \times 7$

Division by prime numbers

Divide the number by the smallest **prime number**. Repeat dividing by larger prime numbers until you reach a prime number.

2	84
2	42
3	21
	7

$84 = 2 \times 2 \times 3 \times 7$

You can write $84 = 2^2 \times 3 \times 7$
This is **index** notation.

- You can find the **highest common factor (HCF)** by using prime factors.

For example, the HCF of 30 and 135:

$30 = 2 \times 3 \times 5 \quad = 2 \times 3 \times 5$

2	30
3	15
	5

$135 = 3 \times 3 \times 3 \times 5 = 3 \times 3 \times 3 \times 5$

3	135
3	45
3	15
	5

$\text{HCF} = 3 \times 5 = 15$

- Write each number as the product of its prime factors.
- Pick out the common factors 3 and 5.
- Multiply these together to get the HCF.

- You can find the **least common multiple (LCM)** by using prime factors.

For example, the LCM of 28 and 126:

$28 = 2^2 \times 7 = 2 \times 2 \times 7$

2	28
2	14
	7

$126 = 2 \times 3^2 \times 7 = 2 \times 3 \times 3 \times 7$

2	126
3	63
3	21
	7

$\text{HCF} = 2 \times 7 = 14$

$\text{LCM} = 2 \times 3 \times 3 \times 14 = 252$

- Write each number as the product of its prime factors.
- Pick out the common factors 2 and 7.
- Multiply these together to get the HCF (14).
- Multiply the HCF by the remaining factors – the remaining factors are 2, 3 and 3.

Exercise N1.6

1 Work out the value of each of these expressions.

a 3×5^2 **b** $2^3 \times 5$ **c** $3^2 \times 7$ **d** $2^2 \times 3^2 \times 5$ **e** $3^2 \times 7^2$

2 Express these numbers as products of their prime factors.

a 18	**b** 24	**c** 40	**d** 39
e 48	**f** 82	**g** 100	**h** 144
i 180	**j** 315	**k** 444	**l** 1350

3 In each of these questions, Jack has been asked to write each of the numbers as the product of its prime factors.

i Mark his work and identify any errors he has made.

ii Correct any of Jack's mistakes.

a 126

2	126
3	63
3	21
	7

Answer: $126 = 2 \times 3^2$

b 210

2	210
3	105
3	21
	7

Answer: $2 \times 3^2 \times 7$

c 221

Answer: 221

4 Find the HCF of

a 9 and 24 **b** 15 and 40 **c** 18 and 24

d 96 and 144 **e** 12, 15 and 18 **f** 425 and 816.

5 Find the LCM of

a 9 and 24 **b** 15 and 40 **c** 18 and 24

d 20 and 30 **e** 12, 15 and 18 **f** 48, 54 and 72.

6 Cancel these fractions to their simplest forms using the HCF of the numerator and denominator to help.

a $\frac{6}{8}$ **b** $\frac{12}{18}$ **c** $\frac{60}{96}$

d $\frac{36}{54}$ **e** $\frac{117}{169}$ **f** $\frac{26}{65}$

7 **a** 48 can be written as $2^x \times y$.
Work out the values of x and y using a method of prime factor decomposition.

b 300 can be written as $2^a \times b \times c^2$.
Work out the values of a, b and c.

N1

Exam review

Key objectives

- Multiply or divide any number by powers of ten
- Understand and use positive and negative numbers, both as positions and translations on a number line
- Add, subtract, multiply and divide integers and then any number
- Understand highest common factor, least common multiple, prime number and prime factor decomposition

1 36 expressed as a product of its prime factors is $2^2 \times 3^2$

a Express 45 as a product of its prime factors.
Write your answer in index form. (3)

b What is the Highest Common Factor (HCF) of 36 and 45? (1)

c What is the Least Common Multiple (LCM) of 36 and 45? (1)

(AQA, 2003)

2 Sally wrote down the temperature at different times on 1st Jan 2003.

Time	Temperature
midnight	−6°C
4 am	−10°C
8 am	−4°C
noon	7°C
3 pm	6°C
7 pm	−2°C

a Write down

i the highest temperature

ii the lowest temperature. (2)

b Work out the difference in the temperature between

i 4 am and 8 am **ii** 3 pm and 7 pm. (2)

At 11 pm that day the temperature had fallen by 5 °C from its value at 7 pm.

c Work out the temperature at 11 pm. (1)

(Edexcel Ltd., 2004)

N2 Estimating and calculating

This unit will show you how to

- Know and use the order of operations, including brackets
- Round numbers to any number of significant figures
- Use rounding to estimate answers to calculations
- Recognise the inaccuracy of rounded measurements
- Multiply and divide by powers of 10 and by decimals between 0 and 1
- Use checking procedures, including approximation to estimate the answer to multiplication and division problems
- Use a range of mental and written methods for calculations with whole numbers and decimals
- Use calculators to carry out more complex calculations
- Give answers to an appropriate degree of accuracy

Before you start ...

You should be able to answer these questions.

1 Calculate

$5 \times 3 + 4 \times 8$

2 Calculate

a $13 \div 10$ **b** 0.03×10

3 Given that $12 \times 35 = 420$, what is the value of

a 12×3.5? **b** $420 \div 12$?

4 Use an appropriate method of calculation to work out

a 15×3.2 **b** $26.6 \div 7$

5 Hugo estimates the value of

$\frac{6.89 \times 12.07}{3.88}$ to be 21.

Write three numbers Hugo could use to get his estimate.

N2.1 Order of operations

This spread will show you how to:

- Know and use the order of operations, including brackets

Keywords
Brackets
Index
Order of operations
Power

When you do long calculations you must work them out according to the **order of operations**.

Order of operations

Brackets
Work out the contents of any **brackets** first.

Powers or Indices
Work out any **powers** or roots.

Division and Multiplication
Work out any multiplications and divisions.

Addition and Subtraction
Finally, work out any additions and subtractions.

To calculate $\frac{(6+4)^2}{5} + 8 \times 2$

$\frac{(6+4)^2}{5} + 8 \times 2 = \frac{(6+4)^2}{5} + 8 \times 2$ (work out the contents of the bracket)

$= \frac{10^2}{5} + 8 \times 2$ (work out any powers)

$= \frac{100}{5} + 8 \times 2$ (work out any divisions and multiplications)

$= 20 + 16$ (work out any additions and subtractions)

$= 36$

- Brackets are used when you need to do an operation in a different order to normal.
- Calculations with brackets need to be thought about carefully.

Example

Put brackets into $5 \times 7 + 8 - 5 = 70$ to make it correct.

$5 \times (7 + 8) - 5 = 5 \times 15 - 5$
$= 75 - 5$
$= 70$

Example

Calculate **a** $\frac{34 \times 8}{5 + 11}$ **b** $30 \div (15 - (12 - 7))$

a $\frac{34 \times 8}{5 + 11} = \frac{(34 \times 8)}{(5 + 11)}$

$= (34 \times 8) \div (5 + 11)$ (work out the brackets)

$= 272 \div 16$ (work out the division)

$= 17$

b $30 \div (15 - (12 - 7)) = 30 \div (15 - (12 - 7))$ (work out the innermost brackets)

$= 30 \div (15 - 5)$ (work out the next set of brackets)

$= 30 \div 10$ (work out the division)

$= 3$

Always write the calculation a line at a time, so you can see each operation clearly.

Exercise N2.1

1 Calculate these using the order of operations.

a $2 + 8 \times 3$ **b** $4 \times 11 - 7$ **c** $4 \times 3 + 5 \times 8$

d $5 + 12 \div 6 + 3$ **e** $(2 + 9) \times 3$ **f** $(1.5 + 18.5) \div 4$

g $(12 + 3) \times (14 - 2)$ **h** $5 + (3 \times 8) \div 6$

2 Calculate these using the order of operations.

a $(4 + 3) \times 2^2$ **b** $3^2 \times (15 - 7)$ **c** $(6^2 - 16) \div 4$

d $2^4 \times (3^2 - 2 \times 4)$ **e** $128 \div (2 + 2 \times 3)^2$ **f** $(8^2 - 7^2) \times 5$

3 Copy each of these calculations.
Insert brackets where necessary to make each of the calculations correct.

a $5 \times 2 + 1 = 15$ **b** $5 \times 3 - 1 \times 4 = 40$ **c** $20 + 8 \div 2 - 7 = 17$

d $2 + 3^2 \times 4 + 3 = 65$ **e** $2 \times 6^2 \div 3 + 9 = 33$ **f** $4 \times 5 + 5 \times 6 = 150$

4 a Karen and Pete answered the same question. Who is correct?
Explain your answer.

Karen wrote:

$5 + (2 \times 9 - 4) = 15$

Pete wrote:

$5 + (2 \times 9 - 4) = 19$

b Duncan said $(5 \times 4)^2$ means the same as 5×4^2.
Is this correct? Explain your answer.

c Use your calculator to work out $(2.4 + 1.65)^2 \times 3.4$.

i Write all the figures on your calculator display.

ii Round your answer to 1 decimal place.

5 Calculate each of these.

a $\dfrac{7^2 - 9}{5 \times 8}$ **b** $\dfrac{4 \times 8}{4^2}$ **c** $\dfrac{15 \times 4}{6 \times 5}$

d $\dfrac{2 \times (3 + 4)^2}{7}$ **e** $\dfrac{(6 + 4)^2}{20} + 7 \times 5$ **f** $\dfrac{6 + (2 \times 4)^2 + 7}{11}$

g $4 + (12 - (3 + 2))$ **h** $120 \div (8 \times (7 - 2))$

Decide whether to use a mental, written or calculator method. Where appropriate give your answer to 2 decimal places.

6 Solve each of these calculations.

a $(15.7 + 1.3) \times (8.7 + 1.3)$ **b** $\dfrac{7^2}{(2.3 \times 4)^2}$ **c** $\dfrac{(7 + 5)^2}{(25 + 7 \times 8)}$

7 Given that $a = 4$ and $b = 3$, calculate $(4a - 3b - 8) + a$.

N2.2 Rounding

This spread will show you how to:

- Round numbers to any number of significant figures
- Use rounding to estimate answers to calculations
- Recognise the inaccuracy of rounded measurements

Keywords
Approximate
Decimal places
Estimate
Rounding
Significant figures

You can round a decimal number to a given accuracy.
To round 718.394 to 2 **decimal places**, look at the **thousandths** digit.

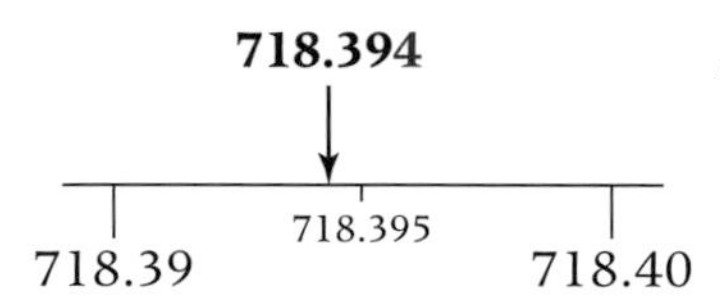

The **thousandths** digit is **4**, so round **down** to 718.39.

718.394 ≈ 718.39 (to 2 decimal places).

When **rounding** numbers to a given degree of accuracy, look at the next digit. If it is **5** or more then round **up**, otherwise round down.

To round 54.76 to 2 **significant figures**:

Look at the 3rd significant figure.

Tens	Units	•	tenths	hundredths
5	4	•	7	6

The **3rd significant** figure is **7**, so the number is rounded up to 55.

54.76 ≈ 55 (to 2 significant figures).

The first **non-zero** digit in the number is called the **1st significant figure** – it has the highest value in the number.

- You can **estimate** the answer to a calculation by rounding the numbers.

Example

Estimate the answer to $\frac{6.23 \times 9.89}{18.7}$.

You can round each of the numbers to 1 significant figure.

$$\frac{6.23 \times 9.89}{18.7} \approx \frac{6 \times 10}{20} = \frac{60}{20} = 3$$

- When a measurement is written, it is always written to a given degree of accuracy. The real measurement can be anywhere within ± half a unit.

Example

A man walks 23 km (to the nearest km). Write the maximum and minimum distance he could have walked.

Because the real measurement has been rounded, it can lie anywhere between 22.5 km (minimum) and 23.5 km (max).

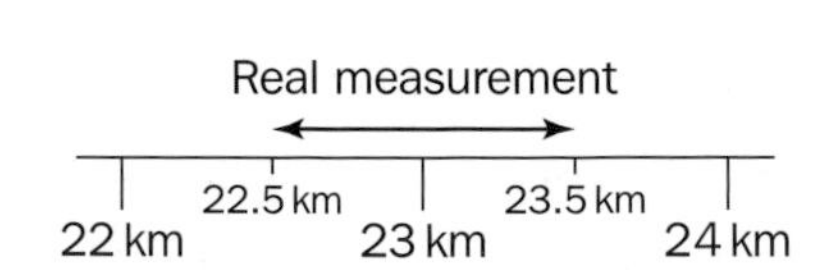

Exercise N2.2

1 Round each of these numbers to the

i nearest 10 **ii** nearest 100 **iii** nearest 1000.

a 3487 **b** 3389 **c** 14 853 m **d** £57 792

e 92 638 kg **f** £86 193 **g** 3438.9 **h** 74 899.36

2 Round each of these numbers to the nearest whole number.

a 3.738 **b** 28.77 **c** 468.63 **d** 369.29

e 19.93 **f** 26.9992 **g** 100.501 **h** 0.001

3 Round each of these numbers to the nearest

i 3 dp **ii** 2 dp **iii** 1 dp.

a 3.4472 **b** 8.9482 **c** 0.1284 **d** 28.3872

e 17.9989 **f** 9.9999 **g** 0.003 987 **h** 2785.5555

dp means decimal places.

4 Round each of these numbers to the nearest

i 3 sf **ii** 2 sf **iii** 1 sf.

a 8.3728 **b** 18.82 **c** 35.84 **d** 278.72

e 1.3949 **f** 3894.79 **g** 0.008 372 **h** 2399.9

i 8.9858 **j** 14.0306 **k** 1403.06 **l** 140 306

sf means significant figures.

5 Write a suitable estimate for each of these calculations. In each case, clearly show how you estimated your answer.

a 4.98×6.12 **b** $17.89 + 21.91$ **c** $\frac{5.799 \times 3.1}{8.86}$

d $34.8183 - 9.8$ **e** $\frac{32.91 \times 4.8}{3.1}$ **f** $\{9.8^2 + (9.2 - 0.438)\}^2$

6 For each of these measurements (given to a specified degree of accuracy), write

i the minimum value it could be **ii** the maximum value it could be.

a 67.0 cm (nearest whole number) **b** 34.7 litres (1 decimal place)

c 8.36 kg (2 decimal places) **d** 0.387 mm (3 decimal places)

7 The length of a car is 2.6 m correct to 1 decimal place.

a Write the maximum value that the length could be.

b Write the minimum value that the length could be.

N2.3 Estimation

This spread will show you how to:

- Multiply and divide by powers of 10 and by decimals between 0 and 1
- Use checking procedures, including approximation to estimate the answer to multiplication and division problems

Keywords
Approximate
Estimate

- You can multiply or divide a number by a power of 10. Move the digits of the number to the left or to the right.

$1.8 \xrightarrow{\times 10 \text{ or } \div 0.1} 18$, $18 \xrightarrow{\div 10 \text{ or } \times 0.1} 1.8$

$12.4 \xrightarrow{\times 100 \text{ or } \div 0.01} 1240$, $1240 \xrightarrow{\div 100 \text{ or } \times 0.01} 12.4$

× 0.1 is the same as ÷ 10.
× 0.01 is the same as ÷ 100.

÷ 0.1 is the same as × 10.
÷ 0.01 is the same as × 100.

- You can multiply and divide by any decimal between 0 and 1 using the mental method of factors.

Example

Calculate **a** 12×0.3 **b** $36 \div 0.04$ **c** $2 \div 0.05$

a $12 \times 0.3 = 12 \times 3 \times 0.1$
$= 36 \times 0.1$
$= 36 \div 10$
$= 3.6$

b $36 \div 0.04 = 36 \div (4 \times 0.01)$
$= 36 \div 4 \div 0.01$
$= 9 \div 0.01$
$= 9 \times 100$
$= 900$

c $2 \div 0.05 = \frac{2}{0.05}$
$= \frac{200}{5} = 40$
$= 40$

You can rewrite a division as a fraction and multiply the numerator and denominator by a power of 10.

- You should **estimate** the answer to a calculation by first rounding the numbers in the calculation.

Example

Estimate the answers to these calculations.

a $\frac{8.93 \times 28.69}{0.48 \times 6.12}$ **b** $\frac{17.4 \times 4.89^2}{0.385}$

a $\frac{8.93 \times 28.69}{0.48 \times 6.12} \approx \frac{9 \times 30}{0.5 \times 6}$
$= \frac{270}{3} = 90$

b $\frac{17.4 \times 4.89^2}{0.385} \approx \frac{20 \times 5^2}{0.4}$
$= \frac{20 \times 25}{0.4} = \frac{500}{0.4}$
$= \frac{5000}{4} = 1250$

A good strategy is to round each number in the calculation to 1 significant figure.

Multiply top and bottom by 10.

Exercise N2.3

1 Round each of these numbers to the nearest **i** 1000 **ii** 100 **iii** 10.

a 1548.9 **b** 5789.47 **c** 17 793.8 kg

d €35 127.35 **e** 236 872

2 Round each of these numbers to
i 3 dp **ii** 2 dp **iii** 1 dp **iv** the nearest whole number.

a 4.3563 **b** 9.8573 **c** 0.9373 **d** 19.4963

e 26.8083 **f** 19.9999 **g** 0.004896 **h** 3896.6567

3 Calculate

a 3×0.1 **b** $15 \div 0.1$ **c** 8×0.01 **d** 2.8×100

e $3.8 \div 0.1$ **f** 0.4×0.1 **g** $9.23 \div 0.1$ **h** $44.6 \div 0.01$

4 Here are five number cards.

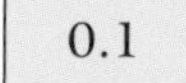

10 | 0.01 | 1000 | 10^2

Fill in the missing numbers in each of these statements using one of these cards.

a $3.24 \times ? = 324$ **b** $14.7 \times ? = 0.147$ **c** $6.3 \div ? = 630$

d $2870 \div ? = 2.87$ **e** $0.43 \div ? = 4.3$ **f** $2.04 \div ? = 204$

5 Work out these calculations using a mental method.

a 12×0.2 **b** 8×0.07 **c** $15 \div 0.3$

d $3 \div 0.15$ **e** 1.2×0.4 **f** $28 \div 0.07$

6 Write a suitable estimate for each of these calculations.
In each case clearly show how you estimated your answer.

a 3.76×4.22 **b** 17.39×22.98 **c** $\frac{4.59 \times 7.9}{19.86}$ **d** $54.31 \div 8.8$

7 Write a suitable estimate for each of these calculations.
In each case clearly show how you estimated your answer.

a $\frac{29.91 \times 38.3}{3.1 \times 3.9}$ **b** $\frac{16.2 \times 0.48}{0.23 \times 31.88}$

c $\{4.8^2 + (4.2 - 0.238)\}^2$ **d** $\frac{63.8 \times 1.7^2}{1.78^2}$

8 Sean estimates the value of $\frac{27.3 \times 11.9}{5.72}$ to be 60.
Write three numbers Sean could use to get his estimate.

N2.4 Mental methods

This spread will show you how to:

- Use a range of mental methods for calculations with whole numbers and decimals

Keywords
Compensation
Mental method
Partitioning
Place value

There are lots of **mental methods** you can use to help you work out calculations in your head.

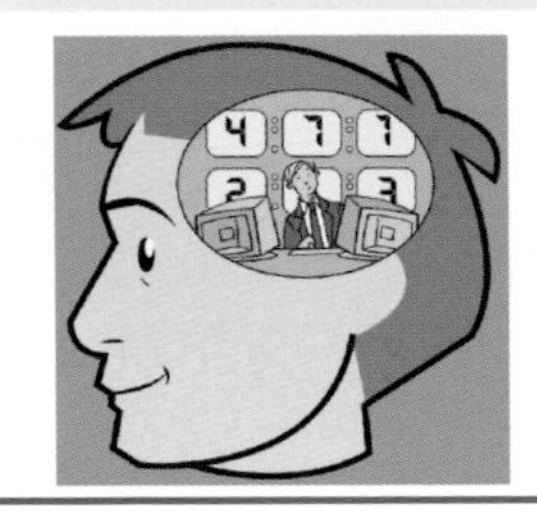

You can use **place value**.

Example

Use the fact that $35 \times 147 = 5145$ to write the value of

a 3.5×1.47 **b** $51.45 \div 3.5$

a $3.5 \times 1.47 = (35 \div 10) \times (147 \div 100)$
$= 35 \times 147 \div 1000$
$= 5145 \div 1000$
$= 5.145$

b $51.45 \div 3.5 = \frac{51.45}{3.5} = \frac{514.5}{35}$
$= \frac{5145 \div 10}{35}$
$= 147 \div 10$
$= 14.7$

You can use **partitioning**.

Example

a Calculate $18.5 - 7.7$. **b** Calculate 6.3×12.

a $18.5 - 7.7 = 18.5 - 7 - 0.7$
$= 11.5 - 0.7$
$= 10.8$

b $12 = 10 + 2$

$6.3 \times 12 = (6.3 \times 10) + (6.3 \times 2)$
$= 63 + 12.6$
$= 75.6$

Split **12** into **10** + **2**. Then work out **10** × 6.3 and **2** × 6.3. Add your two answers together.

You can use **compensation**.

Example

Calculate **a** $12.4 - 4.9$ **b** 23.2×1.9

a $12.4 - 4.9 = 12.4 - 5 + 0.1$
$= 7.4 + 0.1$
$= 7.5$

b $1.9 = 2 - 0.1$

$23.2 \times 1.9 = (23.2 \times 2) - (23.2 \times 0.1)$
$= 46.4 - 2.32$
$= 44.08$

Rewrite **1.9** as **2 − 0.1**. Work out **2** × 23.2 and **0.1** × 23.2. Subtract your two answers.

Example

A rectangular piece of wood measures 3.5 m by 1.9 m. What is its area?

$3.5 \times 1.9 = (3.5 \times 2) - (3.5 \times 0.1)$
$= 7 - 0.35$
$= 6.65 \text{ m}^2$

Use compensation.

Exercise N2.4

1 Calculate these.

a 9×7 **b** $121 \div 10$ **c** 2×2.7 **d** $48.4 \div 2$

e 3.6×100 **f** $430 \div 100$ **g** 23.6×10 **h** $0.78 \div 100$

2 Use an appropriate mental method to calculate these. Show the method you have used.

a 1.4×11 **b** 21×9 **c** 5.3×11 **d** 41×2.8

e 19×7 **f** 12×5.3 **g** $147 \div 3$ **h** $276 \div 4$

i 3.2×11 **j** 31×5.6 **k** 14.9×9 **l** 25.3×31

m 14×8 **n** $51 \div 1.5$ **o** $81 \div 4.5$ **p** 4.4×4.5

3 Use the mental method of partitioning to work out each of these.

a $19.5 - 7.6$ **b** $45.3 + 12.6 + 7.2$ **c** $132.6 - 21.4$

d 7.2×13 **e** 8.4×12 **f** 11×19.2

g $129 \div 3$ **h** $292 \div 4$

4 Use the mental method of compensation to work out each of these.

a $19.5 - 7.9$ **b** $48.4 - 12.8$ **c** $164.5 - 15.9$

d 8.1×19 **e** 36×3.9 **f** 17×5.9

5 Use an appropriate mental method to calculate each of these.

a $27.6 + 21.7$ **b** $1623 - 897$ **c** 32×2.1 **d** 2.9×23

e 19×1.4 **f** 9×7.5 **g** $2.4 \div 0.2$ **h** $\frac{30 \times 0.2}{0.15}$

6 **a** Using the information that $69 \times 147 = 10\,143$, write the value of each of these.

i 69×1470 **ii** 690×1470 **iii** 6.9×147 **iv** 0.69×14.7

v 6.9×0.147 **vi** 690×1.47 **vii** 0.069×14.7 **viii** 0.69×0.147

b Using the information that $37 \times 177 = 6549$, write the value of each of these.

i 3.7×17.7 **ii** 0.37×1770 **iii** $654.9 \div 177$ **iv** $65.49 \div 3.7$

7 Use an appropriate mental method to solve each problem:

a Surinder buys some material. It costs £12.99 per metre. How much does it cost to buy 9 metres?

b Jake sends 29 text messages. Each message costs 4.3p. How much do the 29 text messages cost in total?

N2.5 Written methods

This spread will show you how to:

- Use a range of written methods for calculations with whole numbers and decimals
- Use checking procedures, including approximation to estimate the answer to multiplication and division problems

Keywords
Estimate
Whole number

You can multiply decimals by replacing them with an equivalent **whole-number** calculation that is easier to work out.

Example

Carol is working out the area of carpet she needs for her floor. The floor is in a rectangle with a length of 4.8 m and a width of 3.12 m. What is the area of Carol's floor?

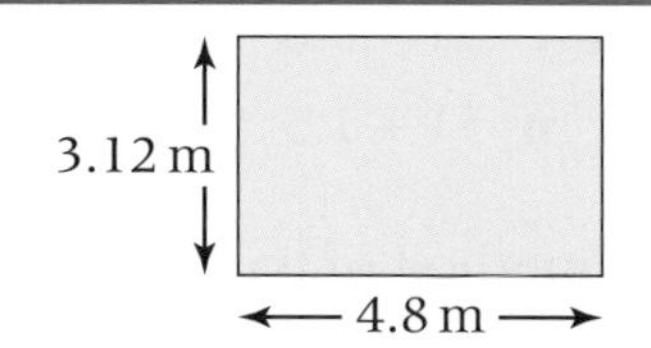

$4.8 \times 3.12 = (48 \div 10) \times (312 \div 100)$
$= 48 \times 312 \div 1000$

×	300	10	2
40	40 × 300 = 12 000	40 × 10 = 400	40 × 2 = 80
8	8 × 300 = 2400	8 × 10 = 80	8 × 2 = 16

Estimate the answer first.
$4.8 \times 3.12 \approx 5 \times 3$
$= 15\text{ m}^2$

$48 \times 312 = 12\,000 + 400 + 80 + 2400 + 80 + 16 = 14\,976$

The area of Carol's floor is $4.8 \times 3.12 = 48 \times 312 \div 1000$
$= 14\,976 \div 1000$
$= 14.98\text{ m}^2$ (2 decimal places)

You can divide a number by a decimal by rewriting the calculation as an equivalent whole-number division.

Example

Mandy has a floor with an area of 91 m^2. She fills the floor with carpet tiles which have an area of 2.8 m^2.
How many tiles does she need to cover the floor?

$91 \div 2.8 = 910 \div 28$

```
28)910
  -840    28 × 30
    70
   -56    28 × 2
   14.0
  -14.0   28 × 0.5
     0
```

$30 + 2 + 0.5 = 32.5$

$910 \div 28 = 91 \div 2.8 = 32.5$

Estimate the answer first.
$9.1 \div 2.8 \approx 90 \div 3$
$= 30$

Mandy needs $91 \div 2.8 = 32.5$ tiles.

Exercise N2.5

1 Use a written method for each of these calculations.

a 16.4 + 9.68 **b** 27.3 + 5.41 **c** 9.51 − 6.7

d 24.3 + 7.69 **e** 34.76 − 8.29 **f** 38.29 − 24.8

2 Use an appropriate method of calculation to work out each of these.

a 15 × 3.4 **b** 5.6 × 18 **c** 8.4 × 13

d 23 × 7.6 **e** 28 × 4.2 **f** 9.7 × 49

3 Use an appropriate method of calculation to work out each of these.

a 27.3 ÷ 7 **b** 36.6 ÷ 6 **c** 70.4 ÷ 8

d 73.8 ÷ 6 **e** 119.7 ÷ 9 **f** 119.2 ÷ 8

g 72 ÷ 1.6 **h** 132 ÷ 2.4 **i** 57.2 ÷ 2.6

For parts g, h and i, rewrite the divisions before performing them.

4 Use a mental or written method to solve each of these problems.

a Oliver sells tomatoes at the market. On Thursday he sells 78.6 kg; on Saturday he sells 83.38 kg. What mass of tomatoes has he sold during the two days?

b A mobile phone without a battery weighs 188.16 g. When the battery is inserted the combined mass of the mobile phone and battery is 207.38 g. What is the mass of the battery?

c A recycling box is full of things to be recycled.
The empty box weighs 1.073 kg.

Bottles	12.45 kg
Cans	1.675 kg
Paper	8.7 kg
Plastic objects	? kg

The total weight of the box and all the objects to be recycled is exactly 25 kg. What is the weight of the plastic objects?

DID YOU KNOW?
At least half the household waste produced in the UK each year could be recycled. Unfortunately in 2005 only 12% was recycled!

5 Use an appropriate method of calculation to work out each of these.

a 2.3 × 1.74 **b** 1.6 × 2.75 **c** 1.7 × 44.3

d 2.5 × 5.88 **e** 8.7 × 4.79 **f** 38 × 4.78

6 **a** Pedro buys 1.8 m of carpet. Each metre costs £1.85. How much does this cost in total?

b Jamal buys 7.8 kg of apples. Each kilogram of apples costs £1.45. How much money does Jamal pay for the apples?

c Brian is a gardener. He plants trees at a rate of 11.8 trees per hour. How many trees does he plant in 6.4 hours?

d Carys works as a car mechanic. She charges £31.70 per hour for her work. How much does Carys charge for working 2.5 hours?

e Des is a plumber. He charges £37.55 as a call out charge, and then £19.60 for each hour he works. How much does Des charge for working $2\frac{1}{2}$ hours?

N2.6 Calculator methods

This spread will show you how to:

- Use calculators to carry out more complex calculations
- Use checking procedures, including approximation to estimate the answer to multiplication and division problems
- Give answers to an appropriate degree of accuracy

Keywords
Appropriate degree of accuracy
Brackets
Order of operations

You can use the bracket keys on a scientific calculator to do calculations where the **order of operations** is not obvious.

Example

a Use a calculator to work out the value of

$$\frac{21.42 \times (12.4 - 6.35)}{(63.4 + 18.9) \times 2.83}$$

Write all the figures on the calculator display.

b Put brackets in this expression so that its value is 45.908.

$1.4 + 3.9 \times 2.2 \times 4.6$

---

a Rewrite the calculation as $(21.42 \times (12.4 - 6.35)) \div ((63.4 + 18.9) \times 2.83)$

Type this into the calculator:

`(21.42×(12.4-6.35))÷((63.4+18.9)×2.83)` → `(21.42×(12.4-` `0.556401856`

So the answer is 0.556 401 856.

Estimate:
$$\frac{20 \times (12 - 6)}{(60 + 20) \times 3} = \frac{120}{240} = 0.5$$

b By inserting a pair of brackets: $(1.4 + 3.9 \times 2.2) \times 4.6$
The calculator should display 45.908. ✓ This is the correct answer.

You can solve multi-step problems using a calculator. You will need to give your answer to an **appropriate degree of accuracy**.

Example

The diagram shows a box in the shape of a cuboid.

a Work out the volume, in m^3, of the box.

b Saleem builds boxes of different sizes.
He charges £7.89 for each m^3 of a box's volume.
Work out Saleem's charge for building this box.

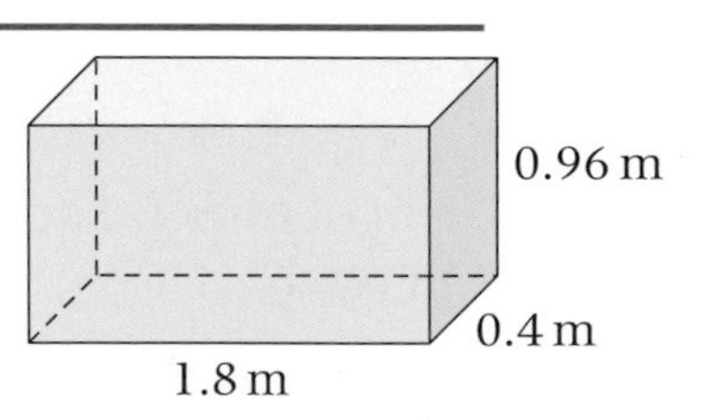

---

a Volume of a cuboid
= length × width × height

Volume ≈ $2 \times 1 \times 0.4 = 0.8\ m^3$
Volume = $1.8 \times 0.4 \times 0.96$
= $0.6912\ m^3$
= $0.7\ m^3$

b Saleem's charge
= cost for each m^3 × number of m^3

Estimate: Saleem's charge ≈ £8 × 0.8
= £6.40

Type: Saleem's charge = £7.89 × 0.6912
= £5.453 568
= £5.45

Exercise N2.6

1 Put brackets into each of these expressions to make them correct.

a $2.4 \times 4.3 + 3.7 = 19.2$

b $6.8 \times 3.75 - 2.64 = 7.548$

c $3.7 + 2.9 \div 1.2 = 5.5$

d $2.3 + 3.4^2 \times 2.7 = 37.422$

e $5.3 + 3.9 \times 3.2 + 1.6 = 24.02$

f $3.2 + 6.4 \times 4.3 + 2.5 = 46.72$

2 Use your calculator to work out each of these. Write all the figures on your calculator.

a $\dfrac{165.4 \times 27.4}{(0.72 + 4.32)^2}$

b $\dfrac{(32.6 + 43.1) \times 2.3^2}{173.7 \times (13.5 - 1.78)}$

c $\dfrac{24.67 \times (35.3 - 8.29)}{(28.2 + 34.7) \times 3.3}$

d $\dfrac{1.45^2 \times 3.64 + 2.9}{3.47 - 0.32}$

e $\dfrac{12.93 \times (33.2 - 8.34)}{(61.3 + 34.5) \times 2.9}$

f $\dfrac{24.7 - (3.2 + 1.09)^2}{2.78^2 + 12.9 \times 3}$

3 Work out each of these using your calculator. In each case give your answer to an appropriate degree of accuracy.

a Véronique lays carpet in her bedroom. The bedroom is in the shape of a rectangle with a length of 4.23 m and a width of 3.6 m. The carpet costs £6.79 per m^2.

i Calculate the floor area of the bedroom.

ii Calculate the cost of the carpet which is required to cover the floor.

b Calculate $\frac{1}{3}$ of £200.

4 Barry sees a mobile phone offer.

Vericheep Fone OFFER
Monthly fee £12.99
FREE - 200 texts every month
FREE - 200 voice minutes every month
Extra text messages 3.2p each
Extra voice minutes 5.5p each

Barry decides to see if the offer is a good idea for him.
His current mobile phone offers him unlimited texts and voice minutes for £22.99 per month.

a In February, Barry used 189 texts and 348 voice minutes. Calculate his bill using the new offer.

b In March, Barry used 273 texts and 219 voice minutes. Calculate his bill using the new offer.

c Explain if the new offer is a good idea for Barry.

N2 Exam review

Key objectives

- Use the hierarchy of operations
- Use an extended range of function keys, relevant across this programme of study
- Make mental estimates of the answers to calculations
- Develop a range of strategies for mental calculation
- Use standard column procedures for multiplication of integers and decimals
- Select and justify appropriate degrees of accuracy for answers to problems

1 **a** Work out 600×0.3 (1)

b Work out $600 \div 0.3$ (1)

c You are told that $432 \times 21 = 9072$
Write down the value of $9072 \div 2.1$ (1)

d Find an approximate value of $\dfrac{2987}{21 \times 49}$
You **must** show all your working. (2)

(AQA, 2003)

2 Use your calculator to work out the value of $\dfrac{6.27 \times 4.52}{4.81 + 9.63}$.

a Write down all the figures on your calculator display. (2)

b Write your answer to part **a** to an appropriate degree of accuracy. (1)

(Edexcel Ltd., 2004)

N3 Fractions, decimals and percentages

This unit will show you how to

- Use fraction notation to describe parts of a shape
- Compare, order and simplify fractions, converting between mixed numbers and improper fractions
- Add, subtract, multiply and divide with fractions
- Recognise and use a unit fraction as a multiplicative inverse
- Recognise the equivalence of fractions, decimals and percentages, ordering them and converting between forms using a range of methods
- Express a number as a percentage of a whole

Before you start ...

You should be able to answer these questions.

1 Here are the results of a survey for a class about their favourite colours.
What fraction of the class chose blue?
Give your answer in its simplest form.

Colour	Frequency
Red	8
Blue	10
Green	7
Black	5
Total	30

2 Copy and complete these equivalent fractions.

a $\frac{2}{3} = \frac{x}{15}$ **b** $\frac{45}{60} = \frac{3}{y}$

3 Calculate $12 \times \frac{1}{4}$

4 Copy and complete.

$10\% = \frac{1}{10} = ?$ $25\% = ? = 0.25$

$50\% = ? = ?$ $? = ? = 0.23$

5 Put these decimals in order from smallest to largest.

0.75 0.8 0.7 0.875

N3.1 Equivalent fractions

This spread will show you how to:

- Compare, order and simplify fractions, converting between mixed numbers and improper fractions

Keywords
Denominator
Equal
Equivalent
Fraction
Improper fraction
Mixed number
Numerator

This pizza is divided into 12 parts.
Adam takes $\frac{3}{12}$ of the whole pizza.

Adam's portion is also $\frac{1}{4}$ of the whole pizza.
$\frac{3}{12}$ and $\frac{1}{4}$ are **equivalent** fractions.

$$\frac{1}{4} \xrightarrow{\times 3} = \frac{3}{12}$$

- **You can find equivalent fractions by multiplying or dividing the numerator and denominator by the same number.**

- **You can simplify a fraction by dividing the numerator and denominator by a common factor.**

This process is called cancelling down.

Example

Write each of these fractions in its simplest form.

a $\frac{18}{30}$ **b** $\frac{64}{80}$ **c** $\frac{13}{27}$

a $\frac{18}{30} \xrightarrow{\div 6} \frac{3}{5}$ (numerator and denominator $\div 6$)

$\frac{18}{30} = \frac{3}{5}$

b $\frac{64}{80} \xrightarrow{\div 4} \frac{16}{20} \xrightarrow{\div 4} \frac{4}{5}$ (numerator and denominator $\div 4$, then $\div 4$)

$\frac{64}{80} = \frac{4}{5}$

c Has no common factors. It cannot be simplified.

- **You can compare and order fractions by writing them as equivalent fractions with the same denominator.**

Example

Which is bigger: $\frac{3}{7}$ or $\frac{4}{9}$?

You need an equivalent fraction for both $\frac{3}{7}$ and $\frac{4}{9}$.

$\frac{3}{7} = \frac{27}{63}$ (numerator and denominator $\times 9$) $\qquad \frac{4}{9} = \frac{28}{63}$ (numerator and denominator $\times 7$)

The common denominator of these equivalent fractions will be $7 \times 9 = 63$.

$\frac{27}{63}$ is less than $\frac{28}{63}$ so $\frac{3}{7}$ is less than $\frac{4}{9}$

Fractions can be used to describe numbers which are bigger than 1, as **mixed numbers** like $1\frac{2}{3}$ and **improper fractions** like $\frac{5}{3}$.

Example

a Change $\frac{13}{8}$ into a mixed number. **b** Change $1\frac{3}{5}$ into an improper fraction.

a $\frac{13}{8} = \frac{8}{8} + \frac{5}{8}$

$= 1 + \frac{5}{8} = 1\frac{5}{8}$

b $1\frac{3}{5} = 1 + \frac{3}{5}$

$= \frac{5}{5} + \frac{3}{5} = \frac{8}{5}$

Exercise N3.1

1 **i** Write the fraction of each of these shapes that is shaded.

ii Write your fraction in its simplest form.

a 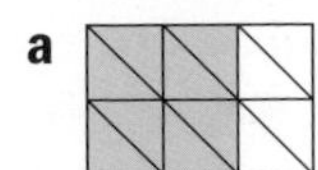

b 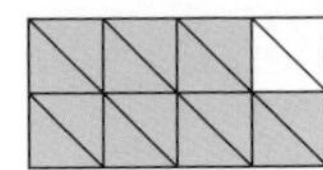

c 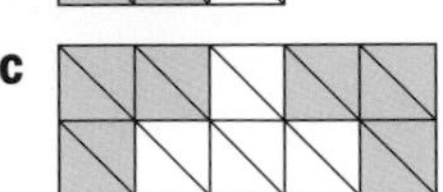

d 

2 Cancel down each of these fractions into its simplest form.

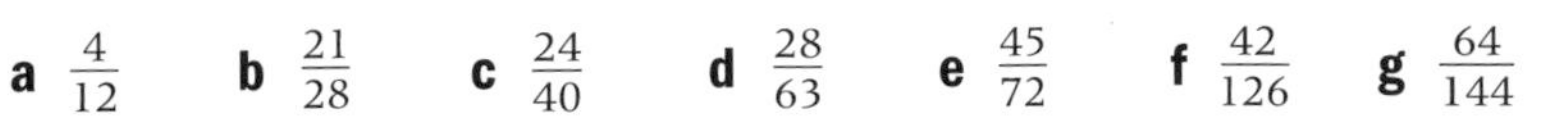

a $\frac{4}{12}$ **b** $\frac{21}{28}$ **c** $\frac{24}{40}$ **d** $\frac{28}{63}$ **e** $\frac{45}{72}$ **f** $\frac{42}{126}$ **g** $\frac{64}{144}$ **h** $\frac{23}{93}$

3 Change each of these fractions to an improper fraction.

a $1\frac{1}{2}$ **b** $3\frac{2}{3}$ **c** $4\frac{3}{8}$ **d** $2\frac{2}{9}$ **e** $5\frac{6}{7}$ **f** $7\frac{4}{5}$ **g** $8\frac{8}{11}$ **h** $12\frac{4}{7}$ **i** $12\frac{7}{13}$

4 Change each of these fractions to a mixed number.

a $\frac{5}{4}$ **b** $\frac{8}{5}$ **c** $\frac{11}{7}$ **d** $\frac{9}{4}$ **e** $\frac{11}{5}$ **f** $\frac{20}{7}$ **g** $\frac{23}{5}$ **h** $\frac{28}{9}$ **i** $\frac{67}{8}$

5 Find the missing number in each of these pairs of equivalent fractions.

a $\frac{2}{3} = \frac{?}{12}$ **b** $\frac{3}{4} = \frac{?}{36}$ **c** $\frac{5}{7} = \frac{40}{?}$ **d** $\frac{7}{8} = \frac{?}{64}$

e $\frac{12}{30} = \frac{?}{5}$ **f** $\frac{6}{7} = \frac{?}{10}$ **g** $\frac{5}{4} = \frac{?}{68}$ **h** $\frac{?}{10} = \frac{154}{220}$

6 **a** Here are two fractions, $\frac{1}{3}$ and $\frac{2}{5}$.

Explain which is the larger fraction.

Use the grids to help with your explanation.

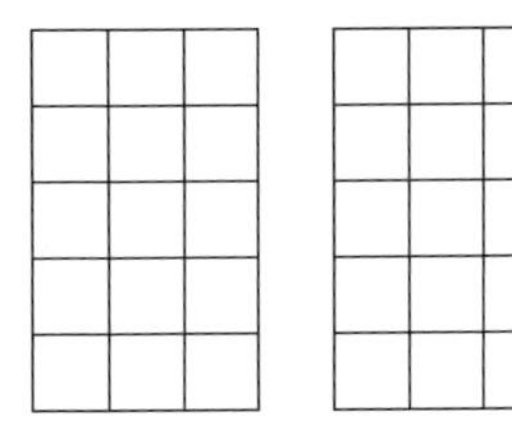

b Write these fractions in order of size.

Start with the smallest fraction.

$\frac{7}{18}$ $\frac{4}{9}$ $\frac{1}{3}$

7 For each pair of fractions, write which is the larger fraction.
Show your working.

a $\frac{3}{8}$ and $\frac{2}{5}$ **b** $\frac{3}{5}$ and $\frac{2}{3}$ **c** $\frac{4}{7}$ and $\frac{2}{5}$

d $\frac{5}{6}$ and $\frac{7}{9}$ **e** $\frac{5}{9}$ and $\frac{4}{7}$ **f** $\frac{7}{5}$ and $\frac{10}{7}$

Convert each of the fractions to an equivalent fraction with the same denominator.

8 Put these fractions in order from smallest to largest.
Show your working.

a $\frac{2}{5}$, $\frac{3}{15}$ and $\frac{1}{3}$ **b** $\frac{4}{7}$, $\frac{15}{28}$ and $\frac{1}{2}$ **c** $\frac{4}{7}$, $\frac{5}{8}$ and $\frac{9}{14}$

N3.2 Adding and subtracting fractions

This spread will show you how to:

- Add and subtract with fractions

Keywords
Common denominator
Equivalent fraction

It is easy to add or subtract **fractions** when they have the same denominator.

+

=

$\frac{3}{8} + \frac{1}{8} = \frac{4}{8}$

- **You can add or subtract fractions with different denominators by first writing them as equivalent fractions with the same denominator.**

Example

Calculate **a** $\frac{3}{5} + \frac{1}{3}$ **b** $1\frac{3}{4} - \frac{5}{7}$ **c** $1\frac{7}{10} + 2\frac{3}{5}$

a $\frac{3}{5} + \frac{1}{3}$

$\frac{3}{5} + \frac{1}{3} = \frac{9}{15} + \frac{5}{15}$

$= \frac{9 + 5}{15}$

$= \frac{14}{15}$

$\frac{3}{5} \overset{\times 3}{=} \frac{9}{15}$ (×3) $\frac{1}{3} \overset{\times 5}{=} \frac{5}{15}$ (×5)

The lowest **common denominator** is the lowest common multiple of 5 and 3, which is 15.

b $1\frac{3}{4} - \frac{5}{7}$

Change the mixed number to an improper fraction:

$1\frac{3}{4} = \frac{7}{4}$

$1\frac{3}{4} - \frac{5}{7} = \frac{7}{4} - \frac{5}{7}$

$= \frac{49}{28} - \frac{20}{28}$

$= \frac{49 - 20}{28}$

$= \frac{29}{28}$

$= 1\frac{1}{28}$

$\frac{7}{4} \overset{\times 7}{=} \frac{49}{28}$ (×7) $\frac{5}{7} \overset{\times 4}{=} \frac{20}{28}$ (×4)

The lowest common denominator is the lowest common multiple of 4 and 7, which is 28.

c $1\frac{7}{10} + 2\frac{3}{5}$

Change the mixed numbers to improper fractions:

$1\frac{7}{10} = \frac{17}{10}$ $2\frac{3}{5} = \frac{13}{5}$

$1\frac{7}{10} + 2\frac{3}{5} = \frac{17}{10} + \frac{13}{5}$

$= \frac{17}{10} + \frac{26}{10}$

$= \frac{17 + 26}{10}$

$= \frac{43}{10}$

$= 4\frac{3}{10}$

$\frac{13}{5} \overset{\times 2}{=} \frac{26}{10}$ (×2)

The lowest common denominator is the lowest common multiple of 5 and 10, which is 10.

An alternative method is to write:

$1 + \frac{7}{10} + 2 + \frac{3}{5}$

$= 3 + \frac{7}{10} + \frac{3}{5}$

$= \ldots$

Exercise N3.2

1 Work out

a $\frac{1}{3} + \frac{1}{3}$ **b** $\frac{3}{8} + \frac{2}{8}$ **c** $\frac{8}{11} - \frac{3}{11}$ **d** $\frac{8}{17} + \frac{5}{17}$

e $\frac{14}{23} - \frac{11}{23}$ **f** $\frac{5}{27} + \frac{8}{27}$

2 Work out each of these, leaving your answer in its simplest form.

a $\frac{2}{3} + \frac{1}{3}$ **b** $\frac{8}{9} - \frac{2}{9}$ **c** $\frac{8}{11} + \frac{5}{11}$ **d** $\frac{15}{13} - \frac{8}{13}$

e $\frac{14}{9} + \frac{1}{9}$ **f** $\frac{17}{12} - \frac{9}{12}$ **g** $1\frac{2}{3} + \frac{2}{3}$ **h** $4\frac{2}{7} - \frac{5}{7}$

3 Work out

Write both fractions as equivalent fractions with the same denominator.

a $\frac{1}{3} + \frac{1}{2}$ **b** $\frac{1}{4} + \frac{3}{5}$ **c** $\frac{3}{5} - \frac{1}{3}$ **d** $\frac{4}{5} - \frac{2}{7}$

e $\frac{5}{8} + \frac{1}{3}$ **f** $\frac{4}{9} + \frac{2}{5}$ **g** $\frac{7}{9} - \frac{2}{11}$ **h** $\frac{7}{15} + \frac{3}{7}$

4 Work out each of these, leaving your answer in its simplest form as appropriate.

a $\frac{2}{5} - \frac{1}{15}$ **b** $\frac{1}{2} - \frac{1}{3}$ **c** $\frac{2}{5} + \frac{7}{20}$ **d** $\frac{1}{2} - \frac{1}{6}$

5 Work out each of these, leaving your answer in its simplest form.

a $\frac{4}{5} + \frac{2}{3}$ **b** $1\frac{1}{2} + \frac{3}{5}$ **c** $1\frac{1}{3} + 1\frac{1}{4}$ **d** $1\frac{2}{7} + \frac{3}{5}$

e $2\frac{2}{5} - \frac{1}{3}$ **f** $3\frac{3}{8} - 1\frac{1}{2}$ **g** $4\frac{1}{3} - 2\frac{3}{4}$ **h** $3\frac{4}{7} - 2\frac{8}{9}$

6 Work out each of these, leaving your answer in its simplest form.

a Pete walked $3\frac{2}{3}$ miles before lunch and then a further $2\frac{1}{4}$ miles after lunch. How far did he walk altogether?

b A bag weighs $2\frac{3}{16}$ lb when it is full. When empty the bag weighs $\frac{3}{8}$ lb. What is the weight of the contents of the bag?

c Henry and Paula are eating pistachios. Henry has a full bag weighing $1\frac{3}{16}$ kg. Paula has a bag that weighs $\frac{4}{5}$ kg. What is the total mass of their two bags of pistachios?

d Simon spent $\frac{2}{3}$ of his wages on a mobile phone. He spent $\frac{1}{5}$ of his wages on a trip to the theatre. Work out the fraction of his wages that he had left.

e Calculate the perimeter of each of these swimming pools.

i

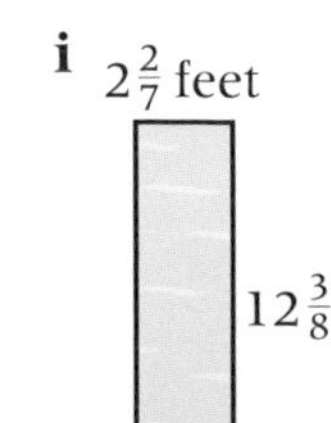

ii

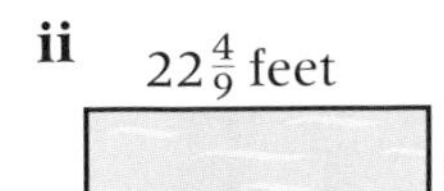

iii

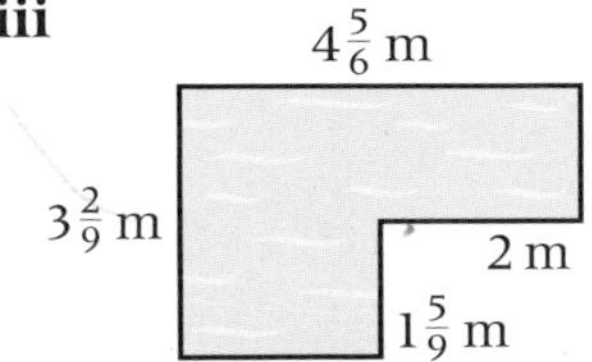

N3.3 Multiplying and dividing fractions

This spread will show you how to:

- Multiply and divide with fractions
- Recognise and use a unit fraction as a multiplicative inverse

Keywords
Equivalent
Integer
Fraction
Multiplicative inverse
Unit fraction

You can multiply a **unit fraction** by an integer using a number line.

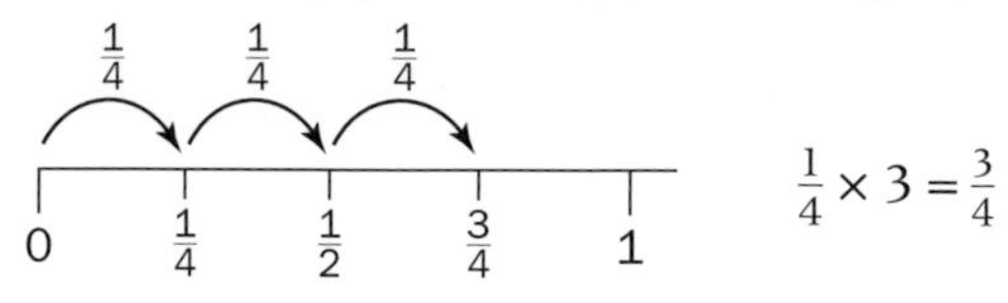

$\frac{1}{4} \times 3 = \frac{3}{4}$

A unit fraction has numerator 1: $\frac{1}{2}, \frac{1}{3}, \frac{1}{4}, \ldots$

Multiplying by $\frac{1}{4}$ is the same as dividing by 4.

$3 \times \frac{1}{4} = \frac{3}{4} \quad \Leftrightarrow \quad 3 \div 4 = \frac{3}{4}$

- **You can multiply any fraction by an integer.**

$$4 \times \frac{3}{8} = \frac{4 \times 3}{8} = \frac{12}{8} = \frac{3}{2} = 1\frac{1}{2}$$

- **You can multiply a fraction by another fraction by multiplying the numerators together and multiplying the denominators together.**

$$\frac{3}{5} \times \frac{5}{8} = \frac{3 \times 5}{5 \times 8} = \frac{15}{40} = \frac{3}{8}$$

You can divide an integer by a unit fraction.

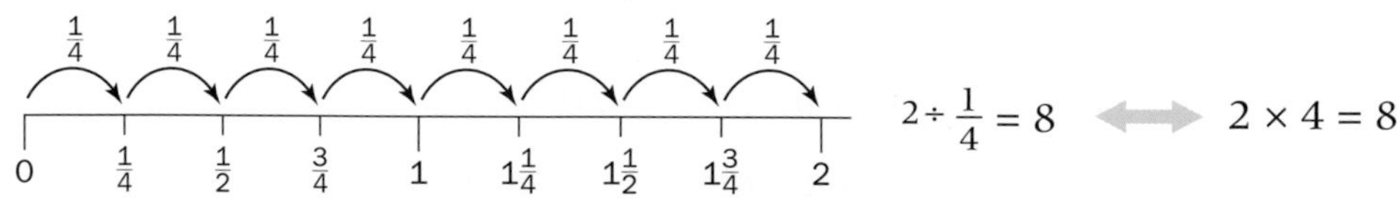

$2 \div \frac{1}{4} = 8$  $2 \times 4 = 8$

Think how many $\frac{1}{4}$s are there in 2 wholes?

1	2	5	6
3	4	7	8

- **You can divide an integer by any fraction using unit fractions.**

$$4 \div \frac{2}{3} = 4 \div 2 \div \frac{1}{3} = 2 \div \frac{1}{3} = 2 \times 3 = 6$$

You can use the relationship between multiplication and division.

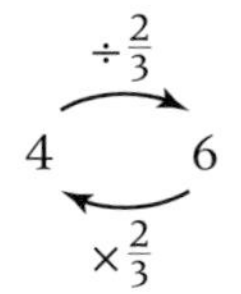

Dividing by $\frac{2}{3}$ is the same as multiplying by $\frac{3}{2}$.

Multiplying by $\frac{2}{3}$ is the same as dividing by $\frac{3}{2}$.

$\times\frac{3}{2}$: 4 → 6; $\div\frac{3}{2}$: 6 → 4

$\times\frac{3}{2}$ is the **multiplicative inverse** of $\div\frac{3}{2}$.

Example

Calculate

a $4 \div \frac{3}{5}$ **b** $\frac{3}{8} \div \frac{5}{6}$ **c** $2\frac{2}{5} \div 1\frac{1}{4}$

a $4 \div \frac{3}{5} = 4 \times \frac{5}{3}$
$= \frac{20}{3}$
$= 6\frac{2}{3}$

b $\frac{3}{8} \div \frac{5}{6} = \frac{3}{8} \times \frac{6}{5}$
$= \frac{3 \times 6}{8 \times 5}$
$= \frac{18}{40}$

c $2\frac{2}{5} \div 1\frac{1}{4} = \frac{12}{5} \div \frac{5}{4}$
$= \frac{12}{5} \times \frac{4}{5}$
$= \frac{12 \times 4}{5 \times 5}$
$= \frac{48}{25} = 1\frac{23}{25}$

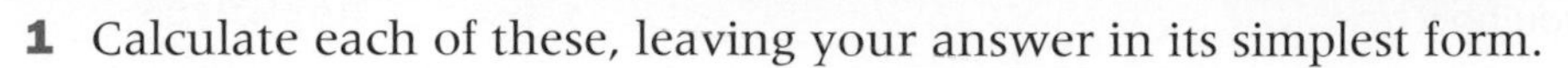

Exercise N3.3

1 Calculate each of these, leaving your answer in its simplest form.

a $3 \times \frac{1}{2}$ **b** $6 \times \frac{1}{3}$ **c** $10 \times \frac{1}{3}$

d $15 \times \frac{1}{7}$ **e** $\frac{1}{10} \times 25$ **f** $\frac{1}{3} \times 13$

2 Calculate each of these, leaving your answer in its simplest form.

a $3 \times \frac{2}{3}$ **b** $6 \times \frac{2}{3}$ **c** $5 \times \frac{2}{3}$ **d** $2 \times \frac{7}{24}$

e $4 \times \frac{3}{20}$ **f** $\frac{4}{5} \times 2$ **g** $\frac{3}{5} \times 10$ **h** $\frac{11}{8} \times 17$

3 Calculate each of these, leaving your answer in its simplest form.

a $4 \div \frac{1}{2}$ **b** $2 \div \frac{1}{5}$ **c** $2 \div \frac{1}{7}$

d $10 \div \frac{1}{2}$ **e** $12 \div \frac{1}{4}$ **f** $22 \div \frac{1}{10}$

4 Calculate each of these, leaving your answer in its simplest form.

a What is the total weight of 7 boxes that each weigh $\frac{2}{5}$ kg?

b What is the total length of 5 pieces of wood that are each $\frac{3}{7}$ of a metre long?

5 Calculate each of these, leaving your answer in its simplest form.

a $4 \div \frac{2}{3}$ **b** $7 \div \frac{2}{5}$ **c** $2 \div \frac{5}{6}$ **d** $12 \div \frac{6}{7}$

e $20 \div \frac{5}{12}$ **f** $5 \div \frac{7}{9}$ **g** $3 \div 1\frac{1}{2}$ **h** $3 \div 1\frac{2}{5}$

6 Calculate each of these, leaving your answer in its simplest form.

a $\frac{2}{5} \times \frac{3}{4}$ **b** $\frac{3}{5} \times \frac{3}{4}$ **c** $\frac{5}{7} \times \frac{3}{4}$ **d** $\frac{4}{7} \times \frac{3}{5}$

e $\frac{5}{6} \times \frac{4}{5}$ **f** $\frac{3}{8} \times \frac{7}{9}$ **g** $\frac{3}{5} \times \frac{10}{9}$ **h** $\frac{15}{16} \times \frac{12}{5}$

i $(\frac{3}{7})^2$ **j** $1\frac{3}{4} \times \frac{2}{7}$ **k** $3\frac{2}{3} \times \frac{7}{11}$ **l** $1\frac{3}{8} \times 1\frac{2}{5}$

7 Calculate each of these, leaving your answer in its simplest form.

a $4 \div \frac{2}{5}$ **b** $\frac{2}{3} \div \frac{4}{5}$ **c** $\frac{4}{5} \div \frac{3}{4}$ **d** $\frac{4}{7} \div \frac{2}{3}$

e $\frac{3}{7} \div \frac{4}{9}$ **f** $\frac{3}{5} \div \frac{1}{2}$ **g** $\frac{3}{4} \div 3$ **h** $\frac{4}{7} \div 5$

i $\frac{4}{11} \div 5$ **j** $\frac{7}{4} \div \frac{2}{3}$ **k** $\frac{7}{4} \div \frac{3}{2}$ **l** $\frac{9}{5} \div \frac{5}{3}$

m $1\frac{1}{2} \div \frac{3}{4}$ **n** $2\frac{1}{4} \div \frac{2}{3}$ **o** $2\frac{2}{5} \div \frac{9}{7}$

8 Calculate each of these, leaving your answer in its simplest form.

a A plank of wood is $4\frac{3}{5}$ metres long. How many pieces of wood of length $1\frac{1}{4}$ metres can be cut from the piece of wood?

b A paint pot can hold $2\frac{3}{4}$ litres of paint. Hector buys $11\frac{1}{5}$ litres of emulsion paint. How many times can Hector fill the paint pot with emulsion paint?

N3.4 Fractions, decimals and percentages

This spread will show you how to:

- Recognise the equivalence of fractions, decimals and percentages, ordering them and converting between forms using a range of methods

Keywords
Decimal
Equivalent
Fraction
Order
Percentage

- **Percentages, fractions and decimals are all ways of writing the same thing.**
- **To change a percentage into a fraction you write it as a fraction out of 100 and then cancel down.**

$$40\% = \frac{4}{10} = \frac{2}{5} = 0.4$$

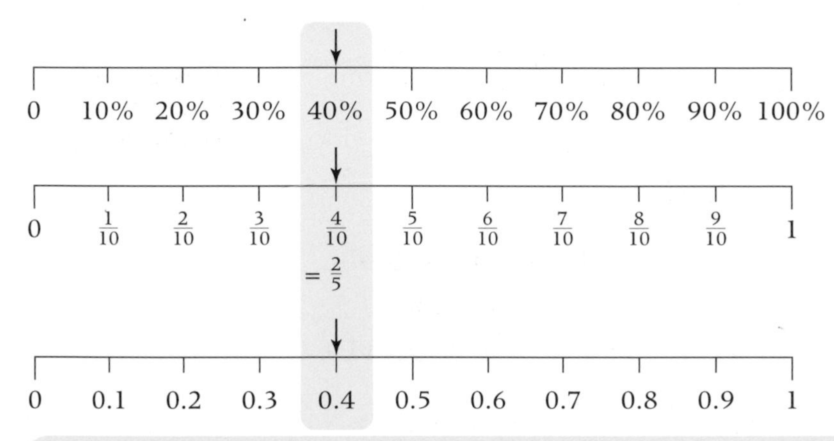

Some useful **equivalents** to remember:

$10\% = \frac{10}{100} = \frac{1}{10} = 0.1$

$20\% = \frac{20}{100} = \frac{1}{5} = 0.2$

$25\% = \frac{25}{100} = \frac{1}{4} = 0.25$

$50\% = \frac{50}{100} = \frac{1}{2} = 0.5$

$75\% = \frac{75}{100} = \frac{3}{4} = 0.75$

- **You can write a terminating decimal as a fraction.**

A terminating decimal ends after a definite number of digits.

$$\begin{aligned} 3.42 &= 3 \text{ units} + \frac{4}{10} + \frac{2}{100} \\ &= 3 \text{ units} + \frac{42}{100} \\ &= 3\frac{42}{100} \\ &= 3\frac{21}{50} \end{aligned}$$

- You can convert a fraction into a decimal ...

... using equivalent fractions

$\frac{3}{5} \xrightarrow{\times 2} \frac{6}{10}$ (numerator and denominator both $\times 2$)

$\frac{7}{40} = \frac{35}{200} = \frac{17.5}{100}$ (numerator and denominator both $\times 5$, then both $\div 2$)

$\frac{3}{5} = \frac{6}{10} = 0.6$

$\frac{7}{40} = \frac{17.5}{100} = 0.175$

... using division

$\frac{7}{40} = 7 \div 40 = 0.175$

You can use short division to convert a fraction into a decimal.

You can convert between percentages and fractions.

$30\% = \frac{30}{100} = \frac{3}{10}$

$145\% = \frac{145}{100} = \frac{29}{20} = 1\frac{9}{20}$

To cancel down a fraction, divide the numerator and denominator by a common factor.

You can convert between percentages and decimals.

$32\% = \frac{32}{100} = 0.32$ (32 ÷ 100)

$5.4\% = \frac{5.4}{100} = 0.054$ (5.4 ÷ 100)

Exercise N3.4

1 Write each of these decimals as a fraction in its simplest form.

a 0.3 **b** 0.6 **c** 0.64 **d** 0.45

e 0.375 **f** 1.08 **g** 3.23

2 Change these fractions to decimals without using a calculator.

a $\frac{3}{10}$ **b** $\frac{11}{25}$ **c** $\frac{26}{25}$ **d** $\frac{124}{200}$

e $\frac{27}{60}$ **f** $\frac{39}{75}$ **g** $\frac{42}{150}$ **h** $3\frac{21}{60}$

3 Change these fractions into decimals using an appropriate method. Give your answers to 2 decimal places where necessary.

You may use a calculator where appropriate.

a $\frac{22}{50}$ **b** $\frac{2}{3}$ **c** $\frac{27}{20}$ **d** $\frac{11}{15}$

e $\frac{8}{7}$ **f** $1\frac{2}{5}$ **g** $2\frac{11}{66}$ **h** $\frac{11}{13}$

4 Write each of these percentages as a fraction in its simplest form.

a 40% **b** 90% **c** 35% **d** 65%

e 1% **f** 362% **g** 15.25% **h** 2.125%

5 Write each of these fractions as a percentage without using a calculator.

a $\frac{27}{50}$ **b** $\frac{2}{5}$ **c** $\frac{17}{20}$ **d** $\frac{13}{25}$ **e** $\frac{2}{3}$ **f** $\frac{48}{200}$ **g** $1\frac{3}{15}$ **h** $\frac{33}{75}$

6 Convert these fractions to decimals using short division.

a $\frac{3}{5}$ **b** $\frac{3}{8}$ **c** $\frac{7}{16}$ **d** $\frac{4}{9}$

e $\frac{5}{7}$ **f** $\frac{7}{12}$ **g** $\frac{15}{23}$ **h** $\frac{8}{35}$

7 Write these percentages as decimals.

a 37% **b** 7% **c** 189% **d** 45%

e 145% **f** 0.8% **g** 250% **h** 123.2%

8 Write these decimals as percentages.

a 0.72 **b** 0.2 **c** 1.25 **d** 0.03

e 1.02 **f** 0.0325 **g** 0.333 ... **h** 1.372

Try converting the fraction into a decimal first!

9 Write these fractions as percentages. Give your answers to 1 decimal place as appropriate.

a $\frac{48}{70}$ **b** $\frac{16}{25}$ **c** $\frac{17}{19}$ **d** $1\frac{11}{12}$ **e** $\frac{5}{19}$

10 Investigate all unit fractions from $\frac{1}{2}$ to $\frac{1}{30}$ using division. Which are recurring decimals?

N3.5 Ordering fractions, decimals and percentages

This spread will show you how to:

- Recognise the equivalence of fractions, decimals and percentages, ordering them and converting between forms using a range of methods
- Express a number as a percentage of a whole

Keywords
Decimal
Equivalent
Fraction
Order
Percentage

You can convert between **fractions**, **decimals** and **percentages** using a range of mental and written methods.

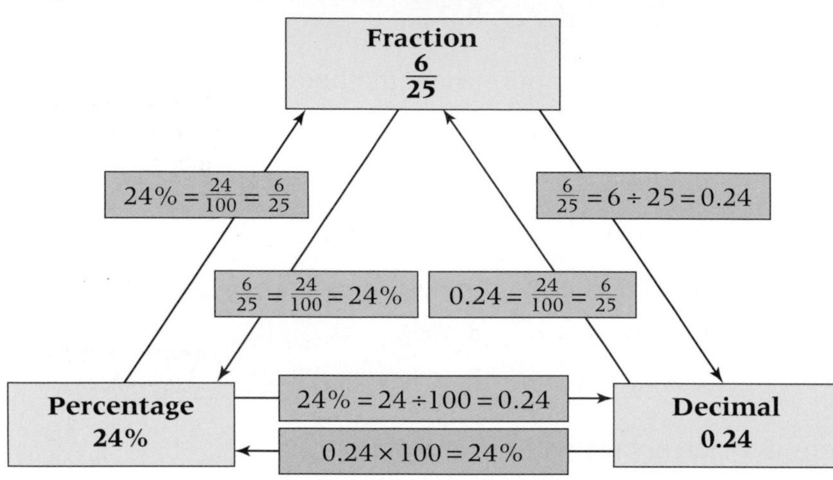

You can write something as a percentage by first finding the fraction.

Example

a What percentage of this shape is shaded?

b In a class at Clarendon College there are 40 students. 24 of them are men. What percentage of the class are men?

---

a There are 16 equal parts.
7 of the parts are shaded.
The fraction shaded is $\frac{7}{16}$.
% shaded = $\frac{7}{16} = 0.4375 = 43.75\%$

b There are 40 students in the class.
24 of the students are men.
The fraction of men is $\frac{24}{40}$.
% of men = $\frac{24}{40} = \frac{12}{20} = \frac{60}{100} = 60\%$

- **You can order fractions, decimals and percentages by converting them into decimals.**

Example

Write these numbers in order of size. Start with the smallest number.

0.8 70% $\frac{7}{8}$ $\frac{3}{4}$

$70\% = \frac{70}{100} = 0.7$ $\frac{7}{8} = \frac{35}{40} = \frac{175}{200} = \frac{875}{1000} = 0.875$ $\frac{3}{4} = 0.75$

Rewrite $\frac{7}{8}$ as an equivalent fraction with a denominator of 1000.

Place the decimals in order: 0.7 0.75 0.8 0.875
70% $\frac{3}{4}$ 0.8 $\frac{7}{8}$

Exercise N3.5

1 Copy and complete this table.

Fraction	Decimal	Percentage
$\frac{3}{8}$		
	0.28	
		15%
	0.375	
$\frac{4}{5}$		
		17.5%

2 For each pair of fractions, write which is the larger fraction. Show your working.

Convert each of the fractions into a decimal.

a $\frac{5}{8}$ and $\frac{3}{5}$ **b** $\frac{4}{5}$ and $\frac{2}{3}$ **c** $\frac{5}{7}$ and $\frac{3}{5}$ **d** $\frac{3}{8}$ and $\frac{4}{11}$

e $\frac{10}{7}$ and $\frac{16}{11}$ **f** $\frac{14}{9}$ and $\frac{17}{11}$ **g** $1\frac{2}{7}$ and $1\frac{7}{23}$ **h** $2\frac{7}{12}$ and $2\frac{8}{11}$

3 Copy each pair of fractions, decimals and percentages and insert '>' or '<' in between them. Show your working out clearly for each question.

$>$ = greater than
$<$ = less than

a $\frac{3}{5}$ 0.7 **b** $\frac{7}{15}$ $\frac{30}{65}$ **c** $\frac{5}{9}$ $\frac{7}{13}$ **d** $2\frac{2}{3}$ 265%

4 Put these fractions, decimals and percentages in order from smallest to largest. Show your working.

a 47%, $\frac{12}{25}$ and 0.49 **b** $\frac{4}{5}$, 78% and 0.81

c $\frac{5}{8}$, 66% and $\frac{7}{12}$ **d** $\frac{5}{16}$, 0.3, 29% and $\frac{7}{22}$

5 In each of these questions express the answer first as a fraction, then convert the fraction to a percentage using an appropriate method.

a In a class there are 28 students. 19 of the students are right-handed. What percentage of the class are right-handed?

b In a survey of 80 people, 55 said they would prefer full-time education to be compulsory until the age of 18. What percentage of the 80 people preferred full-time education to be compulsory until the age of 18?

c In a football squad of 24 players, 5 of the players are goalkeepers. What percentage of the football squad are not goalkeepers?

d In a mixed packet of 54 biscuits, 36 of the biscuits are covered in chocolate. What percentage of the biscuits in the packet are not covered in chocolate?

6 a Leon scores 68% in his French exam and gets $\frac{37}{54}$ in his German exam. In which subject did he do the best? Explain your answer.

b In a college survey 23% of the students said they did not like eating meat. In Sarah's class $\frac{7}{31}$ students said they did not like eating meat. How do the results of Sarah's class compare with the rest of the college?

N3

Exam review

Key objectives

- Order fractions by rewriting them with a common denominator
- Convert simple fractions of a whole to percentages of a whole and vice versa
- Multiply and divide a given fraction by an integer, by a unit fraction and by a general fraction

1 Work out:

a $\frac{2}{5} \times \frac{1}{4}$

Give your answer as a fraction in its simplest form. (2)

b $\frac{2}{5} - \frac{1}{4}$

(2)

(OCR, 2003)

2 a Write these five fractions in order of size.

Start with the smallest fraction. (2)

$\frac{3}{4}$ $\frac{1}{2}$ $\frac{3}{8}$ $\frac{2}{3}$ $\frac{1}{6}$

b Write these numbers in order of size.

Start with the smallest number. (2)

65% $\frac{3}{4}$ 0.72 $\frac{2}{3}$ $\frac{3}{5}$

(Edexcel Ltd., 2004)

N4 Ratio and proportion

This unit will show you how to

- Use decimals, fractions and percentages to describe proportions
- Use proportions to make simple comparisons
- Understand and use ratio, proportion and direct proportion in problem solving
- Calculate missing amounts when two quantities are in direct proportion
- Simplify a ratio and express it in the form 1 : *n*
- Calculate exchange rates and solve problems involving exchange rates
- Solve problems involving compound measures

Before you start ...

You should be able to answer these questions.

1 What proportion of this shape is shaded?

2 Neil buys 2 pizzas at a cost £7.00. What is the cost of 8 pizzas?

3 John weighs 50 kg. Kevin weighs 75 kg. How many times heavier than John is Kevin?

4 What is 56 ÷ 14?

5 Krishna earns £8.50 per hour. How much does he get paid for 8 hours work?

6 Beatrice is driving her car. She travels 200 miles in 4 hours. What is her average speed?

N4.1 Proportion

This spread will show you how to:

- Use decimals, fractions and percentages to describe proportions
- Use proportions to make simple comparisons

Keywords
Decimal
Equivalent
Fraction
Percentage
Proportion

- A **proportion** is a part of the whole.
  You can use **percentages, fractions** and **decimals** to describe proportions.

Example

a What proportion of this shape is shaded?

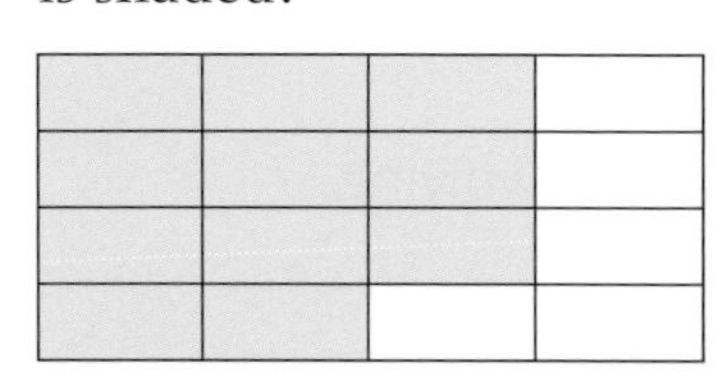

b What proportion of cars are blue?

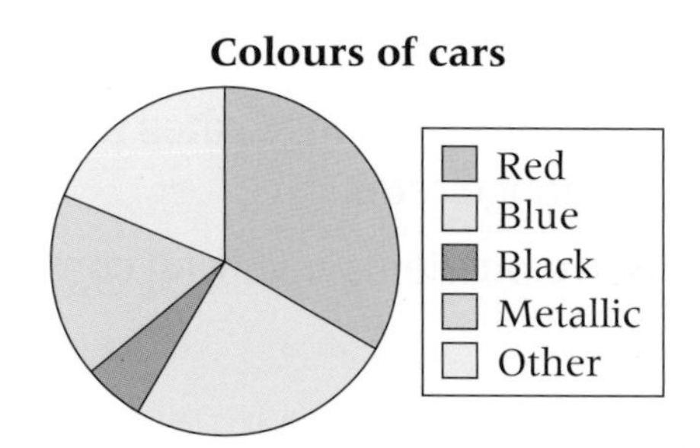

a The proportion shaded is $\frac{11}{16}$

$= 11 \div 16$

$= 0.6875$

$= 68.75\%$

b The sector representing blue cars is 90 degrees.

Fraction of blue cars $= \frac{90}{360} = \frac{1}{4}$

$= 1 \div 4$

$= 0.25$

$= 25\%$

Change the fraction to a decimal by division.
Change the decimal to a percentage by multiplying by 100.

- **You can use proportions to make simple comparisons.**

Example

John scores $\frac{26}{40}$ in his Maths exam and 63% in his English exam.
In which exam did he score the highest mark?

Change the Maths mark into a percentage:

$\frac{26}{40} = 26 \div 40$

$= 0.65$

$= 0.65 \times 100\%$

$= 65\%$

Maths mark = 65% English mark = 63%

John scored a higher mark in his Maths exam.

It is easier to compare proportions by converting them to percentages.

Exercise N4.1

1 Write the proportion of each of these shapes that is shaded. Write each of your answers as

i a fraction in its simplest form

ii a percentage (to 1 decimal place as appropriate).

a

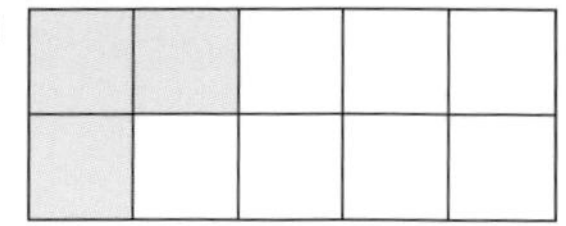

b

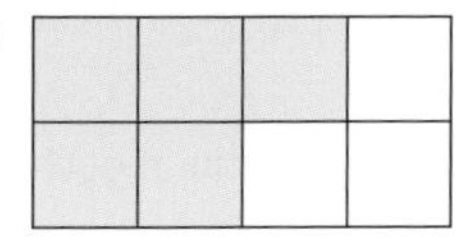

c

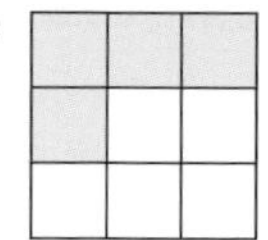

d

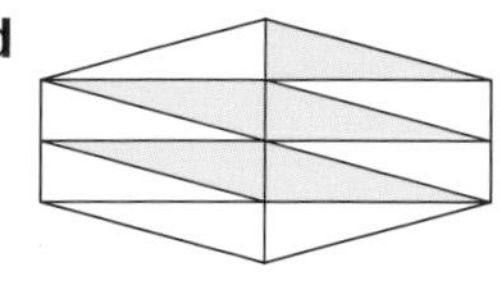

e

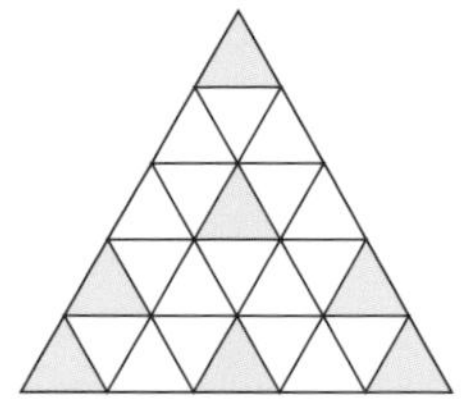

2 Put these quantities in order from lowest to highest.
Do these without a calculator.

a $\frac{3}{5}$, 61%, 0.63 **b** $\frac{7}{10}$, $\frac{17}{25}$, 69%, 0.71 **c** $\frac{7}{20}$, $\frac{2}{5}$, $\frac{3}{8}$, 0.36, 34%

Do these with a calculator.

d $\frac{3}{7}$, 42%, $\frac{2}{5}$ **e** 15%, $\frac{3}{19}$, 0.14, $\frac{1}{5}$ **f** 81%, $\frac{8}{9}$, 0.93, 0.9, $\frac{19}{20}$

3 Solve each of these without using a calculator.
Express each of your answers

i as a fraction in its lowest form **ii** as a percentage.

a Brian's mark in a Geography test was 42 out of 70.
What proportion of the test did he answer correctly?

b Hannah collects ornaments. She has 25 ornaments altogether.
13 of the 25 ornaments are from Russia.
What proportion of Hannah's ornaments are from Russia?

c A restaurant makes a service charge of 3p in every 20p.
Work out 3p as a proportion of 20p.

d A class at Clarendon College has 30 students.
21 of these students are men.
What proportion of the class are men?

e Sakarako makes cakes. On Friday she makes 60 cakes.
Sakarako puts icing on 48 of the cakes.
What proportion of the cakes have icing on them?

4 These are Samina's end of year exam results:

English $\frac{17}{25}$ French $\frac{29}{40}$ Maths 79%

History $\frac{23}{30}$ Science $\frac{15}{20}$ Technology $\frac{28}{45}$

a Put these results in order from lowest to highest.

b In which subject did Samina do the best?

N4.2 Unitary method

This spread will show you how to:

- Calculate missing amounts when two quantities are in direct proportion

Keywords
Direct proportion
Unitary method

- A **ratio** tells you how many times bigger one number is compared to another number. You can calculate a ratio using division.

Ratio is covered in greater detail on pages 48–51.

- Numbers or quantities are in **direct proportion** when the ratio of each pair of corresponding values is the same.

Example

Here is a price list for the cost of different-sized tins of paint.

Number of litres	Total cost (£)
2	9.00
3	13.50
5	22.50
10	45.00

Is the number of litres of paint in proportion to the total cost?

For 2 tins of paint — Ratio of 'number of litres' to 'total cost' = $\frac{9}{2} = 4.5$

For 3 tins of paint — Ratio of 'number of litres' to 'total cost' = $\frac{13.5}{5} = 4.5$

For 5 tins of paint — Ratio of 'number of litres' to 'total cost' = $\frac{22.5}{5} = 4.5$

For 10 tins of paint — Ratio of 'number of litres' to 'total cost' = $\frac{45}{10} = 4.5$

The ratio is the same, so the numbers are in direct proportion.

Remember: to find the ratio of two numbers you divide them.

- You can use the **unitary method** to solve problems involving direct proportion. In this method you always find the value of one unit of a quantity.

Example

Here is a recipe for mushroom soup.
Work out the number of grams of mushrooms needed to make mushroom soup for eight people.

Mushroom Soup
(for 3 people)

270 g of button mushrooms
45 ml of white wine
450 ml of vegetable stock
3 tbsp of olive oil
3 garlic cloves

Number of people	Grams of mushrooms
3	270
÷ 3 → 1	÷ 3 → 90
× 8 → 8	× 8 → 720

You need 720 g of mushrooms for eight people.

The number of grams of mushrooms is in direct proportion to the number of people.

Exercise N4.2

1 How many times bigger than

a 12 is 60 **b** 3 is 24 **c** 12 is 84

d 20 is 90 **e** 18 is 27 **f** 9 m is 30 m

g 16 km is 50 km **h** £10 is £72 **i** 36 kg is 240 kg

j $25 is $215 **k** 10 mm is 18 mm **l** 220p is 300p?

(Give your answers to 1 dp as appropriate.)

2 What proportion of

a 15 is 5 **b** 20 is 10 **c** 10 cm is 3 cm

d 12 kg is 3 kg **e** 4 cm is 15 mm **f** £30 is £12

g 3 m is 50 cm **h** 40 km is 60 km **i** £8 is £16?

3 Which of these sets of numbers are in direct proportion? Explain your reasoning.

a

A	B
3	12
30	120
6	24
1.5	6

b

C	D
4	13
35	115
7	21
15	47

c

E	F
2	3.6
7	12.6
15	27
13	23.4

d

G	H
12	37.2
8	24.8
19	58.9
45	139.5

4 Steve works as a manager. He is paid by the hour. Copy and complete this table to help him work out his pay.

Hours worked	6	1	3	30	38	48
Pay (£)	£300					

5 **a** 2 pizzas cost £7.50. What is the cost of 5 pizzas?

b 4 erasers cost £1.40. What is the cost of 17 erasers?

c 11 packets of seeds cost £17.49. What is the cost of 7 packets of seeds?

d 5 tennis balls cost £2.60. What is the cost of 12 tennis balls?

e 5 kg of apples cost £3.95. What is the cost of 12 kg of apples?

f There are 2640 megabytes of memory on 11 identical memory sticks. How much memory is there on 3 memory sticks?

g 30 protractors cost £3.30. What is the cost of 17 protractors?

h It takes 10 litres of orange juice to fill 25 cups. Work out how many litres of orange juice are needed to fill 30 cups.

i Three 1 litre tins of paint cost a total of £23.85. Find the cost of seven of the 1 litre tins of paint.

j 12 cans of tomato soup weigh 5040g. What is the weight of 15 cans?

In questions **a–i** the items are all the same price.

N4.3 Introducing ratio

This spread will show you how to:

- Simplify a ratio and express it in the form 1 : n
- Solve simple problems involving ratio

Keywords
Ratio
Scale
Simplest form
Unitary form

You can compare the size of two quantities using a **ratio**.

- You can simplify a ratio by dividing both parts by the same number. When a ratio cannot be simplified any further it is in its **simplest form**.

Example

Express each of these ratios in its simplest form.

a 55 : 65 **b** 3 m : 120 cm **c** 2.5 : 4.5

a 55 : 65

55 : 65 (÷ 5) → 11 : 13

Answer 11 : 13

b 3 m : 120 cm

300 : 120 (÷ 10) → 30 : 12 (÷ 6) → 5 : 2

Answer 5 : 2

c 2.5 : 4.5

2.5 : 4.5 (× 10) → 25 : 45 (÷ 5) → 5 : 9

Answer 5 : 9

In part **b**, convert the measurements to the same unit.

Convert decimals to whole numbers by multiplying everything by the same number.

- **A ratio can be expressed in the form 1 : *n*. This is called the unitary form.**

Example

Write these ratios in the form 1 : n.

a 6 cm : 9 cm **b** 4 cm : 2 m

a 6 cm : 9 cm

6 : 9 (÷ 6) → 1 : 1.5

Answer 1 : 1.5

b 4 cm : 2 m Change 2 m into 200 cm

4 : 200 (÷ 4) → 1 : 50

Answer 1 : 50

Maps and plans are drawn to **scale**.

Example

On a plan, a real-life measurement of 4 m is drawn as a length of 25 cm. What is the scale of the plan?

Ratio of plan : real life
= 25 cm : 4 m

Change 4 m into 400 cm

25 : 400 (÷ 25) → 1 : 16

The scale of the plan is 1 : 16.

Exercise N4.3

1 Write each of these ratios in its simplest form.

a 4 : 8 **b** 16 : 10 **c** 40 : 25 **d** 36 : 24

e 95 : 45 **f** 28 : 168 **g** 4 : 8 : 6 **h** 20 : 25 : 40

2 Write each of these ratios in its simplest form.

a 40 cm : 1 m **b** 55 mm : 8 cm **c** 3 km : 1200 m

d 4 m : 240 cm **e** 700 mm : 42 cm **f** 12 mins : 450 secs

3 Express each of these pairs of measurements as a ratio in its simplest form.

a A table width of 40 cm to a table length of 1.2 m. What is the ratio of width to length?

b A lorry is 8.4 m long. A van is 360 cm long. What is the ratio of the lorry length to the van length?

c An alloy contains 24 kg of tin and 3.2 kg of zinc. What is the ratio of tin to zinc in the alloy?

d A bag of newly dug potatoes contains 6.4 kg of potatoes and 576 g of earth. What is the ratio of earth to potatoes in the bag?

4 Write each of these ratios in its simplest form.

a 1.5 : 4.5 **b** 2.25 : 1.5 **c** $\frac{1}{2} : \frac{1}{4}$ **d** $\frac{3}{8} : \frac{1}{2}$

5 Write each of these ratios in the form 1 : n.

a 4 : 10 **b** 39 : 150 **c** 30 : 125 **d** 720 g : 30 kg

e 95p : £22.50 **f** 3.2 : 480 **g** 0.36 : 4.5

6 In each of these questions, work out the scale of the map, plan or model. This is a ratio expressed in the form 1 : n.

a On a plan, a length of 4 cm represents a real-life measurement of 2 m. What is the scale of the plan?

b On a map, a length of 3.2 cm represents a real-life measurement of 640 m. What is the scale of the map?

c On a plan, a length of 7.2 cm represents a real-life measurement of 2.592 m. What is the scale of the plan?

7 Solve each of these problems.

a In a batch of concrete the ratio of sand to cement is 5 : 3. How much sand is needed to mix with 21 kg of cement?

b In a metal alloy the ratio of aluminium to zinc is 7 : 2.5. How much aluminium is needed to mix with 10 kg of zinc?

N4.4 Calculating with ratio

This spread will show you how to:

- Solve problems involving ratio and proportion
- Divide an amount in a given ratio

Keywords
Ratio
Scale

You can use ratios to solve problems.

Example

An alloy is made from zinc and copper in the ratio 3 : 8. How much zinc would you need to mix with 40 kg of copper? Give your answer to an appropriate degree of accuracy.

Zinc : Copper

3 : 8

×5 ×5

? 40 kg

? = 3 × 5 = 15 kg

Look for the scaling factor : 40 ÷ 8 = 5

Example

A map has a **scale** of 1 : 2500. A distance in real life is 50 m. What is this distance on the map?

map : real life

1 : 2500

×2 ×2

? 5000 cm

? = 1 × 2 = 2 cm

Look for the scaling factor : 5000 ÷ 2500 = 2

You can divide a quantity in a given ratio.

Example

Sean and Patrick share £350 in the ratio 3 : 7.
How much money do they each receive?

Sean receives 3 parts for every 7 parts that Patrick receives.

Total number of parts = 3 + 7 = 10 parts
Each part = £350 ÷ 10 = £35

Sean will receive 3 parts = 3 × £35 = £105
Patrick will receive 7 parts = 7 × £35 = £245

Check your answer by adding up the two parts.
They should add up to the amount being shared!
£105 + £245 = £350

Exercise N4.4

1 In each question, simplify the ratio, draw a diagram and write two statements. The first one is done for you.

a A length of 50 cm : length of 80 cm

50 : 80

5 : 8

a length of 80 cm = $\frac{8}{5}$ × a length of 50 cm

a length of 50 cm = $\frac{5}{8}$ × a length of 80 cm

$\times\frac{8}{5}$

50 cm 80 cm

$\times\frac{5}{8}$

b Andrew's height of 144 cm : Andrew's width of 48 cm

c A limousine of length 6.4 m : a car of length 280 cm

d A can containing 330 ml : a can containing 0.44 litres

2 Solve each of these problems.

a The ratio of men to women in a waiting room is 4 : 5. There are 12 men in the waiting room. How many women are there?

b In a metal alloy the ratio of aluminium to tin is 8 : 5. How much aluminium is needed to mix with 55 kg of tin?

c The ratio of the number of purple flowers to the number of white flowers in a garden is 5 : 11. There are 132 white flowers. How many purple flowers are there?

d The ratio of male students to female students in a college is 7 : 6. There are 588 male students. How many female students are there at the college?

3 **a** A map has a scale of 1 : 400. A distance in real life is 4.8 m. What is this distance on the map?

b In a college the ratio of lecturers to students is 1 : 22.5. If there are 990 students at the college, how many lecturers are there?

c The model of an aircraft is in the scale 1 : 32. If the real aircraft is 12.48 m long, how long is the model?

4 A map has a scale of 1 : 5000.

a What is the distance in real life of a measurement of 6.5 cm on the map?

b What is the distance on the map of a measurement of 30 m in real life?

5 Solve each of these problems.

a Divide £90 in the ratio 3 : 7.

b Divide 369 kg in the ratio 7 : 2.

c Divide 103.2 tonnes in the ratio 5 : 3.

d Divide 35.1 litres in the ratio 5 : 4.

e Divide £36 in the ratio 1 : 2 : 3.

N4.5 Exchange rates

This spread will show you how to:

- Calculate exchange rates and solve problems involving exchange rates

Keywords
Rate
Ratio
Scale

- A **rate** is a way of comparing two quantities. It tells you how many units of one quantity there are compared to one unit of another quantity. You can calculate a rate using division.

Example

Tariq is paid £374 a week. Each week he works for 44 hours.
What is his hourly rate of pay?

The rate of pay = £374 for every 44 hours

$= \frac{£374}{44}$ for every 1 hour

= £8.50 for every hour

= £8.50 per hour

- An exchange rate is a way of comparing two currencies. It tells you how many units of one currency there are compared to one unit of another currency.

Example

Katherine went to France. She changed £300 into €480.

a What was the exchange rate of euros to the pound?
b Katherine had €200 left over and changed this back into pounds at the same exchange rate. How many pounds did she get?

a Rate of euros to pounds $= \frac{€480}{£300}$

$= 480 \div 300$ euros for every pound

$= 1.6$ euros for every pound

Exchange rate is £1 = €1.6 (£1 will buy you €1.6).

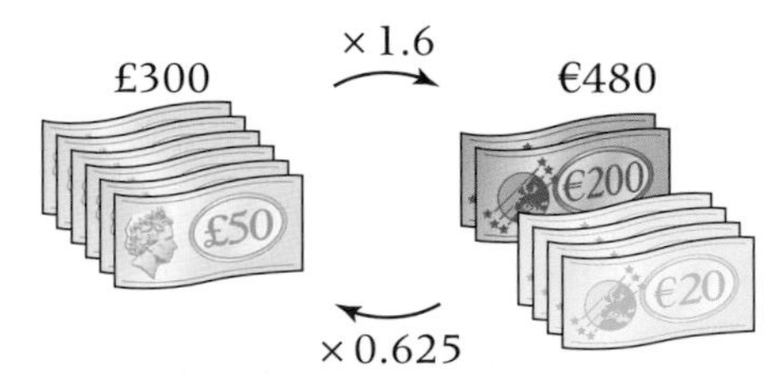

b

? $\xrightarrow{\times 1.6}$ €200

? $\xleftarrow{\div 1.6}$ €200

€200 ÷ 1.6 = £125
Katherine got back £125.

Exercise N4.5

1 Work out the hourly rate of pay for each person.

Person	Money earned (£)	Hours worked (hours)	Hourly rate of pay (£ per hour)
Leonard	£210	30	
Pavel	£216	32	
Andy	£554.40	48	

2 Work out the rate for each of these.

a Wendy works for 3 hours. She gets paid £22.26. What is her hourly rate of pay?

b Brian is a bricklayer. On average, he lays 680 bricks in 4 hours. What is his hourly rate of laying bricks?

c Gustav travels in his car for 6 hours. He travels 330 km. What is his average speed?

How far does he travel each hour?

3 Work out the exchange rate into pounds for each of these currencies.

Country	Number of pounds (£)	Number of other currency	Exchange rate (=£1)
Lithuania	£13	65 litas	£1 = ? litas
Namibia	£260	3003 dollars	£1 = ? dollars
Qatar	£82.40	519.12 riyals	£1 = ? riyals

4 Each of these people change amounts of money from pounds into euros. The exchange rate is £1 = €1.44.
Work out the number of euros each person receives.

Person	Amount (£)	Exchange rate (£1 = €1.44)	Amount (€)
Bernice	£240	£1 = €1.44	
Ingeborg	£720	£1 = €1.44	
Andrew	£6300	£1 = €1.44	

5 Nina went to Spain. She changed £500 into euros. The exchange rate was £1 = €1.64.

a Work out how many euros Nina received.

b Nina came home. She had €149 left. The new exchange rate was £1 = €1.58. Work out how much, in pounds, Nina received back for €149.

N4.6 Compound measures

This spread will show you how to:

- Round answers to an appropriate degree of accuracy
- Solve problems involving compound measures, including speed and density

Keywords
Compound
Density
Speed

Compound measures involve a combination of measurements and units.

- The **density** of a material is its mass divided by its volume.
- **Speed** is the distance travelled divided by the time taken.

The units for density can be grams per cubic centimetre (g/cm^3), or kilograms per cubic metre (kg/m^3).

- **A formula triangle can be a useful way of remembering the relationships between the different parts of a compound measure.**
  For example:
  speed = distance ÷ time
  distance = speed × time
  time = distance ÷ speed

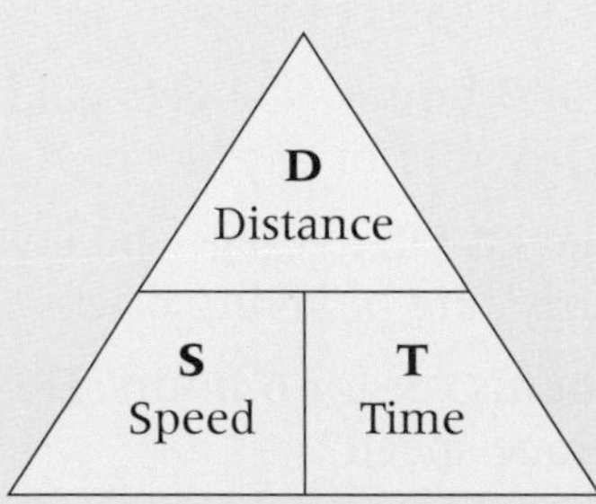

Speed can be measured in miles per hour (mph), kilometres per hour (kph) or metres per second (m/s).

Example

a A cube of side 3.5 cm has a mass of 600 g. Find the density of the cube in g/cm^3, correct to 3 significant figures.
b A car travels 240 miles in 3 hours 45 minutes. Find the average speed of the car in miles per hour.
c Lubricating oil has a density of 0.58 g/cm^3. Find
i the mass of 2.5 litres of this oil
ii the volume of 10 grams of the oil.

Volume of a cube = length^3.

a Volume = 3.5^3 cm^3 = 42.875 cm^3
Density = 600 g ÷ 42.875 cm^3 = 13.994 ... g/cm^3 = 14.0 g/cm^3 to 3 sf

Density = $\frac{\text{mass}}{\text{volume}}$

b Average speed $= \frac{\text{total distance}}{\text{total time}}$

$= \frac{240}{3.75}$

= 64 mph

Put the time in hours.

c i Mass = density × volume = 0.58 g/cm^3 × 2500 cm^3 = 1450 g
ii Volume = mass ÷ density = 10g ÷ 0.58g/cm^3 = 17.2 cm^3

1 litre ≡ 1000 cm^3

Example

A metal cuboid has a length of 2 cm, width 4 cm and height 6 cm.
a Find the volume of the cuboid.
b If the mass of the cuboid is 7.8 kg, find its density in g/cm^3.

a Volume of cuboid = 2 × 4 × 6 = 48 cm^3

b Density of cuboid $= \frac{\text{mass}}{\text{volume}} = \frac{7800}{48}$ = 163 g/cm^3 to 3 sf

Volume of a cuboid = length × width × height.

Change the mass into grams.

Exercise N4.6

1 Rod cycles 18 miles in 2 hours. Find his average speed, in miles per hour (mph).

2 If 4 metres of fabric costs £8.40, find the price of the fabric in pounds per metre.

3 A car travels 24 miles in 45 minutes. Find the average speed of the car in miles per hour (mph).

4 A train leaves Euston at 8:57 a.m. and arrives at Preston at 11:37 a.m. If the distance is 238 miles find the average speed of the train.

5 Copy and complete the table to show speeds, distances and times for five different journeys.

Speed (kph)	Distance (km)	Time
105		5 hours
48	106	
	84	2 hours 15 minutes
86		2 hours 30 minutes
	65	1 hour 45 minutes

6 A cube of side 2 cm weighs 40 grams.

Volume of cube = length^3.

a Find the density of the material from which the cube is made, giving your answer in g/cm^3.

b A cube of side length 2.6 cm is made from the same material. Find the mass of this cube, in grams.

7 A box has a length and width of 22.50 mm, and a height of 3.15 mm. It has a mass of 9.50 g.

Volume of cuboid = length × width × height.

a Find the density of the metal from which the box is made, giving your answer in g/cm^3.

b How many boxes can be made from 1 kg of the material?

8 Emulsion paint has a density of 1.95 kg/litre. Find

a the mass of 4.85 litres of the paint.

b the number of litres of the paint that would have a mass of 12 kg.

9 A steel cable weighs 2450 kg.
The cable has a uniform circular cross-section of radius 0.85 cm.
The steel from which the cable is made has a density of 7950 kg/m^3.
Find the length of the cable.

Volume of a cylinder = πr^2 × length.

l

0.85 cm

N4 Exam review

Key objectives

- Use knowledge of operations and inverse operations, and of methods of simplification, in order to select and use suitable strategies and techniques to solve problems and word problems, including those involving ratio and proportion, fractions, percentages and measures and conversion between measures, and compound measures defined within a particular situation

1 **a** Tom has £2200.
He gives $\frac{1}{4}$ to his son and $\frac{2}{5}$ to his daughter.
How much does Tom keep for himself?
You **must** show all your working. (3)

b Mrs Jones inherits £12 000.
She divides the £12 000 between her three children Laura, Mark and Nancy in the ratio 7 : 8 : 9, respectively.
How much does Laura receive? (2)

(AQA, 2003)

2 A group of students visited the USA. The table shows information about the numbers of hamburgers the students bought on the visit:

Number of hamburgers	Number of students
0	1
1	1
2	4
3	8
4	8
5	7

a Work out the total number of hamburgers that these students bought. (3)

One of these students bought a pair of sunglasses in the USA.
He paid \$35.50.
In England, an identical pair of sunglasses costs £26.99.
The exchange rate is £1 = \$1.42.

b In which country were the sunglasses cheaper, and by how much?
Show all your working. (3)

(Edexcel Ltd., 2004)

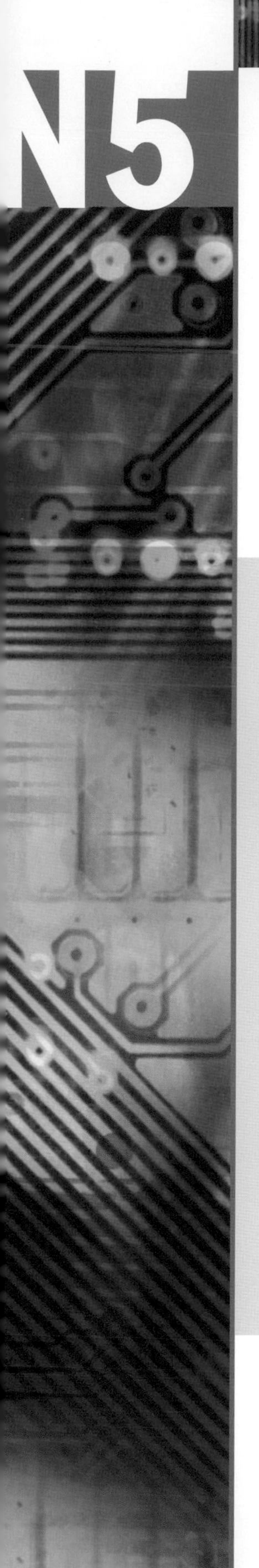

N5 Integers, powers and roots

This unit will show you how to

- Use powers and index notation for small positive integer powers
- Use the square, square root, cube and cube root functions of a scientific calculator
- Understand and use reciprocals
- Multiply and divide by powers of 10
- Use simple index laws including negative indices
- Understand and use standard form in calculations with large and small numbers
- Use calculators to calculate in standard form

Before you start ...

You should be able to answer these questions.

1 Calculate

a 12^2 **b** $\sqrt{81}$

2 Calculate

a 5^3 **b** $\sqrt[3]{8}$

3 Calculate

a 2^4 **b** 10^6

4 Write all the factors of 24.

5 Write the first 6 prime numbers.

6 Work out the value of each of these expressions.

a $3^2 \times 5$ **b** $2^2 \times 5^2$

N5.1 Squares and square roots

This spread will show you how to:

- Use powers and index notation for small positive integer powers
- Use the square and square root functions of a scientific calculator

Keywords
Index
Power
Square number
Square root

- **A square number is the result of multiplying a whole number by itself.**

Your calculator should have an x^2 function key.

Square numbers can be written using **index** notation.

$1^2 = 1 \times 1 = 1$ $2^2 = 2 \times 2 = 4$ $3^2 = 3 \times 3 = 9$

- **A square root is a number that when multiplied by itself it is equal to a given number. Square roots are written using √ notation.**

$\sqrt{225} = 15$ and -15 because $15 \times 15 = 225$ and $-15 \times -15 = 225$

You can write $\sqrt{225} = \pm 15$

Example

a Calculate the value of $\sqrt{300}$ using a calculator.

b Find $\sqrt{900}$ using a calculator.

Use a calculator to find a square root using the $\sqrt{x}$ function key.

a Using the calculator you would type

$\sqrt{x}$ 3 0 0 = √300 17.32050808

So $\sqrt{300} = 17.32$ (2 dp)

Check: $17.32^2 \approx 17^2$
$17^2 < 20^2 = 400$
$17^2 > 15^2 = 225$
17.32^2 is greater than 225 but less than 400.

b $\sqrt{900} = \sqrt{9 \times 100}$
$= \sqrt{9} \times \sqrt{100}$
$= 3 \times 10$
$= 30$

You can use trial and improvement to **estimate** the square root of a number to a given number of decimal places.

Example

Use trial and improvement to find $\sqrt{30}$ to 1 decimal place.

Estimate	Check (square of estimate)	Answer	Result
5	5^2	25	Too small
6	6^2	36	Too big
5.5	5.5^2	30.25	Too big
5.4	5.4^2	29.16	Too small
5.45	5.45^2	29.7025	Too small

So $\sqrt{30} = 5.5$ (1 decimal place)

Examiner's tip: In these GCSE questions you get most of the marks for your working out.

Exercise N5.1

1 Write

a the 5th square number

b the 11th square number

c the 15th square number

d the 17th square number.

2 In each of these lists of numbers, identify the square numbers.

a 8, 16, 24, 30, 36

b 49, 59, 69, 79, 99

c 140, 121, 135, 144, 136

d 214, 218, 223, 225, 222

3 Use your calculator to work out each of these. Give your answer to 2 decimal places as appropriate.

a 16^2 **b** 3.7^2 **c** 50^2 **d** 6.7^2

e 17.8^2 **f** $(-4.2)^2$ **g** 1.9^2 **h** 0.1^2

i $(-3.9)^2$ **j** 2.1^2 **k** $(-0.7)^2$ **l** 13.25^2

4 Calculate each of these using a calculator, giving your answer to 2 dp as appropriate. Remember to give your answer as both a positive and a negative square root.

a $\sqrt{529}$ **b** $\sqrt{157}$ **c** $\sqrt{41}$ **d** $\sqrt{0.16}$

e $\sqrt{6.76}$ **f** $\sqrt{800}$ **g** $\sqrt{1345}$ **h** $\sqrt{38.6}$

i $\sqrt{7093}$ **j** $\sqrt{234.652}$

5 Without a calculator, write the whole number that is closest in value to

a $\sqrt{50}$ **b** $\sqrt{80}$ **c** $\sqrt{30}$ **d** $\sqrt{40}$

e $\sqrt{120}$ **f** $\sqrt{150}$ **g** $\sqrt{8}$ **h** $\sqrt{5}$

6 Use your calculator to work out these problems.

a $\sqrt{7} = 2.645751$
Calculate $(2.645751)^2$.
Explain why the answer is not 7.

b Two consecutive numbers are multiplied together. The answer is 3192. What are the two numbers?

7 **a** Use a trial and improvement method to find the square root of 20 to 1 decimal place.

Estimate	Check	Answer	Result
4	4^2	16	Too small
5	5^2	25	
4.5			

b Use a similar method to find

i $\sqrt{40}$ **ii** $\sqrt{60}$ **iii** $\sqrt{95}$

N5.2 Cubes and cube roots

This spread will show you how to:

- Use powers and index notation for small positive integer powers
- Use the square, square root, cube and cube root functions of a scientific calculator

Keywords
Cube number
Cube root
Index
Power

- A **cube number** is the result of multiplying a whole number by itself and then multiplying by that number again.

Your calculator should have a x^3 function key.

Cubes of numbers are written using **index** notation.

$1^3 = 1 \times 1 \times 1 = 1$ $2^3 = 2 \times 2 \times 2 = 8$ $3^3 = 3 \times 3 \times 3 = 27$

- A **cube root** is a number that when multiplied by itself and then multiplied by itself again is equal to a given number. Cube roots are written using $\sqrt[3]{\ }$ notation.

$\sqrt[3]{4913} = 17$ because $17^3 = 17 \times 17 \times 17 = 4913$

- A positive number has a positive cube root and a negative number has a negative cube root.

$\sqrt[3]{-125} = -5$ because $(-5)^3 = -5 \times -5 \times -5 = -125$

Example

Calculate the value of $\sqrt[3]{200}$.

Using a calculator you might type

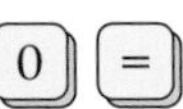

$\sqrt[3]{x}$ 2 0 0 =

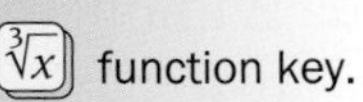

So $\sqrt[3]{200} = 5.85$ (2 dp)

Check: $5.85^3 \approx 6^3$
$5.85^3 < 6^3 = 216$
$5.85^3 > 5^3 = 125$
5.85^3 is greater than 125 but less than 216.

Use a calculator to find a cube root using the $\sqrt[3]{x}$ function key.

You can use trial and improvement to estimate cube roots.

Example

Use trial and improvement to find $\sqrt[3]{18}$ to 1 decimal place.

Estimate	Check (cube of estimate)	Answer	Result
2	2^3	8	Too small
3	3^3	27	Too big
2.5	2.5^3	15.625	Too small
2.6	2.6^3	17.576	Too small
2.7	2.7^3	19.683	Too big
2.65	2.65^3	18.609 625	Too big

So $\sqrt[3]{18} = 2.6$ (1 decimal place)

Examiner's tip
In these GCSE questions you get most of the marks for your working out.

Note you can also use this technique for **square** roots.

Exercise N5.2

1 Write

a the 7th cube number

b the 10th cube number

c the 13th cube number

d the 19th cube number.

2 In each of these lists of numbers, identify the square and cube numbers.

a 4, 11, 16, 27, 35

b 24, 44, 64, 84, 124, 144

c 156, 196, 216, 256, 286

d 700, 800, 900, 1000, 1200

3 Use your calculator to work out each of these. Give your answer to 2 decimal places as appropriate.

a 8^3 **b** 2.4^3 **c** 20^3 **d** 3.9^3

e 11.7^3 **f** $(-2.8)^3$ **g** 8.9^3 **h** 0.5^3

i $(-5.4)^3$ **j** 9.9^3 **k** $(-0.1)^3$ **l** 16.85^3

4 Calculate these using a calculator, giving your answers to 2 dp as appropriate.

a $\sqrt[3]{729}$ **b** $\sqrt[3]{100}$ **c** $\sqrt[3]{64}$ **d** $\sqrt[3]{86}$

e $\sqrt[3]{7.6}$ **f** $\sqrt[3]{2.7}$ **g** $\sqrt[3]{1.331}$ **h** $\sqrt[3]{56.3}$

i $\sqrt[3]{12\,167}$ **j** $\sqrt[3]{-216}$ **k** $\sqrt[3]{-70}$ **l** $\sqrt[3]{0.015\,625}$

5 **a** Use a trial and improvement method to find the cube root of each of these numbers to 1 decimal place.

i $\sqrt[3]{20}$

Estimate	Check (cube of estimate)	Answer	Result
2	2^3	8	Too small
3	3^3	27	
2.5			

ii $\sqrt[3]{50}$

Estimate	Check (cube of estimate)	Answer	Result
3	3^3	27	Too small
4	4^3		

iii $\sqrt[3]{80}$ **iv** $\sqrt[3]{150}$ **v** $\sqrt[3]{300}$ **vi** $\sqrt[3]{500}$ **vii** $\sqrt[3]{900}$ **viii** $\sqrt[3]{1500}$

b Use the cube root key on your calculator to check your answers.

N5.3 Powers and reciprocals

This spread will show you how to:

- Use powers and index notation for small positive integer powers
- Understand and use reciprocals
- Multiply and divide by powers of 10

Keywords
Index
Power
Powers of 10
Reciprocal

- You can use **index** notation to describe **powers** of any number.

$4^5 = 4 \times 4 \times 4 \times 4 \times 4$

To work out 4^5 you might type: [4] [y^x] [5] [=]

The calculator display should read 1024
So $4^5 = 1024$.

Some calculators may have a different key, for example [x^y] or [^] or [y^x].

- Powers of the same number can be multiplied and divided.

When multiplying, you add the indices.

$5^3 \times 5^4 = (5 \times 5 \times 5) \times (5 \times 5 \times 5 \times 5) = 5^7$

$5^{3+4} = 5^7$

When dividing, you subtract the indices.

$3^5 \div 3^2 = \frac{3 \times 3 \times 3 \times 3 \times 3}{3 \times 3} = 3 \times 3 \times 3 = 3^3$

$3^{5-2} = 3^3$

- **Simplified, the index laws are**

$a^m \times a^n = a^{(m+n)}$ $\qquad a^m \div a^n = a^{(m-n)}$

- **Any number raised to the power of zero is equal to 1.**

$7^0 = 1$ $\qquad 10^0 = 1$

- **The reciprocal of a number is 1 divided by that number.**

Reciprocal of $10 = \frac{1}{10} = 0.1$ $\qquad$ Reciprocal of $4^2 = \frac{1}{4^2} = \frac{1}{16} = 0.0625$

To calculate reciprocals use the [$\frac{1}{x}$] or [x^{-1}] key on your calculator.

- **A negative power represents the reciprocal of a number.**

$8^{-1} = \frac{1}{8} = 0.125$ $\qquad 10^{-2} = \frac{1}{10^2} = \frac{1}{100} = 0.01$

The decimal system is based upon **power of ten**. This makes it easy to multiply or divide by powers of 10.

Example

Calculate **a** 6.3×10^3 **b** $120 \div 10^4$

a

Thousands	Hundreds	Tens	Units	•	tenths	hundredths
			6	•	3	
6	3	0	0	•		

$\times 10^3$

$6.3 \times 10^3 = 6300$

b

Hundreds	Tens	Units	•	tenths	hundredths	thousandths
1	2	0	•			
			•	0	1	2

$\div 10^4$

$120 \div 10^4 = 0.012$

When you multiply by 10^3, all the digits move **three** places to the left.

When you divide by 10^4, all the digits move **four** places to the right.

Exercise N5.3

1 Calculate these without using a calculator.

a 4^2 **b** 2^5 **c** 5^3 **d** 7^4 **e** 9^3

2 Use the x^y function key on your calculator to work out these, giving your answers to 2 decimal places where appropriate.

a 15^3 **b** 3^6 **c** 2^{10} **d** 21.6^4 **e** 13^3

3 Use your calculator to work out each of these.

a $2^4 + 3^2$ **b** $10^3 \div 5^2$ **c** $8^6 - 13^3$ **d** $10^6 \div 5^3$ **e** $(\frac{1}{4})^4 + (\frac{1}{8})^2$

4 Use the x^y function key on your calculator to find the value of x.

a $3^x = 27$ **b** $5^x = 625$ **c** $10^x = 10\,000$ **d** $4^x = 16\,384$ **e** $x^2 = 529$

5 Calculate these without using a calculator.

a 3.4×10^2

Thousands	Hundreds	Tens	Units	•	tenths	hundredths	
			3	•	4		$\times 10^2$
				•			

b 76.6×10^3 **c** $85 \div 10^3$ **d** 2.3×10^4 **e** 0.312×10^6

f 5.62×10^4 **g** $2960 \div 1000$

6 Simplify each of these, leaving your answer as a single power of the number.

a $3^2 \times 3^2$ **b** $7^3 \times 7^2$ **c** $2^7 \times 2^5$ **d** $10^7 \div 10^4$

e $3^{10} \div 3^6$ **f** $4^2 \times 4^3 \times 4^2$ **g** $10^5 \times 10^2 \times 10^3$

h $7^4 \times 7^1 \times 7$ **i** $\frac{2^3 \times 2^5}{2^2}$ **j** $\frac{10^3 \times 10^4}{10^2}$ **k** $\frac{4^2 \times 4^2 \times 4^2}{4^6}$

7 Use your calculator to work out the reciprocal of each of these numbers.

a 10 **b** 8 **c** 1000 **d** 3 **e** 7 **f** 13

8 Calculate each of these without using a calculator.

a 4^0 **b** 5^{-1} **c** 13^0

9 Copy these and fill in the missing numbers.

a $540 \div 10^2 =$ ___ **b** $6850 \div 10^? = 6.85$

c $3.12 \times$ ___ $= 31\,200$ **d** $1.73 \times 10^6 =$ ___

10 Simplify each of these, leaving your answer as a single power of the number or letter where appropriate.

a $y^2 \times y^3$ **b** $4^8 \times 4^2$ **c** $w^{12} \div w^7$ **d** $4^y \div 4^2$ **e** $\frac{g^6 \times g^3}{g^5}$ **f** $3^2 \times 4^3$

N5.4 Standard index form for large numbers

This spread will show you how to:

- Understand and use standard form in calculations with large and small numbers
- Use calculators to calculate in standard form

Keywords
Standard form

You can use **standard form** to represent large numbers.

- In **standard form**, a number is written as $A \times 10^n$.
  - A is a number between 1 and 10 (but not including 10). Using algebra, $1 \leqslant A < 10$.
  - The value of n is an integer.

  For example, $856 = 8.56 \times 10^2$ and $43\,994 = 4.3994 \times 10^4$.

Example

Write these numbers in standard form.

a 235 **b** 12 492 **c** 15×10^4 **d** 0.23×10^6

a $235 = 2.35 \times 10^2$ **b** $12\,492 = 1.2492 \times 10^4$
c $15 \times 10^4 = 1.5 \times 10^5$ **d** $0.23 \times 10^6 = 2.3 \times 10^5$

You can use a calculator to input standard form. On a standard Casio you use the EXP key.

Example

The Andromeda Galaxy has a radius of about 1 040 700 000 000 000 000 km.
Write this in standard form.

$1\,040\,700\,000\,000\,000\,000 = 1.0407 \times 10^{18}$

13×10^5 is *not* in standard form, because 13 is larger than 10.

The correct version is 1.3×10^6

0.75×10^4 is *not* in standard form, because 0.75 is less than 1.

The correct version is 7.5×10^3

You can calculate with numbers in standard form.

- Multiplication works like this:
  $(3 \times 10^5) \times (4 \times 10^3) = (3 \times 4) \times 10^{(5+3)} = 12 \times 10^8 = 1.2 \times 10^9$
- Division works like this:
  $(1.4 \times 10^8) \div (7 \times 10^5) = (1.4 \div 7) \times 10^{(8-5)} = 0.2 \times 10^3 = 2 \times 10^2$

Multiplication – add the indices

Division – subtract the indices

Example

Calculate

a $(4.2 \times 10^3) \times (2 \times 10^2)$ **b** $(3.6 \times 10^5) \div (1.2 \times 10^3)$ **c** $(5.4 \times 10^4) \times (2 \times 10^3)$

a $(4.2 \times 10^3) \times (2 \times 10^2) = (4.2 \times 2) \times (10^3 \times 10^2)$
$= 8.4 \times 10^{(3+2)} = 8.4 \times 10^5$

b $(3.6 \times 10^5) \div (1.2 \times 10^3) = (3.6 \div 1.2) \times (10^5 \div 10^3)$
$= 3 \times 10^{(5-3)} = 3 \times 10^2$

c $(5.4 \times 10^4) \times (2 \times 10^3) = (5.4 \times 2) \times (10^4 \times 10^3)$
$= 10.8 \times 10^7 = 1.08 \times 10^8$

To do part **a** on a calculator:

4.2 EXP 3 × 2 EXP 2

However, your calculator may give the answer 840 000. Why?

Exercise N5.4

1 Write these numbers as powers of 10.

a 100 **b** 10 **c** 100 000 **d** 1

2 Write these numbers in standard form.

a 200 **b** 800 **c** 9000 **d** 650

e 6500 **f** 952 **g** 23.58 **h** 255.85

3 These numbers are in standard form. Write each of them as an 'ordinary' number.

a 5×10^2 **b** 3×10^3 **c** 1×10^5 **d** 2.5×10^2

e 4.9×10^3 **f** 3.8×10^6 **g** 7.5×10^{11} **h** 8.1×10^{18}

4 Although they are written as multiples of powers of 10, these numbers are not in standard form. Rewrite each of them correctly in standard form.

a 60×10^1 **b** 45×10^3 **c** 0.65×10^1 **d** 0.05×10^8

5 Work out these calculations, giving your answers in standard form. Do not use a calculator.

a $(2 \times 10^2) \times (2 \times 10^3)$ **b** $(3 \times 10^4) \times (3 \times 10^3)$

c $(5 \times 10^3) \times (5 \times 10^4)$ **d** $(8 \times 10^7) \times (3 \times 10^5)$

6 Evaluate these, showing your working.
Do not use a calculator; give your answers in standard form.

a $(4 \times 10^4) \div (2 \times 10^2)$ **b** $(8.4 \times 10^9) \div (4.2 \times 10^5)$

c $(2 \times 10^6) \div (4 \times 10^4)$ **d** $(3 \times 10^5) \div (4 \times 10^2)$

7 Use a calculator to evaluate these.
Give your answers in standard form, to 3 significant figures.

a $(2.5 \times 10^5) \times (3.9 \times 10^4)$ **b** $(4.1 \times 10^6) \div (3 \times 10^2)$

c $(4.95 \times 10^3) \times (8.11 \times 10^7)$ **d** $(3.7 \times 10^{11}) \div (1.8 \times 10^3)$

8 The speed of light is approximately 3×10^8 metres per second. Copy and complete the table to show the time taken for light from the Sun to reach the various planets. (Hint: divide distance by speed)

Planet	Mean distance from Sun (m)	Light travel time
Mercury	5.79×10^{10}	
Earth	1.50×10^{11}	
Mars	2.28×10^{11}	
Jupiter	7.78×10^{11}	
Pluto	5.90×10^{12}	

DID YOU KNOW?

In 2006, astronomers decided that Pluto could no longer be regarded as a planet, due to its erratic orbit and small size. There are now considered to be only 8 planets in our solar system.

N5.5 Standard form for small numbers

This spread will show you how to:

- Understand and use standard form in calculations with large and small numbers
- Use calculators to calculate in standard form

Keywords
Standard form

It is often useful to write small numbers, such as 0.000415, in standard form.

- Negative powers of 10, such as 10^{-4}, represent small numbers.

  $10^{-4} = \frac{1}{10^4} = \frac{1}{10\,000} = 0.0001$

- You can write any small number in standard form.

  $0.00312 = 3.12 \times 0.001 = 3.12 \times 10^{-3}$

- You can calculate with small numbers expressed in standard form.

  $(4.25 \times 10^{-3}) \div (3.75 \times 10^{4}) = (4.25 \div 3.75) \times 10^{(-3-4)} = 1.13 \times 10^{-7}$

- You can obtain a small number as a result of a calculation involving large numbers.

  $(3 \times 10^{5}) \div (4 \times 10^{8}) = 0.75 \times 10^{-3} = 7.5 \times 10^{-4}$

Small numbers are often used to describe microscopic or subatomic distances.

You can also work with numbers in standard form on a scientific calculator. The button for entering the power of 10 is often marked EXP or EE; so for 4.5×10^{-3}, you might enter

Example

Write these numbers in standard form.

a 0.003 **b** 0.000 000 416 **c** 0.45

a $0.003 = 3 \times 10^{-3}$
b $0.000\,000\,416 = 4.16 \times 10^{-7}$
c $0.45 = 4.5 \times 10^{-1}$

Example

Calculate. **a** $(4.8 \times 10^{4}) \times (3.6 \times 10^{-5})$ **b** $(4.5 \times 10^{3}) \div (9.7 \times 10^{8})$

These can be done directly, using a scientific calculator.
The answers, to 3 significant figures, are **a** 1.73 **b** 4.64×10^{-6}

Example

What is wrong with this calculation? $10^4 \div 10^{-5} = 10^{-1}$ ✗

Because you are **dividing**, you need to **subtract** the indices. The correct power of 10 for the answer is $4 - (-5) = 4 + 5 = 9$, so $10^4 \div 10^{-5} = 10^9$

Be careful with the signs of the indices in examples like this.

Exercise N5.5

1 Write these numbers in standard form.

a 0.3 **b** 0.0047 **c** 0.000 078 **d** 0.4485

2 Although these numbers are written as multiples of powers of 10, they are not in standard form. Write each of the numbers correctly in standard form.

a 28×10^{-2} **b** 0.4×10^{-1} **c** 13.5×10^{-4} **d** 12×10^{-8}

3 Write these measurements using standard form.

a One hundredth of a kilometre **b** Two thousandths of a gram

c Five millionths of a metre **d** 11 thousandths of a litre

4 Work out these calculations without a calculator.
Give your answers in standard form.

a $(2.5 \times 10^{-3}) \times (2 \times 10^{2})$ **b** $(4.6 \times 10^{-6}) \times (2 \times 10^{-2})$

c $(4 \times 10^{4}) \div (2 \times 10^{6})$ **d** $(8.4 \times 10^{-2}) \div (2 \times 10^{6})$

5 Use a scientific calculator to work out these. Give your answers in standard form, and to three significant figures.

a $(3.7 \times 10^{-4}) \times (3.1 \times 10^{-4})$ **b** $(5.3 \times 10^{5}) \div (2.9 \times 10^{8})$

c $(3.18 \times 10^{2}) \div (6.55 \times 10^{7})$ **d** $(1.79 \times 10^{5}) \times (2.8 \times 10^{-6})$

6 Work out these calculations without using a calculator, giving your answers in standard form.

a $(5 \times 10^{-1}) + (2 \times 10^{-2})$ **b** $(4 \times 10^{-2}) + (6 \times 10^{-3})$

c $(2 \times 10^{-2}) + (9 \times 10^{-4})$ **d** $(1.5 \times 10^{-2}) - (2 \times 10^{-3})$

You may find it easier to convert the numbers to 'ordinary' numbers first, and then convert the answers back to standard form.

7 Use a scientific calculator to check your answers to question 6.

8 Use a scientific calculator to find the volume of a cube of side length 4.5×10^{-3} metres. Give your answer in m^3, to 3 sf, in standard form.

9 A pack of 500 sheets of A4 paper weighs 2.65 kg. Find the mass in kg of a single sheet of paper, giving your answer in standard form.

10 A sheet of gold leaf is one ten thousandth of a millimetre thick.
The diameter of an atom of gold is about 0.26 nanometres.
(One nanometre is 10^{-9} metres.)
Approximately how many atoms thick is the sheet of gold leaf?
Show your working.

N5 Exam review

Key objectives

- Use standard index form expressed in conventional notation and on a calculator display
- Use index laws for multiplication and division of integer powers

1 Calculate $\sqrt{9.61} + 2.91^2$ (2)
Give your answer to 3 significant figures.

(AQA, 2004)

2 Here are six numbers written in standard form.

2.6×10^5 1.75×10^6 5.84×10^0
8.2×10^{-3} 3.5×10^{-1} 4.9×10^{-2}

a Write down the largest number. (1)

b Write down the smallest number. (1)

c Write 4.9×10^{-2} as an ordinary number. (1)

d Work out $2.6 \times 10^5 \div 0.1$
Give your answer in standard form. (1)

(AQA, 2004)

N6 Fraction and percentage calculations

This unit will show you how to

- Express one number as a fraction (or proportion) of another number
- Use fractions and percentages as operators
- Use equivalent fractions and mental methods to calculate simple percentages
- Calculate a fraction and percentage of an amount using a variety of methods
- Calculate percentage increase and decrease using a range of methods
- Solve percentage problems
- Calculate simple and compound interest

Before you start ...

You should be able to answer these questions.

1 Calculate

$\frac{3}{5}$ of £55

2 Calculate

10% of $430

3 Decrease 76 kg by 25%.

4 A shop increases all of its prices by 20%.

A cup normally costs £4.50.

What is the new price of the cup?

5 Increase £350 by 5%.

N6.1 Fraction of a quantity

This spread will show you how to:

- Express one number as a fraction (or proportion) of another number
- Use fractions as operators

Keywords
Denominator
Fraction
Numerator

You can express one number as a **fraction** of another number.

Example

In a class there are 32 students. 20 of the students are female. What fraction of the class are male? Give your answer in its simplest form.

There are 32 students in the class altogether.
Number of males = 32 − 20 = 12
Fraction who are male = $\frac{12}{32} = \frac{3}{8}$

You can calculate a fraction of a quantity using a variety of methods.

Mental method
To calculate $\frac{3}{8}$ of 32 people:
$\frac{1}{8}$ of 32 people $= 32 \div 8$
$= 4$ people
$\frac{3}{8}$ of 32 people $= 3 \times 4$
$= 12$ people

Written method
To calculate $\frac{3}{7}$ of 63 kg:
$\frac{3}{7}$ of 63 kg $= \frac{3}{7} \times 63$
$= 3 \times \frac{1}{7} \times 63$
$= \frac{3 \times 63}{7}$
$= \frac{189}{7}$
$= 27$ kg

Calculator method
To calculate $\frac{7}{11}$ of £90:
Decimal equivalent of $\frac{7}{11}$
$= 7 \div 11 = 0.636\,363\ldots$
$\frac{7}{11}$ of £90 $= \frac{7}{11} \times$ £90
$= 0.636\,363\ldots \times 90$
$=$ £57.2727
$=$ £57.27

Example

Morgan is 150 cm tall. Peter is $\frac{6}{5}$ of the height of Morgan.

a Calculate the height of Peter.
b What fraction of Peter's height is Morgan?

a Peter's height $= \frac{6}{5}$ of Morgan's height
$= \frac{6}{5} \times 150$ cm
$= 6 \times 150 \times \frac{1}{5}$
$= \frac{900}{5} = 180$ cm

$\times\frac{6}{5}$
150 cm → 180 cm

b Morgan's height as a fraction of Peter's height $= \frac{150\text{ cm}}{180\text{ cm}}$
$= \frac{150}{180}$
$= \frac{5}{6}$

Morgan is $\frac{5}{6}$ of the height of Peter.

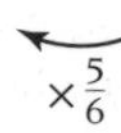
$\times\frac{5}{6}$

Exercise N6.1

1 Calculate each of these, leaving your answer in its simplest form.

a $5 \times \frac{1}{2}$ **b** $8 \times \frac{1}{4}$ **c** $8 \times \frac{1}{3}$ **d** $13 \times \frac{1}{7}$ **e** $\frac{1}{12} \times 24$ **f** $\frac{1}{3} \times 4$

2 Calculate each of these, leaving your answer in its simplest form.

a $6 \times \frac{2}{3}$ **b** $5 \times \frac{3}{4}$ **c** $6 \times \frac{2}{3}$ **d** $4 \times \frac{7}{6}$

e $5 \times \frac{9}{20}$ **f** $\frac{4}{5} \times 28$ **g** $\frac{4}{9} \times 30$ **h** $\frac{11}{18} \times 14$

i $1\frac{2}{3} \times 6$

3 Calculate each of these, leaving your answer in its simplest form.

a What is the total mass of four packets that each weigh $\frac{1}{5}$ kg?

b A cake weighs $\frac{7}{20}$ of a kg. What is the mass of 10 cakes?

c What is the total capacity of 12 jugs that each have a capacity of $\frac{3}{5}$ of a litre?

d What is the total mass of 16 bags of flour that each weigh $\frac{9}{10}$ of a kilogram?

4 Use a mental or written method to work out these. Leave your answers as fractions in their simplest form where appropriate.

a $\frac{3}{10}$ of €40 **b** $\frac{2}{5}$ of £70 **c** $\frac{3}{4}$ of 50 m **d** $\frac{4}{7}$ of 64 km

e $\frac{3}{8}$ of £1000 **f** $\frac{5}{6}$ of 70 mm **g** $\frac{11}{12}$ of 1500 m **h** $\frac{4}{13}$ of 60 g

5 Use a suitable method to calculate each of these. Where appropriate round your answer to 2 decimal places.

a $\frac{8}{15}$ of 495 kg **b** $\frac{9}{10}$ of \$5000 **c** $\frac{5}{9}$ of 8 kg

d $\frac{7}{9}$ of 1224 cups **e** $\frac{13}{18}$ of 30 tonnes **f** $\frac{4}{15}$ of 360°

g $\frac{12}{31}$ of 360° **h** $\frac{13}{15}$ of 1 hour **i** $\frac{17}{15}$ of £230

j $\frac{5}{6}$ of 24 hours

6 Express each of these as proportions. Give each answer as a fraction in its simplest form.

a 40 kg as a fraction of 60 kg **b** 15 m as a fraction of 25 m

c 40 cm as a proportion of 2 m **d** 55p as a fraction of £3

7 **a** A shirt normally costs £40. In a sale the price is reduced by $\frac{2}{5}$. What is the new price of the shirt?

b Benito receives £20 a week pocket money. He saves $\frac{3}{8}$ of it. How much money does he save each week?

c Karen rents out her holiday cottage to tourists for $\frac{4}{5}$ of the 365 day year. For how many days is her cottage empty?

N6.2 Percentage of a quantity

This spread will show you how to:

- Use equivalent fractions and mental methods to calculate simple percentages
- Calculate a fraction and percentage of an amount using a variety of methods

Keywords
Percentage

You can calculate simple **percentages** of amounts in your head using equivalent fractions.

Example

Calculate **a** 10% of £83 **b** 5% of 164 m.

a 10% of £83 is the same as working out $\frac{1}{10}$ of £83 $= \frac{1}{10} \times 83$

$= 83 \div 10$

$= £8.30$

b 10% of 164 m $= \frac{1}{10}$ of 164 m

$= \frac{1}{10} \times 164$

$= 164 \div 10$

$= 16.4$ m

5% of 164 m $= \frac{1}{2}$ of (10% of 164 m)

$= 16.4 \div 2$

$= 8.2$ m

$10\% = \frac{10}{100} = \frac{1}{10}$

$\times \frac{1}{10}$ is the same as $\div 10$.

To find 5% of something:

- Find 10% of it.
- ÷ 2.

Harder percentages of amounts can be worked out using a written or calculator method.

Written method

Change the percentage to its equivalent fraction and multiply by the amount.

To calculate 9% of 24 m:

9% of 24 m $= \frac{9}{100} \times 24$

$= \frac{9 \times 1 \times 24}{100}$

$= \frac{9 \times 24}{100}$

$= \frac{216}{100}$

$= 2.16$ m

9% of an amount
$= \frac{9}{100}$ of it
$= 9 \times \frac{1}{100}$ of it.
This is the same as × 9 and ÷ 100, so you work out $9 \times 24 \div 100$.

Calculator method

Change the percentage to its equivalent decimal and multiply by the amount.

To calculate 37% of £58:

37% of £58 $= \frac{37}{100} \times 58$

$= 0.37 \times 58$

$= £21.46$

$37\% = \frac{37}{100} = 37 \div 100 = 0.37$
$2.7\% = \frac{2.7}{100} = 2.7 \div 100 = 0.027$

Exercise N6.2

1 Calculate these percentages without using a calculator.

a 50% of £300 **b** 50% of 4 kg **c** 50% of £80

d 50% of 37 kg **e** 1% of £30 **f** 10% of 342.8 m

2 Calculate these percentages without using a calculator.

a 5% of £180 **b** 20% of 410 kg **c** 20% of $25

d 25% of £3 **e** 75% of £42 **f** 5% of 3.8 m

3 Calculate these percentages using a mental or written method.

a 15% of £340 **b** 60% of 120 Mb **c** 60% of £75

d 80% of £50 **e** 30% of 455 m **f** 2.5% of £880

g 70% of 1570 mm **h** 15% of 42 kg **i** 45% of 70 mm

4 Write the method you would use to calculate each of these without using a calculator.

a 15% of anything **b** 5% of anything **c** 35% of anything

d 17.5% of anything **e** 95% of anything

5 **a** Paul downloads a file from the internet. The file is 16 Mb. After 2 minutes he has downloaded 70% of the file. How much of the file has Paul downloaded?

b Winston has to move 24 tonnes of gravel. In the morning he moves 55% of the gravel. How many tonnes of gravel has he moved?

c A train journey is 395 km long. Lola is travelling on a train that has completed 23% of the journey. How many kilometres has Lola's train travelled?

d At Herbie's school there is a charity race night. The evening raises £234. 58% of the money raised goes to charities and the rest is given to the school. How much money is given to the school?

6 Calculate these using a mental or written method. Show all the steps of your working out. Give your answers to 2 decimal places as appropriate.

a 12% of £17 **b** 16% of 87 km **c** 8% of £38

d 32% of €340 **e** 17% of 65 m **f** 73% of 46 cm

g 85% of 148 m **h** 2% of £76.40 **i** 25% of £85

N6.3 Percentage increase and decrease

This spread will show you how to:

- Calculate percentage increase and decrease using a range of methods

Keywords
Decrease
Increase
Percentage

Percentages are used in real life to show how much an amount has increased or decreased.

- To calculate a **percentage increase**, work out the increase and add it to the original amount.
- To calculate a **percentage decrease**, work out the decrease and subtract it from the amount.

Example

a Alan is paid £940 a month. His employer increases his wage by 3%. Calculate the new wage Alan is paid each month.

b A new car costs £19 490. After one year the car depreciates in value by 8.7%. What is the new value of the car?

a Calculate 3% of the amount.
Add to the original amount.
Increase in wage = 3% of £940 = $\frac{3}{100} \times £940$
$= \frac{3 \times 940}{100} = \frac{2820}{100}$
Increase in wage = £28.20 per month
Alan's new wage = £940 + £28.20 = £968.20

The percentage calculation has been worked out using a written method.

b Calculate 8.7% of the amount.
Subtract from the original amount.
Depreciation = 8.7% of £19 490
$= \frac{8.7}{100} \times £19\,490$
$= 0.087 \times £19\,490$
Value reduction = £1695.63
New value of car = £19 490 − £1695.63 = £17 794.37

The percentage calculation has been worked out using a calculator.

You can calculate a percentage increase or decrease in a single calculation.

Example

a In a sale all prices are reduced by 16%. A pair of trousers normally costs £82. What is the sale price of the pair of trousers?

b Last year, Leanne's Council Tax bill was £968. This year the local council have raised the bill by 16%. How much is Leanne's new bill?

a Sale price = (100 − 16)% of the original price
= 84% of £82
$= \frac{84}{100} \times 82$
$= 0.84 \times 82$
= 68.88
= £68.88

b New bill = (100 + 16)% of the original bill
= 116% of £968
$= \frac{116}{100} \times £968$
$= 1.16 \times £968$
= £1122.88

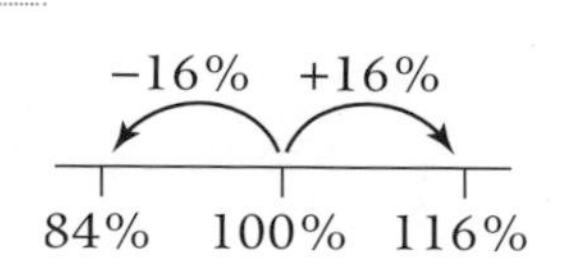

Exercise N6.3

1 Calculate these amounts using an appropriate method.

a 25% of 18 kg **b** 20% of 51 m **c** 15% of 360°

d 2% of 37 cm **e** 65% of 510 ml **f** 17.5% of 360°

g 28% of 65 kg **h** 31% of 277 kg **i** 3.6% of 154 kg

j 0.3% of 1320 m^2

2 Calculate each of these using a mental or written method.

a Increase £350 by 10% **b** Decrease 74 kg by 5%

c Increase £524 by 5% **d** Decrease 756 km by 35%

e Increase 960 kg by 17.5%

3 Calculate these. Give your answers to 2 decimal places as appropriate.

a Increase £340 by 17% **b** Decrease 905 kg by 42%

c Increase £1680 by 4.7% **d** Decrease 605 km by 0.9%

e Increase $2990 by 14.5%

4 These are the weekly wages of five employees at Suits-U clothing store. The manager has decided to increase all the employees' wages by 4%. Calculate the new wage of each employee. Give your answers to 2 decimal places as appropriate.

Employee	Original wage	Increase	New wage
Hanif	£350	×1.04	
Bonny	£285.50		
Wilf	£412.25		
Gary	£209.27		
Marielle	£198.64		

5 Use an appropriate method to work out each of these. Give your answers to 2 decimal places as appropriate.

a A drink can contains 440 ml. The size is increased by 12%. How much drink does it now contain?

b The price of a coat was £185. The price is reduced by 10% in a sale. What is the sale price of the coat?

c A house is bought for £195 000. During the next year, the house increases in price by 28.3%. What is the new value of the house?

d The number of students in Clarendon College is 940. Next year the college expects the number of students to increase by 15%. How many students does the college expect next year?

N6.4 Percentage problems

This spread will show you how to:

- Calculate percentage increase and decrease using a range of methods
- Solve percentage problems

Keywords
Decrease
Increase
Percentage
Profit

You will often encounter percentages in real life.

- VAT (Value Added Tax) is a tax which is added to bills for services and purchases. VAT is always given as a **percentage**.

Example

Helen has a contract for her home phone.
She pays £38.29 for calls and a quarterly charge of £19.60.
VAT has to be added at 17.5%.
Calculate the cost of Helen's bill + VAT.

Estimate:
Bill = £40 + £20 = £60   VAT = 20% of £60 = $\frac{20}{100} \times 60$
= 0.2×60
= £12

Bill + VAT = £60 + £12
= £72

Bill = £38.29 + £19.60   VAT = 17.5% of £57.89 = $\frac{17.5}{100} \times 57.89$
= £57.89   = 0.175×57.89
= 10.130 75
= £10.13

Bill + VAT = £57.89 + £10.13
= £68.02

You should always estimate your answers when working with a calculator to solve percentage problems.

Calculate VAT at 17.5% of the amount.
Add this to the original amount.

When people buy and sell things they try to sell for more than they paid.

- The difference between the selling price and cost price is called the **profit**. A profit is normally written as a percentage of the cost price.

Example

Danii buys and sells protractors. She buys each protractor for 12p and sells them for 15p. What is her percentage profit?

Profit = 15p − 12p   Calculate the profit.
= 3p

Percentage profit = $\frac{3\text{p}}{12\text{p}}$ Write the profit as a proportion of the cost price.

= $\frac{3}{12}$

= 0.25

= 25%   Express this fraction as a percentage.

$\%\text{ profit} = \frac{\text{profit}}{\text{cost price}}$

To change a decimal into a percentage just × 100.

Exercise N6.4

1 Calculate the selling price of each of these items. In each case give the answer to 2 decimal places.

Item	Cost price	% profit or % loss	Selling price
DVD	£8.50	Profit 20%	
DVD player	£29.50	Loss 12%	
Pack 5 CD-RW	£4.60	Profit 73%	
USB hub 4-way	£21.30	Profit 32%	
TFT monitor	£185	Loss 4.6%	

2 Here are the prices of various objects without VAT. Calculate the real price of each item including VAT at a rate of 17.5%.

a

b

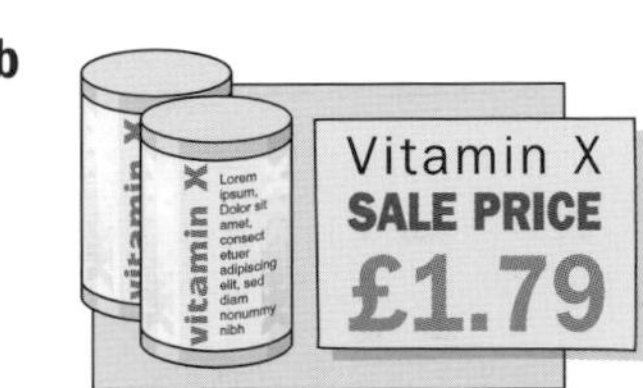

c

d

3 For each of these questions

i make an estimate

ii work out your answer using a suitable method

iii give your answer to an appropriate degree of accuracy.

a A carton of milk contains 580 ml. The size of the carton is increased by 28%. How much milk does it now contain?

b A piece of wood is 7.8 m long. It is reduced in length by 37%. What is the new length of the piece of wood?

c A lorry carries a load of sand that weighs 2.8 tonnes. The lorry loses 2.3% of its load as it travels. What mass of sand does the lorry now carry?

d A kitten weighs 300 g. After a month the kitten has grown by 73.2%. What is the new mass of the kitten?

4 Samir can buy a Games Console for one cash payment of £189, or pay a deposit of 24% and then 12 equal monthly payments of £12. Which is the better option?
Explain and justify your answer.

N6.5 Interest

This spread will show you how to:

- Calculate percentage increase and decrease using a range of methods
- Calculate simple and compound interest

Keywords

Appreciation
Compound interest
Decrease
Depreciation
Increase
Percentage
Simple interest

When you borrow money from a bank or building society you have to pay interest. When you save money, you earn interest.

People sometimes choose to have the interest they earn at the end of each year taken out of their bank account. This is called **simple interest**.

- **To calculate simple interest you multiply the interest earned at the end of the year by the number of years.**

Example

Calculate the simple interest on investing £7650 for 3 years at an interest rate of 4.3%.

Interest each year = 4.3% of £7650 = $\frac{4.3}{100} \times 7650$
= 0.043 × 7650 = £328.95

Total amount of simple interest after 3 years = 3 × £328.95
= £986.85

Don't forget to estimate:
4.3% of £7650
≈ 5% of £7000
= 10% of £7000 ÷ 2
= £700 ÷ 2
= £350

People usually choose to leave the interest in their bank account. This is called **compound interest**.

Compound interest is not examined in the Foundation tier.

- **To calculate compound interest you work out the total amount of money in the bank account at the end of each year.**

Example

Ben puts £1200 into a bank account. Each year the bank pays a rate of interest of 10%. Work out the amount of money in Ben's bank account after 3 years.

Year 1 amount
= (100 + 10)% of £1200
= 110% × £1200
= 1.1 × £1200 = £1320

Year 2 amount
= 110% of £1320
= 1.1 × £1320 = £1452

Year 3 amount
= 110% of £1452
= 1.1 × £1452 = £1597.20

- Some items grow in value over time. This is called **appreciation**.
- Other items reduce in value over time. This is called **depreciation**.

Example

A company buys a van at a cost of £15 000. Each year the van depreciates in value by 17%. Work out the value of the van after 2 years.

End of Year 1 amount
= (100 − 17)% of £15 000
= 83% × £15 000
= 0.83 × £15 000 = £12 450

End of Year 2 amount
= 83% of £12 450
= 0.83 × £12 450 = £10 333.50

Exercise N6.5

1 **a** Louise puts £8750 into a bank account. The bank pays interest of 7% on any money she keeps in the account for 1 year. Calculate the interest received by Louise at the end of the year.

b Jermaine puts £45 800 into a savings account. The account pays interest of 5.1% on any money he keeps in the account for 1 year. Calculate the interest received by Jermaine at the end of the year.

c Vicky takes out a loan of £24 800 for 1 year from a building society. The building society charges interest on the loan of 7.6% for 1 year. Calculate the total amount of money that Vicky must pay back at the end of the year.

2 Calculate the simple interest paid on £13 582

a at an interest rate of 5% for 2 years

b at an interest rate of 13% for 5 years

c at an interest rate of 4.9% for 7 years

d at an interest rate of 4.85% for 4 years.

3 Calculate the simple interest paid on

a an amount of £3950 at an interest rate of 10% for 2 years

b an amount of £6525 at an interest rate of 8.5% for 2 years

c an amount of £325 at an interest rate of 2.4% for 7 years

d an amount of £239.70 at an interest rate of 4.25% for 13 years.

4 **a** Nanette buys an antique wall covering for £430. At the end of the year the wall covering has risen in value by 11%. Calculate the new value of the wall covering.

b A new car costs £15 500. After 1 year the car depreciates in value by 9%. What is the value of the car after 1 year?

c A house is bought for £128 950. After 1 year, the house increases in value by 16%. What is the new value of the house?

5 **a** Patricia puts £8000 into a bank account. Each year the bank pays a compound interest rate of 5%. Work out the amount of money in Patricia's bank account after 2 years.

b Simone puts £12 500 into a savings account. Each year the building society pays a compound interest rate of 6.5%. Work out the amount of money in Simone's bank account after 3 years.

c Antonio invests £3400 into a Super Saver account. Each year the account pays a compound interest rate of 6.2%. Work out the amount of money in Antonio's account after 4 years.

N6 Exam review

Key objectives

- Calculate a given fraction of a given quantity
- Select and use suitable strategies and techniques to solve problems and word problems, including those involving ratio, proportion, fractions and percentages

1 a Paul invested £1200 for 2 years at a rate of 3.7% per year simple interest. Calculate the amount of interest Paul received. (3)

b Paul wants to buy a new computer.

PC ESSENTIALS

£890 plus VAT

COMPUTERS FOR ALL

£999 including VAT

At PC Essentials Paul needs to pay £890 plus VAT at $17\frac{1}{2}$%.
The total price is £999 at Computers For All.

Find the difference in the price of the computers. (3)

(OCR, 2003)

2 Ben bought a car for £12 000.

Each year the value of the car depreciated by 10%.

Work out the value of the car two years after he bought it. (3)

(Edexcel Ltd., 2003)

A1 Expressions

This unit will show you how to

- Use letters to represent unknown numbers in algebraic expressions
- Simplify algebraic expressions by collecting like terms
- Use index notation and simple laws of indices
- Expand single and double brackets within algebraic expressions
- Transform algebraic expressions using the rules of arithmetic
- Factorise an algebraic expression

Before you start ...

You should be able to answer these questions.

1 $3 + 3 + 3 + 3 = 4$ lots of $3 = 4 \times 3 = 12$.

a Write these addition sums as multiplications.

i $2 + 2 + 2$

ii $5 + 5 + 5 + 5 + 5 + 5$

iii $10 + 10 + 10 + 10$

iv $7 + 7 + 7$

b Work out the answer of the multiplications.

2 a Work out.

i $3 + 2 + 4$ **ii** $4 + 2 + 3$ **iii** $2 + 4 + 3$

b What do you notice?

3 a Work out.

i $2 \times 3 \times 5$ **ii** $5 \times 2 \times 3$ **iii** $3 \times 2 \times 5$

b What do you notice?

4 Follow the order of operations to work out these calculations.

a $3 \times 5 - 12 + 2$ **b** $4 + 2 \times 3 - 1$

c $6 \div 3 + 2$ **d** $3 \times 4 - 4 \div 2$

5 a Write all the factors of

i 18 **ii** 12 **iii** 24

b Write all the common factors of 18, 12 and 24.

c What is the highest common factor of 18, 12 and 24?

A1.1 Algebraic expressions

This spread will show you how to:

- Use letters to represent unknown numbers in algebraic expressions
- Simplify algebraic expressions by collecting like terms

Keywords
Expression
Like terms
Simplify
Term

You can describe everyday situations using algebra.

- **In algebra, you use letters to represent unknown numbers.**

These boxes hold n pens each.

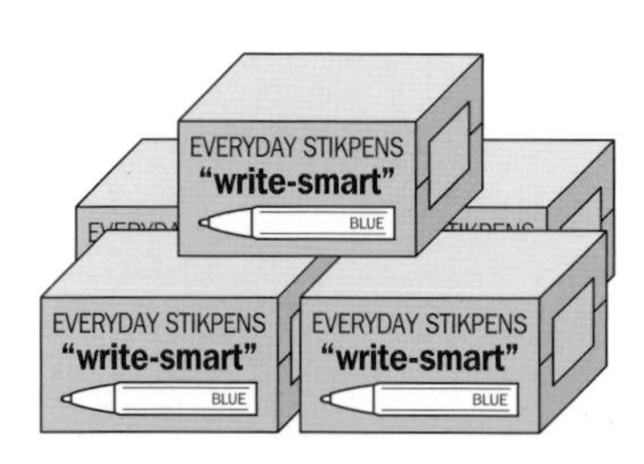

In 5 boxes there are
$n + n + n + n + n = 5 \times n = 5n$ pens

These boxes hold s pens each.

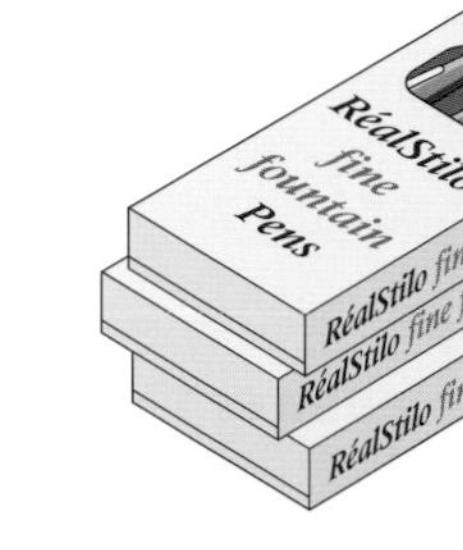

In 3 boxes there are $3s$ pens.

There are $5n + 3s$ pens in total.

$5n$ and $3s$ are **terms.**

$5n + 3s$ is an **expression.**

- **You can simplify an algebraic expression by collecting like terms.**
  Like terms have exactly the same letters.

Example

Simplify these expressions:

a $4x + 2y - 2x + 3y$ **b** $7p - 3q + 5q - p$ **c** $5c - 2b + 2c - 3b$

---

a $4x + 2y - 2x + 3y$
$= 4x - 2x + 2y + 3y$
$= 2x + 5y$

b $7p - 3q + 5q - p$
$= 7p - p + 5q - 3q$
$= 6p + 2q$

c $5c - 2b + 2c - 3b$
$= 5c + 2c - 2b - 3b$
$= 7c - 5b$

Rearrange, keeping terms and their signs together.

Example

In a fruit shop,
apples cost 20p each and
oranges cost 15p each.
Write an expression for the cost
of x apples and y oranges.

Cost of x apples: $x \times 20 = 20x$
Cost of y oranges: $y \times 15 = 15y$
Total cost: $20x + 15y$

Write numbers before letters.

Exercise A1.1

1 Simplify these expressions.

a $b + b + b + b$ **b** $y + y + y - y$ **c** $a - a + a + a + a + a - a$

d $3p + 6p$ **e** $5x - 2x$ **f** $4z - z + 3z$

2 Simplify these expressions.

a $2p + 5q + 3p + q$ **b** $6x + 2y + 3x + 5y$

c $4m + 2n - 2m + 6n$ **d** $5x + 3y - 4x + 2y$

e $7r - 4s + r - 2s$ **f** $2f - 3g + 5g - 6f$

g $3a + 2b + 5c - a + 4b$ **h** $7u - 5v + 3w + 3v - 2u$

i $5x - 3y - 2x + 4z - y + z$ **j** $4r + 6s - 3t + 2r + 5t - s$

3 **a** One guitar has 6 strings. How many strings are there on t guitars?

b One cookie has 3 peanuts on top. How many peanuts are needed for n cookies?

c One horse has 4 horseshoes. How many horseshoes are needed for x horses?

4 In one month, Dan sends x texts.

a Alix sends 4 times as many texts as Dan. How many is this?

b Kris sends 8 more texts than Alix. How many is this?

5 A factory makes bags.

a The factory makes m small bags. A small bag has 2 zips. How many zips do they need?

b The factory makes three times as many large bags as small bags. How many large bags do they make?

c Each large bag has 4 buttons. How many buttons do they need for the large bags?

6 In a pizza takeaway

- a medium pizza has 6 slices of tomato
- a large pizza has 10 slices of tomato.

How many slices of tomato are needed

a for c medium pizzas

b for d large pizzas?

c Write an expression for the total number of slices of tomato needed for c medium and d large pizzas.

7 Write algebraic expressions for the cost in pence of

a f teas and g scones

b j fruit juices and k flapjacks

c x teas, y milks and z scones

d r milks, s fruit juices and t flapjacks.

Café price list	
Tea	50p
Fruit juice	80p
Milk	60p
Scone	30p
Flapjack	40p

A1.2 Indices

This spread will show you how to:

- Use index notation and simple laws of indices

Keywords
Base
Index
Indices
Power
Simplify

- **You can use index notation to write repeated multiplication.**

$5 \times 5 = 5^2$ $\quad m \times m = m^2$ $\quad$ You say '*m* squared'

$5 \times 5 \times 5 = 5^3$ $\quad m \times m \times m = m^3$ $\quad$ You say '*m* cubed'

$5 \times 5 \times 5 \times 5 = 5^4$ $\quad m \times m \times m \times m = m^4$ $\quad$ You say '*m* to the **power** of 4'

'index 4' and 'power 4' mean the same.

5 is the **base**, 4 is the index.

You can simplify expressions with **indices** and numbers.

Indices is the plural of index.

Example

Simplify

a $y \times y \times y \times y \times y$ $\quad$ **b** $4 \times r \times r$

c $3 \times p \times p \times p \times q \times q$ $\quad$ **d** $2 \times s \times s \times 3 \times t \times t \times t$

a $y \times y \times y \times y \times y = y^5$

y is multiplied by itself 5 times, so index is 5.

b $4 \times r \times r = 4 \times r^2 = 4r^2$

c $3 \times p \times p \times p \times q \times q$
$= 3 \times p^3 \times q^2$
$= 3p^3q^2$

d $2 \times s \times s \times 3 \times t \times t \times t$
$= 2 \times s^2 \times 3 \times t^3$
$= 2 \times 3 \times s^2 \times t^3$
$= 6s^2t^3$

Rearrange so the numbers are together.

You can simplify expressions with powers of the same base.

$n^2 \times n^2 = n \times n \times n \times n = n^4$

$t^5 \div t^2 = \frac{t^5}{t^2} = \frac{{}^1\cancel{t} \times {}^1\cancel{t} \times t \times t \times t}{{}^1\cancel{t} \times {}^1\cancel{t}} = t^3$

- **To multiply powers of the same base, add the indices.**
$a^m \times a^n = a^{(m+n)}$

- **To divide powers of the same base, subtract the indices.**
$a^m \div a^n = a^{(m-n)}$

These are the **index laws.** They are described for numbers on page 60.

Example

Simplify

a $s^2 \times s \times s^3$ $\quad$ **b** $\frac{d^2 \times d^4}{d^3}$

$s = s^1$

a $s^2 \times s \times s^3 = s^{(2+1+3)} = s^6$

b $\frac{d^2 \times d^4}{d^3} = \frac{d^6}{d^3} = d^{(6-3)}$
$= d^3$

- **Any value raised to the power of zero is equal to 1.**

$x^0 = 1$ $\quad (4y)^0 = 1$ $\quad 3p^0 = 3 \times 1 = 3$

Exercise A1.2

1 Write these terms in the simplest form.

a $y \times y \times y \times y$ **b** $m \times m \times m \times m \times m \times m$ **c** $x \times x \times x$ **d** $p \times p$

2 Simplify these terms.

a $3 \times t \times t$ **b** $4 \times p \times q \times q$

c $6 \times v \times v \times w \times w \times w$ **d** $2 \times r \times r \times r \times r \times s$

3 Simplify these terms.

a $2 \times m \times m \times 3 \times n$ **b** $4 \times y \times y \times y \times 2 \times z \times z$

c $3 \times g \times 4 \times h \times h \times h$ **d** $5 \times x \times 2 \times y \times y \times y \times y$

4 Simplify these terms.

a $3m^2 \times 2$ **b** $3 \times 4p^3$ **c** $2x \times 3y^2$ **d** $5r^2 \times 2s^2$

5 Simplify these terms.

a $n^2 \times n^3$ **b** $s^3 \times s^4$ **c** $p^3 \times p$ **d** $t \times t^3$

6 Write each of these terms as a single power in the form x^n.

a $x^2 \times x^2 \times x^3$ **b** $x \times x^5 \times x^2$ **c** $x^3 \times x^2 \times x^4$ **d** $x^5 \times x \times x$

7 Write each of these terms as a single power in the form r^n.

a $r^4 \div r^2$ **b** $r^5 \div r^4$ **c** $r^7 \div r^2$ **d** $r^8 \div r^5$

8 Simplify these terms.

a $\frac{m^6}{m^2}$ **b** $\frac{x^4}{x^3}$ **c** $\frac{t^7}{t^5}$ **d** $\frac{y^4}{y}$ **e** $\frac{n^3}{n^3}$

9 Simplify these terms.

a $\frac{x^2 \times x^3}{x^4}$ **b** $\frac{m^3 \times m}{m^2}$ **c** $\frac{s^2 \times s^4}{s^3}$ **d** $\frac{v \times v^3 \times v^2}{v^4}$

e $\frac{q^2 \times q^3 \times q^2}{q^4}$ **f** $\frac{t^3 \times t \times t^2}{t^2}$ **g** $\frac{p^4 \times p^2 \times p}{p^7}$ **h** $\frac{y^2 \times y^4 \times y}{y^3 \times y^2}$

10 Match each of the pairs.

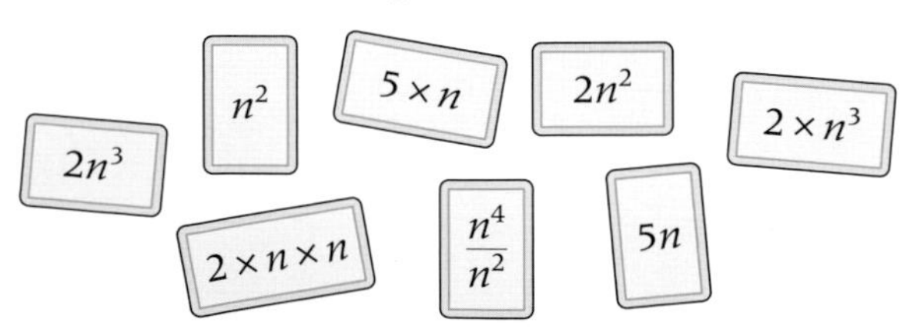

A1.3 Brackets in algebra

This spread will show you how to:

- Expand single and double brackets within algebraic expressions

Keywords
Brackets
Expand
Simplify

You can use **brackets** in algebraic equations.

- **You can multiply out brackets.**
  - **You multiply each term inside the bracket by the term outside.**

$$2(x+4) = 2 \times x + 2 \times 4 = 2x + 8$$

$2 \times (x+4) = 2(x+4)$
You don't write the ×.

To **simplify** expressions with brackets, expand the brackets and collect like terms.

Example

Expand these expressions.

a $2(3x+1)$ **b** $n(n+5)$

a $2(3x+1) = 2 \times 3x + 2 \times 1$
$= 6x + 2$

b $n(n+5) = n \times n + 5 \times n$
$= n^2 + 5n$

Expand means 'multiply out'.

$n \times n = n^2$

Example

Simplify these expressions.

a $m(m+2) + m$ **b** $3(x+1) + 2(4x+2)$

a $m(m+2) + m = m^2 + 2m + m$
$= m^2 + 3m$

b $3(x+1) + 2(4x+2)$
$= 3x + 3 + 8x + 4$
$= 11x + 7$

Like terms have the same power of the same letter. m^2 and m are **not** like terms.

You can **expand** a double bracket by multiplying pairs of terms.

Each term in the first bracket multiplies each term in the second bracket.

$$(p+7)\ (p+3) \longrightarrow p^2 + 3p + 7p + 21 \longrightarrow p^2 + 10p + 21$$

$p^2 + 10p + 21$ is the product of $(p+7)$ and $(p+3)$.

F... **F**irsts
O ... **O**uters
I ... **I**nners
L ... **L**asts

Example

Expand and simplify.

$(y+3)(y+4)$

$(y+3)(y+4) = y^2 + 4y + 3y + 12$
$= y^2 + 7y + 12$

F: $y \times y = y^2$
O: $y \times 4 = 4y$
I: $3 \times y = 3y$
L: $3 \times 4 = 12$

Exercise A1.3

1 Expand the brackets in these expressions.

a $3(m+2)$ **b** $4(p+6)$ **c** $2(x+4)$ **d** $5(q+1)$
e $2(6+n)$ **f** $3(2+t)$ **g** $4(3+s)$ **h** $2(4+v)$

2 Expand these expressions.

a $3(2q+1)$ **b** $2(4m+2)$ **c** $3(4x+3)$ **d** $2(3k+1)$
e $5(2+2n)$ **f** $3(4+2p)$ **g** $4(1+3y)$ **h** $2(5+4z)$

3 At a pick-your-own fruit farm, Lucy picks n apples.

Mary picks 5 more apples than Lucy.

a Write in terms of n, the number of apples Mary picks.

Nat picks 3 times as many apples as Mary.

b Write in terms of n, the number of apples Nat picks.

4 Expand and simplify each of these expressions.

a $3(p+3)+2p$ **b** $2(m+4)+5m$ **c** $4(x+1)-2x$
d $2(5+k)+3k$ **e** $4(2t+3)+t-2$ **f** $3(2r+1)-2r+4$

5 On Monday a shop sells s DVDs.
On Tuesday the shop sells 6 more DVDs than on Monday.

a Write an expression for the number of DVDs it sells on Tuesday.

On Wednesday the shop sells twice as many DVDs as on Tuesday.

b Write an expression for the number of DVDs it sells on Wednesday.

On Thursday the shop sells 7 more DVDs than on Wednesday.

c Write an expression for the number of DVDs it sells on Thursday.

Give your answer in its simplest form.

6 Expand and simplify each of these expressions.

a $2(n+3)+3(n+2)$ **b** $4(p+1)+2(3+p)$

c $4(2x+1)+2(x+3)$ **d** $2(3n+2)+3(4n+1)$

7 A small box contains 12 chocolates.
Sam buys y small boxes of chocolates.

a Write an expression for the number of chocolates Sam buys.

A large box contains 20 chocolates.
Sam buys 2 more of the large boxes than the small ones.

b Write an expression for the number of large boxes of chocolates he buys.

c Find, in terms of y, the total number of chocolates in the large boxes that Sam buys.

d Find, in terms of y, the total number of chocolates Sam buys.
Give your answer in its simplest form.

8 Expand the brackets in these expressions.

a $x(4x+1)$ **b** $m(m^2+2)$ **c** $2t(t^2+4)$ **d** $3p(p^2+1)$

9 Expand and simplify.

a $(p+2)(p+1)$ **b** $(5w+1)(3w+9)$ **c** $(2m+1)(3-m)$ **d** $(y+1)^2$

A1.4 Simplifying expressions

This spread will show you how to:

- Transform algebraic expressions using the rules of arithmetic

Keywords
Brackets
Indices

You can add, subtract, multiply or divide algebraic terms.

$3n + 5n + 8n = 16n$ $4p - p = 3p$ $2 \times 6p = 12p$ $8r \div 4 = 2r$

You can use the acronym BIDMAS to remember the order. Look back at page 16 for a reminder.

- **To simplify an expression, you follow the same order of operations as in arithmetic.**

  Brackets ⟹ Indices ⟹ Division or Multiplication ⟹ Addition or Subtraction

Example

Simplify these expressions.

a $4n + 2 \times 5n$ **b** $3r \times 2s$ **c** $4t^2 - 3 \times t^2 + t$

a $4n + 2 \times 5n$ multiplication before addition
$= 4n + 10n$
$= 14n$

b $3r \times 2s = 3 \times 2 \times r \times s$
$= 6rs$

c $4t^2 - 3 \times t^2 + t = 4t^2 - (3 \times t^2) + t$
$= 4t^2 - 3t^2 + t$
$= t^2 + t$

Collect like terms.

To simplify an expression with brackets, expand the brackets first.

$$3(n - 2) = 3 \times n + 3 \times -2 = 3n - 6$$

$3 \times -2 = -6$

When you expand brackets, keep each term with its sign.

$$4(x - 1) - 2(x + 3) = 4 \times x + 4 \times -1 + -2 \times x + -2 \times 3$$
$$= 4x \quad - 4 \quad - 2x \quad - 6$$
$$= 4x - 2x - 4 - 6$$
$$= 2x - 10$$

Example

Expand and simplify these expressions.

a $3(5y - 2)$ **b** $2(3p + 1) - 3(p + 2)$ **c** $2(3m + 1)(2m - 2)$

a $3(5y - 2) = 3 \times 5y + 3 \times -2$
$= 15y - 6$

b $2(3p + 1) - 3(p + 2) = 6p + 2 - 3p - 6$
$= 3p - 4$

c $2(3m + 1)(2m - 2) = 2(6m^2 - 6m + 2m - 2)$
$= 2(6m^2 - 4m - 2)$
$= 12m^2 - 8m - 4$

In part **c**, multiply the brackets together first and then multiply by the 2 outside the brackets.

Exercise A1.4

1 Simplify these expressions.

a $3r + 3 \times 2r$ **b** $2m^2 + 2m \times m$ **c** $6x \div 2 + x$

d $2t \times 4v$ **e** $5m \times 2n$ **f** $3x \times 2y^2$

g $x^2 + x^2 + x$ **h** $3 \times 3w - 2 \times 4$ **i** $z \times z^2 + 3z + 1$

2 Expand these expressions.

a $4(2y + 3)$ **b** $2(3x - 2)$ **c** $3(2k - 2)$ **d** $4(1 - n)$

3 Write these expressions in their simplest form.

a $3k^2 - 2 \times k^2 + k$ **b** $4m + 6m \div 2 + m^2$ **c** $6t - (4 \times -t) + 5$

4 Expand these expressions.

a $2m(m - 3)$ **b** $4p(2p - 1)$ **c** $r(r^2 + 3)$ **d** $2s(s^2 - 4)$

5 Expand and simplify each of these expressions.

a $3(r + 2) + 2(r - 1)$ **b** $4(s + 1) - 2(s + 2)$

c $3(2j + 3) - 2(j + 2)$ **d** $3(4t - 2) + 3(t - 1)$

6 Simplify these expressions.

a $3(4m + 1)(m - 1)$ **b** $2(3p - 1)(p - 2)$

c $-5(2q + 3)(q - 3)$ **d** $4(2v - 3)(v - 1)$

7 Jake is n years old.
Jake's sister is 4 years older than Jake.
Jake's mother is 3 times older than his sister.
Jake's father is 4 times older than Jake.
Jake's uncle is 2 years younger than Jake's father.
Jake's grandmother is twice as old as Jake's uncle.

a Copy the table and write each person's age in terms of n.

Jake	Sister	Mother	Father	Uncle	Grandmother
n					

b Find, in terms of n, how much older Jake's grandmother is than his mother. Give your answer in its simplest form.

A1.5 Factorising

This spread will show you how to:

- Factorise an algebraic expression

Keywords
Common factor
Factor
Factorise

- When you **factorise** a quantity, you break it up into factors.

$$2(x+4) \underset{\text{factorise}}{\overset{\text{expand}}{=}} 2x+8$$

In number...
a factor is a number that exactly divides into another number.

2, 3 and 4 are factors of 12.

In algebra...
a factor is a number or letter that exactly divides into another term.

3 and $2x$ are factors of $6x$.

- To factorise an expression, look for a **common factor** for all the terms.

A common factor divides into all the terms.

$3x+9$: $\div 3$ each term gives $3(x+3)$

Write the common factor outside the bracket.

a^2-a: $\div a$ each term gives $a(a-1)$

Sometimes the common factor is a letter.

a divides into a^2 and a.

Example

a Find the common factors of $12x$ and 8.
b Factorise $12x + 8$.

a 2 and 4 are common factors of $12x$ and 8.
b $12x + 8 = 4 \times 3x + 4 \times 2$
$= 4(3x + 2)$

To factorise completely, use the highest common factor. The highest common factor of $12x$ and 8 is 4.

Example

Factorise

a $3y - 9$
b $x^2 - 3x$
c $2m^3 + m$
d $4p^3 - 2p$

a $3y - 9 = 3 \times y - 3 \times 3$
$= 3(y - 3)$
b $x^2 - 3x = x \times x - 3 \times x$
$= x(x - 3)$
c $2m^3 + m = m \times 2m^2 + m \times 1$
$= m(2m^2 + 1)$
d $4p^3 - 2p = 2p \times 2p^2 - 2p \times 1$
$= 2p(2p^2 - 1)$

You can check your answer by expanding:
$x(x - 3) = x^2 - 3x$

Exercise A1.5

1 Find all the common factors of

a $2x$ and 6 **b** $4y$ and 12 **c** 10 and $20j$ **d** 6 and $12p$

e 9 and $6q$ **f** $6t$ and 4 **g** $4x$ and 10 **h** $24t$ and 8

2 Find the highest common factor of

a $3x$ and 9 **b** $12r$ and 10 **c** $6m$ and 8 **d** 4 and $4z$

3 Find the highest common factor of

a y^2 and y **b** $4s^2$ and s **c** $7m$ and m^3 **d** $2y^2$ and $2y$

4 Factorise these expressions.

a $2x + 10$ **b** $3y + 15$ **c** $8p - 4$ **d** $6 + 3m$

e $5n + 5$ **f** $12 - 6t$ **g** $14 + 4k$ **h** $9z - 3$

5 Factorise these expressions.

a $w^2 + w$ **b** $z - z^2$ **c** $4y + y^2$ **d** $2m^2 - 3m$

e $4p^2 + 5p$ **f** $7k - 2k^2$ **g** $3n^3 - 2n$ **h** $5r + 3r^2$

6 The cards show expansions and factorisations.
Match the cards in pairs.

$4(x+3)$	$4x^2 - 3x$	$3(x-4)$	$4x + 3x^2$
$3x - 12$	$x(4+3x)$	$4x + 12$	$x(4x-3)$

7 Factorise these expressions.

a $4y - 12$ **b** $2x^2 + 3x$ **c** $3y^2 - y$ **d** $15 + 5t^2$

e $3m + 9m^2$ **f** $2r^2 - 2r$ **g** $4v^3 + v$ **h** $3w^2 + 3w$

Check your answers by expanding.

8 Debbie, Kate and Bryn factorise $16x^2 + 4x$.
Here are their answers.

Debbie	Kate	Bryn
$16x^2 + 4x = 2(x^2 + 2x)$	$16x^2 + 4x = 4x(4x + 1)$	$16x^2 + 4x = 2x(8x^2 + 2)$

a Who is correct?

b Explain where the other two have gone wrong.

A2.3 Equations with the unknown on both sides

This spread will show you how to:

- Solve linear equations using the balance method, including equations with fractional or negative solutions
- Solve equations with the unknown on both sides

Keywords
Solve
Unknown

When you **solve** an equation, you find the value of the letter term or **unknown**.

In some equations the unknown is on both sides of the equals sign.

- **You can solve equations with the unknown on both sides using the balance method.**

Here is an example:

$4x + 2 = 3x + 5$ The $3x$ term has the smallest number of x.

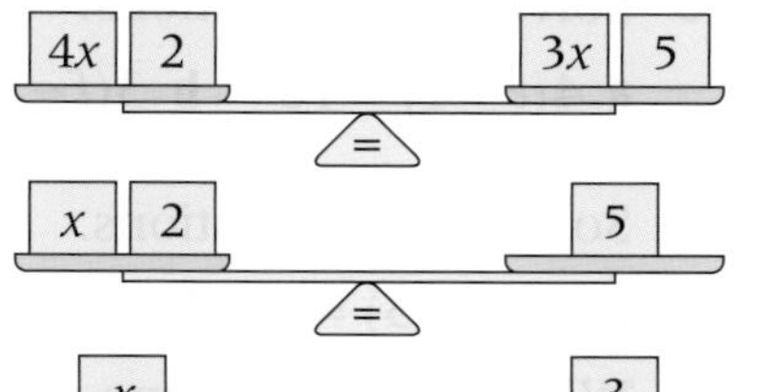

$4x - 3x + 2 = 3x - 3x + 5$ Subtract $3x$ from both sides.

$x + 2 = 5$

$x - 2 = 5 - 2$ Subtract 2 from both sides.

$x = 3$

Example

Solve the equation

$3y - 7 = 2y + 3$

$3y - 2y - 7 = 2y - 2y + 3$ The smallest term in y is $2y$. Subtract $2y$ from both sides.

$y - 7 = 3$ Add 7 to both sides.

$y = 3 + 7 = 10$

Example

Here is an equation:

$4t + 10 = 2t + 4$

Find the value of t.

$4t - 2t + 10 = 2t - 2t + 4$ The smallest term in t is $2t$. Subtract $2t$ from both sides.

$2t + 10 = 4$

$2t + 10 - 10 = 4 - 10$

$2t = -6$ Divide both sides by 2.

$t = -3$

Exercise A2.3

1 Solve these equations.

a $4m + 2 = 3m + 7$
b $6p - 5 = 5p - 2$
c $3t + 2 = 2t + 5$
d $3n - 11 = 2n - 4$
e $4q + 2 = 5q - 6$
f $5s - 2 = 4s + 6$

2 Solve these equations.

a $2s + 5 = 3s + 8$
b $4t - 2 = 5t + 2$
c $6u + 10 = 5u + 8$
d $4v - 6 = -15 - 5v$

3 Find the value of the unknown in each of these equations.

a $2a + 14 = 6a - 6$
b $4b - 2 = 6b + 6$
c $3c - 4 = c + 1$
d $5d + 15 = -6 - 2d$

4 Solve these equations.

a $4x + 3 = 18 + 2x$
b $4x + 10 = 2x + 4$
c $8x + 15 = 12x + 14$
d $6x - 4 = 10x + 2$

5 Mae doubles a number and adds 5 to get 21.

a Write an expression for Mae's calculation.
Use n to represent the number.

$21 = \square n + \square$

b Tim multiplies the same number by 4 and subtracts 11 to get 21.
Write an expression for Tim's calculation, using n to represent the number.

$\square n - \square = 21$

c Write your expressions from parts **a** and **b** as an equation.

expression **a** = 21, expression **b** = 21
expression **a** = expression **b**

d Solve your equation to find the value of n.

6 Write equations for these 'think of a number' problems as you did in question **5**.

Solve them to find the value of the number.

a I think of a number, multiply by 2 and add 7. I get the same answer when I multiply the number by 4 and subtract 13.

$2n + 7 = 4n - 13$

b I think of a number, multiply by 5 and subtract 8. I get the same answer when I double the number and add 10.

$\square n - \square = \square n + \square$

c I think of a number, multiply by 3 and add 4. I get the same answer when I multiply by 5 and add 12.

A2.4 More equations with brackets

This spread will show you how to:

- Solve equations with the unknown on both sides
- Solve equations involving brackets

Keywords
Brackets
Expand

This square pattern is made from rectangular tiles. Each tile has length $(x+3)$ cm and width x cm.

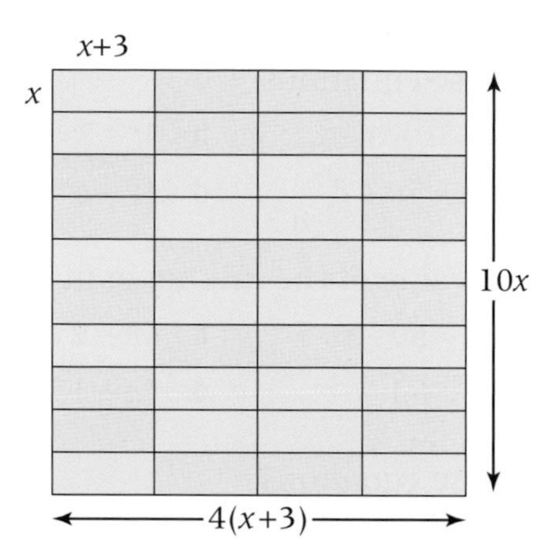

The pattern is 10 tiles wide: $10x$
The pattern is 4 tiles long: $4(x+3)$

The pattern is a square, so length = width: $4(x+3) = 10x$

Expand the **brackets**: $4x + 12 = 10x$

Subtract $4x$ from both sides: $12 = 10x - 4x$
$12 = 6x$

Divide both sides by 6: $2 = x$
So $x = 2$.

The square is 20 cm by 20 cm.

- **To solve equations with the unknown on both sides and brackets:**
  - **Expand the brackets**
  - **Use the balance method.**

Example

Solve

a $2(y+4) = 4y$

b $6r - 2 = 4(r+3)$

---

a $2(y+4) = 4y$ — Expand the brackets.
$2y + 8 = 4y$ — Subtract $2y$ from both sides.
$8 = 4y - 2y$
$8 = 2y$ — Divide both sides by 2.
$4 = y$

b $6r - 2 = 4(r+3)$
$6r - 2 = 4r + 12$ — Subtract $4r$ from both sides.
$6r - 4r - 2 = 4r - 4r + 12$
$2r - 2 = 12$ — Add 2 to both sides.
$2r = 12 + 2$
$2r = 14$ — Divide both sides by 2.
$r = 7$

Exercise A2.4

1 Solve these equations.

a $2(r+6) = 5r$ **b** $6(s-3) = 12s$

c $4(2t+8) = 24t$ **d** $5(v-1) = 6v$

2 Solve these equations.

a $2(a+5) = 7a-5$ **b** $3(b-2) = 5b-2$

c $2(c+6) = 5c-3$ **d** $3d+8 = 2(d+2)$

3 Solve these equations.

a $3(2x-4) = 7x-18$ **b** $2(3y+2) = 5y-2$

c $4(2z+1) = 6z+15$ **d** $-4(6m+1) = -17m-18$

4 Solve these equations.

a $2(e+3) = 4e-1$ **b** $4f+3 = 2(f+2)$

c $4(2g+1) = 6g+1$ **d** $3(2h+3) = 5h+8$

5 **a** Choose one expression from each set of cards.

b Write them as an equation:
expression from set 1 = expression from set 2

c Solve your equation to find the value of x.

d Repeat for different pairs of expressions.

Set 1

$2(x+3)$	$4(2x-1)$	$3(4x+1)$

Set 2

$3x-2$	$4x+1$	$8x-3$

6 The triangle and the square have equal perimeter.

a Write an expression for the perimeter of the triangle.

b Write an expression for the perimeter of the square.

c Use your two expressions to write an equation.

d Solve your equation to find the value of x.

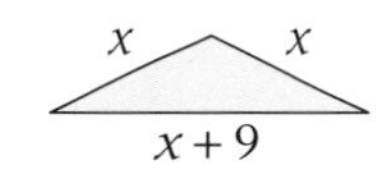

$x+1$

$x+1$

perimeter of triangle = perimeter of square

7 **a** Write an expression for the area of square A.

b Square B and square A have equal area.
Write an equation in x to show this.

c Solve your equation to find the value of x.

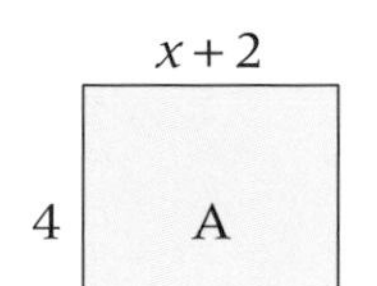

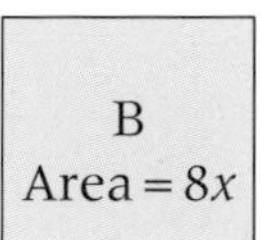

8 A blouse has m buttons. A shirt has $m+2$ buttons.

a Write an expression for the number of buttons on four blouses.

b Write an expression for the number of buttons on three shirts.

c Three shirts have the same number of buttons in total as four blouses.
Write an equation and solve it to find the value of m.

d How many buttons are there on a shirt?

A2.5 Equations with fractions

This spread will show you how to:

- Solve equations involving fractions and negative signs

Keywords
Fraction

Problems in algebra often contain fractions.

Tom and Lauren share a box of chocolates.
Tom has half the chocolates.
He counts them.
There are 15.

How many chocolates were there in the box?

To solve this problem, you multiply 15×2 to get 30.

You can write the problem in algebra like this: $\frac{x}{2} = 15$

$x \div 2 = \frac{x}{2}$

Using the balance method, you do the same to both sides: $\frac{x}{2} \times 2 = 15 \times 2$

$x = 30$

The opposite of ÷ is ×

- **You can solve equations involving fractions using the balance method.**

Example

Solve these equations.

a $\frac{x}{4} = -3$ **b** $\frac{x}{3} + 2 = 7$ **c** $\frac{x+3}{2} = 5$ **d** $\frac{5-2x}{3} = 7$

a

$$\frac{x}{4} = -3$$
$$4 \times \frac{x}{4} = -3 \times 4$$
$$x = -12$$

b

$$\frac{x}{3} + 2 = 7$$
$$\frac{x}{3} + 2 - 2 = 7 - 2$$
$$\frac{x}{3} = 5$$
$$3 \times \frac{x}{3} = 5 \times 3$$
$$x = 15$$

c

$$\frac{x+3}{2} = 5$$
$$2 \times \frac{x+3}{2} = 5 \times 2$$
$$x + 3 = 10$$
$$x = 10 - 3$$
$$x = 7$$

d

$$\frac{5-2x}{3} = 7$$
$$3 \times \frac{5-2x}{3} = 7 \times 3$$
$$5 - 2x = 21$$
$$5 - 2x - 5 = 21 - 5$$
$$-2x = 16$$
$$\frac{-2x}{-2} = \frac{16}{-2}$$
$$x = -8$$

Exercise A2.5

1 Solve these equations.

a $\frac{x}{3} = 3$ **b** $\frac{m}{4} = -2$ **c** $\frac{-n}{3} = 6$ **d** $\frac{m}{5} - 4$

2 Find the value of the unknown in each of these equations.

a $\frac{s}{3} + 5 = 8$ **b** $4 - \frac{t}{2} = 1$ **c** $\frac{u}{5} + 7 = 5$ **d** $16 = \frac{v}{4} + 13$

3 Solve these equations.

a $\frac{2x}{3} + 5 = 9$ **b** $\frac{3y}{2} - 5 = 4$ **c** $3 - \frac{2z}{5} = -3$ **d** $\frac{3q}{2} + 5 = -7$

4 Solve these equations.

a $\frac{x + 5}{3} = 2$ **b** $\frac{x - 3}{4} = 2$ **c** $\frac{x + 9}{2} = -4$ **d** $\frac{10 - x}{4} = 1$

5 Solve these equations.

a $\frac{2x + 1}{5} = 5$ **b** $\frac{3x - 2}{4} = 4$ **c** $\frac{11 - 2x}{3} = -1$ **d** $\frac{31 - 3x}{4} = 4$

6 I think of a number.
I divide my number by 4 and add 6.

a Write an expression for 'I divide my number by 4 and add 6'.
Use n to represent the number.

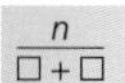

b My answer is 10.
Using your expression from part **a**, write an equation to show this.

Expression **a** = 10

c Solve your equation to find the number, n.

7 Use the method in question **6** to write an equation and find the missing number in these problems.

a I think of a number.
I divide it by 3 and subtract 4.
The answer is 7.

b I think of a number.
I half it and add 8.
The answer is 3.

8 The perimeter of this equilateral triangle is $2x + 6$.

a Write an expression for the length of one side of the triangle.

b The length of one side of the triangle is 8 cm.
Find the value of x.

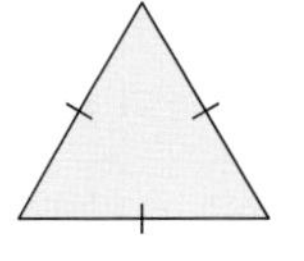

Length of side = $\frac{\text{perimeter}}{3}$

9 The perimeter of this square is $4 + x$.
The length of one side is 10 cm.
Write an equation and solve it to find the value of x.

A2.6 Trial and improvement

This spread will show you how to:

- Use systematic trial and improvement to estimate the solutions of an equation

Keywords
Estimate
Trial and improvement

- To solve equations with powers, you can use **trial and improvement**.
  - You **estimate** a solution and **try** it in the equation.
  - If your estimate doesn't fit, you **improve** it and try again.

Example

Use trial and improvement to find the value of x in this equation.

$x^2 = 87$

Give your answer to 1 dp.

$x^2 = 87$

An estimate for x is 9.2

Try 9.2 in the equation: $9.2^2 = 84.64$ too small

Try 9.3 in the equation: $9.3^2 = 86.49$ too small

Try 9.4 in the equation: $9.4^2 = 88.36$ too big

So the solution is between 9.3 and 9.4.

Try the halfway value, 9.35: $9.35^2 = 87.42$ too big

So the solution is between 9.3 and 9.35

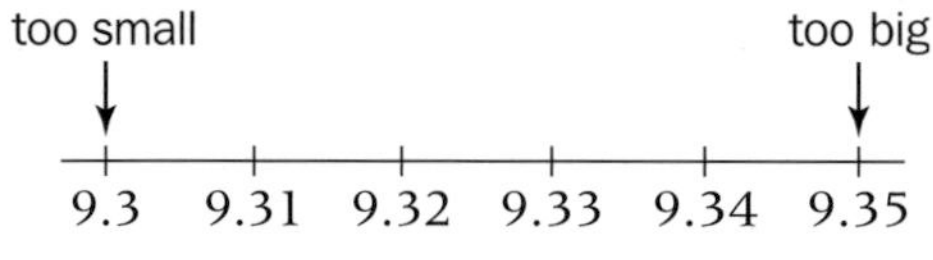

The solution is 9.3 to 1 dp.

$9^2 = 81$, so estimate that x is a bit bigger than 9.

When the answer is too small, improve your estimate by choosing a slightly bigger value.

All the values between 9.3 and 9.35 round to 9.3

- **When using trial and improvement you need to work systematically.** **You can show your trials in a table.**

Example

The equation

$x^3 + x = 33$

has a solution between 3 and 4.

Use trial and improvement to find the solution.

Give your answer correct to 1 decimal place.

x	x^3	$x^3 + x$	Too big or too small?
3.5	42.875	46.375	too big
3.2	32.768	35.968	too big
3.1	29.791	32.891	too small
3.15	31.255…	34.405…	too big

The solution is between 3.1 and 3.15.

The solution is 3.1 to 1 dp.

Draw a table.

The solution is between 3.1 and 3.2. Try the halfway value.

All the values between 3.1 and 3.15 round to 3.1.

Exercise A2.6

1 Bina is using trial and improvement to find a solution to $x^2 = 29$.
She draws this table.

x	x^2	**Too big or too small?**
5.5	30.25	
5.4		

a Copy the table and fill in the rest of the rows for the values 5.5 and 5.4.

b What value could you try next?
Write this value in your table and complete the row.

c Improve your estimate and write the value in the table.
Complete the row for this estimate.

d Continue in this way until you have found a solution to 1 decimal place.

You may need to add extra rows to your table.

2 The equation $x^2 - x = 13$ has a solution between 4 and 5.
Copy and complete the table to find this solution to 1 decimal place.
Draw as many rows as you need.

x	x^2	$x^2 - x$	**Too big or too small?**
4.5			

3 The equation $x^3 + x = 146$ has a solution between 5 and 6.
Copy and complete the table to find this solution to 1 decimal place.
Draw as many rows as you need.

x	x^3	$x^3 + x$	**Too big or too small?**

4 **a** Substitute $x = 1$, $x = 2$ and $x = 3$ into the equation

$$x^3 - x = 9$$

b Use your answers from part **a** to help you estimate a solution to the equation $x^3 - x = 9$.

c Draw up a table for this equation.

d Use your answer from part **b** as your first estimate in your table.

e Find the solution to 1 decimal place.

Your table will be similar to the one in question **3**.

5 Use trial and improvement to find a solution to

$$x^3 + x = 73$$

Give your answer to 2 decimal places.

A2

Exam review

Key objectives

- Solve linear equations that require prior simplification of brackets, including those that have negative signs occurring anywhere in the equation, and those with a negative solution
- Solve linear equations in one unknown, with integer or fractional coefficients, in which the unknown appears on either side or on both sides of the equation
- Use systematic trial and improvement to find approximate solutions of equations where there is no simple analytical method of solving them

1 **a** Solve

$$\frac{x}{2} = 10$$

(1)

b A cream cake costs x pence.
A chocolate cake costs 50 pence more than a cream cake.

i Write down, in terms of x, the cost of one chocolate cake. (1)

ii Three cream cakes and one chocolate cake cost £4.30 in total.
Form an equation in x and solve it to find the cost of one cream cake. (3)

(OCR, 2004)

2 **a** Solve $7x + 18 = 74$. (2)

b Solve $4(2y - 5) = 32$. (2)

c Solve $5p + 7 = 3(4 - p)$. (3)

(Edexcel Ltd., 2003)

A3 Sequences

This unit will show you how to

- Generate and describe sequences using a term-to-term rule
- Understand the difference between increasing and decreasing sequences
- Generate and describe sequences using a position-to-term rule
- Describe and find the general term of a linear sequence
- Explain how the formula for the general term of a linear expression works

Before you start ...

You should be able to answer these questions.

1 Work out the difference between

a 5 and 8 **a** 3 and 9

c 6 and 11 **d** −4 and +2

2 Find the difference between

a 10 and 7 **b** 9 and 5

c 12 and 7 **d** 3 and −1

3 Write down the first six multiples of

a 4 **b** 3

c 5 **d** 6

4 Copy and complete

a $2 \times 7 = ?$ **b** $4 \times ? = 36$

c $8 \times ? = 24$ **d** $3 \times ? = 21$

5 Work out

a 2^2 **b** 4^2

c 5^2 **d** 3^2

e 6^2 **f** 9^2

A3.1 Term-to-term rules

This spread will show you how to:

- Generate and describe sequences using a term-to-term rule
- Understand the difference between increasing and decreasing sequences

Keywords
Common difference
Decreasing
Increasing
Linear
Rule
Sequence
Term

You can describe a **sequence** by giving the first **term** and the term-to-term **rule**.
The rule tells you how to work out each term from the one before.

Example

Write the first five terms in the sequence with first term 4 and term-to-term rule 'add 3'.

4, 7, 10, 13, 16

Start with 4; add 3 each time.

- **In an increasing sequence, the terms are getting larger.** For example 5, 9, 13, 17, 21, ...
- **In a decreasing sequence, the terms are getting smaller.** For example 12, 10, 8, 6, 4, ...

You can work out the term-to-term rule for a sequence and use it to find more terms.

Example

For each sequence, work out the two missing terms.
Describe the sequence in words.

a 19, 9, −1, −11, ... **b** 2, ___, 14, ___, 26, 32

a 19, 9, −1, −11, ___, ___
The first term is 19 and the terms decrease by 10 each time.
The next two terms are −21 and −31.
b From 26 to 32 is an increase of 6.
The sequence is 2, 8, 14, 20, 26, 32.
The first term is 2 and the terms increase by 6 each time.

A **linear** sequence increases or decreases in equal-sized steps.
The size of the 'step' is called the **common difference**.
For example:

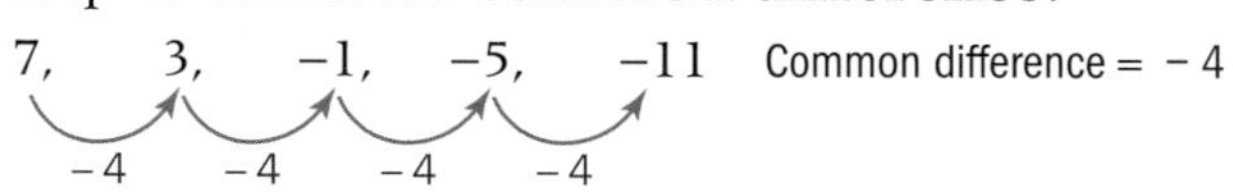

In some sequences, the 'steps' from one term to another are not equal.

Example

Describe how this sequence is increasing: 3, 4, 7, 12, 19, ...
Work out the next two terms in the sequence.

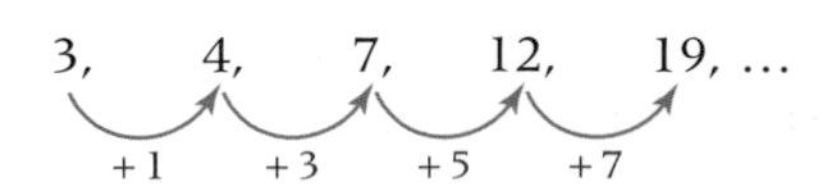

The first difference is 1 and the difference increases by 2 each time.
The next two terms are 28 and 39.

To find the next term, +9:
$19 + 9 = 28$
To find the one after, +11:
$28 + 11 = 39$

Exercise A3.1

1 Write out the first five terms of these sequences.

1st term	Rule
6	Increase by 4 each time
26	Increase by 5 each time
10	Decrease by 3 each time
−6	Increase by 2 each time
−23	Increase by 7 each time

2 Write down the first five terms in each of these sequences.

a Even numbers

b Odd numbers larger than 16

c Multiples of 4

d Multiples of 6 greater than 20

e Two more than the 5 times table

f Square numbers

g One more than square numbers

h Powers of 2.

3 Write out the first five terms of these sequences.

a 2nd term 7, increases by 3 each time.

b 2nd term 19, decreases by 6 each time.

c 3rd term 12, increases by 4 each time.

d 3rd term 14, decreases by 8 each time.

e 5th term is 8, increases by 6 each time.

4 For each sequence

i work out the two missing terms

ii describe the sequence.

a 3, 10, 17, 24, ___, ___

b −7, 1, 9, 17, ___, ___

c −19, −16, −13, −10, ___, ___

d 7, ___, 15, 19, 23, ___

e 3, ___, ___, 27, 35, 43

f 16, 11, 6, ___, ___, −9

g ___, −8, −5, −2, ___, 4

h ___, 6, ___, −8, −15

5 For each sequence

i write the terms in order from smallest to largest

ii describe the sequence.

a 8, 3, 5, 12, 2

b 25, 18, 27, 22, 28

c −3, 0, −6, 4, −5

d 20, 8, 6, 15, 11

e −21, −9, −23, −17, −24

f −1, 5, −15, −7, 3

6 **Challenge**
In each set of numbers there are two sequences mixed together.
Write out each pair of sequences.

a 1, 3, 4, 5, 7, 7, 9, 10, 11, 13

b 2, 4, 7, 8, 11, 15, 16, 19, 23, 32

A3.2 The general term

This spread will show you how to:

- Generate and describe sequences using a position-to-term rule

Keywords
General term
*n*th term
Position-to-term

- **A position-to-term rule links a term with its position in the sequence.**

For example, the 4 times table: 4, 8, 12, 16, 20, 24, ... can be written as

1st term =T(1)	2nd term =T(2)	3rd term =T(3)	4th term =T(4)	5th term =T(5)	6th term =T(6)
4	8	12	16	20	24

$T(1) = 1 \times 4$
$T(2) = 2 \times 4$
$T(3) = 3 \times 4$
$T(10) = 10 \times 4$

The **general term** or ***n*th term** of the 4 times table is $T(n) = n \times 4$ or $4n$.

You can generate a sequence from the general term.

Example

Find the first three terms and the 10th term of the sequence with general term $3n + 2$.

1st term $3 \times 1 + 2 \rightarrow 5$
2nd $3 \times 2 + 2 \rightarrow 8$
3rd $3 \times 3 + 2 \rightarrow 11$
10th $3 \times 10 + 2 \rightarrow 32$

To find a term, substitute its position number for n.

Example

The nth term of a sequence is $n^2 - 8$.
Copy and complete the table of results.

Term number	Term
1	
2	
3	
4	
10	
n	$n^2 - 8$

Term number	Term
1	$1^2 - 8 = 1 - 8 = -7$
2	$2^2 - 8 = 4 - 8 = -4$
3	$3^2 - 8 = 9 - 8 = 1$
4	$4^2 - 8 = 16 - 8 = 8$
10	$10^2 - 8 = 100 - 8 = 92$
n	$n^2 - 8$

Squared comes first then subtract 8.

Exercise A3.2

1 Find the first three terms and the 10th term of these sequences.

a $5n + 1$ **b** $3n + 8$ **c** $8n - 4$ **d** $6n - 8$

e $24 - 2n$ **f** $15 - 5n$ **g** $7n - 20$ **h** $4n - 6$

2 Copy and complete the table of results for each sequence.

a $3n + 8$

Term number	Term
1	11
2	
3	
5	
10	
n	$3n+8$

b $6n - 15$

Term number	Term
1	−9
2	
3	
5	
10	
n	$6n-15$

3 Write the first five terms of the sequences with nth term

a $n^2 + 4$ **b** $n^2 - 2$ **c** $2n^2$ **d** $12 - n^2$

4 Find the 2nd, 5th and 10th terms of the sequences with nth term

a $n^2 + 8$ **b** $n^2 - 6$ **c** $2n^2 + 7$

5 The general term of a sequence is given by $4n + 1$.

a Find the first three terms.

b Copy and complete:
Each term is _____ more than a multiple of _____

c Is 222 in the sequence? Explain.

6 The general term of a sequence is $2n^2 + 1$.

a Find the first three terms.

b Manjit worked out that the 5th term was 101. Explain why he was wrong.

7 The general term of the sequence of square numbers is n^2.

a Write the first five terms of this sequence.

b Work out the differences between consecutive terms.

c Copy and complete this table.

1st square number	1	1
2nd square number	4	1 + 3
3rd square number	9	1 + 3 + 5
4th square number		
5th square number		
6th square number		

d What square number is equal to
$1 + 3 + 5 + 6 + 7 + 9 + 11 + 13 + 15 + 17 + 19$?
Do not work out the addition!

A3.4 Pattern sequences

This spread will show you how to:

- Explain how the formula for the general term of a linear expression works

Keywords
Common difference
*n*th term
Pattern

You can find the **nth term** for a sequence of **patterns**.

Here is a pattern made of pencils:

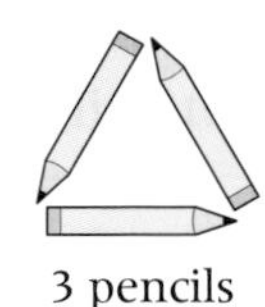

3 pencils

The next pattern in the sequence is:

5 pencils

Examiner's tip: The techniques described here are useful in many investigations.

You add 2 pencils each time.

The nth term is an expression for the number of pencils in the nth pattern.
To find the nth term, write the numbers of pencils in a table.

Pattern number	1	2	3	4	5
Number of pencils	3	5	7	9	11

You can work out the number of pencils in pattern 5 by continuing the number sequence.

The number sequence 3, 5, 7, 9, 11 has **common difference** 2.
So the nth term contains the term $2n$.

$2n$	2	4	6	8	10
Sequence	3	5	7	9	11

Each term is 1 more than a multiple of 2.
The nth term is $2n + 1$.

The nth term is the same as the general term.

Example

Here are some patterns made up of dots.

○ ○
○ ○ ○

Pattern number 1

○ ○ ○
○ ○ ○ ○

Pattern number 2

○ ○ ○ ○
○ ○ ○ ○ ○

Pattern number 3

Pattern number	1	2	3	4	5
Number of dots	5	7			

a Draw the next pattern in the sequence.
b Copy and complete the table for the first five terms.
c How many dots will there be in pattern number 10?

---

a

○ ○ ○ ○ ○
○ ○ ○ ○ ○ ○

Pattern number 4

b

Pattern number	1	2	3	4	5
Number of dots	5	7	9	11	13

Each time you add 2 dots – one to each row

c The number sequence continues:

Pattern number	6	7	8	9	10
Number of dots	15	17	19	21	23

There will be 23 dots in the 10th pattern.

Exercise A3.4

1 Here is a sequence of patterns made from squares.

Pattern number 1   Pattern number 2   Pattern number 3

Pattern number	1	2	3	4	5
Number of squares					

a Draw the next pattern in the sequence.

b Copy and complete the table for the first five terms.

c How many squares will there be in pattern number 10?

2 Here are some patterns made from crosses.

Pattern number	1	2	3	4	5
Number of crosses					

a Draw the next pattern in the sequence.

b Copy and complete the table for the first five terms.

c Describe in words how the sequence grows.

d Work out the common difference for the number sequence.

e Find the nth term for the number sequence.

3 Here is a sequence of patterns of squares.

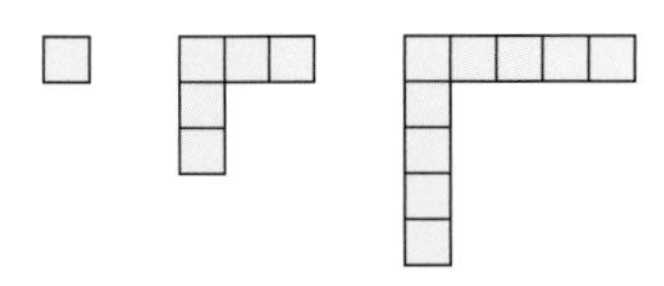

a Write the number of squares in the next two patterns.
Explain how you worked them out.

b Find, in terms of n, an expression for the number of squares in the nth pattern.

c Find the number of squares in the 50th pattern.

4 Repeat question **3** for these sequences of patterns.

a

b

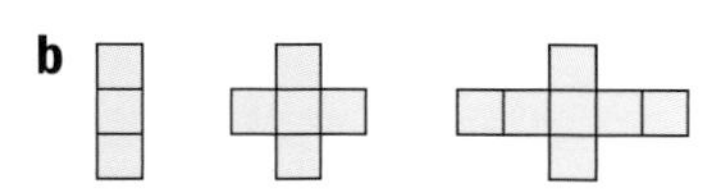

A3.5 More pattern sequences

This spread will show you how to:

- Explain how the formula for the general term of a linear expression works

Keywords
Justify

A farmer uses hurdles to build pens for his sheep.
For the first pen he needs four hurdles.

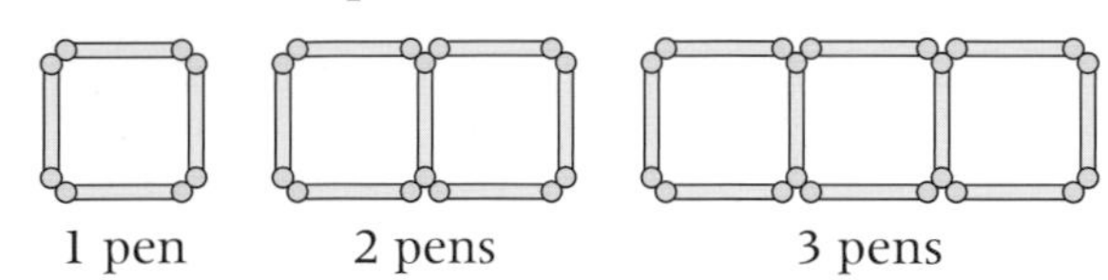

1 pen 2 pens 3 pens

To make each new pen, he adds three more hurdles.
The sequence for the number of hurdles is

4, 7, 10, 13, 16, ...

Common difference = 3

nth term = $3n + 1$

n	1	2	3	4	5
$3n$	3	6	9	12	15
Sequence	4	7	10	13	16

You can **justify** the nth term by looking at the pattern it comes from.

Justify means explain why it is correct.

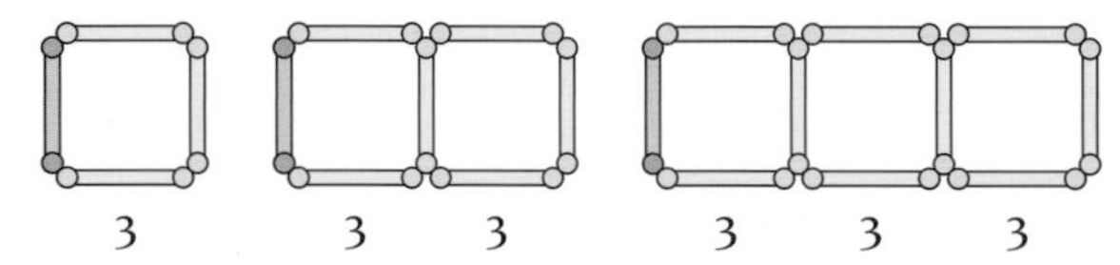

3 3 3 3 3 3

The nth pattern will have n lots of three hurdles (or $3n$), plus the one on the left-hand side.

Example

Here is a sequence of patterns of dots.

a Write the number of dots in the first five patterns.
b Find, in terms of n, an expression for the number of dots in the nth pattern.
c Justify your expression in n by comparing it with the dot patterns.

a 5, 7, 9, 11, 13
b Common difference = 2
Number of dots in nth pattern is $2n + 3$.

n	1	2	3	4	5
$2n$	2	4	6	8	10
Sequence	5	7	9	11	13

Each term is 3 more than a multiple of 2.

c The nth pattern will have $2 \times n$ dots, arranged with n above and n below the horizontal row, plus the horizontal row of three dots.

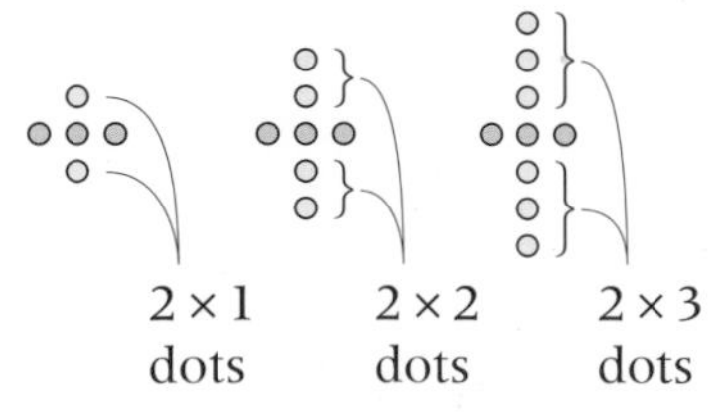

Look for what changes (the $2n$ term) and what stays the same (the +3 term).

Exercise A3.5

1 Here is a sequence of patterns of dots.

a Write the number of dots in the first five patterns.

b Find, in terms of n, an expression for the number of dots in the nth pattern.

c Justify your expression in n by comparing it with the dot patterns.

2 Kim is designing bead patterns for different-sized cushions. Here are her patterns for the first three sizes.

Size 1 Size 2 Size 3

a Write the numbers of beads in the patterns for the first five sizes.

b Find, in terms of n, an expression for the number of beads for the nth size pattern.

c Justify your expression in n by comparing it with the bead patterns.

3 Jas is building a house of cards.

Stage 1 Stage 2 Stage 3

a How many cards does he add each time?

b Write down the number of cards for the first five stages.

c Find, in terms of n, an expression for the number of cards in the nth stage.

d Justify your expression in n by comparing it with the diagrams.

e There are 52 cards in a pack. If Jas continues his pattern, can he use them all? Explain your answer.

4 Dave is building a fence from vertical and horizontal posts.
The fence grows like this:

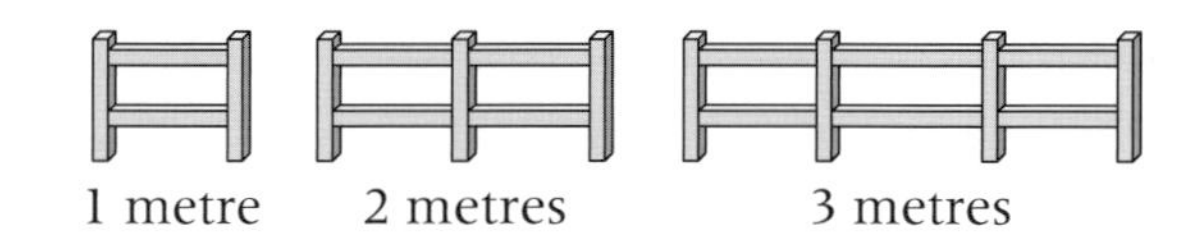

a Describe in words how the fence grows.

b How many posts will he need for 5 metres of fence?

c Find an expression in n for the number of posts for an n metre fence.

d Justify your expression in n by comparing it with the fence diagrams.

e Dave's garden is 26 metres long. How many posts will he need?

5 A car hire company charges.

> Small car = £20 per day + £50   Medium car = £25 per day + £60
> Large car = £30 per day + £70

a Copy and complete this table of charges for 1 to 7 days.

b How would you work out the charge for

i large car for 10 days
ii medium car for 14 days
iii small car for n days?

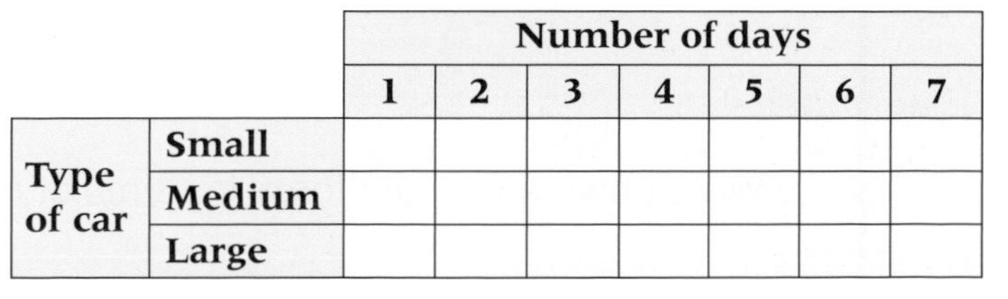

		Number of days						
		1	**2**	**3**	**4**	**5**	**6**	**7**
Type of car	**Small**							
	Medium							
	Large							

A3 Exam review

Key objectives

- Generate terms of a sequence using term-to-term and position-to-term definitions of the sequence
- Use linear expressions to describe the nth term of an arithmetic sequence, justifying its form by reference to the activity or context from which it was generated

1 A sequence of numbers is shown.

2 5 8 11 14

a Find an expression for the nth term of the sequence. (2)

b Explain why 99 will not be a term in this sequence. (2)

(AQA, 2004)

2 Here are some patterns made with sticks.

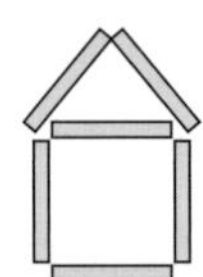

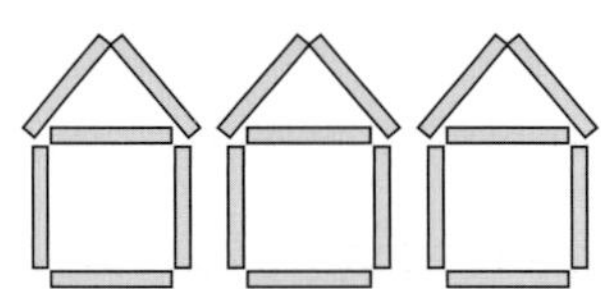

The graph shows the number of sticks m used in pattern number n:

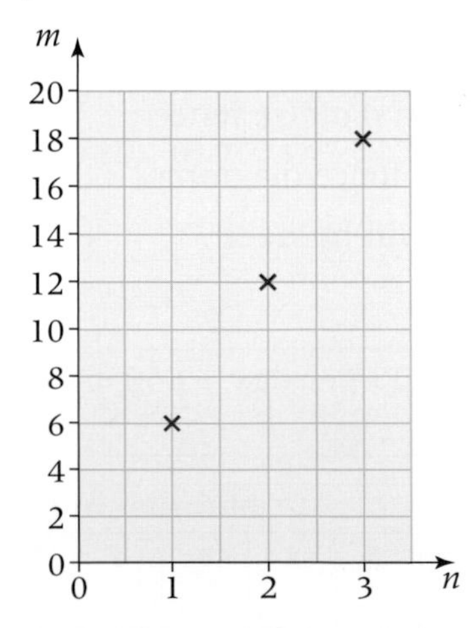

Write down a formula for m in terms of n. (2)

(Edexcel Ltd., 2003)

A4 Formulae and inequalities

This unit will show you how to

- Understand and use the words equation, formula, identity, and expression
- Substitute values into expressions, functions and formulae
- Use formulae from mathematics and other subjects
- Write formulae to represent everyday situations
- Change the subject of simple formulae
- Understand the difference between a practical demonstration and a proof
- Use a counter-example to show that a statement is false
- Solve simple one-sided and two-sided inequalities, representing the solution on a number line

Before you start ...

You should be able to answer these questions.

1 Work out the value of each expression when $y = 3$.

a $2y$ **b** $4y - 1$ **c** y^2

2 Multiply out

a $3(x + 1)$ **b** $2(x - 1)$

c $4(2x + 3)$ **d** $3(4x - 2)$

3 Solve, using the balance method

a $2x + 3 = 11$ **b** $3y - 4 = 14$

4 Factorise

a $4x + 8$ **b** $6x + 2$ **c** $3y - 9$

5 Write the prime numbers from this list.

2 3 5 7 8

11 13 15 17 19

6 Insert > or < to make these statements true.

a 2☐5 **b** 8☐3

c 42☐40 **d** 12☐15

A4.1 Formulae, equations and identities

This spread will show you how to:

- Understand and use the words equation, formula, identity, and expression

Keywords
Equation
Expression
Formula
Function
Identity
Substitute

In algebra you use letters to represent numbers.

- An **expression** is made up of algebraic terms. It has no equals sign.

$2x + 3b$ and $2(l + w)$ are expressions.

- A **function** links two variables. When you know one, you can work out the other.

$x \longrightarrow 3x + 2$ or $y = 3x + 2$ is a function.

- A **formula** is a rule linking two or more variables.

$P = 2l + 2w$ is a formula for the perimeter of a rectangle.

- An **equation** is only true for one unique value of a variable. You can solve the equation to find this value, the solution.

$x + 4 = 10$ is an equation.
Its solution is $x = 6$.

- An **identity** is true for any values of the variables.

$a + a + a + a \equiv 4a$ is an identity.

The sign $\equiv$ means identically equal to.

You can **substitute** values into expressions, functions and formulae.

Formulae is the plural of formula.

Example

Work out the value of each expression when $x = 2$, $y = 4$, $z = 3$.

a $xy - z$ **b** $3x + 5y$ **c** $\dfrac{3x^2y^3}{z}$

$xy = x \times y$

a $xy - z = 2 \times 4 - 3$
$= 8 - 3 = 5$

b $3x + 5y = 3 \times 2 + 5 \times 4$
$= 6 + 20 = 26$

c $\dfrac{3x^2y^3}{z} = \dfrac{3 \times 2^2 \times 4^3}{3}$
$= \dfrac{\cancel{3}^1 \times 4 \times 64}{\cancel{3}^1}$
$= 256$

Write in the multiplication sign.

You can solve equations using the balance method.

Example

Solve $3x + 5 = 17$.

$3x + 5 = 17$
$3x + 5 - 5 = 17 - 5$ Subtract 5 from both sides.
$3x = 12$
$3x \div 3 = 12 \div 3$ Divide both sides by 3.
$x = 4$

Do the same to both sides.

Exercise A4.1

1 Work out the value of each expression when
$a = 6$, $b = 3$, $c = \frac{1}{2}$, $d = 4$.

a ad **b** $2b$ **c** ab **d** $a + cd$ **e** $ad + b$
f $3b - d$ **g** $2dc - a$ **h** $ab - dc$ **i** $a^2b - d$ **j** $b + c^2d$

2 Work out the value of each expression when $e = -2$, $f = 3$, $g = -6$.

a $ef + g$ **b** $eg + f$ **c** $2e^2 - g$
d $\frac{fg}{e}$ **e** $\frac{3e^3}{g}$ **f** $\frac{f^2}{g} + e$

3 For each function, work out the value of y
i when $x = 4$
ii when $x = -2$.

a $y = 4x + 3$ **b** $y = 2x - 6$ **c** $y = \frac{1}{2}x + 10$
d $y = 6x - 1$ **e** $y = x^2$ **f** $y = 2x^2 + 4$

4 Solve these.

a $2x + 3 = 15$ **b** $2y - 5 = 17$ **c** $4x + 15 = 7$
d $3y - 13 = -10$ **e** $23 = 6b + 5$ **f** $7 = 19 + 3c$
g $15 - 7f = 1$ **h** $60 - 3g = 72$

5 Multiply out the brackets and write each of these as an identity.
The first one has been done for you.

a $7(x + 4) \equiv 7 \times x + 7 \times 4 \equiv 7x + 28$
b $3(x - 2)$ **c** $2(3 + x)$ **d** $5(2 - x)$

6 Copy and complete these identities by factorising.

a $8m + 4 \equiv 4(\quad)$ **b** $12n - 9 \equiv \square(\quad)$ **c** $15p + 55 \equiv \square(\quad)$
d $q^2 + 2q \equiv q(\quad)$ **e** $16r - 28 \equiv \square(\quad)$ **f** $4pq - 10q \equiv \square(\quad)$

7 Copy the table.

Expressions	Equations	Functions	Formulae	Identities

Write these under the correct heading in your table.

a $a + bc = bc + a$ **b** $y = 3x + 2$ **c** $a + bc = d$
d $3a + 5 = -4$ **e** $4xy + 3x - z$ **f** $E = mc^2$
g $s = ut$ **h** $x - 1 = y$

A4.2 Substituting into formulae

This spread will show you how to:

- Substitute values into expressions, functions and formulae

Keywords
Formula
Substitute

- You can **substitute** numbers into a **formula** written in words.

The formula for the perimeter of a regular hexagon is

Perimeter = 6 × length of one side

This regular hexagon has sides of length 3 cm.
Its perimeter = 6 × 3 cm = 18 cm.

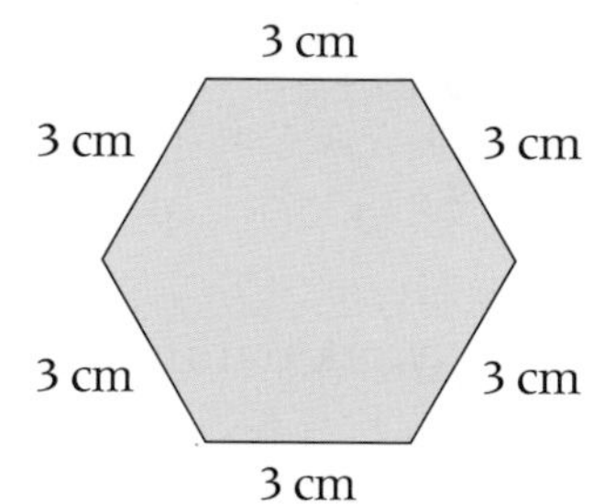

- You can substitute numbers into a formula written using letters.

Example

Apples cost 20p each and bananas cost 25p each.

The formula for the cost (in pence) of a apples and b bananas is

$$\text{cost} = 20a + 25b$$

Find the cost of 8 apples and 3 bananas.

a apples cost $a \times 20$p.
b bananas cost $b \times 25$p.

Cost = 20 × 8 + 25 × 3
= 160 + 75
= 235 pence or £2.35

Substitute $a = 8$ and $b = 3$

- You can solve an equation to find a value from a formula.

Example

In the formula $v = u + at$

a find v when $u = 2$, $a = 5$, $t = 20$ **b** find u when $v = 50$, $a = 2$, $t = 15$

a $v = u + at$
$v = 2 + 5 \times 20$
$= 2 + 100 = 102$

b $v = u + at$
$50 = u + 2 \times 15$
$50 = u + 30$
$50 - 30 = u + 30 - 30$
$20 = u$

Subtract 30 from both sides.

Exercise A4.2

1 The area of card needed to make an open cube-shaped box is calculated by

Area of card = 5 × area of one side of box

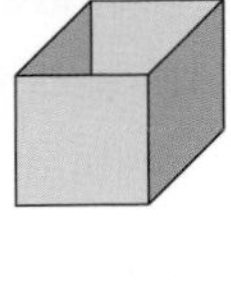

Use this formula to work out the area of card needed for a box with

a area of one side 12 cm^2 **b** area of one side 20 cm^2

c area of one side 1 m^2 **d** area of one side 2.3 m^2

2 The bill for a mobile phone is calculated using the formula

Cost in pounds = 0.05 × number of texts + 0.10 × minutes of calls

a Use the formula to work out the bills for

i Nadia: 40 texts and 20 minutes of calls

ii Saleem: 5 texts and 70 minutes of calls

iii Marcus: 32 texts and 15 minutes of calls.

b What is the cost for a call lasting 1 minute? Give your answer in pence.

3 The cost of hiring a van is given by the formula $C = 25d + 40$ where C is the cost in pounds and d is the number of days. Work out the cost of

a hiring the van for 3 days **b** hiring the van for 10 days.

4 In the formula $v = u + at$ find v when

a $u = 3, a = 5, t = 2$ **b** $u = 12, a = 4, t = 9$

5 In the formula $V = lwh$ find V when $l = 3, w = 6, h = 8$.

6 Use the formula $s = ut + \frac{1}{2}at^2$ to find the value of s when

a $u = 2, a = 3, t = 5$ **b** $u = -3, t = 4, a = 12$

7 Use the formula $t = \frac{v - u}{a}$ to find t when

a $v = 35, u = 27, a = 4$ **b** $v = 12, u = 16, a = -2$

8 In the formula $A = 2b + c$, find c if $A = 14$ and $b = 5$.

9 $P = 2l + 2w$

a Find w when $P = 18, l = 6$. **b** Find l when $P = 20, w = 8$.

10 Using the simple interest formula, $I = \frac{PRT}{100}$

a find P when $I = 12, R = 20, T = 30$.

b find R when $I = 3.6, T = 18, P = 10$.

11 Gemma and Paul evaluate $2x^2$ when $x = 6$.

Who is right? Explain why.

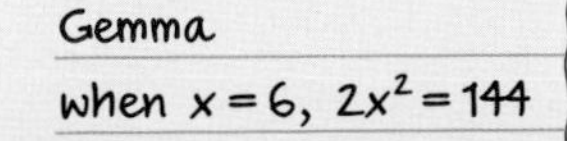

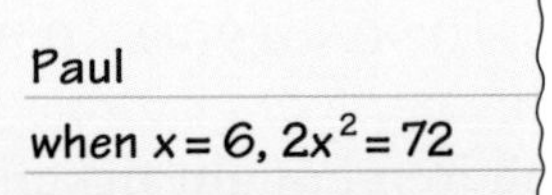

A4.3 Writing formulae

This spread will show you how to:

- Write formulae to represent everyday situations
- Use formulae from mathematics and other subjects

Keywords
Formula

A **formula** can save time when you have to work out similar calculations over and over again.

- **You can write a formula to represent an everyday situation.**
  - **Write the formula in words and then using letters.**
  - **Explain what the letters represent.**

Example

A plumber charges £25 for a callout and £30 per hour of work.
Write a formula for the plumber's charge.

Charge in pounds = 25 + 30 × number of hours of work

$C = 25 + 30h$

where C = charge in pounds, h = number of hours of work.

Example

A bus ticket to town costs £3 for an adult and 90p for a child.

a Write a formula to work out the cost in pounds of bus tickets for different numbers of adults and children.

b Use your formula to work out the cost of tickets for 3 adults and 5 children.

c Mr and Mrs Karim and their children paid £8.70 for bus tickets.
How many children bought tickets?

a Cost in pounds = 3 × number of adults + 0.90 × number of children

$C = 3n + 0.90m$

where C = cost in pounds, n = number of adults, m = number of children.

90p = £0.90

b $C = 3n + 0.90m$
$= 3 \times 3 + 0.90 \times 5$
$= 9 + 4.50 = £13.50$

$n = 3$, $m = 5$

Write the answer in pounds.

c $C = 3n + 0.90m$
$8.70 = 3 \times 2 + 0.90m = 6 + 0.90m$
$8.70 - 6 = 6 - 6 + 0.90m$
$2.70 = 0.90m$
$2.70 \div 0.90 = 0.90m \div 0.90$
$3 = m$
3 children bought tickets.

Substitute $C = 8.70$ and $n = 2$.

Exercise A4.3

1 An electrician charges £35 for each job + £20 per hour.

a Write a formula for the electrician's charge in pounds.

b Use your formula to find the charge of a job that takes 3 hours.

2 The cost of a taxi is £2 for a callout + 60p for each mile.

a Write a formula for the cost of a taxi in pounds.

b Work out the cost for a journey of

i 5 miles **ii** 15 miles.

3 In Spain a hire car costs €75 plus €35 a day.

a Write a formula for the cost of hiring a car in euros.

b How much does it cost to hire a car for 7 days?

c Louise paid €495 to hire a car.
How many days did she hire it for?

4 Pencils are arranged in rectangles.
The number of pencils needed to make a rectangle is
$2 \times$ number of pencils along the bottom + 2

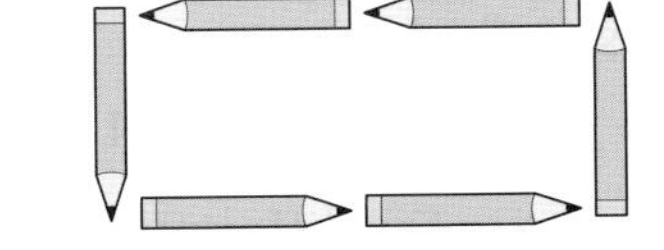

a Write this formula using letters.

b Check that your formula gives the correct answer for a rectangle of length 5.

c Work out how many pencils are needed for a rectangle of length 8.

d A rectangle uses 48 pencils. What is its length?

5 Cinema tickets cost £6.80 for adults and £4.50 for children.
Five adults took a group of children to the cinema.
The tickets cost £97 in total.
How many children went to the cinema?

Write a formula first.

6 The cost of hiring a steam cleaner is

- £32.50 for the first day
- £24.75 for each extra day.

Tariq paid £131.50 to hire the steam cleaner.
How many days did he hire it for?

7 Tickets for a fun day cost £3 for adults and £1.50 for children.
The total cost of tickets for a group of adults and children was £54.

a Write a formula for the cost for n adults and m children.

In the group there were 4 children for every adult.

b Write this information using algebra: $m = 4 \times ...$

c Substitute your expression from part **b** into your formula from part **a**.

d Use your formula from part **c** to work out the number of adults in the group.

A4.7 Two-sided inequalities

This spread will show you how to:

- Solve simple two-sided inequalities, representing the solution on a number line

Keywords
Greater than
Inequality
Integer
Less than
Solution set

You can read inequalities in two directions.

If $x > 5$, then $5 < x$ If $y \leqslant -2$, then $-2 \geqslant y$

You can use two-sided inequalities to show upper and lower limits.

$2 < x < 5$ means that x is greater than 2 and less than 5.

$x > 2$ and $x < 5$.

On a number line:

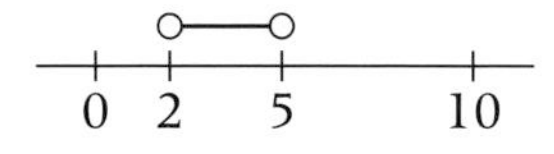

You can combine two inequalities to show a range of values:

$y > 3$ and $y < 10$
combine to give $3 < y < 10$

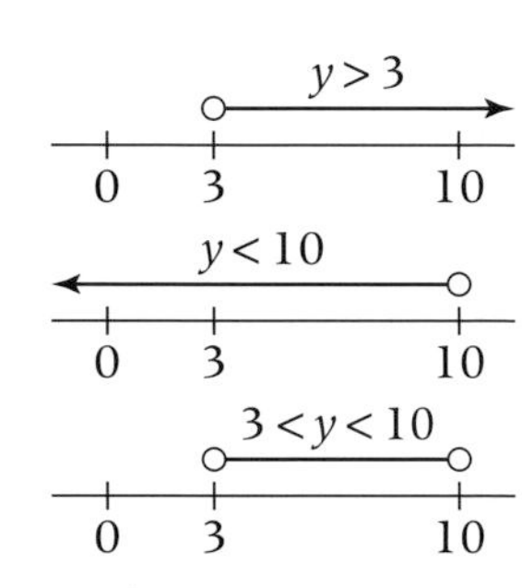

Example

Split each two-sided inequality into two single inequalities.

a $-3 < x \leqslant 1$ **b** $8 > y \geqslant 2$

---

a $x > -3$, $x \leqslant 1$ **b** $y < 8$, $y \geqslant 2$

- **You can solve two-sided inequalities to find the solution set.**
  - **You can give the integer values in the solution set.**

An integer is a whole number such as −1, 4, 10.

Example

$-4 < 2n \leqslant 6$
n is an integer.

a Show all the possible values of n on a number line.
b Write all the possible values of n.

---

a $-4 < 2n$, so $2n > -4$ $2n \leqslant 6$
$n > -2$ $n \leqslant 3$

Divide both sides by 2 to get n on its own.

The solution is $-2 < n \leqslant 3$:

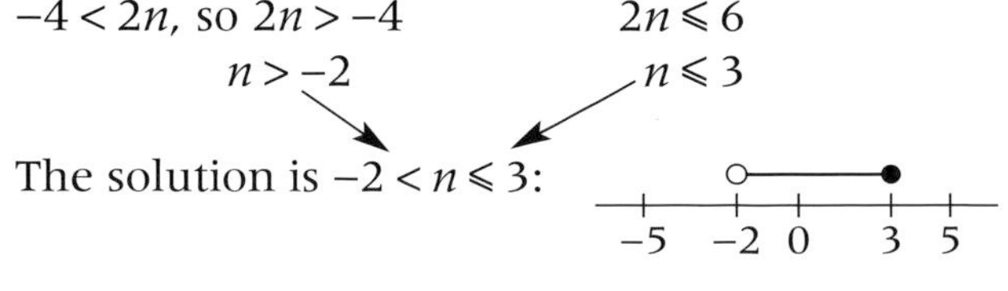

−2 is not included.
3 is included.

b The possible integer values of n are −1, 0, 1, 2, 3.

Exercise A4.7

1 If $7 < x$ (7 is less than x) then you can say x is greater than 7 or $x > 7$.
Write each of these in another way.

a $5 > x$ **b** $6 \leqslant y$ **c** $4 \geqslant y$ **d** $9 < r$

e $15 \geqslant w$ **f** $2 < s$ **g** $-12 > u$ **h** $-4 \leqslant v$

2 Split each two-sided inequality into two single inequalities.

a $1 < x < 5$ **b** $-1 < x < -5$ **c** $-2 < x < 4$
d $-6 \leqslant x \leqslant -1$ **e** $2 < x < 7$ **f** $2 \geqslant x \geqslant -1$

3 Write the inequalities shown by these number lines.

a

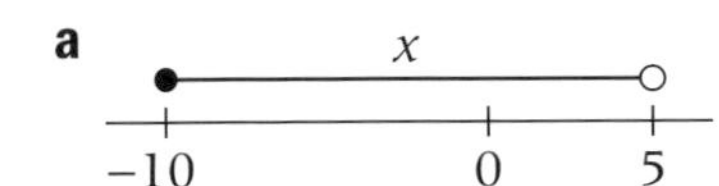

b

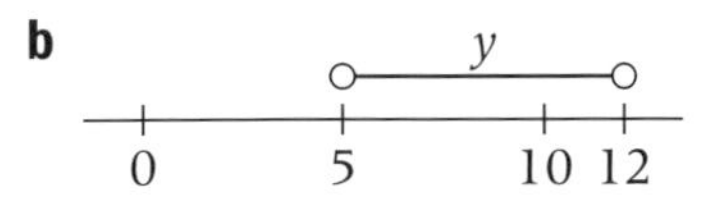

c

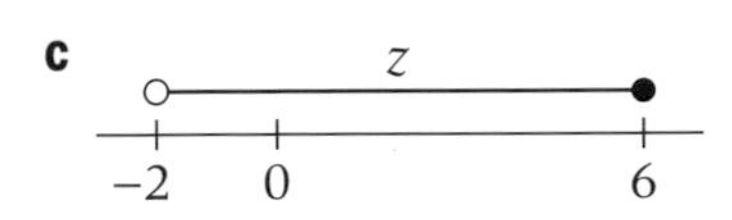

d

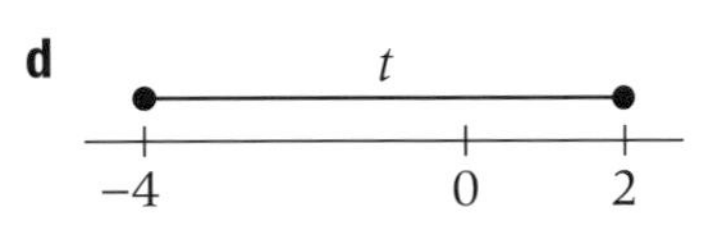

4 Show these inequalities on number lines like this.

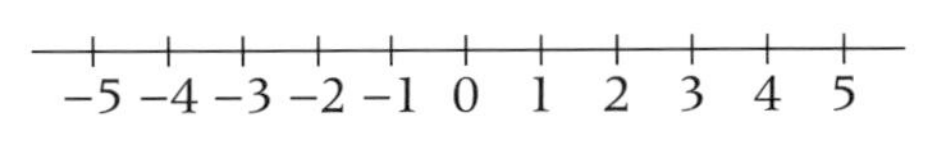

a $-3 \leqslant x < 2$ **b** $4 \geqslant n > -1$ **c** $1 < y \leqslant 3$ **d** $-4 \leqslant m < 0$

5 For each of the inequalities in question 4, list the integer values.

6 If x can take the possible integer values $-1, 0, 1, 2, 3$, which of these could be true?

a $x > -2$ **b** $-1 \leqslant x \leqslant 3$ **c** $-2 < x < 4$

d $-1 \leqslant x < 3$ **e** $-2 < x \leqslant 3$

7 n is an integer.

$-4 \leqslant n < 5$

a Show all the possible values of n on a number line.

b Write all the possible values of n.

> Using a number line may be helpful.

8 $2 \leqslant 2x < 10$
x is an integer.
Write all the possible values of x.

9 In these inequalities, y is an integer.
For each inequality, write all the possible values of y.

a $-6 < 2y \leqslant 4$

b $-3 \leqslant 3y < 15$

c $3 < 2y \leqslant 10$

A4 Exam review

Key objectives

- Know the meaning of and use the words equation, formula, identity and expression
- Use formulae from mathematics and other subjects
- Substitute numbers into a formula
- Change the subject of a simple formula
- Understand the difference between a practical demonstration and a proof
- Solve simple inequalities

1 An approximate conversion between temperatures in Fahrenheit (F) and Celsius (C) is given by the formula

$$F = 2C + 30.$$

Use this formula to find the value of C when F = 68. (3)

(OCR, 2004)

2 Tayub said, 'When $x = 3$, then the value of $4x^2$ is 144'.

Bryani said, 'When $x = 3$, then the value of $4x^2$ is 36'.

a Who was right?

Explain why. (2)

b Work out the value of $4\,(x + 1)^2$ when $x = 3$. (1)

(Edexcel Ltd., 2003)

Graphical solutions

This unit will show you how to

- Plot straight line graphs
- Change the subject of an equation
- Recognise the general equation forms of horizontal, vertical and diagonal line graphs
- Understand that parallel lines have the same gradient
- Read x and y values from a graph
- Use graphs to find solutions to equations
- Recognise the form of and plot simple quadratic graphs

Before you start ...

You should be able to answer these questions.

1 Plot these points on a coordinate grid.

a (3, 6) **b** (−4, 2)

c (5, −3) **d** (−3, −1)

2 Copy and complete this table of values for the equation

$y = 2x + 5$

x	−2	−1	0	1	2
y					

3 Write the letters of the parallel lines.

a $y = x$

b $y = x + 2$

c $y = 2x$

d $y = -x$

e $y = x - 2$

f $y = x + 4$

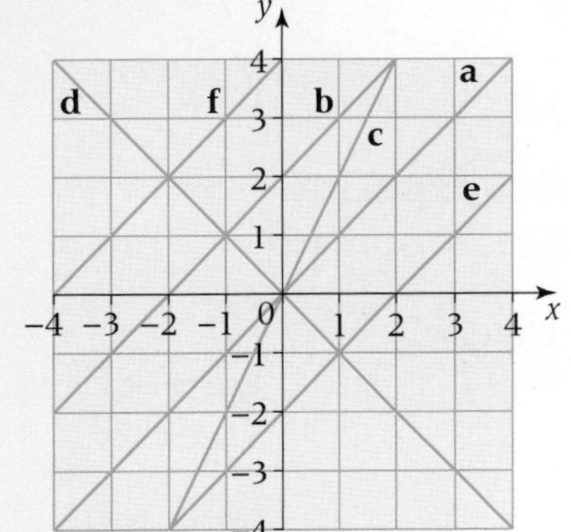

4 Which line in question 3 is the steepest?

5 Rearrange these equations to make y the subject.

a $2x - y = 5$ **b** $4x + 2y = 8$

A5.1 Drawing linear graphs

This spread will show you how to:

- Plot straight line graphs

Keywords
Explicit
Implicit

You can represent a function by drawing its **graph**.

- **To plot a graph of a function:**
  - **Draw up a table of values**
  - **Calculate the value of *y* for each value of *x***
  - **Draw a suitable grid**
  - **Plot the (*x*, *y*) pairs and join them with a straight line.**

Example

Draw the graph of $y = 3x - 4$.

First construct a table of values.

x	−2	−1	0	1	2
y	−10	−7	−4	−1	2

Choose four or five values, including negative values and zero.

Then plot the points and draw the line.

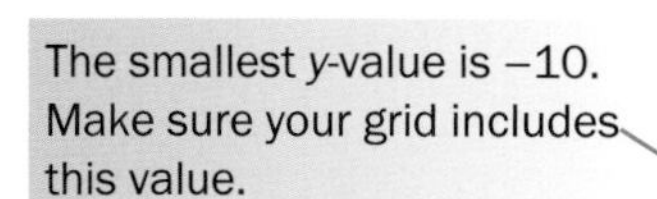

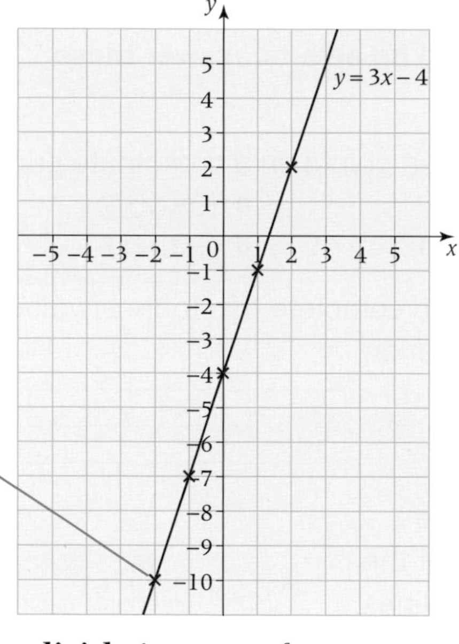

The equation $y = 3x - 4$ gives y **explicitly** in terms of x.
The equation $2x + 3y = 6$ gives y **implicitly** in terms of x.

An **explicit** function has the variables separated by the = sign.
An **implicit** function can have the variables on the same side of the = sign.

- **To draw a graph of an implicit function:**
  - **Draw up a table of values for $x = 0$ and $y = 0$.**

  or

  - **Rearrange the equation to make y the subject.**

For $2x + 3y = 6$:

x	0	$2x = 6$ $x = 3$
y	$3y = 6$ $y = 2$	0

For $2x + 3y = 6$:

$3y = 6 - 2x$

$y = \dfrac{6 - 2x}{3}$

x	−2	−1	0	1	2
y	$\frac{10}{3}$	$\frac{8}{3}$	2	$\frac{4}{3}$	$\frac{2}{3}$

Exercise A5.1

1 **a** Copy and complete the table of values for $y = 2x + 5$.

x	-3	-1	0	1	3
y	-1			7	

b Draw the graph of $y = 2x + 5$ on a copy of the grid.

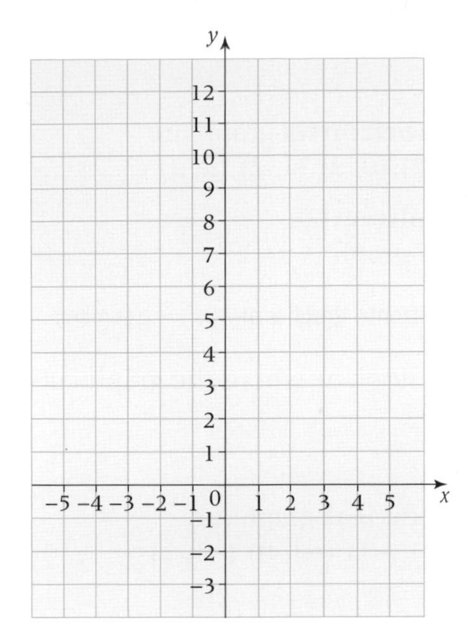

2 Draw the graphs of these functions.

a $y = 3x - 2$ **b** $y = -2x + 4$ **c** $y = \frac{1}{2}x + 3$ **d** $y = 5 - x$

3 Draw the graphs of these functions.

a $x + y = 4$ **b** $2x - y = 3$ **c** $2y + x = 4$ **d** $3y + 6x = 9$

4 Draw the graphs of these functions.

a $y = 5x - 3$ **b** $4y - x = 2$ **c** $y = 3 - 4x$ **d** $2x - 3y = 6$

5 Rearrange these equations to make y the subject.
Which is the odd one out?

$y = 2x + \frac{1}{2}$

$2y - 4x = 1$

$4x + 2y = 1$

$8x - 4y = -2$

6 **a** Draw the graph of $y = 3x + 4$.

b Use your graph to find

i the value of y when $x = \frac{1}{2}$ **ii** the value of x when $y = -\frac{1}{2}$.

7 Draw the graphs of these functions on the same axes.

a $y = 2x + 1$ **b** $y = 2x - 2$ **c** $y = 2x + 5$ **d** $y = 2x$

What do you notice?

8 Draw the graphs of these functions on the same axes.

a $y = -x$ **b** $y = -x + 3$ **c** $y = -x - 2$

What do you notice?
Where do you think the graph of $y = -x + 1$ would be on your grid?

A5.2 Horizontal and vertical graphs

This spread will show you how to:

- Understand the equation forms of horizontal and vertical graphs

Keywords
Horizontal
Vertical
y-intercept

The points on this **horizontal** graph are
(−3, 2) (−2, 2) (−1, 2) (0, 2) (1, 2) (2, 2) (3, 2)

The y-coordinate is always 2.
The equation of this graph is $y = 2$.

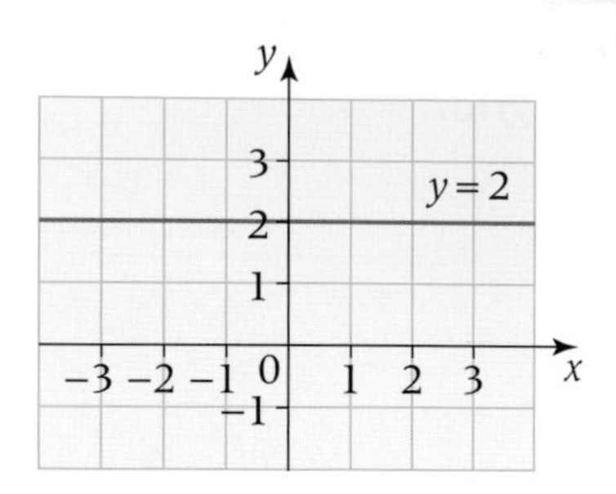

Where a graph cuts the y-axis is called the **y-intercept**.

- **The equation of a horizontal graph is always y = a number.**

To find the number, look at where the graph cuts the y-axis.

The points on this **vertical** graph are
(−1, −3) (−1, −2) (−1, −1) (−1, 0) (−1, 1)
(−1, 2) (−1, 3)

The x-coordinate is always −1.
The equation of this graph is $x = -1$.

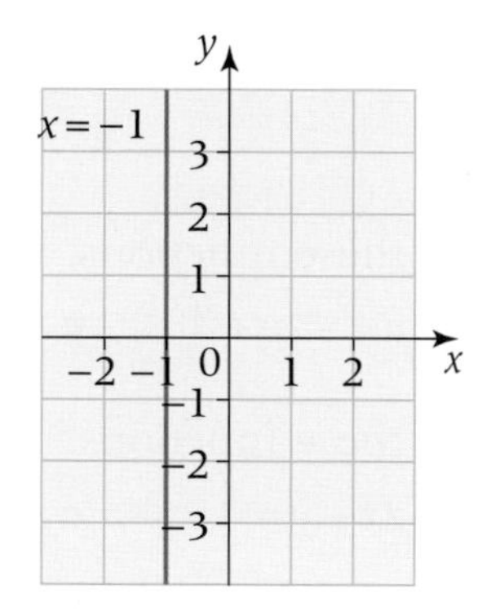

- **The equation of a vertical graph is always x = a number.**

To find the number, look at where the graph cuts the x-axis.

Example

a Draw the graphs of $y = 3$ and $x = 2$ on the same pair of axes.

b Find the coordinates of the point P where the two graphs cross.

a

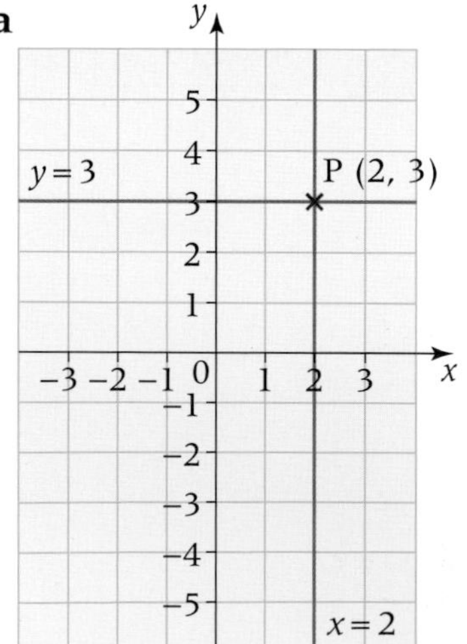

$y = 3$ is a horizontal graph.
It cuts the y-axis at 3.
$x = 2$ is a vertical graph.
It cuts the x-axis at 2.

b The coordinates of P are (2, 3).

Exercise A5.2

1 Write the equations of these horizontal and vertical graphs.

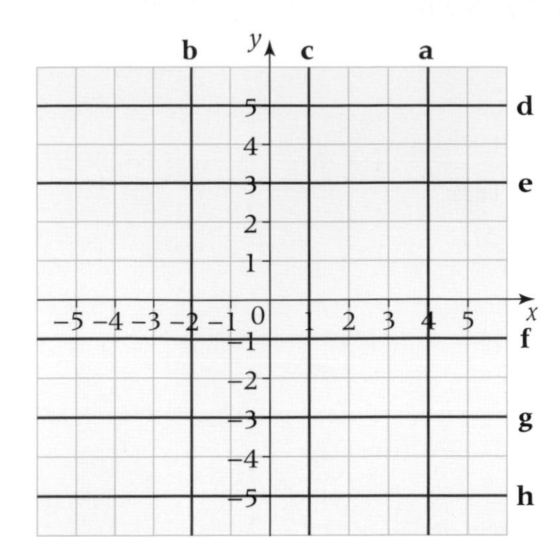

2 Copy the grid from question **1** without the graphs. Draw these graphs on the same grid.

a $y = 4$ **b** $x = 5$ **c** $y = -2$ **d** $x = -2$ **e** $x = -4$

3 Draw a table like this:

Horizonal lines	Vertical lines

Write these equations into the correct columns of your table:

$y = 4$ $x = 5$ $y = -10$ $x = 15$ $y = -2$ $y = -6$ $x = -3$ $x = +4$

4 Use your graphs from question **2** to find the coordinates of the points where these pairs of graphs cross.

a $x = -2$ and $y = 4$ **b** $x = 5$ and $y = -2$ **c** $x = -4$ and $y = 4$

5 Write the coordinates of the points where these pairs of lines cross.

a $x = 2$ and $y = -3$ **b** $x = 1$ and $y = 6$ **c** $x = 3$ and $y = -1$

6 **a** Copy the grid from question **1** without the graphs.
b Draw a square on your grid using two vertical and two horizontal lines.
c Write the equations of the four lines.

7 **a** Draw a grid with the x-axis and the y-axis going from −5 to +5.
b Draw the line $y = 0$ on your grid. What is another name for this line?
c Draw the line $x = 0$ on your grid. What is another name for this line?

8 **a** Draw the graph of $y = 3x + 1$.
b On the same grid, draw the graph of $y = 7$.
c Write the coordinates of the point P where the graphs cross.

A5.3 The equation of a straight line

This spread will show you how to:

- Understand the equation form of a general straight line graph
- Understand that parallel lines have the same gradient

Keywords
Gradient
Intercept
Parallel

- **Straight line graphs can be vertical, horizontal or diagonal.**

Examples of vertical lines: $x = 1, x = 5, x = -2, \ldots$

Examples of horizontal lines: $y = 2, y = 7, y = -1, \ldots$

The line $y = 2x + 1$ is a diagonal line.

You can describe a diagonal line by

- how steep it is and
- where it crosses the y-axis (the y-**intercept**).

Gradient measures steepness.

The graph $y = 2x + 1$ has gradient 2.
For every 1 unit across, the graph goes up 2 units.

The graph $y = 2x + 1$ has y-intercept 1.
It crosses the y-axis at (0, 1).

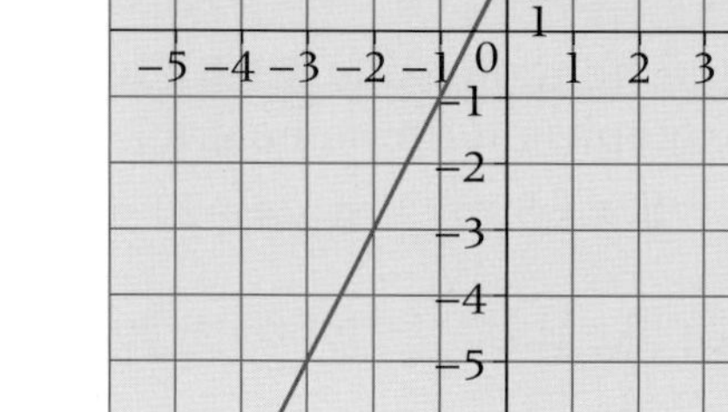

- **Diagonal lines have an equation of the form**

$y = mx + c$ where m is the gradient and c is the y-intercept.

Example

Match the equations to the graphs.

a $x = 2$ **b** $y = 3$
c $y = 3x + 2$ **d** $y = 3x - 1$

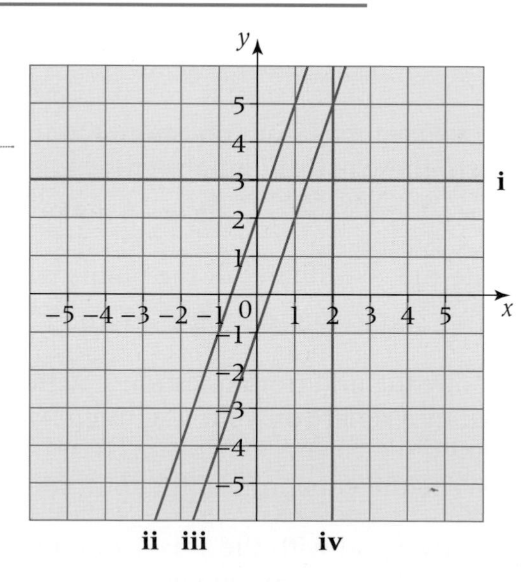

a → **iv** ($x = 2$ is a vertical graph)

b → **i** ($y = 3$ is a horizontal graph)

c → **ii** ($y = 3x + 2$ has y-intercept 2)

d → **iii** ($y = 3x - 1$ has y-intercept -1)

In the example above, the graphs of $y = 3x + 2$ and $y = 3x - 1$ are **parallel**.

- **Parallel lines have the same gradient.**

Exercise A5.3

1 Write the gradient and y-intercept for each of these equations.

a $y = 3x - 1$ **b** $y = 2x + 5$ **c** $y = 4x - 3$ **d** $y = \frac{1}{2}x + 2$

e $y = 5x + 1$ **f** $y = -3x + 7$

2 Match the equations to the graphs.

a $y = 3x + 1$

b $y = 3$

c $y = 2x + 4$

d $x = -2$

e $y = x + 3$

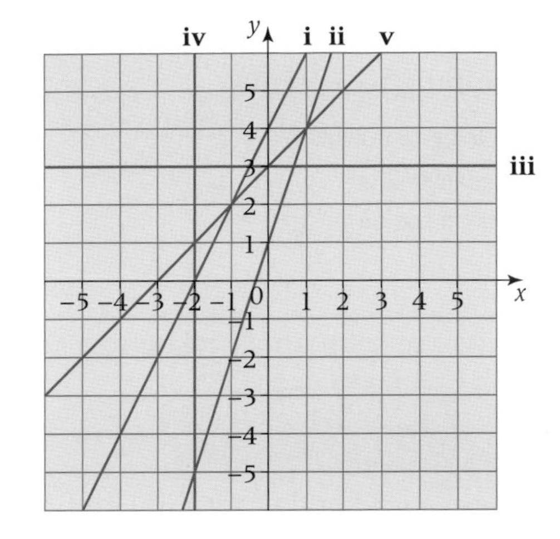

3 Rearrange these equations in the form $y = mx + c$.

a $x + y = 5$ **b** $y - x = 3$ **c** $x + y = -2$ **d** $x - y = 3$

e $2x + y = 6$ **f** $5x + y = 9$ **g** $3x + y = -2$ **h** $y - 2x = 5$

i $2y + x = 4$ **j** $2y - x = 8$ **k** $2x + 4y = 16$ **l** $6x + 2y = 8$

4 List the equations from question **3** whose graphs are parallel to

a $y = x$ **b** $y = -x$ **c** $y = -3x$ **d** $y = -\frac{1}{2}x$

5 Here are the equations of five straight lines.
Which three are parallel?

a $y = 2x + 4$ **b** $2x - y = 3$ **c** $y + 2x = 7$ **d** $4y + 2x = 8$ **e** $y - 2x = 5$

6 Write these equations for straight lines in order of steepness, starting with the least steep.

$y = 4x + 3$ | $y = 2x - 2$ | $y = 3x + 1$ | $y = \frac{1}{2}x - 9$ | $y = x + 11$

7 Write the equation of a straight line that is parallel to $y = 3x - 1$.

8 Write the equation of a straight line that is parallel to $y + 4x = 2$.

9 Write the equation of the straight line that is parallel to $y = \frac{1}{2}x - 3$ and passes through the point $(0, 4)$.

10 Draw the graph of $y = -x$.
What can you say about the slope of the graph?
Is this true for all graphs with negative gradient?
Draw more graphs to test your idea.

A5.4 Finding the equation of a graph

This spread will show you how to:

- Draw a line of best fit through a set of linearly related points and find its equation

Keywords
Gradient
Intercept

You can deduce the equation of a straight line from its graph.
You need to find the y-**intercept** and the **gradient**.
For a straight line joining two points,

- **For a straight line joining two points,**

  Gradient = $\dfrac{\text{increase in y-value}}{\text{increase in x-value}}$

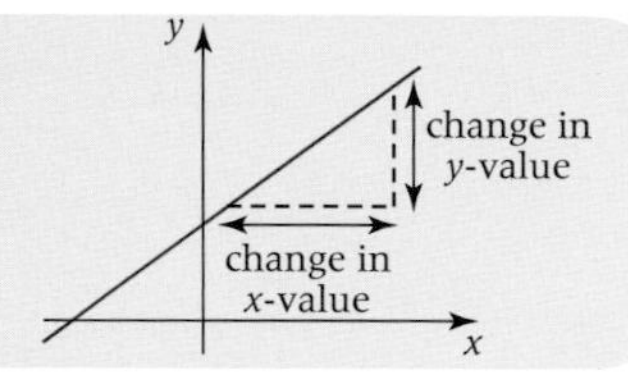

Example

The table shows how a set of points y relate to a set of points x.

x	−2	−1	0	1	2
y	−5	−3	−1	1	3

a Plot these points on a grid and draw a line of best fit through them.
b Find the equation of your line.

a

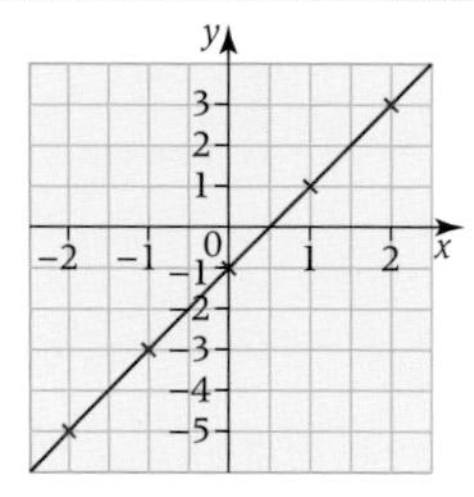

b The equation of a straight line is $y = mx + c$. c is the y-intercept.
The graph crosses the y-axis at $(0,-1)$ $\Rightarrow$ $c = -1$. m is the gradient.

$$\text{Gradient} = \frac{\text{vertical change}}{\text{horizontal change}}$$

$$= \frac{2}{1} = 2$$

So $m = 2$

The equation of the line is $y = 2x - 1$.

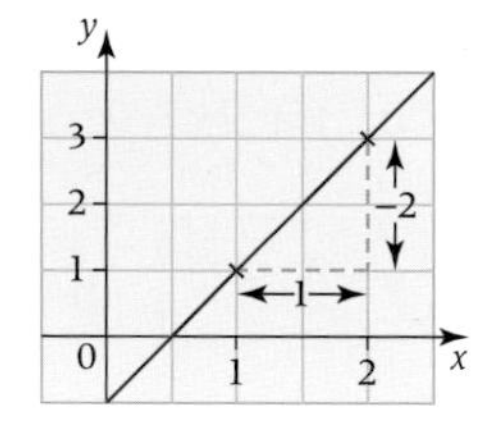

Lines that slope downwards have negative gradients.

Example

Find the equation of the line on this graph.

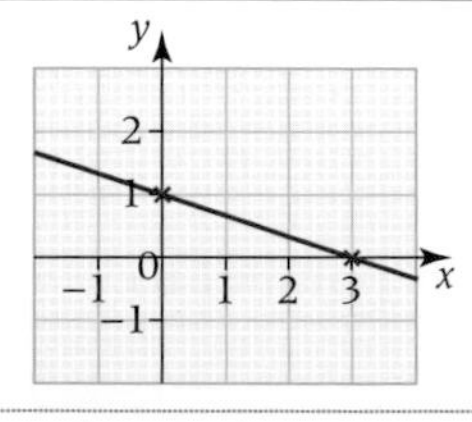

The line crosses the y-axis at $(0,1)$ $\Rightarrow$ $c = 1$

$$m = \frac{\text{increase in } y\text{-value}}{\text{increase in } x\text{-value}} = \frac{-1}{3}$$

So $y = -\frac{1}{3}x + 1$ is the equation of the line.

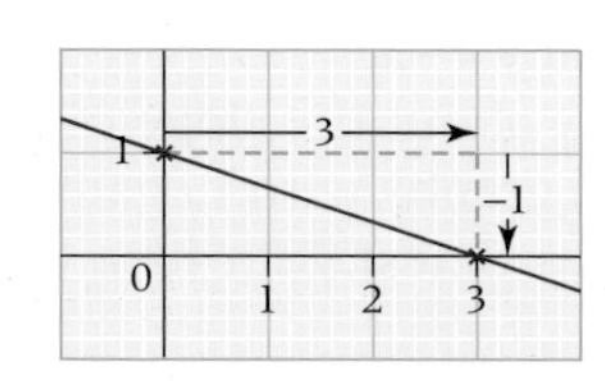

$y = mx + c$

Exercise A5.4

1 The table shows how a set of points y relate to a set of points x.

a Plot these points on a grid and draw a line of best fit through them.

x	−2	−1	0	1	2
y	−3	−1	1	3	5

b Find the equation of the line.

2 Find the equation of each of these straight line graphs.

a

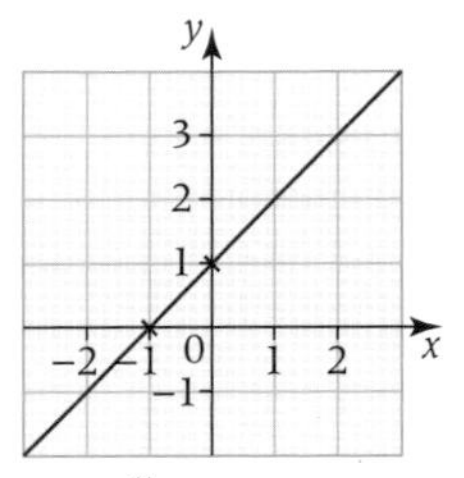

b

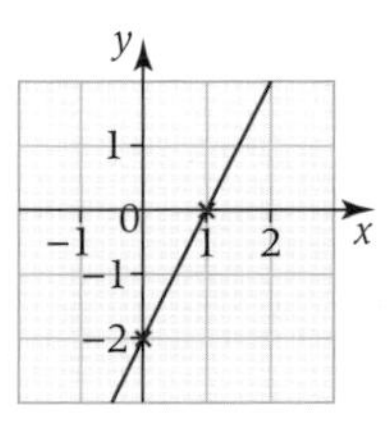

c

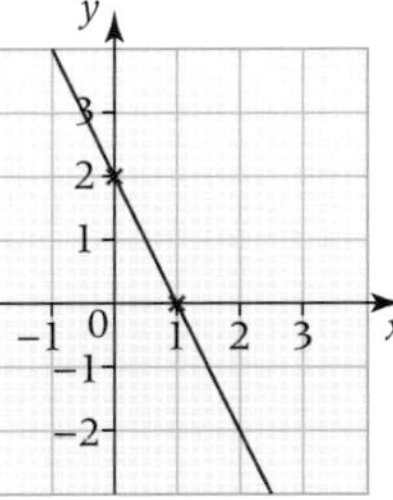

d

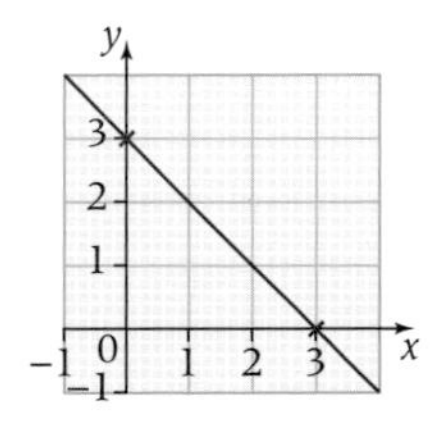

3 Match the graphs with the equations.

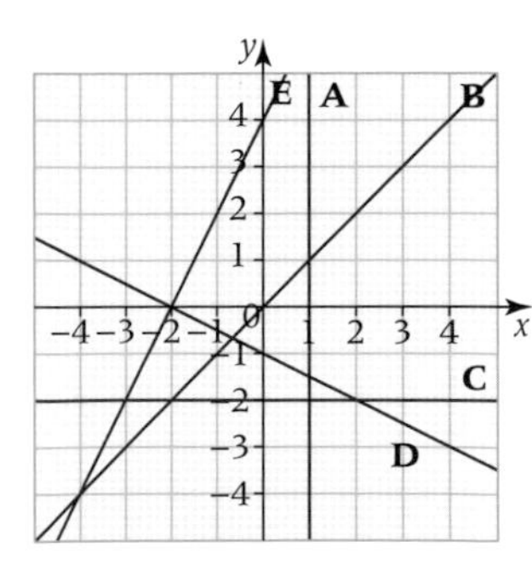

i $y = x$

ii $x = 1$

iii $y = -2$

iv $y = -\frac{1}{2}x - 1$

v $y = 2x + 4$

4 Shelly is a plumber. She has a call-out charge of £30, and then charges £40 per hour.

a Copy and complete the table for different lengths of time.

Duration of job (hours), x	0	1	2	3	4
cost (£), y	30				

b Using suitable axes, plot a graph of y against x.

c Find the equation of your line.

d Work out how much Shelly would charge for a job lasting $2\frac{1}{2}$ hours.

A5.5 Finding solutions from graphs

This spread will show you how to:

- Plot straight line graphs
- Use graphs to find solutions to equations

Keywords
Solution

- **All the points on a graph line fit the equation of the line.**
  You could extend the graph an infinite distance.

You draw the graph line right to the edge of the grid, to show it continues.

- **You can read x and y values from a graph.**

Example

Here is the graph of $y = 3x - 2$.

a Find the value of y when $x = \frac{1}{2}$.
b Find the value of x when $y = 7$.

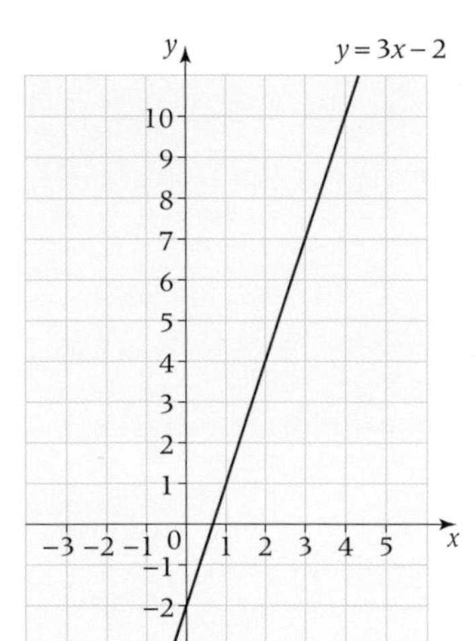

a Draw a vertical line from $x = \frac{1}{2}$ up to the graph.
Draw a horizontal line from the graph to the y-axis.
Read off the value of y.
When $x = \frac{1}{2}$, $y = -\frac{1}{2}$.

b Draw a horizontal line from $y = 7$ to the graph.
Draw a vertical line from the graph to the x-axis.
Read off the value of x.
When $y = 7$, $x = 3$.

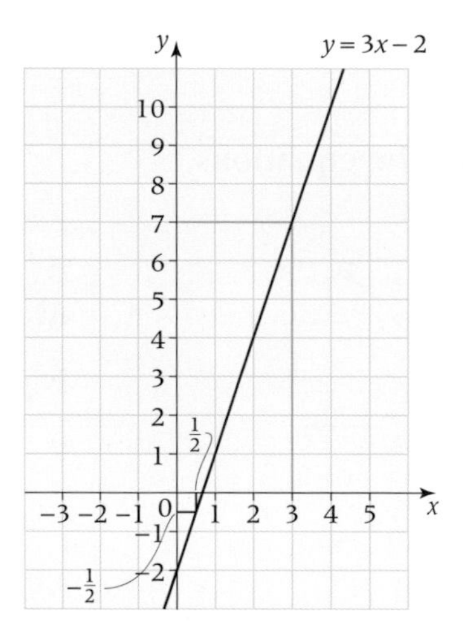

- **You can use a graph to find solutions to equations.**

Example

For each equation, **describe** how you could draw a graph to find the value of x.

You do not need to draw the graph.

a $2x + 6 = 10$

b $-5x + 1 = 11$

a Draw the graph of $y = 2x + 6$.
Read off the x-value when $y = 10$.

b Draw the graph of $y = -5x + 1$.
Read off the x-value when $y = 11$.

Exercise A5.5

1 Here is the graph of $y = -2x + 3$.
Use the graph to find

a the value of y when $x = \frac{1}{2}$

b the value of y when $x = -2\frac{1}{2}$

c the value of x when $y = -1$

d the value of x when $y = 9$.

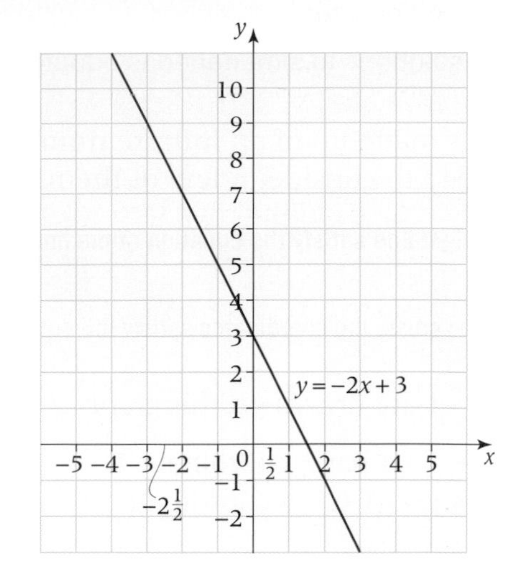

2 **a** Draw the graph of $y = \frac{1}{2}x + 3$.

b Point P on this line has y-coordinate 7.
Use your line to find the x-coordinate of point P.

c Which of these points lie on the line?
$(2, 4)$ $(3, 7)$ $(-3, 1\frac{1}{2})$ $(-4, 2)$

3 **a** What graph would you draw to find the solution to the equation $4x - 7 = -15$?

b Draw the graph and use it to find the value of x.

4 Write down three equations that you could solve using the graph in question **1**.
Use the graph to find a solution to each of your equations.

5 Draw the graph of $y = 3x + 2$.

a Use your graph to solve

i $3x + 2 = 6.5$ **ii** $3x + 2 = 17$ **iii** $3x + 2 = -7$.

b You can rewrite the equation $3x + 6 = 5$ as $3x + 2 + 4 = 5$.
You can use your graph from part **a** to solve $3x + 2 = 1$.
Rewrite the equation $3x + 7 = 16$ as $3x + 2 = \ldots$
Find the value of x from your graph.

Subtract 4 from each side:
$3x + 2 = 5 - 4 = 1$
So $3x + 2 = 1$ is equivalent to $3x + 6 = 5$.

6 **a** Draw a graph to solve the equation $2x + 6 = 10$.
What is the value of x?

b Use your graph from **a** to solve
i $2x + 6 = 14$ **ii** $2x + 6 = -3$ **iii** $2x + 9 = 12$.

A5.6 Solutions of more than one equation

This spread will show you how to:

- Use graphs to find solutions to simultaneous equations

Keywords
Solution
Simultaneous equations

A straight line graph is made up of an infinite number of points. You normally only need to consider a few of them.

- **All the points on a straight line satisfy the equation of the line.**

- **Where two straight lines cross, the coordinates satisfy the equations of both lines.**

The lines $x = 4$ and $y = -1$ are drawn on this graph.

Every point on the line $x = 4$ has x-coordinate 4.
Every point on the line $y = -1$ has y-coordinate -1.

So the point where they cross is $(4, -1)$.

The point P satisfies both equations:
$x = 4$ **and** $y = -1$. It has coordinates $(4, -1)$.

P is the **solution** to the **simultaneous equations** $x = 4$ and $y = -1$.

- **Simultaneous equations are both true at the same time.**

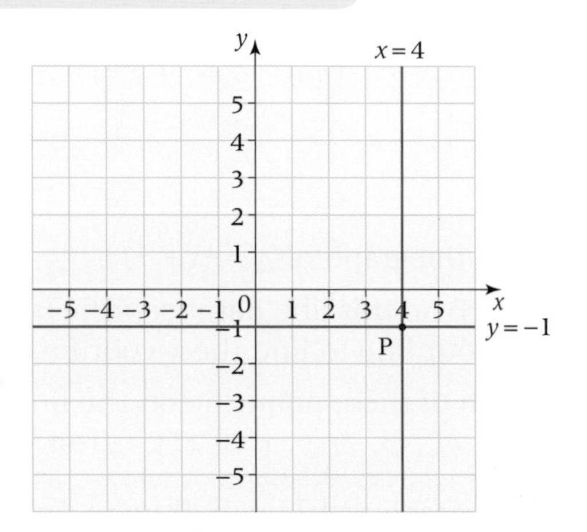

Example

a Draw the graphs of $y = x + 6$ and $y = 2x$ on the same pair of axes.
b Write the coordinates of the point where they cross.
c What can you say about this point?

a $y = x + 6$

x	−2	−1	0	1	2
y	4	5	6	7	8

$y = 2x$

x	−2	−1	0	1	2
y	−4	−2	0	2	4

b Graphs cross at $(6, 12)$

c $(6, 12)$ satisfies both equations.

$x = 6$ is the solution to $x + 6 = 12$ **and** $2x = 12$.

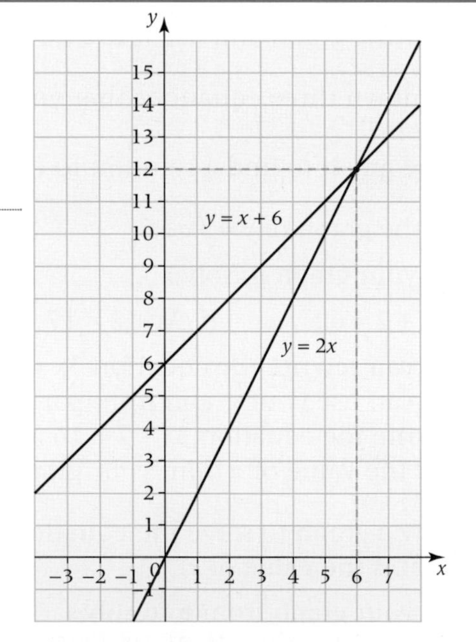

- **To solve a pair of simultaneous equations**
  - **Draw the graphs of both equations on the same axes**
  - **Locate the point where they cross**
  - **The coordinates are the solutions**

Exercise A5.6

1 Write the coordinates of the points where these lines cross. Draw graphs to check your answers.

a $x = 2$ and $y = 3$ **b** $x = -1$ and $y = -4$ **c** $x = 3$ and $y = 7$

d $x = -2$ and $y = -4$ **e** $x = 7$ and $y = -2$ **f** $y = 1$ and $x = 4$

2 For each pair of equations, decide whether the lines will cross.

a $y = 2x + 1$ and $y = 4x + 2$ **b** $y = 3$ and $y = x + 1$

c $y = 3x + 2$ and $y = 3x - 1$ **d** $y = x$ and $y = -x$

Consider the gradient of each line.

3 **a** Draw graphs of the pairs of lines from question **2** that cross.

b For each pair, write the coordinates of the point where the lines cross.

4 **a** Draw the graphs of $y = -x + 2$ and $y = 2x - 1$ on the same axes.

b Write the coordinates of the point where the two lines cross.

c Does the point (1, 2) satisfy both of these equations? Explain how you know.

5 **a** Draw the graphs of $y = 2x - 4$ and $y = x - 1$ on the same axes.

b Write the coordinates of the point where they cross.

6 **a** Draw the graphs of these two equations on the same axes:

$x + 2y = 8 \qquad x - y = 2$

b Where the two lines cross, the x and y values satisfy both these equations.
Write these x and y values.

7 Jenny buys 2 cakes and 3 sandwiches. The total cost is £8.
She writes this as an equation:

$2x + 3y = 8$

where x is the cost of a cake and y is the cost of a sandwich.
Tim buys 5 cakes and 2 sandwiches. The total cost is £9.

a Copy and complete this equation for Tim:

$5\square + \square y = \square$

b Draw the graph of $2x + 3y = 8$.

c Draw the graph for Tim's equation on the same axes.

d Write the coordinates of the point where the two lines cross.

e The x-coordinate of the point where the two lines cross gives the cost of a cake in pounds (£). How much does a cake cost?

f How much does a sandwich cost?

Graphs of quadratic functions

This spread will show you how to:

- Recognise the form of and plot simple quadratic graphs

Keywords
Quadratic
Solution

A **quadratic** function includes a 'squared' term as the highest power, for example x^2.

These are all quadratic functions:

x^2 $\quad x^2 + 3 \quad x^2 + 3x - 1 \quad 2x - x^2$

- **To draw a graph of a quadratic function:**
  - **Draw a table of values**
  - **Calculate the value of y for each value of x**
  - **Draw a suitable grid**
  - **Plot the (x, y) pairs and join them with a smooth curve.**

Example

a Draw the graph of $y = x^2 + 1$.
b Label the minimum point.
c What are the values of x and y at the minimum point?

a Draw a table of values.

$y = x^2 + 1$

x	−3	−1	0	1	3
y	10	2	1	2	10

Draw a suitable grid.
Then plot the coordinate pairs and join the points.

b

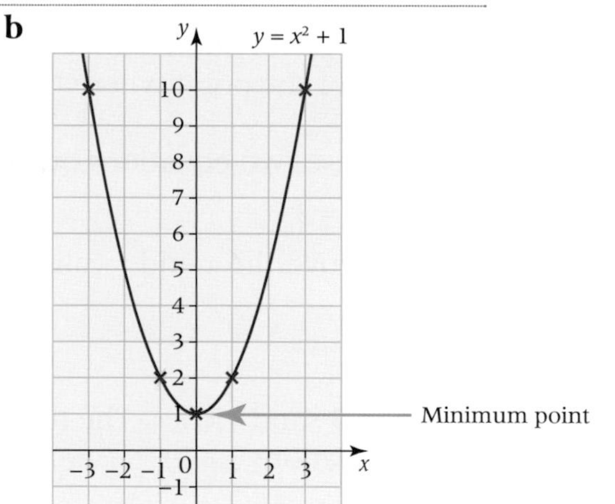

c At the minimum point, $x = 0$, $y = 1$.

Graphs of quadratic functions have a distinctive shape.

- **The graph of a quadratic function**
  - **is always a U-shaped curve**
  - **is symmetrical about a vertical line**
  - **always has a maximum point or a minimum point.**

Quadratic graphs can be this way up :

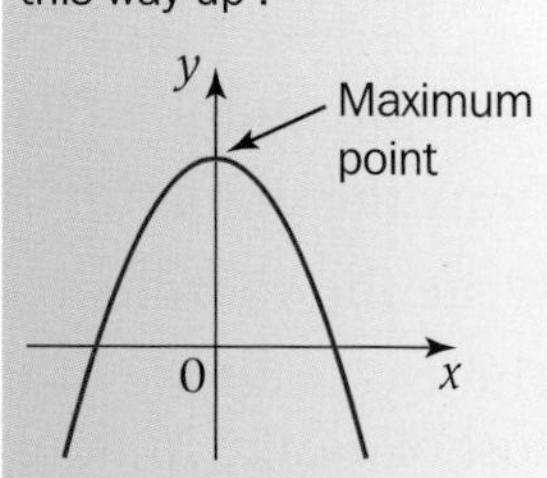

Exercise A5.7

1 **a** Copy and complete the table of values for $y = x^2$.

x	-3	-1	0	1	3
y					

b Draw a pair of axes from 0 to 10 on the y-axis and from -5 to $+5$ on the x-axis.

c Plot the coordinate pairs on the grid.

d Join the points with a smooth curve.

e Label the minimum point. What is the value of y at the minimum point?

2 Draw the graphs of $y = x^2 + 2$ and $y = x^2 + 3$ on the same axes as your graph from question **1**.
What do you notice?

3 Draw the graphs of $y = x^2 - 1$ and $y = x^2 - 3$ on the same pair of axes.
What do you notice?

4 Match these graphs to their equations.

a $y = x^2 + 1$

b $y = x^2 - 2$

c $y = 2 + x^2$

d $y = x^2 - 1$

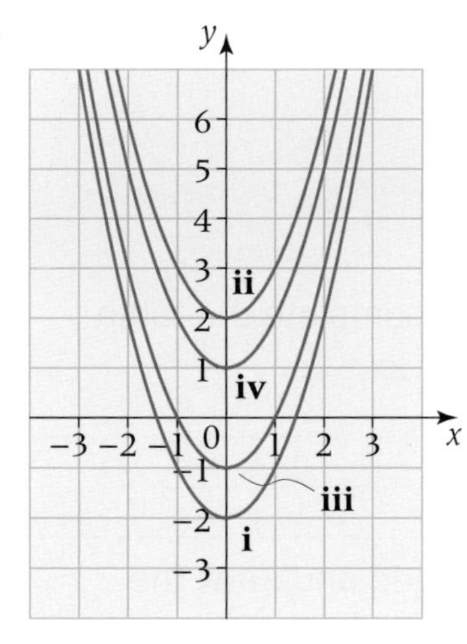

You can use the table of values you drew up in question **1**.

5 **a** Draw the graph of $y = x^2$ on square grid paper.

b Draw the line $y = 9$ on your graph.
Write the coordinates of the points where the graphs cross.

c Your two pairs of coordinates from part **b** both give solutions to the equation $x^2 = 9$.
What are the two x-values that satisfy this equation?

6 **a** Copy and complete the table for $y = x^2 - 2x + 1$.

x	-2			1	2		4
y	9	4	1			4	

b On a suitable grid, draw the graph of $y = x^2 - 2x + 1$.

c Draw a line of symmetry on your graph.

d Use your graph to find two approximate solutions to the equation $x^2 - 2x + 1 = 5$.

A5 Exam review

Key objectives

- Plot graphs of functions in which y is given explicitly in terms of x or implicitly
- Recognise that equations of the form $y = mx + c$ correspond to straight-line graphs in the coordinate plane
- Generate points and plot graphs of simple quadratic functions

1 **a** On a copy of the grid below, draw the graph of $y = 2x - 3$ for values of x from -2 to $+3$. (3)

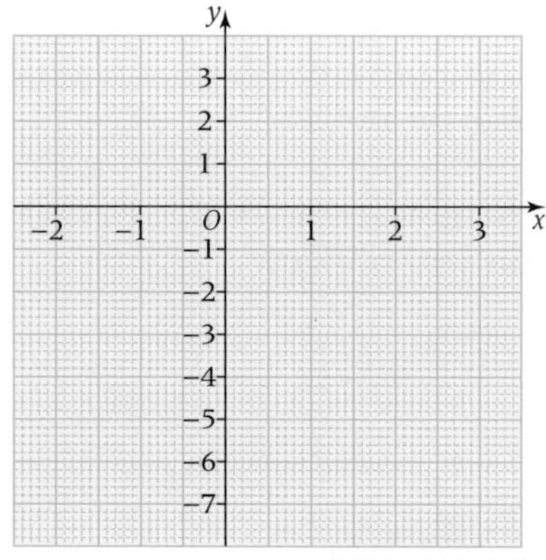

b The line $y = 2$ crosses $y = 2x - 3$ at P.
Write down the coordinates of P. (1)

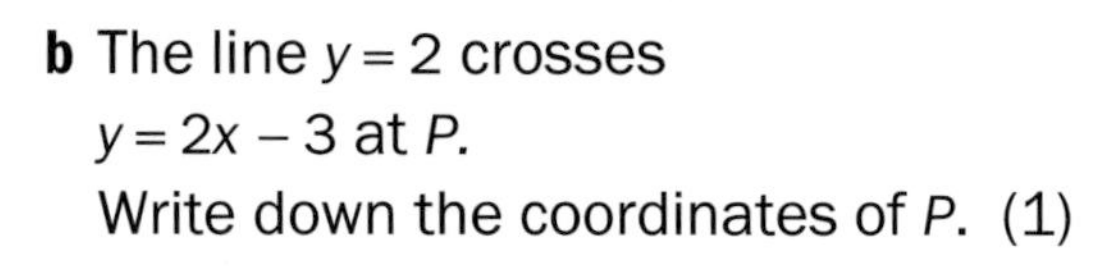

(AQA, 2004)

2 **a** Copy and complete the table for $y = x^2 - 3x + 1$:

x	−2	−1	0	1	2	3	4
y	11		1	−1		1	5

b Copy the grid and draw the graph of $y = x^2 - 3x + 1$. (2)

c Use your graph to find an estimate for the minimum value of y. (2)

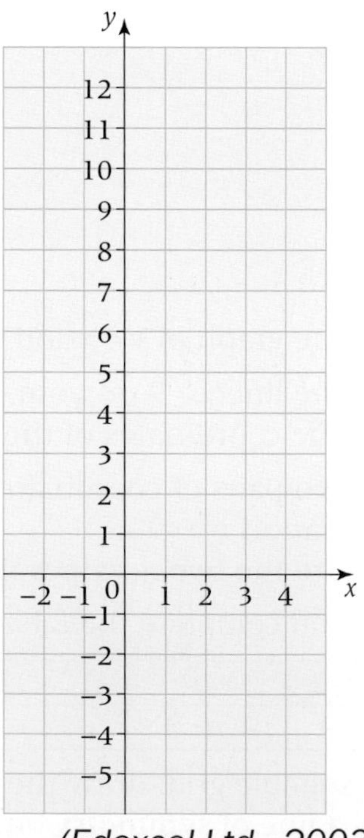

(Edexcel Ltd., 2003)

Real-life graphs

This unit will show you how to

- Draw and use conversion graphs
- Draw a straight-line graph by plotting and joining three points
- Draw, discuss and interpret graphs arising from real-life situations
- Understand and use compound measures, including speed
- Work out an average speed from a distance–time graph
- Understand the trend shown by a graph
- Read values off graphs and scales

Before you start ...

You should be able to answer these questions.

1 Copy and complete, by following the pattern

5 miles = 8 km
10 miles = __ km
20 miles = __ km.

2 Copy and complete, by following the pattern

1 m = 100 cm
__ m = 1000 cm
20 m = __ cm.

3 Write the value shown on this scale.

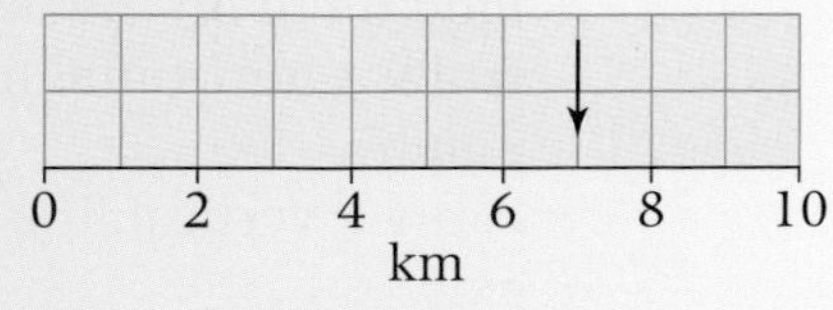

4 Write the value shown on this scale.

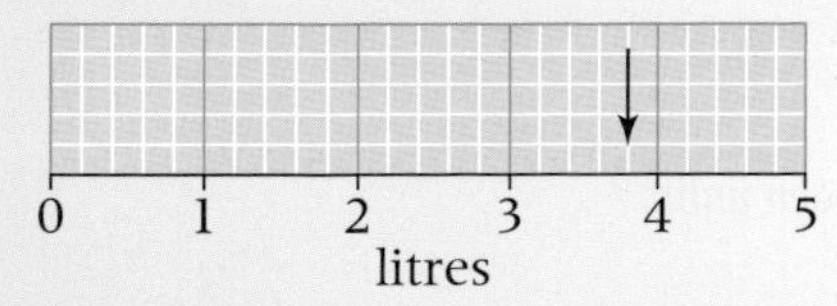

5 Write the value shown on this scale.

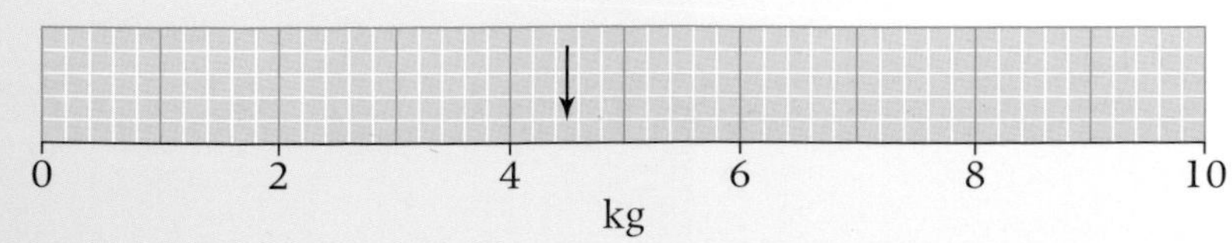

A6.1 Conversion graphs

This spread will show you how to:

- Draw and use conversion graphs
- Read values off graphs and scales

Keywords
Convert
Scale
Units

You can use a conversion graph to **convert**

- a distance in miles to a distance in kilometres
- a distance in kilometres to a distance in miles.

Example

Use the conversion graph to convert these distances.

a 2.5 miles to kilometres
b 6 km to miles

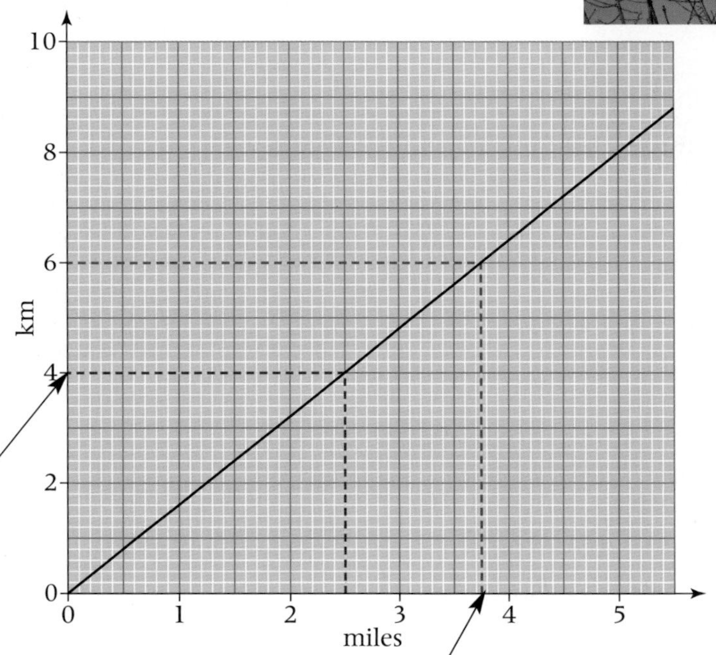

Read the **scale** on the axes carefully. The vertical axis goes up in 2s.

a To convert 2.5 miles to kilometres:

- Find 2.5 miles on the 'miles' axis
- Draw a vertical line up to the graph
- Draw a horizontal line across to the 'kilometres' axis
- Read off the value: 4 km.

b To convert 6 km to miles:

- Find 6 km on the 'kilometres' axis
- Draw a horizontal line across to the graph
- Draw a vertical line down to the 'miles' axis
- Read off the value: 3.75 miles.

- **You can use a conversion graph to convert between units:**
  - distance (miles ↔ km)
  - weight (pounds ↔ kg)
  - temperature (°C ↔ °F)
  - currency (£ ↔ €)

Exercise A6.1

1 Use the conversion graph on page 150 to convert

a 3 miles to kilometres **b** 10 km to miles

c 4.5 miles to km **d** 2 km to miles.

Which is longer: 1 mile or 1 km?

Use the conversion graph to help you decide.

2 Use this kilograms to pounds (lb) conversion graph to convert

a 6 lbs to kg

b 10 lbs to kg

c 4 kg to lbs

d 1.8 kg to lbs

e 7 lbs to kg

f 2.2 kg to lbs

g 12 lbs to kg

h 5 kg to lbs.

Which is heavier: 1 kg or 1 lb? Explain your reasons.

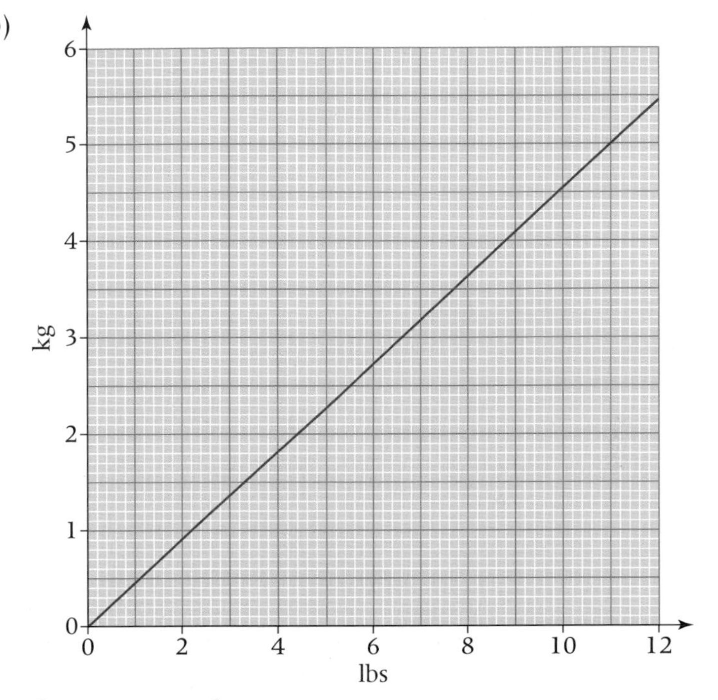

3 **a** Use this °C to °F conversion graph to convert these temperatures.

i 20 °C to °F

ii 58 °F to °C

iii 30 °C to °F

iv 20 °F to °C

b The freezing point of water is 0 °C. What is the freezing point of water in °F?

c Copy this table. Use the graph to convert the temperatures in this table to °F.

Reykjavik	Oslo	Paris	Madrid	Sydney
−2 °C	2 °C	6 °C	12 °C	24 °C
°F	°F	°F	°F	°F

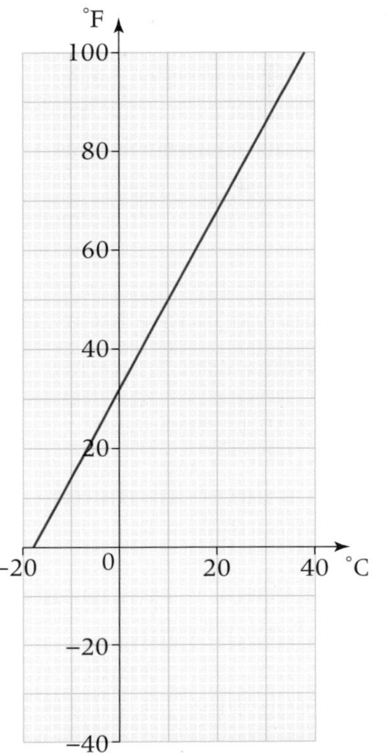

A6.2 Drawing conversion graphs

This spread will show you how to:

- Draw, discuss and interpret graphs arising from real-life situations

Keywords
Conversions
Exchange rate

Davina buys some euros for a trip to France.
The **exchange rate** is £1 = €1.60.

She draws a conversion graph to help her convert prices.

First she works out some simple **conversions** to plot on the graph:

£1 = €1.60 | £0 = €0
£10 = €16
£20 = €32

Two points is enough to plot a straight line. The third point checks the line is accurate.

She wants to include prices up to £40.
£20 = €32, so £40 = €64.
The euros scale needs to go up to at least €64.

She chooses the scale so her graph fits her paper.

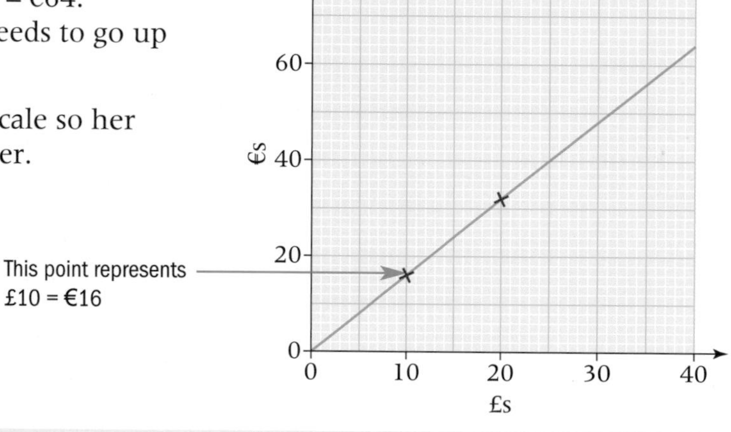

The graph has £s on the horizontal axis and euros on the vertical axis. You could use either axis for either currency.

- **To draw a conversion graph:**
  - **Work out three simple conversions**
  - **Decide on a suitable scale**
  - **Plot your three points**
  - **Join the points with a straight line, right to the edge of the grid.**

Example

The conversion rate for miles to kilometres is

5 miles = 8 kilometres

a Work out two simple conversions you could plot for a miles to kilometres conversion graph.

b The graph needs to convert distances up to 50 miles. What is the highest value needed on the km scale?

a You know: 5 miles = 8 km
Try the zero value
0 miles = 0 km
Use doubling: 10 miles = 16 km

b 5 miles = 8 km
So 5 × 10 miles = 8 × 10 km
50 miles = 80 km
The highest value needed on the km scale is 80 km.

Exercise A6.2

1 The conversion rate for millimetres to centimetres is

1 cm = 10 mm

a Work out two simple conversions you could plot for a millimetres to centimetres conversion graph.

b The graph needs to convert distances up to 10 cm.
What is the highest value needed on the mm scale?

2 The conversion rate for pounds (lb) to kilograms (kg) is

1 kg = 2.2 lb

a Copy and complete these conversions.

0 kg = ____ lb

10 kg = ____ lb

5 kg = ____ lb

b Copy the axes on to graph paper.

c Complete the labelling of the axes.

d Use your conversions from part **a** to draw a conversion graph from pounds to kilograms on your grid.

e Use your graph to convert

i 10 lb to kg **ii** 5 lb to kg

iii 3 kg to lb **iv** 2.5 kg to lb.

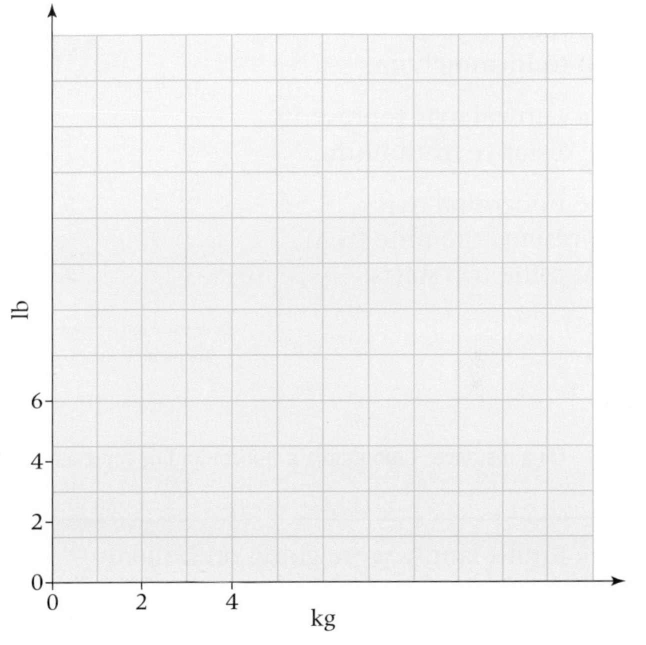

3 One day the exchange rate for pounds (£) to US dollars ($) is

£1 = $1.70

a Work out three simple conversions you could plot for a pounds to dollars conversion graph.

b Max needs a graph to convert amounts up to £30 to dollars.
What is the highest value needed on the dollars scale?

c Draw a conversion graph for pounds to dollars.

d Use your graph to find

i £5 in dollars **ii** $30 in pounds.

4 The exchange rate for pounds to New Zealand dollars (NZ$) is

£1 = NZ$2.40

a Draw a conversion graph to convert amounts up to £20 to New Zealand dollars.

b Use your graph to decide which cap is cheapest.

A6.3 Distance–time graphs

This spread will show you how to:

- Draw, discuss and interpret graphs arising from real-life situations

Keywords
Distance
Time

- **A distance–time graph represents a journey.**
- **It shows how the distance from the starting point changes over time.**

This distance–time graph illustrates Ayesha's shopping trip to Birmingham.

The vertical axis represents the distance from home.

The horizontal axis represents the time from when the trip starts.

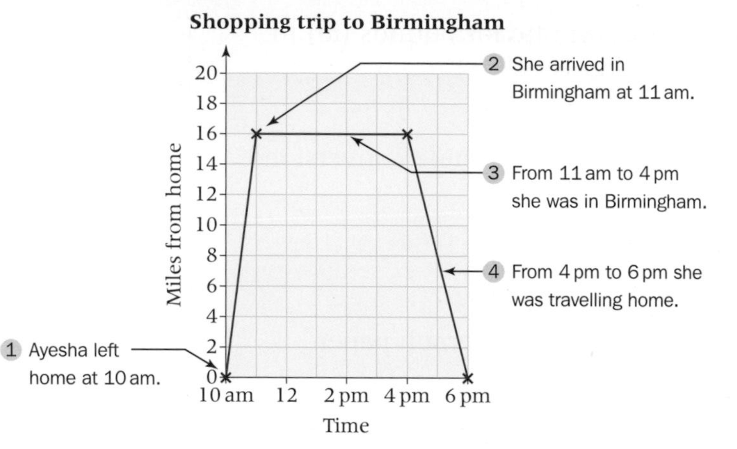

- **On a distance–time graph, a horizontal line represents a stay in one place.**

Example

The Smith family were going on holiday. The distance–time graph shows the first part of their journey.

a What time did they set off?
b How far did they drive before they stopped for a break?
c How long did they stop for?
d The journey to Southend-on-Sea is 240 km. The Smiths arrived at 4 pm. Copy and complete the graph to show the last part of their journey.

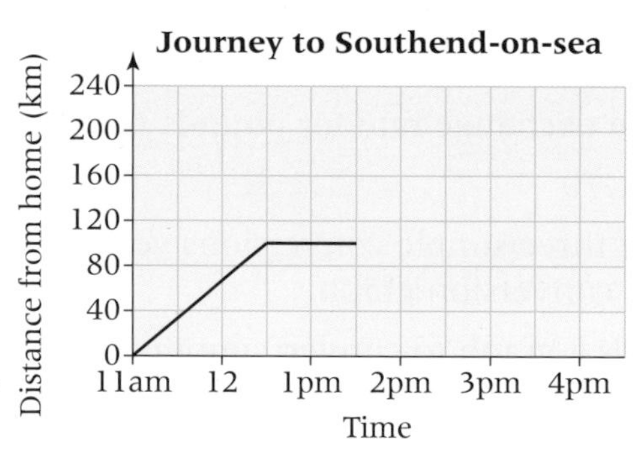

a They set off at 11 am.
b They drove 100 km before stopping.
c 12.30 to 1.30 = 1 hour
d Journey ends at 4 pm, 240 km.

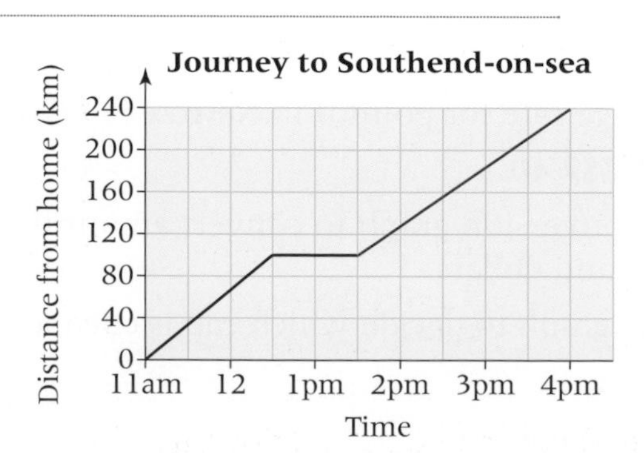

Exercise A6.3

1 The distance–time graph illustrates the first part of a coach tour.

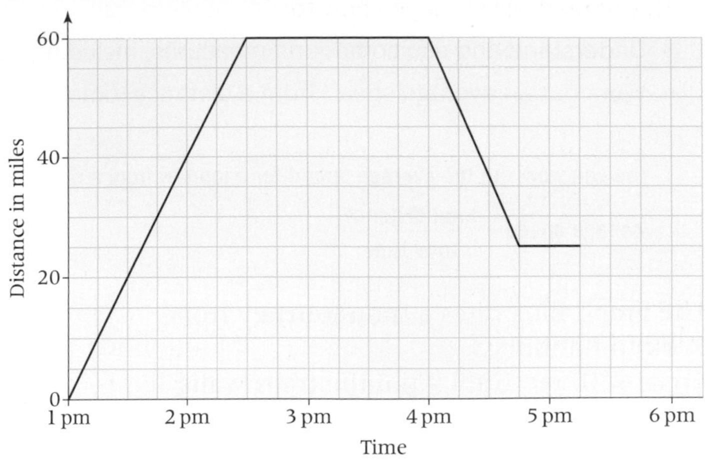

a What time did the coach tour set off?

b The coach arrived at Kinross Castle at 2.30 pm.
How many miles did the coach drive to the castle?

c How long did the coach stop at the castle?

d The tour stopped for a tea break at 4.45 pm.
How far from home were they?

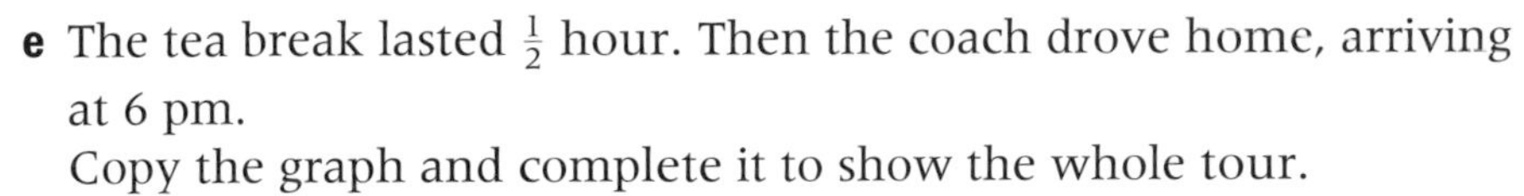

e The tea break lasted $\frac{1}{2}$ hour. Then the coach drove home, arriving at 6 pm.
Copy the graph and complete it to show the whole tour.

2 The graph illustrates a cycle ride.
Match each section of the graph to a part of the description below.

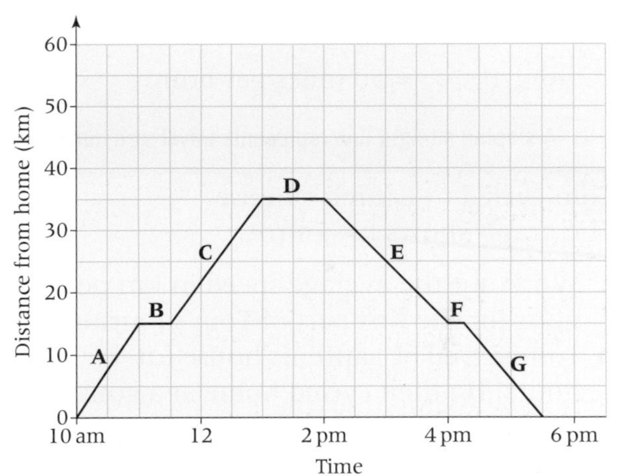

a 15 minute tea break

b Cycle 20 km in $1\frac{1}{2}$ hours

c 1 hour lunch break

d Cycle 15 km in 1 hour

e Cycle 15 km in $1\frac{1}{4}$ hours

f $\frac{1}{2}$ hour rest

g Cycle 20 km in 2 hours

3 The graph represents Karim's car journey to Cornwall.

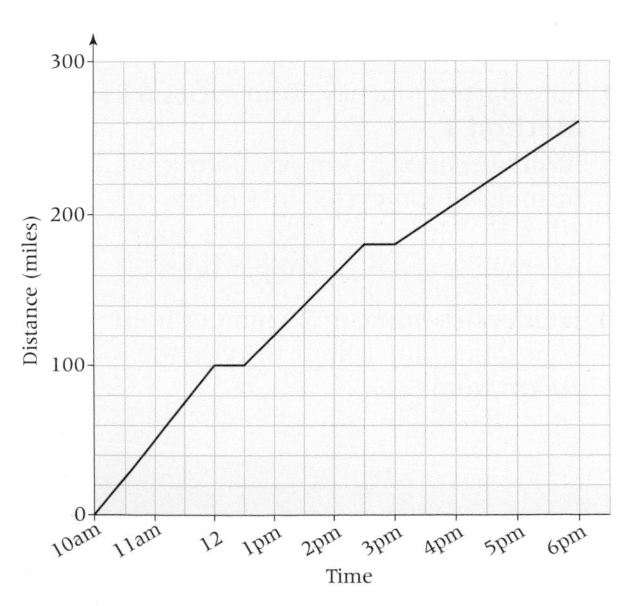

a How long did the journey take him altogether?

b How many miles was the journey altogether?

c How long did it take him to drive the first 100 miles?

d How long did he stop for at 12 pm?

e After his break at 12 pm, he drove another 80 miles before he stopped again.

How long did it take him to drive this 80 miles?

A6.4 Average speed

This spread will show you how to:

- Understand and use compound measures, including speed
- Work out an average speed from a distance–time graph

Keywords
Average speed
Constant speed

- **You can work out the average speed for a journey from a distance–time graph.**

  Average speed $= \dfrac{\text{total distance}}{\text{total time}}$

The graph illustrates a train journey from York to Banbury.
From 8.30 pm until 9 pm the train waits at Birmingham.
The whole journey is 200 miles.
The whole journey takes 4 hours.

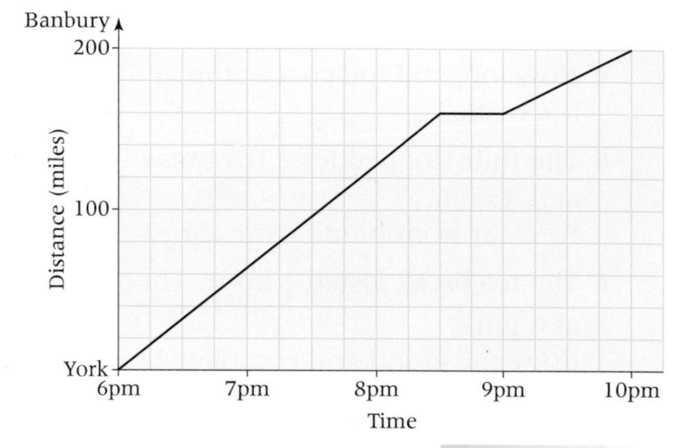

The average speed for the whole journey is

Average speed $= \dfrac{\text{total distance}}{\text{total time}}$

$= \dfrac{200}{4} = 200 \div 4$

$= 50$ miles per hour

- **A sloping straight line represents travel at a constant speed.**

Constant speed means no stops or change in speed.

Example

Tom cycled to Lauren's house.
The graph shows his journey.

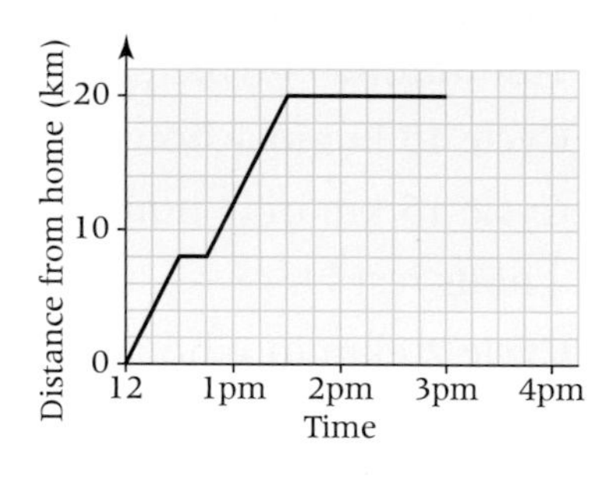

a Work out the average speed in km per hour for the first part of Tom's journey.

b Tom stayed at Lauren's house for $1\frac{1}{2}$ hours. He then cycled home at a constant speed of 20 km per hour.
Copy and complete the graph to show this information.

---

a First part of journey, Tom cycles 8 km in $\frac{1}{2}$ hour.

Method 1
Average speed in km per hour = number of km cycled in 1 hour
8 km in $\frac{1}{2}$ hour $\rightarrow$ 16 km in 1 hour
Average speed = 16 km per hour

Method 2
Using the formula:
Average speed $= 8 \div \frac{1}{2}$
$= 16$ km per hour

For speeds in km per hour, use distances in km and times in hours.

b Tom cycles home at 20 km per hour.
So it takes him 1 hour to cycle the 20 km home.

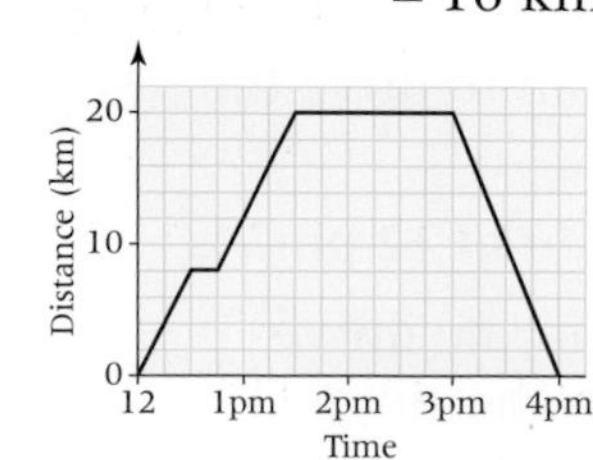

Exercise A6.4

1 The graph shows Shani's journey to her grandmother's house.

a What was the total distance she travelled?

b What was the total time for the journey?

c Work out the average speed for the complete journey using the formula

$$\text{Average speed} = \frac{\text{total distance}}{\text{total time}}$$

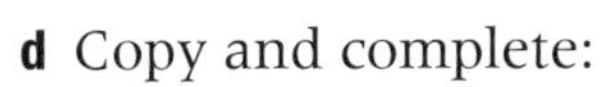

d Copy and complete:

Average speed for first part of journey $= \frac{}{2} =$ ____ km per hour

e Work out the average speed for the second part of her journey.

f Which part of the journey was fastest, the first or second?

g Copy and complete:

The steeper the graph, the __________ the speed.

2 The graph illustrates a boat trip to two islands and then back to port.

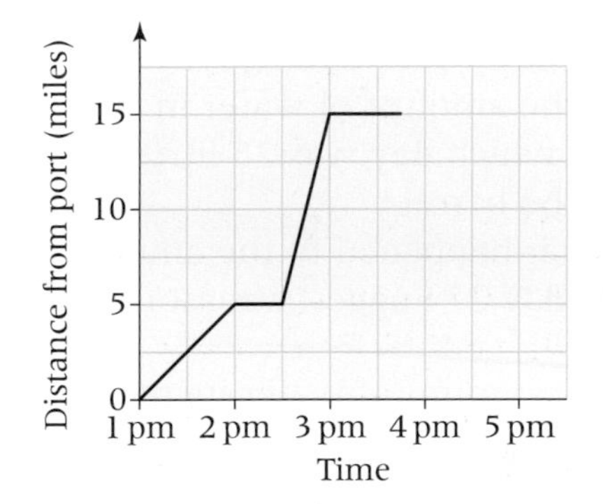

a How far from port was the first island?

b How long did the boat stop at the first island?

c From the graph, on which part of the trip was the boat travelling fastest? Explain how you know.

d Work out the average speed for the trip to the first island.

e How many miles is it from the first island to the second?

f How long did it take to travel from the first island to the second?

g Work out the average speed for the second part of the trip, to the second island.

Use your answers to parts **e** and **f**.

h How far is the second island from the port?

i The boat waited for $\frac{3}{4}$ hour at the second island and then sailed back to port at a steady speed of 15 miles per hour. Copy and complete the graph for the trip.

j What time did the boat arrive back in port?

3 Here is a graph showing part of Dave's trip to the dentist.

a How far is the dentist's from Dave's home?

b Dave took 30 minutes to walk to the dentist's. Work out his walking speed in km/h.

c Dave was at the dentist's for 30 minutes. Then he jogged home at 6 km/h. Copy and complete the travel graph.

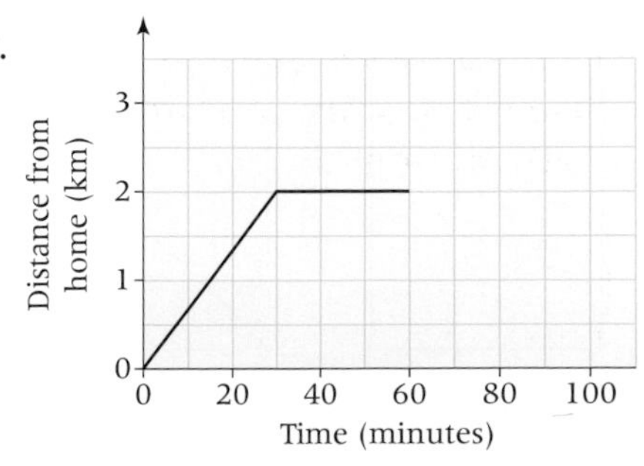

A6.5 Real-life graphs

This spread will show you how to:

- Understand the trend shown by a graph

Keywords
Decrease
Increase
Trend

- **The shape of a graph shows the trend.**

The graph shows the numbers of video recorders sold over a 10 year period.

The trend is that the number of video recorders sold is **decreasing**.

You can read information from a graph, but read the axis labels and scale carefully.

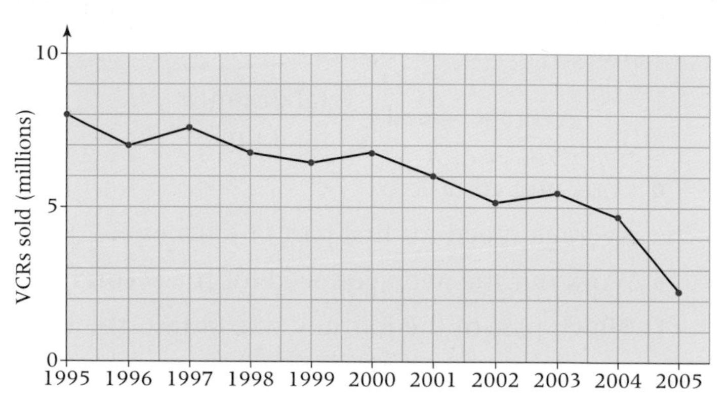

Example

The graph shows the amount of rainwater in a barrel over a few days.

a On day 1 it rained heavily. What happened to the amount of water in the barrel?
b On which day was 25 litres poured out of the barrel?
c What happened to the amount of water on day 2? Suggest a reason for this.

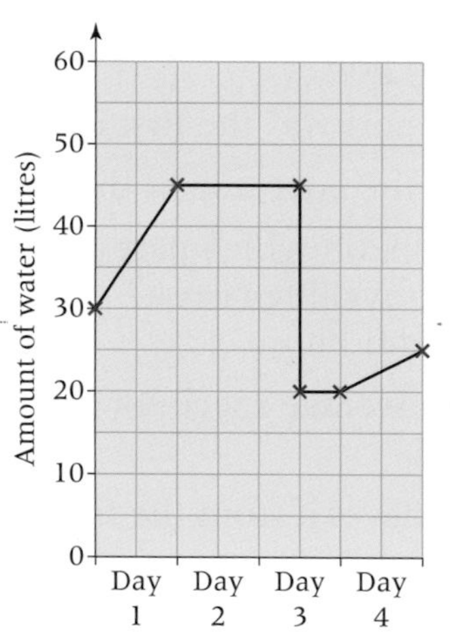

a The amount of water increased.
b Day 3, as the amount suddenly reduced by 25 litres.
c Amount of water stayed the same. It probably did not rain on day 2 and no water was poured out.

Think what could affect the amount of water.

- **A straight line shows that a quantity is changing at a steady rate. The steeper the slope, the faster the change.**

quantity increasing | no change | quantity decreasing

The graphs show the water level as two tanks fill with water.

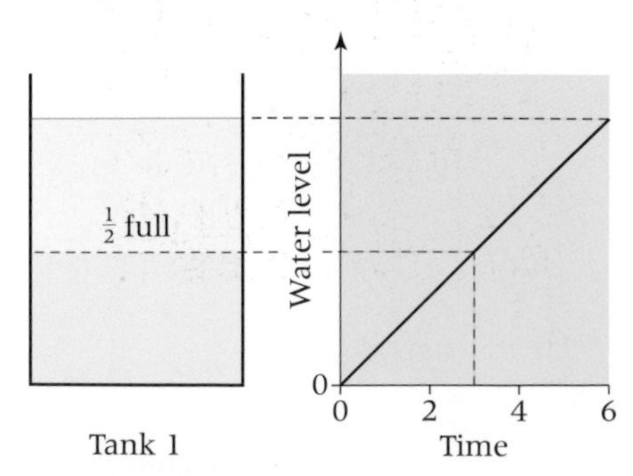

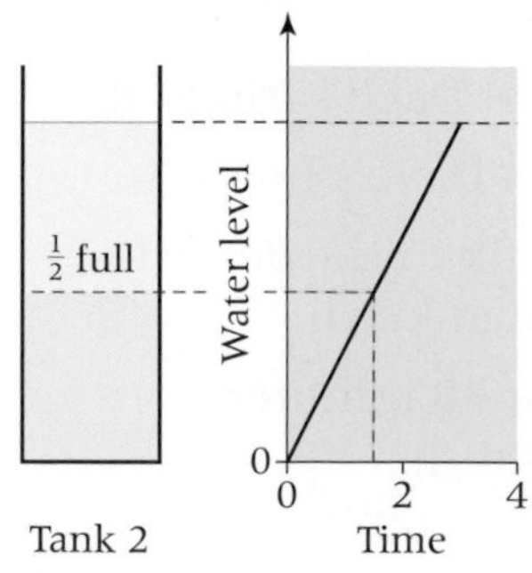

Water is poured into both tanks at a steady rate.

The first tank fills more slowly, as it is wider.

The second tank fills more quickly, as it is narrower.

The steeper the slope, the faster the change in water level.

Exercise A6.5

1 The graph shows sales of 'Time 2 Chat' mobile phones.

a How many phones were sold in March?

b How many phones were sold in June?

c How many more phones were sold in June than in January?

d Here are the sales figures for the next three months.

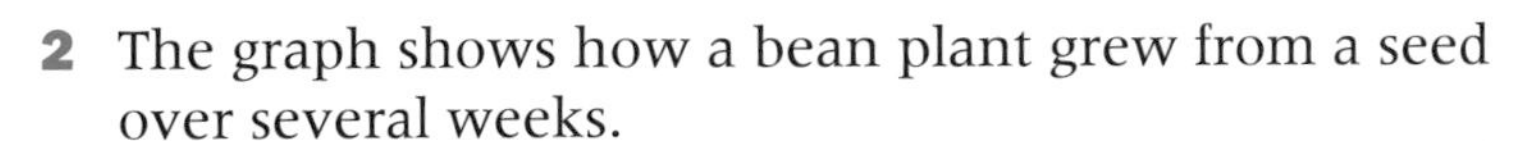

Month	October	November	December
Number of phones sold	250	325	400

Copy the graph and complete it for this information.

e What happened to sales in November and December? Suggest a reason for this.

f What overall trend does the graph show?

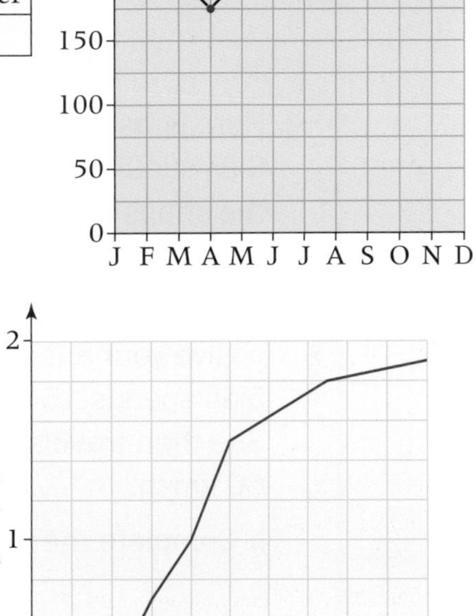

2 The graph shows how a bean plant grew from a seed over several weeks.

a How tall was the plant after 6 weeks?

b How much did the plant grow between weeks 8 and 10?

c How tall did the plant grow in total?

d How much did the plant grow between 15 and 20 weeks.

e Is the plant likely to reach a height of 3 metres? Explain your answer.

3 Rain rushes into these rain barrels at a steady rate. The graphs show how the water level changes. Match each graph to a rain barrel.

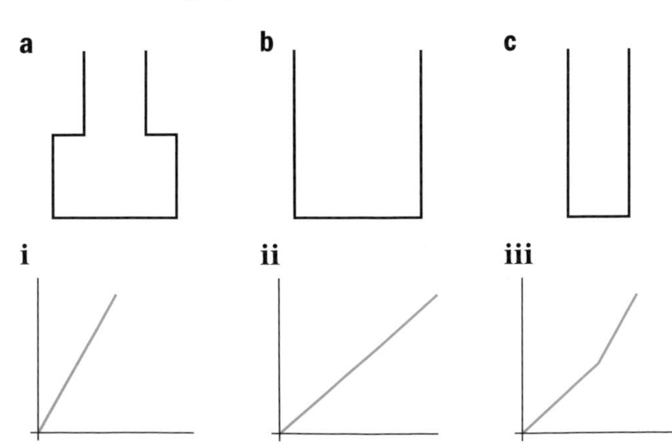

A6 Exam review

Key objectives

- Construct linear functions and plot the corresponding graphs arising from real-life problems
- Discuss and interpret graphs arising from real situations

1 Here is part of a travel graph of Siân's journey from her house to the shops and back.

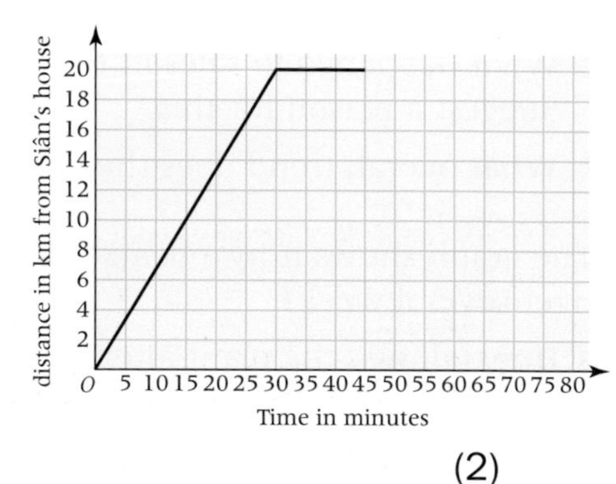

a Work out Siân's speed for the first 30 minutes of her journey.
Give your answer in km/h.

Siân spends 15 minutes at the shops.
She then travels back to her house at 60 km/h. (2)

b Complete the travel graph. (2)

(Edexcel Ltd., 2003)

2 Anil cycled from his home to the park.

Anil waited in the park.

Then he cycled back home.

Here is a distance–time graph for Anil's complete journey:

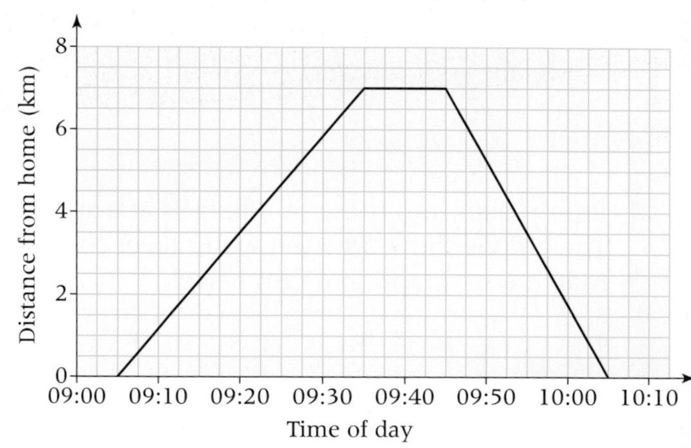

a At what time did Anil leave home? (1)

b What is the distance from Anil's home to the park? (1)

c How many minutes did Anil wait in the park? (1)

d Work out Anil's average speed on his journey home. (3)
Give your answer in kilometres per hour.

(Edexcel Ltd., 2004)

S1

Length and area

This unit will show you how to

- Know rough metric equivalents to imperial measures
- Convert measurements from one unit to another
- Calculate the area and perimeter of rectangles, triangles and shapes made from rectangles and triangles
- Use the formula to find the area of any parallelogram
- Use formulae of rectangles, triangles and parallelograms to find the area of any trapezium
- Calculate the circumference and area of a circle

Before you start ...

You should be able to answer these questions.

1 Evaluate.

a 60×10 **b** 71×1000 **c** 4.8×10

d 26.3×100 **e** 4.5×1000 **f** $600 \div 100$

g $750 \div 10$ **h** $6500 \div 1000$ **i** $32 \div 10$

2 Evaluate.

a $6 \times \frac{8}{5}$ **b** $9 \times 2\frac{1}{2}$ **c** $6 \times 1\frac{3}{4}$

3 Evaluate.

a $5.8 + 2$ **b** $14.8 + 0.7$ **c** $6.4 + 2.6$

4 Measure this line

a in millimetres **b** in centimetres

5 Calculate the perimeter and area of this rectangle.

8 cm

5 cm

S1.1 Metric and imperial measures

This spread will show you how to:

- Know rough metric equivalents to imperial measures
- Convert measurements from one unit to another

Keywords
Capacity
Convert
Equivalents
Imperial
Length
Mass
Metric

You can measure **length**, **mass** and **capacity** using **metric** and **imperial units.**

You can **convert** between metric units by multiplying or dividing by 10, 100, 1000, . . .

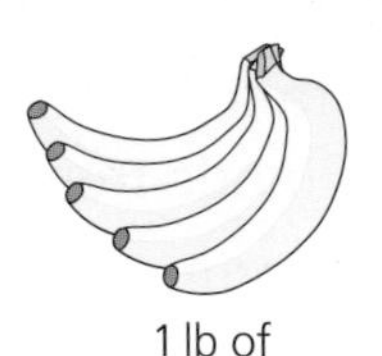

1 lb of bananas

Paris
40 km

- **Length** is a measure of distance.

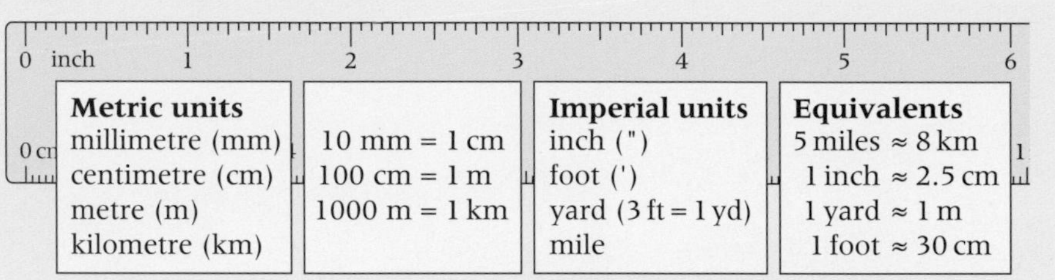

Metric units		Imperial units	Equivalents
millimetre (mm)	10 mm = 1 cm	inch (")	5 miles ≈ 8 km
centimetre (cm)	100 cm = 1 m	foot (')	1 inch ≈ 2.5 cm
metre (m)	1000 m = 1 km	yard (3 ft = 1 yd)	1 yard ≈ 1 m
kilometre (km)		mile	1 foot ≈ 30 cm

≈ means approximately equal to.

1 metre is a bit longer than 1 yard.

- **Mass** is a measure of the amount of matter in an object. Mass is linked to weight.

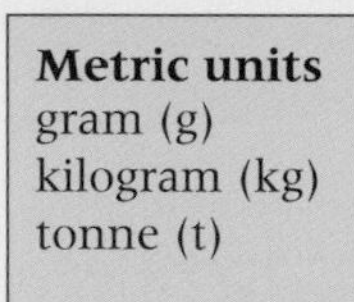

Metric units		Imperial units	Equivalents
gram (g)	1000 g = 1 kg	ounce (oz)	1 ounce ≈ 30 g
kilogram (kg)	1000 kg = 1 tonne	pound (lb)	1 kg ≈ 2.2 lb
tonne (t)		stone	
		ton	

1 lb of jam

Sugar 1 kg

1 kg of sugar

- **Capacity** is a measure of the amount of liquid a 3-D shape will hold.

Metric units		Imperial units	Equivalents
millilitre (ml)	1000 ml = 1 litre	pint	1 pint ≈ 600 ml
centilitre (cl)	100 cl = 1 litre	gallon	1.75 pints ≈ 1 litre
litre			1 gallon ≈ 4.5 litres

1 pint of milk 1 litre of lemonade

Example

Calculate the approximate length of a 12 inch ruler in

a centimetres **b** millimetres.

0 1 2 3 4 5 6 7 8 9 10 11 12
inches

a $1'' \approx 2\frac{1}{2}$ cm
$12'' \approx 2\frac{1}{2} \times 12 = 30$ cm

b 1 cm = 10 mm
$30 \text{ cm} = 30 \times 10 = 300$ mm

cm → × 10 → mm
mm → ÷ 10 → cm

Exercise S1.1

1 Choose one of these metric units to measure each of these items.

millimetre	gram	millilitre	centimetre
kilogram	centilitre	metre	tonne
litre	kilometre		

a your height
b amount of tea in a mug
c your weight
d length of a suitcase
e weight of a suitcase
f distance from Paris to Madrid
g quantity of drink in a can
h amount of petrol in a car
i weight of an elephant
j weight of an apple.

Write the appropriate abbreviation next to your answers.

2 Convert these measurements to the units shown.

a 20 mm = ___ cm
b 400 cm = ___ m
c 450 cm = ___ m
d 4000 m = ___ km
e 0.5 cm = ___ mm
f 4.5 kg = ___ g
g 6000 g = ___ kg
h 6500 g = ___ kg
i 2500 kg = ___ t
j 3 litres = ___ ml

3 Convert these distances to miles.

a Berlin 16 km
b Dusseldorf 40 km
c Bonn 88 km
d Dresden 84 km

4 Convert these measurements to centimetres.

a 1 inch **b** 5 inches **c** 6 inches **d** 12 inches **e** 36 inches

5 Use 1 kg ≈ 2.2 lb to convert these weights to pounds.

a 2 kg **b** 40 kg **c** 50 kg **d** 0.5 kg **e** 2.5 kg

6 Use 1 oz ≈ 30 g to convert ounces to grams in these recipes.

a
Lemon Curd
6 oz butter
12 oz caster sugar
6 lemons
6 eggs

b
Cumberland Pudding
8 oz rice
4 oz raisins
3 oz sugar
4 oz currants
1 egg
beef marrow

c
Chocolate Crunchies
6 oz self-raising flour
2 oz cornflour
2 oz cornflakes
1 oz drinking chocolate
6 oz margarine
3 oz sugar

7 The speed limit on a motorway in the UK is 70 miles per hour. Calculate the speed limit in kilometres per hour.

S1.2 Perimeter and area of a rectangle and a triangle

This spread will show you how to:

- Calculate the area and perimeter of rectangles, triangles and shapes made from rectangles and triangles

Keywords
Area
Base
Perimeter
Perpendicular height
Square units

- **The perimeter of a shape is the distance around it.**

Perimeter is a length, so it is measured in mm, cm, m or km.

- **The area of a shape is the amount of space it covers.**

Area is measured in **square units**: mm^2, cm^2, m^2 or km^2.

You can find the area of a rectangle using:

- **Area of a rectangle = length × width**

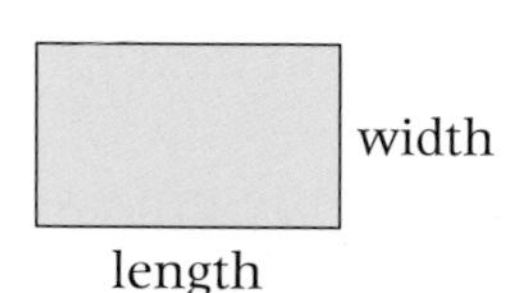

This formula also works for a square.

You can find a formula for the area of any triangle.

For this triangle ...

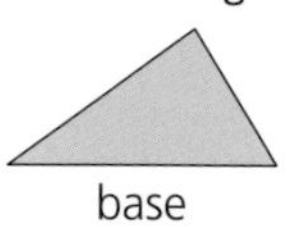

complete the rectangle ...

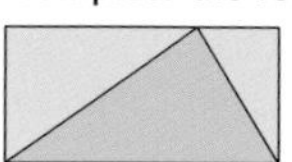

the area has doubled.

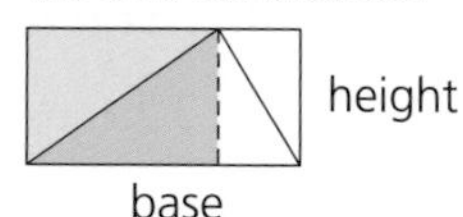

The area of a triangle is half the area of the surrounding rectangle.

- **Area of triangle = $\frac{1}{2}$ × base × height**

The height must be perpendicular to the base.

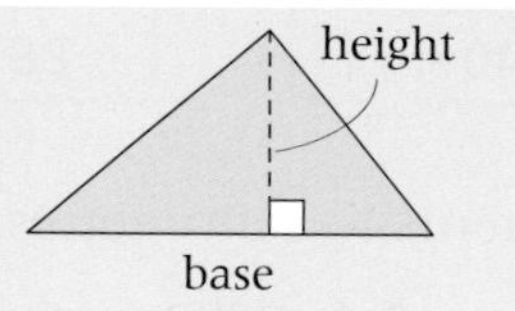

Perpendicular means at right angles.

You can find the area of a compound shape by splitting it into rectangles and triangles.

Example

Calculate the perimeter and area of each shape.

a

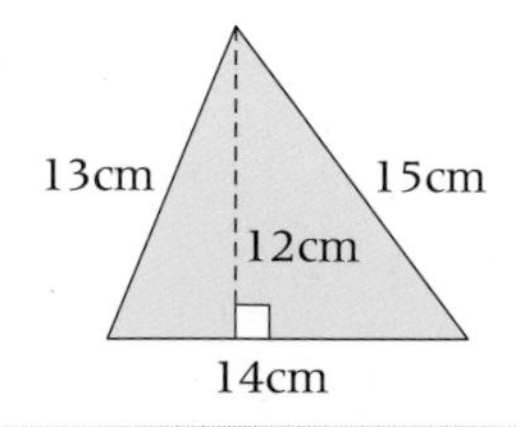

b

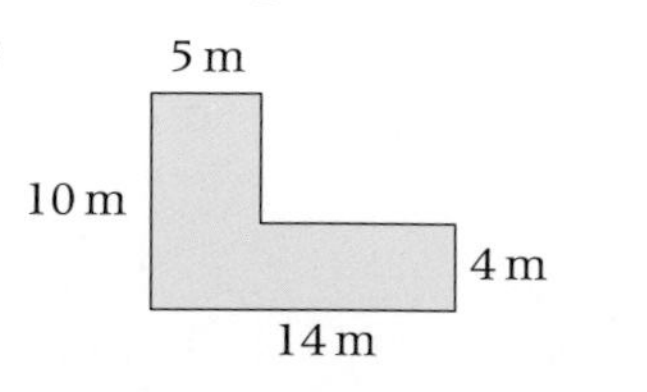

a Perimeter = 15 + 13 + 14
= 42 cm

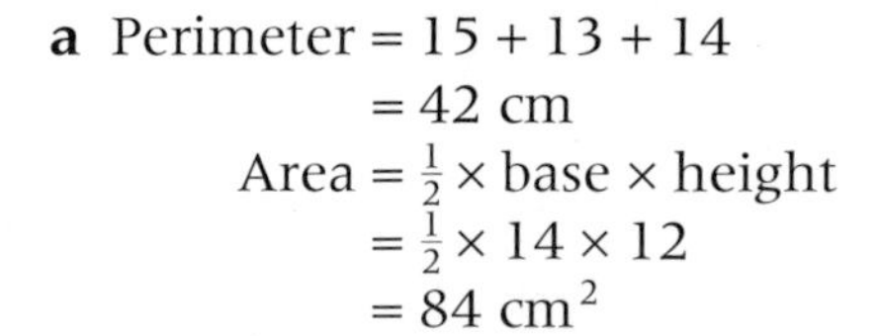

Area = $\frac{1}{2}$ × base × height
= $\frac{1}{2}$ × 14 × 12
= 84 cm^2

b Perimeter = 5 + 6 + 9 + 4 + 14 + 10
= 48 m
Area = area of green rectangle
+ area of orange rectangle
= 10 × 5 + 9 × 4
= 50 + 36 = 86 m^2

The two missing lengths are:
10 − 4 = 6 m
and
14 − 5 = 9 m

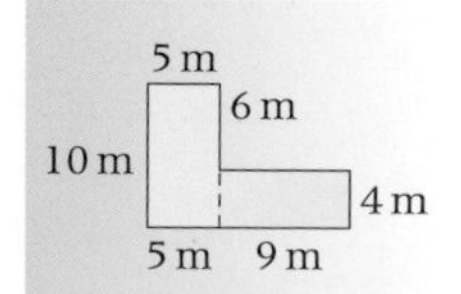

Exercise S1.2

1 Calculate the perimeter and area of each rectangle.
Give the units of your answers.

a 4 m, 2 m

b 8 cm, 1.5 cm

c 13.5 mm, 6 mm

d 5.4 cm, 8 cm

e 12 m, 3.2 m

2 Calculate the area of each triangle.

a

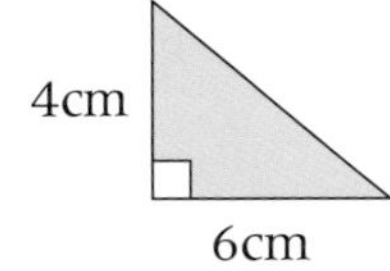

b

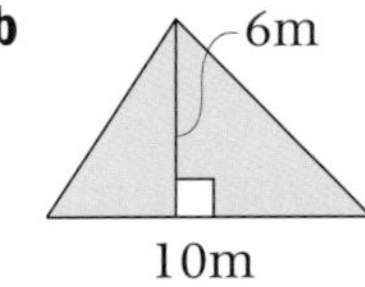

c

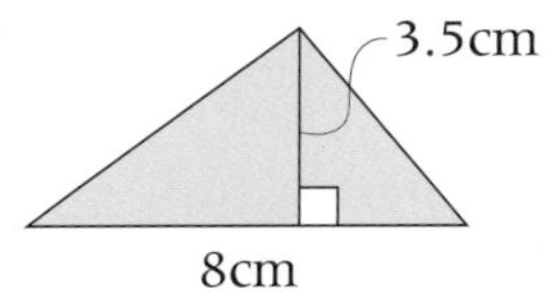

d

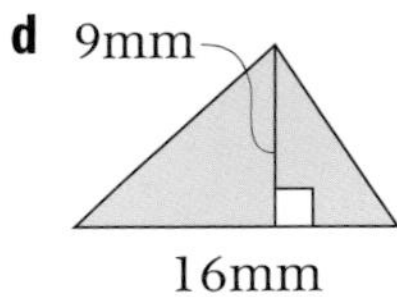

e

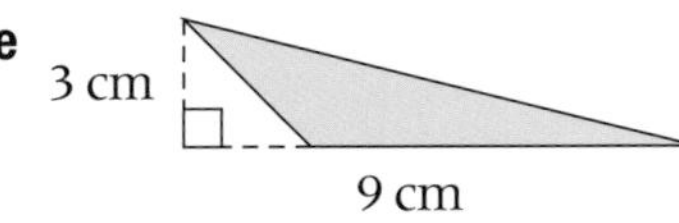

3 Calculate the missing lengths. Give the units of your answers.

a

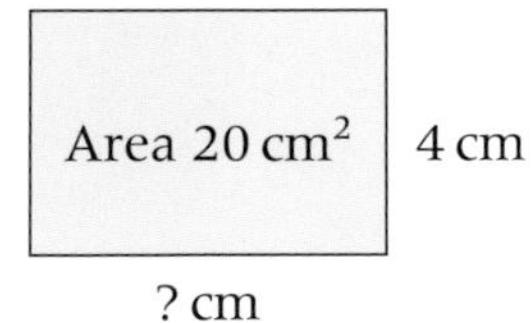

b

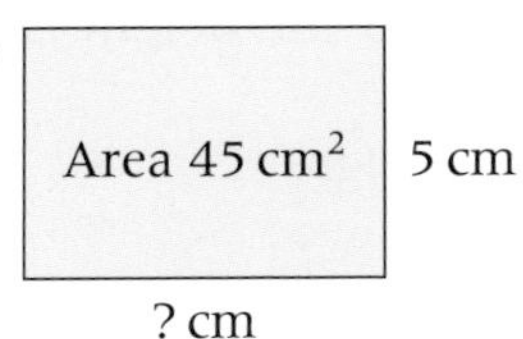

c

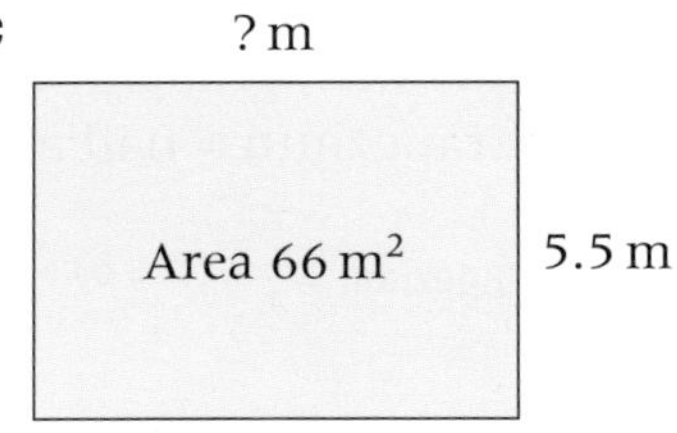

4 Calculate the perimeter and area of each shape.
State the units of your answers.

a

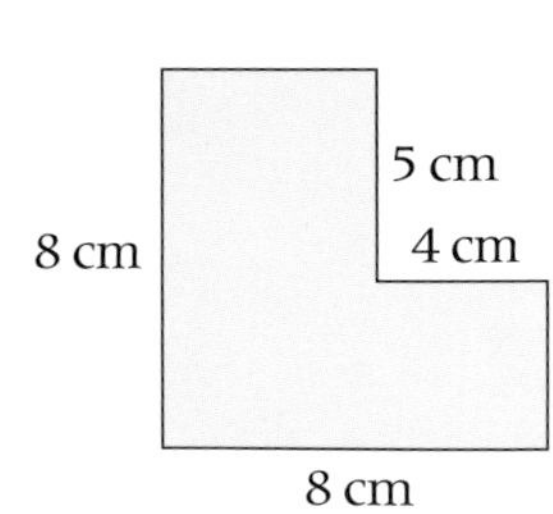

b

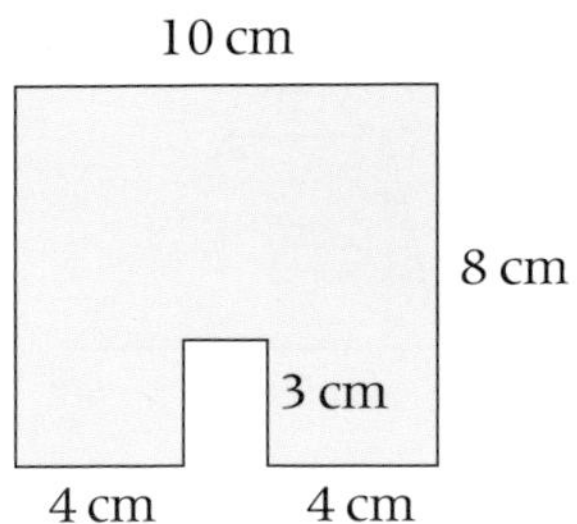

c

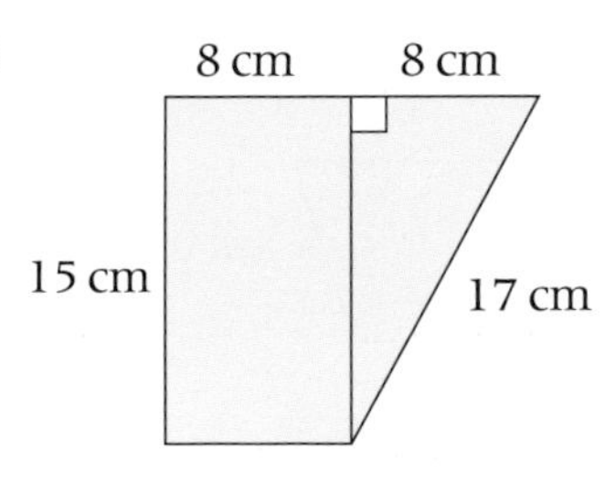

S1.3 Area of a parallelogram and a trapezium

This spread will show you how to:

- Use the formula to find the area of any parallelogram
- Use formulae of rectangles, triangles and parallelograms to find the area of any trapezium

Keywords
Area
Base
Congruent
Parallelogram
Perpendicular height
Trapezium

You can find the formula for the **area** of any **parallelogram**.

For this parallelogram ...

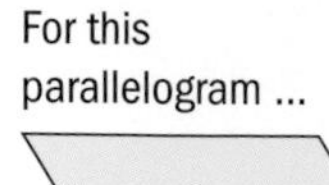

cut off one triangle ...

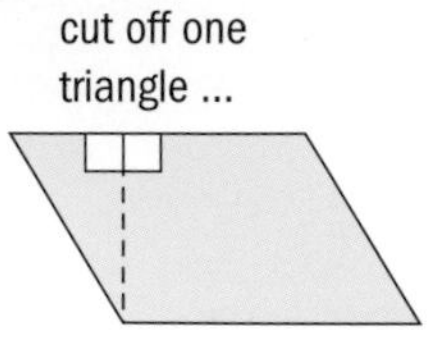

and fit it on the other end ... to make a rectangle.

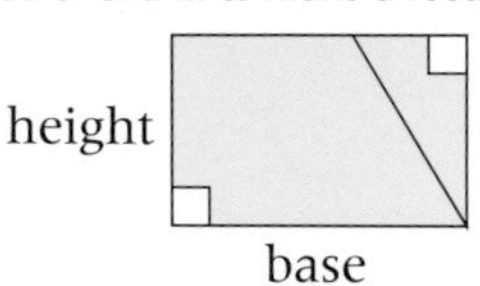

- **Area of parallelogram = base × perpendicular height.**

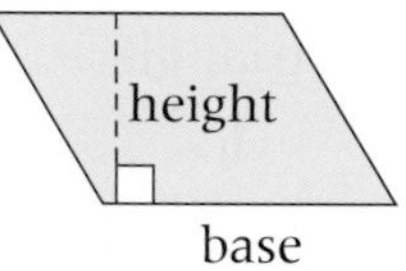

The height must be perpendicular to the base.

You can find the formula for the area of any **trapezium**.

You can fit two **congruent** trapeziums together to make a parallelogram.

Congruent means identical.

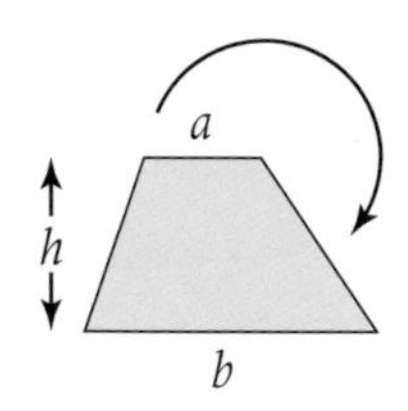

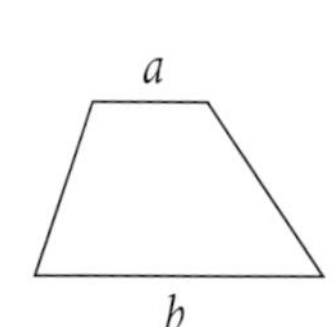

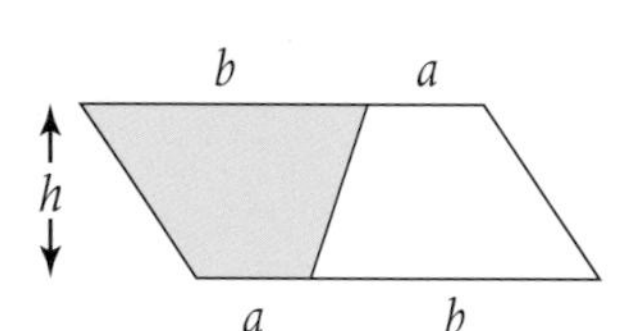

The base of the parallelogram is $a + b$ and the height is h.
Area of parallelogram = $(a + b) \times h$
Area of trapezium = half area of parallelogram.

- **Area of trapezium** = $\frac{1}{2} \times (a + b) \times h$

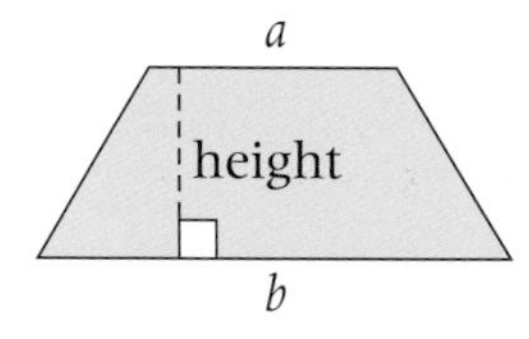

The height is the perpendicular distance between the parallel sides.

Example

Calculate the area of each shape.

a

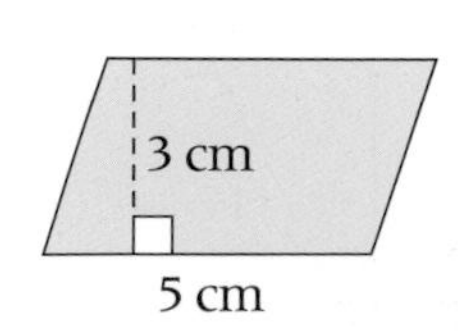

b

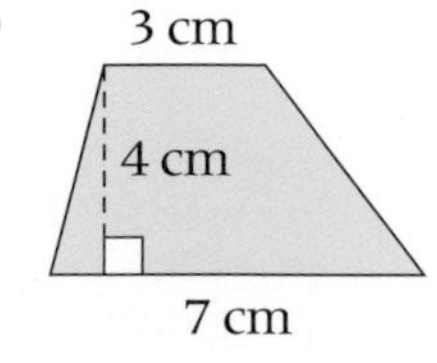

a Area of parallelogram = 5×3
$= 15\text{ cm}^2$

b Area of trapezium = $\frac{1}{2}(3 + 7) \times 4$
$= 5 \times 4$
$= 20\text{ cm}^2$

Exercise S1.3

1 Calculate the area of each parallelogram.

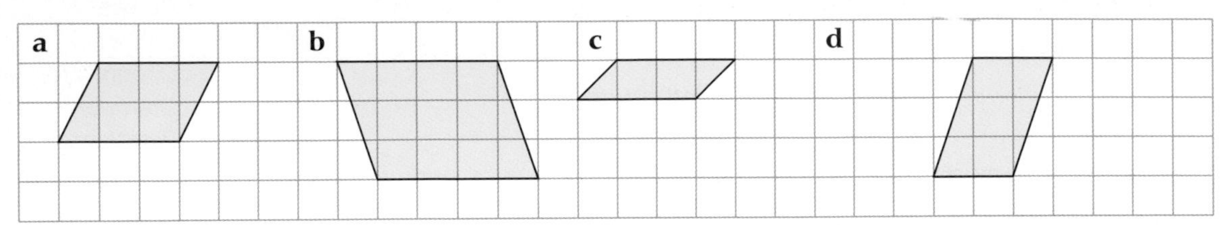

2 Calculate the area of each trapezium.

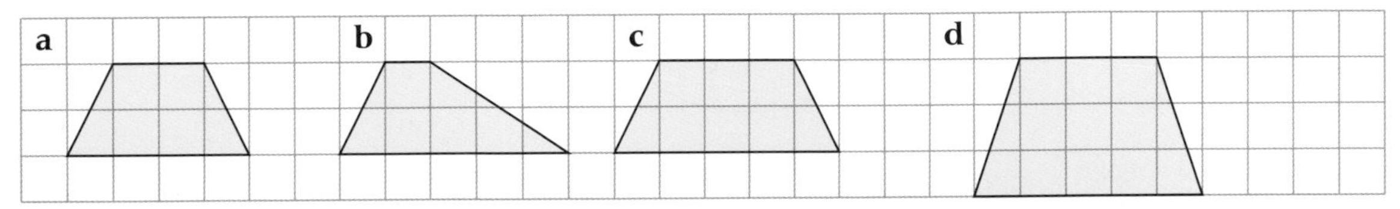

3 Calculate the area of each parallelogram. State the units of your answers.

a 8cm, 10cm **b** 20m, 40m **c** 8mm, 15mm **d** 16cm, 24cm

4 Calculate the area of each trapezium. State the units of your answers.

a 8cm, 5cm, 12cm **b** 20mm, 15mm, 30mm **c** 5m, 4m, 9m **d** 14cm, 10cm, 18cm

5 The areas of these shapes are given. Calculate the unknown lengths.

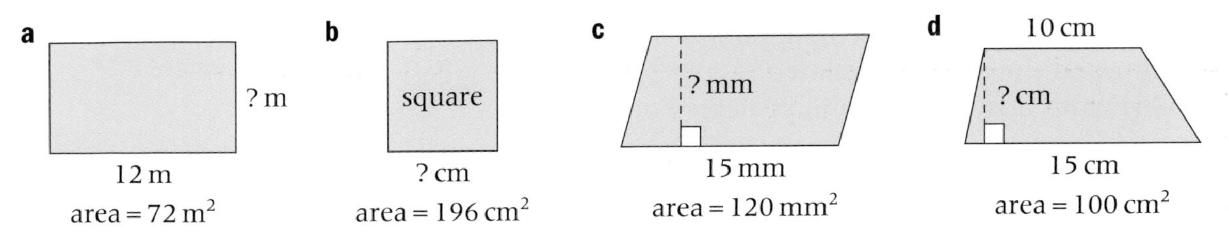

6 **a** Calculate the area of this shape using the formula for the area of a trapezium.

b Calculate the area by adding the areas of the triangles and the square.

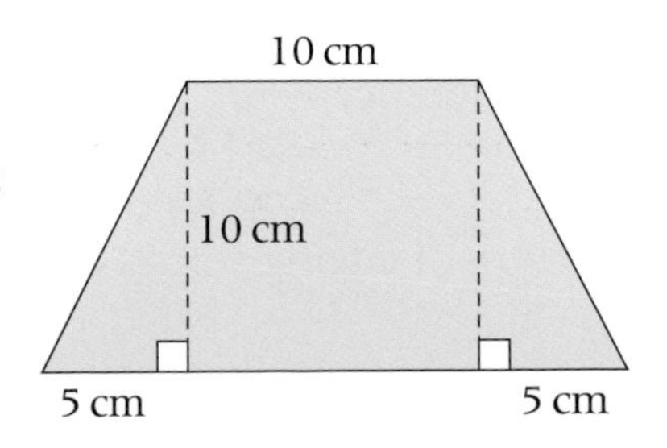

S1.4 Circumference and area of a circle

This spread will show you how to:
- Calculate the circumference and area of a circle

Keywords
Centre
Circle
Circumference
Diameter
Pi (π)
Radius

In a **circle**:
- the **radius** is r
- the **diameter** is d
- the **circumference** is C.

C, d and r are all measures of length.

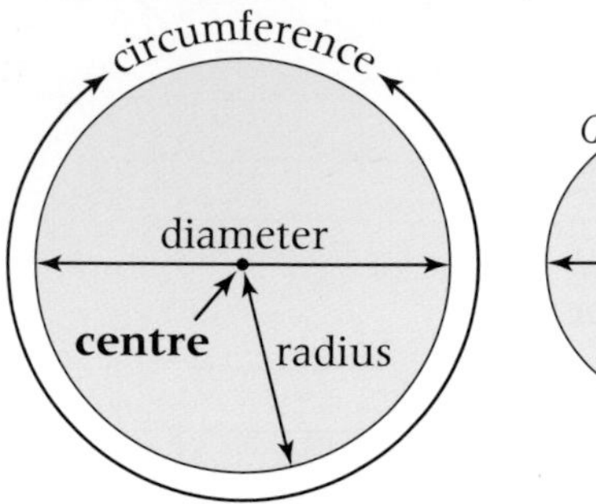

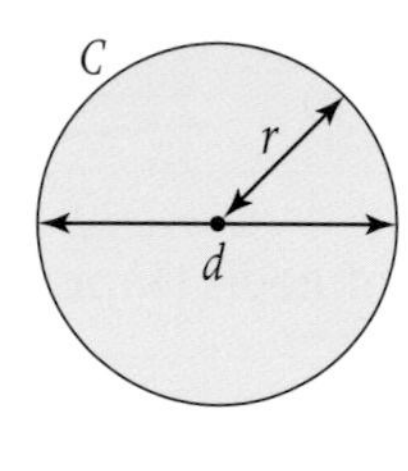

The perimeter of a circle is called the circumference.

- **Diameter = 2 × radius**
- **$C = \pi \times$ diameter $= \pi d = 2\pi r$**

$d = 2 \times r$

$\pi = 3.14 \ldots$

Example

Calculate the circumference of this circle.

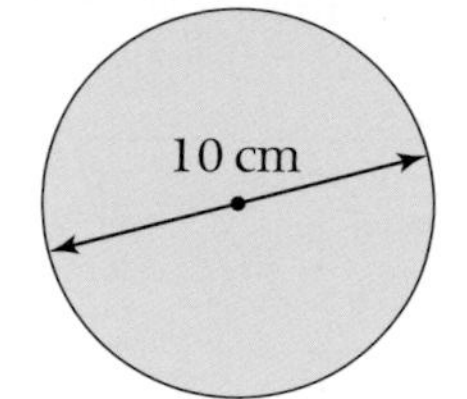

$C = \pi \times d$
$= 3.14 \times 10$
$= 31.4$ cm   Remember to state the units.

Circumference is measured in units of length.

- **Area of a circle $= \pi \times$ radius × radius**
  $= \pi \times r \times r$ or πr^2

r^2 means $r \times r$

A circular lawn has radius 3 metres.

a Calculate the area of the lawn. State the units of your answer.

b Calculate the length of edging stones needed to fit all round the edge of the lawn.
Give your answer to a suitable degree of accuracy.

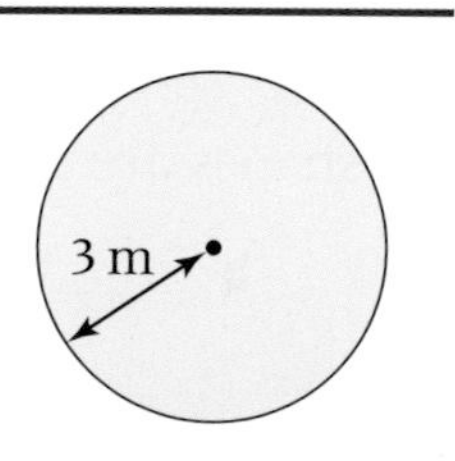

a Area $= \pi r^2$
$= 3.14 \times 3 \times 3$
$= 3.14 \times 9$
$= 28.26$ m^2

b Circumference $= \pi d$
$= 3.14 \times 6$
$= 18.84$ m
So 19 m of edging stones are needed.

Area is measured in square units.

Exercise S1.4

Take $\pi = 3.14$ for all questions on this page.

1 Calculate the circumferences of these circles. State the units of your answers.

a diameter = 10 cm

b diameter = 8 m

c diameter = 12 cm

d 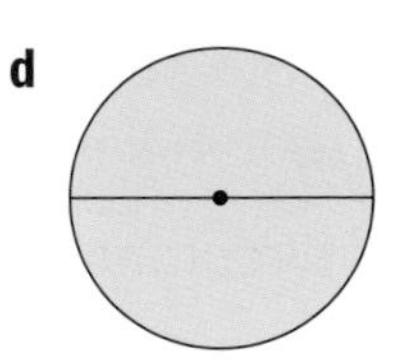

diameter = 20 m

e radius = 2 m

f 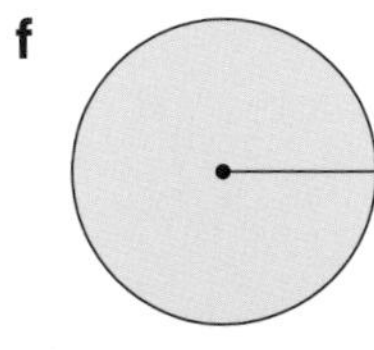

radius = 8 cm

g 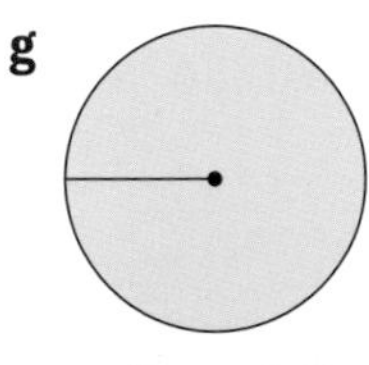

radius = 1.5 m

h 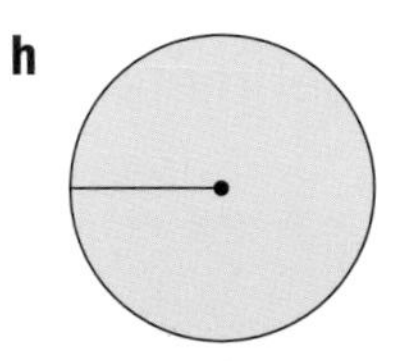

radius = 3.5 cm

2 Calculate the diameter of a circle, if its circumference is

a 18.84 cm **b** 15.7 m **c** 28.26 cm **d** 47.1 m **e** 314 cm

3 Calculate the areas of these circles. State the units of your answers.

a 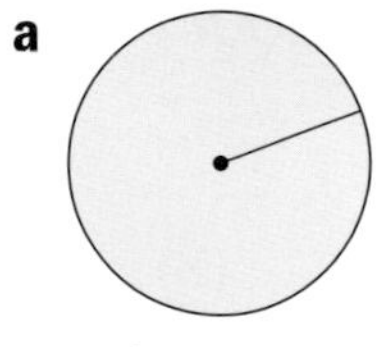

radius = 7 cm

b 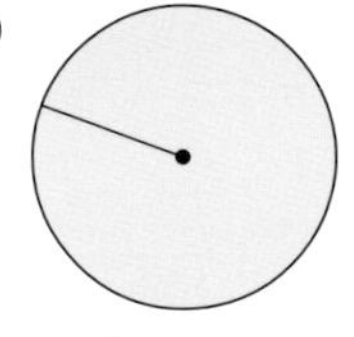

radius = 5 m

c 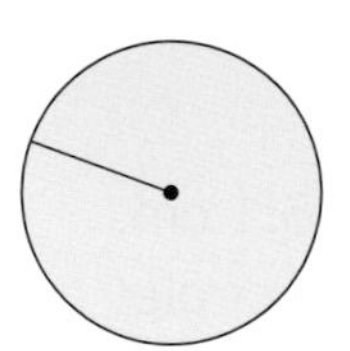

radius = 4 cm

d 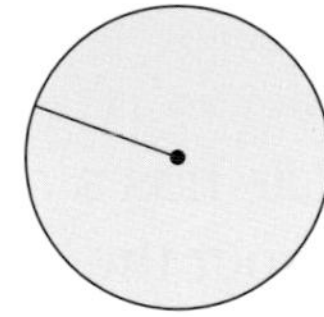

radius = 3 m

e 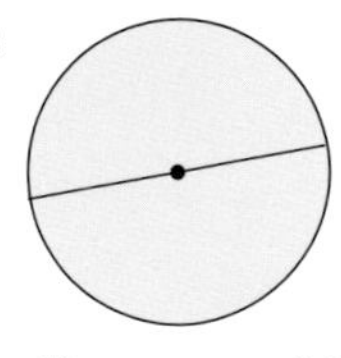

diameter = 20 m

f 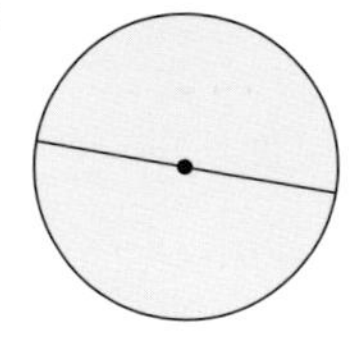

diameter = 16 cm

g 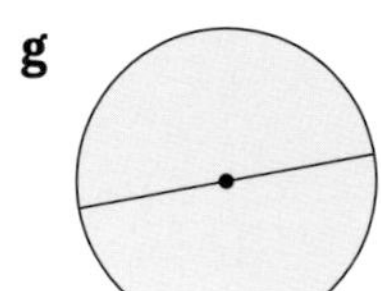

diameter = 12 mm

h 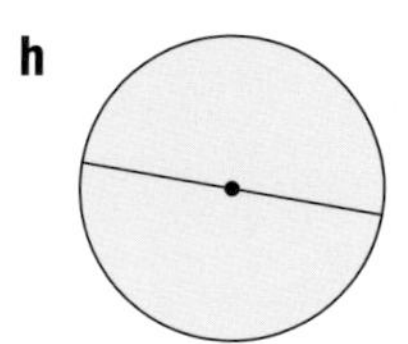

diameter = 18 cm

4 A garden pond is circular.
The radius of the pond is 1.5 m.

a Calculate the diameter of the pond.

b Calculate the circumference of the pond.

c Calculate the area of the pond.

Give your answers to a suitable degree of accuracy.

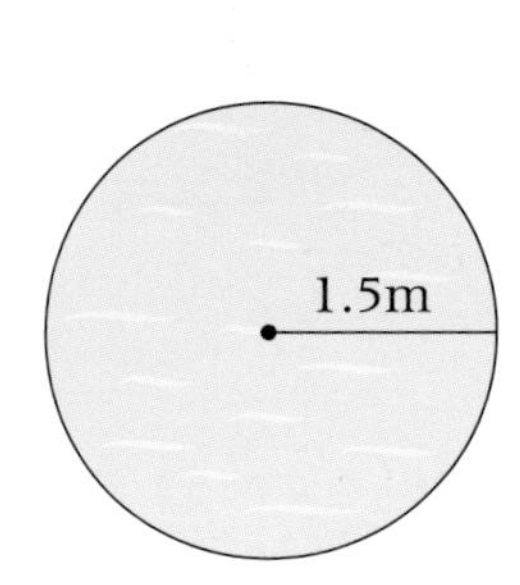

S1 Exam review

Key objectives

- Calculate perimeters and areas of shapes made from triangles and rectangles
- Find circumferences of circles and areas enclosed by circles

1 The length of a rectangle is 10.8 cm.
The perimeter of the rectangle is 28.8 cm.

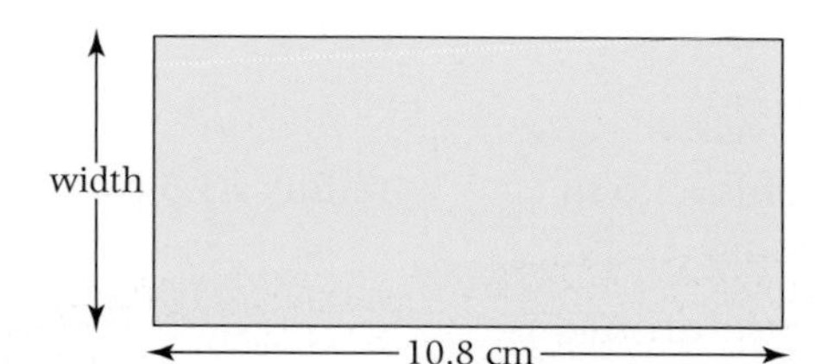

Not to scale

Calculate the width of the rectangle. (3)

(AQA, 2004)

2 A circle has a radius of 6.1 cm.
Work out the area of the circle. (3)

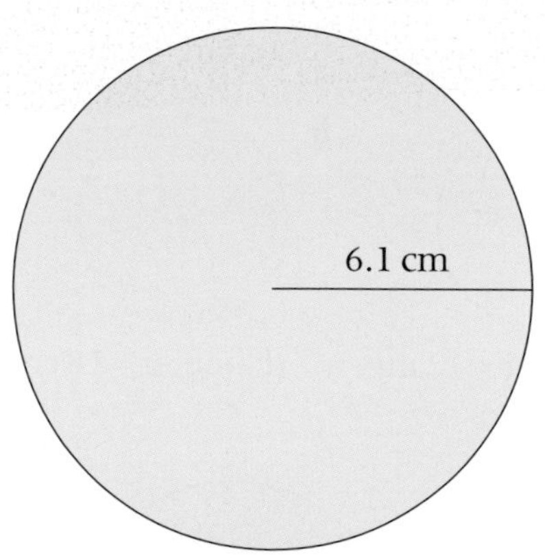

(Edexcel Ltd., 2003)

S2 Angles

This unit will show you how to

- Recall and use properties of lines and angles
- Recall the geometric properties of triangles
- Explain why the angle sum of any quadrilateral is 360°
- Calculate and use interior and exterior angles of polygons
- Use parallel lines, alternate angles and corresponding angles
- Understand and use angle properties

Before you start ...

You should be able to answer these questions.

1 Evaluate.

a 180 – 107 **b** 180 – 38

c 360 – 183 **d** 360 – 217

e 360 – 197

2 Evaluate.

a 180 ÷ 2 **b** 360 ÷ 3

c 360 ÷ 4

3 Use a protractor to measure these angles.

a

b

S2.1 Angle properties

This spread will show you how to:

- Recall and use properties of lines and angles
- Recall the geometric properties of triangles

Keywords
Angle
Degree (°)
Straight line
Triangle
Vertically opposite

You should know these facts:

There are 360° at a point.

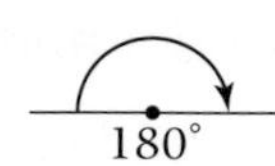

There are 180° on a straight line.

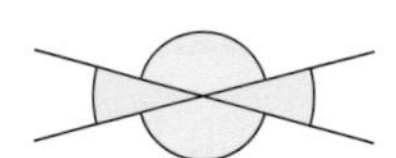

Vertically opposite angles are equal.

Example

Calculate the values of x, y and z. Give a reason for each of your answers.

∟ means the angle is a right angle.

a

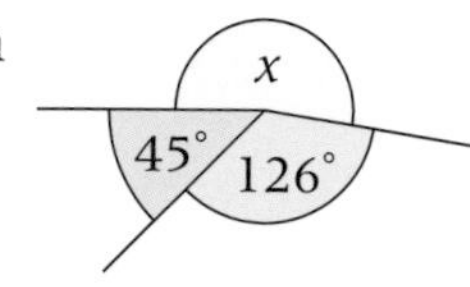

b

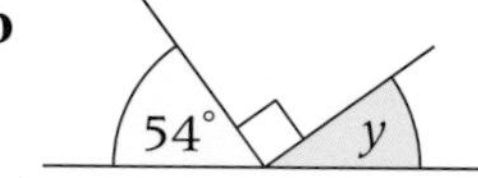

c

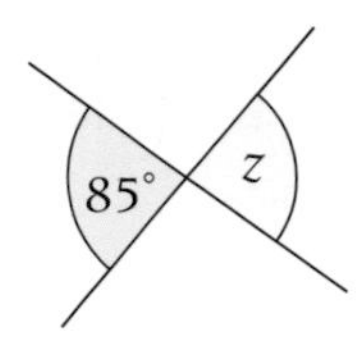

a $126° + 45° = 171°$
$360° - 171° = 189°$
$x = 189°$

(angles at a point add to 360°)

b $54° + 90° = 144°$
$180° - 144° = 36°$
$y = 36°$

(angles on a straight line add to 180°)

c $z = 85°$

(vertically opposite angles are equal)

You should know the names of these triangles:

Right-angled

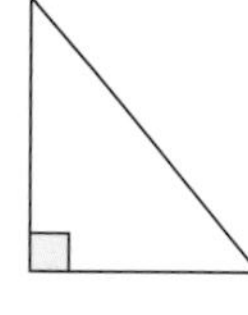

One 90° angle, marked ∟

Equilateral

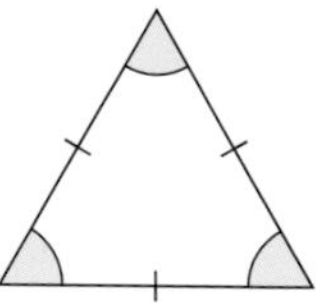

3 equal angles
3 equal sides

Isosceles

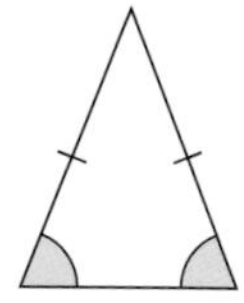

2 equal angles
2 equal sides

Scalene

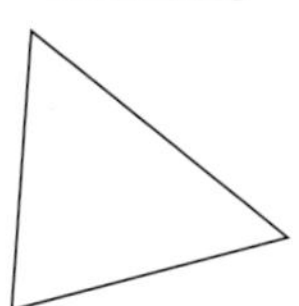

No equal angles
No equal sides

A triangle is a 2-D shape with 3 sides and 3 angles.

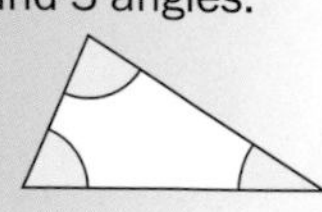

- **The angles in a triangle add to 180°.**

Draw any triangle, tear off the corners,

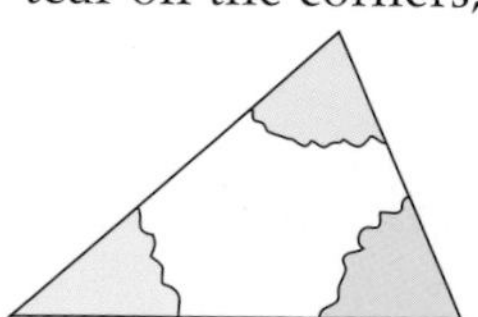

and put them together to make a straight line.

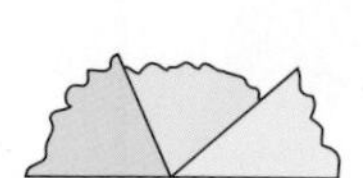

Example

Calculate the value of p.
Give a reason for your answer.

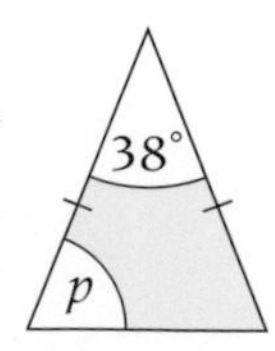

$180° - 38° = 142°$ (angles in a △ add to 180°)
$142° \div 2 = 71°$ (two equal angles in an isosceles △)
$p = 71°$

Exercise S2.1

1 Calculate the size of the unknown angles in each diagram.
Give a reason for each answer.

The diagrams are not drawn to scale.

a

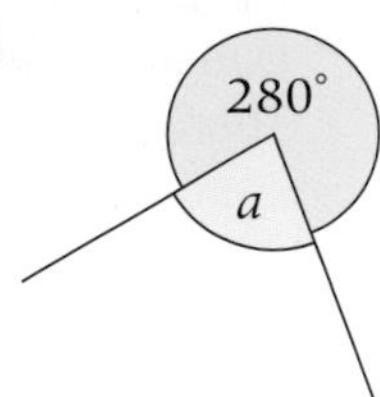

b

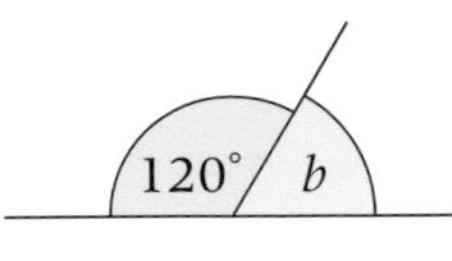

c

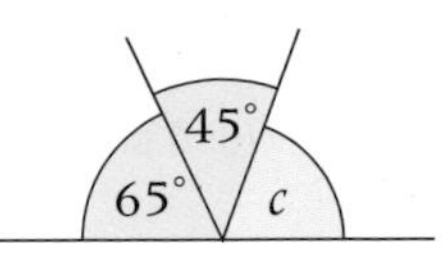

d

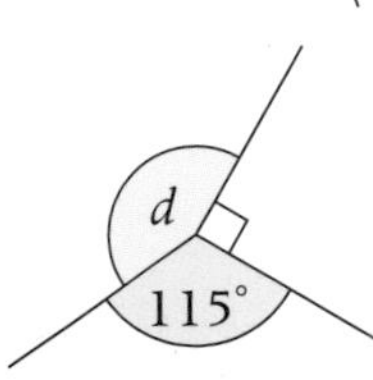

e

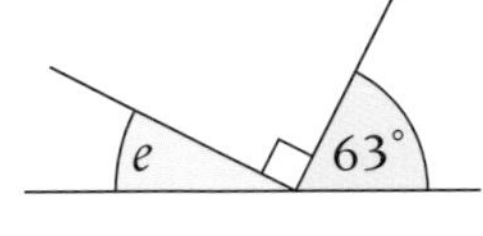

f

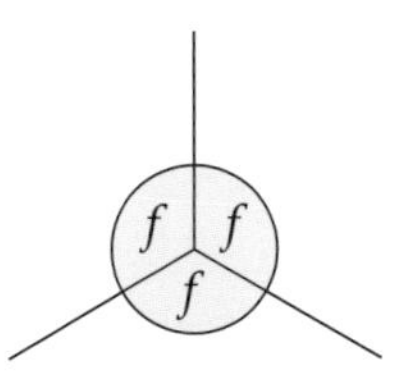

2 Calculate the size of the angles marked by letters in each diagram.

a

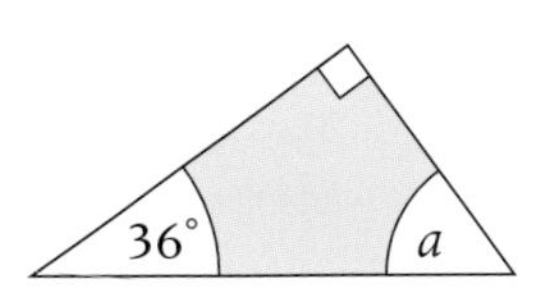

b

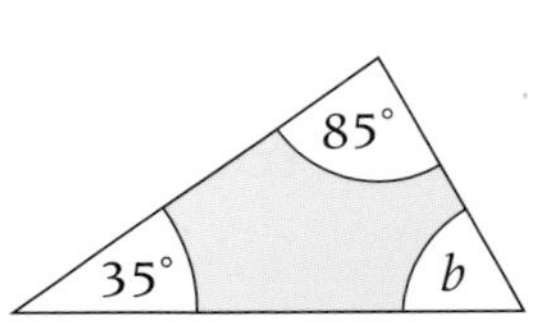

c

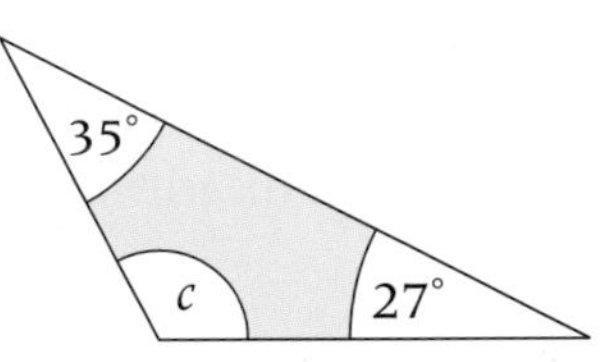

d

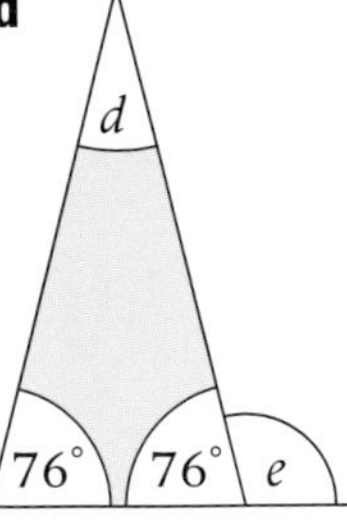

e

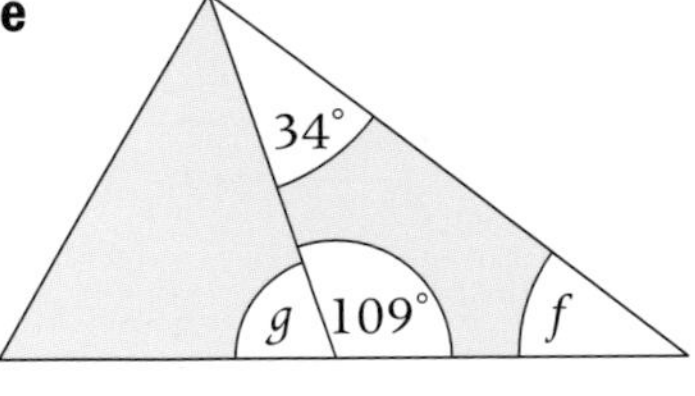

3 Find the unknown angle and state the type of triangle.

a

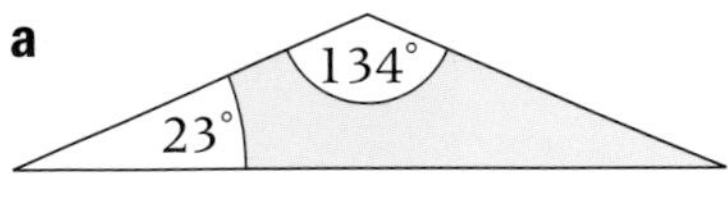

b

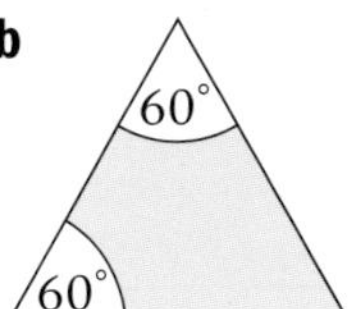

c

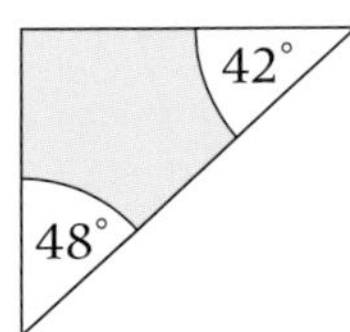

4 Calculate the size of the angles marked with letters in each diagram.

a

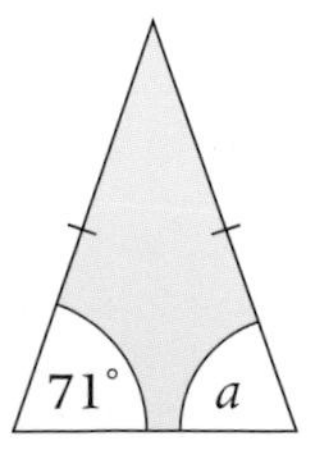

b

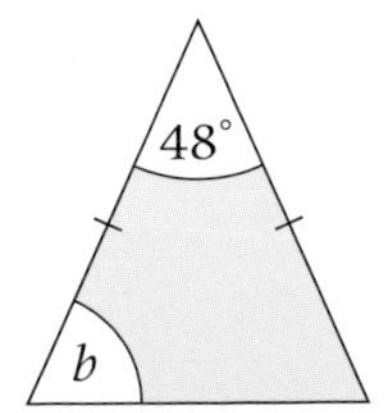

c

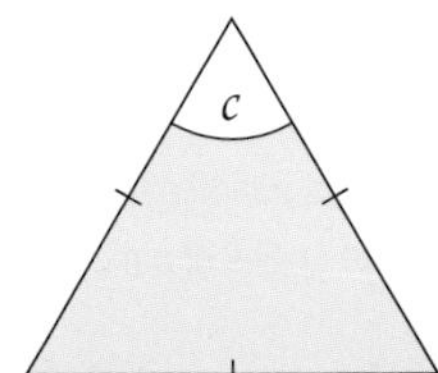

S2.2 Angles in quadrilaterals

This spread will show you how to:

- Explain why the angle sum of any quadrilateral is 360°

Keywords
Angle
Degree (°)
Quadrilateral
Triangle

- **A quadrilateral is a 2-D shape with 4 sides and 4 angles.**

You should know the names of these **quadrilaterals**:

Square

Rectangle

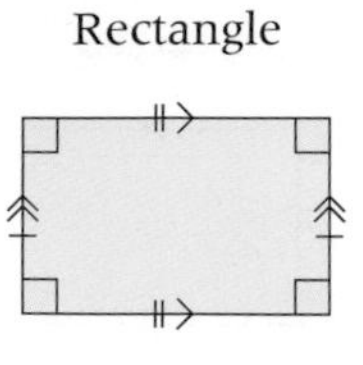

Rhombus

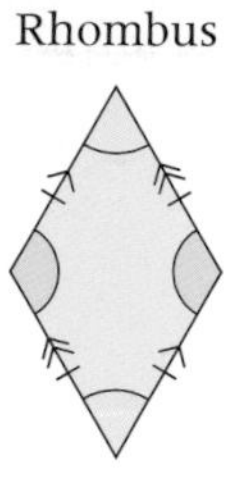

Parallelogram

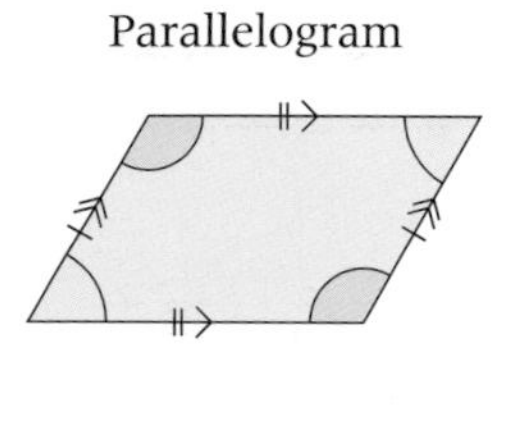

The equal angles are the same colour.

→ means the lines are parallel.

Trapezium

Isosceles trapezium

Kite

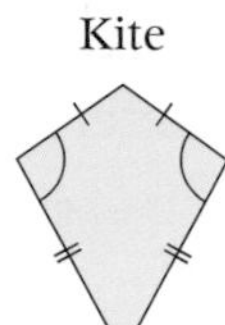

Arrowhead

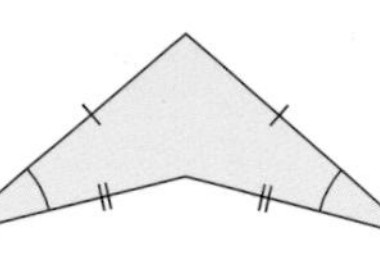

You can draw a diagonal in a quadrilateral to form two **triangles**.
$2 \times 180° = 360°$

The angles in each triangle add to 180°.

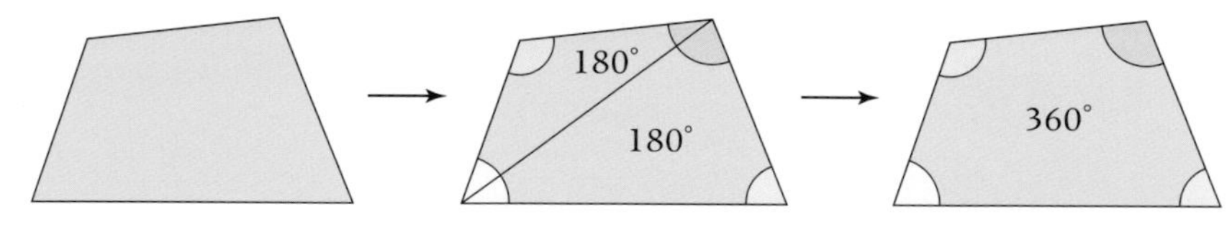

- **The angles in a quadrilateral add to 360°.**

Example

Calculate the values of x, y and z. Give a reason for each of your answers.

a

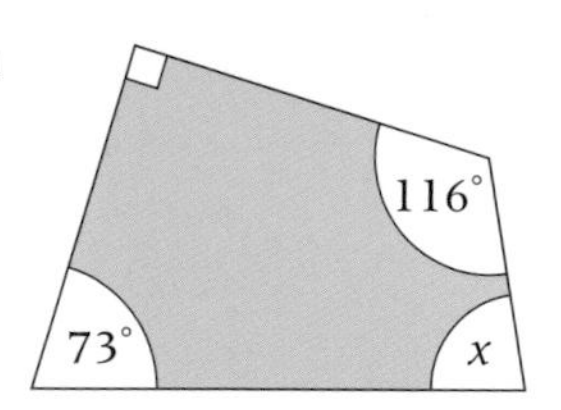

b

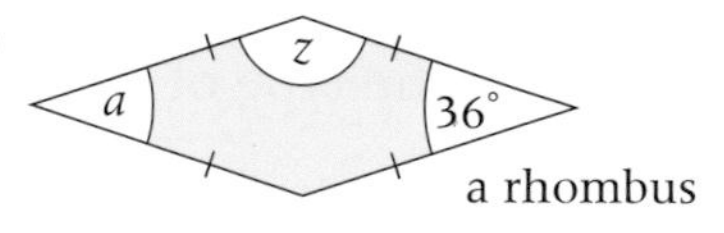

a rhombus

a $90° + 116° + 73° = 279°$

$360° - 279° = 81°$

$x = 81°$ (angles in a quadrilateral add to 360°)

b $a = 36°$ (opposite angles of a rhombus are equal)

$36° + 36° = 72°$

$360° - 72° = 288°$ (angles in a quadrilateral add to 360°)

$288° \div 2 = 144°$ (opposite angles of a rhombus are equal)

$z = 144°$

Exercise S2.2

The diagrams are not drawn accurately.

1 Calculate the size of the unknown angles in each diagram.

a

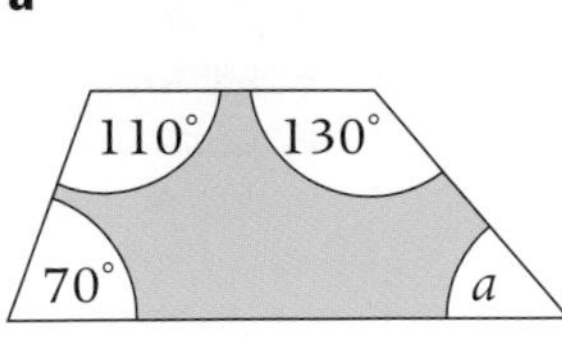

b

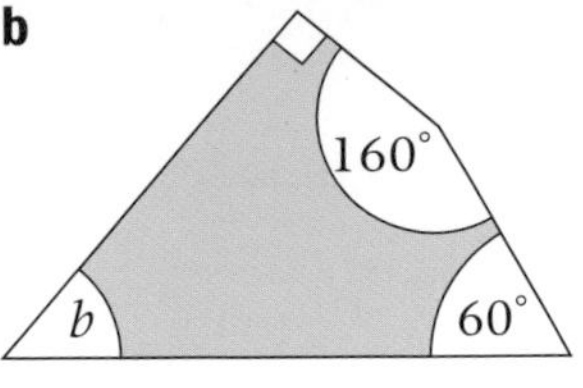

c

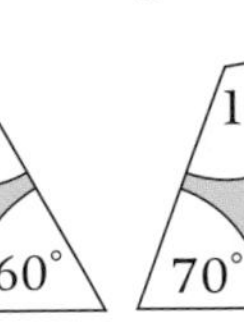

d

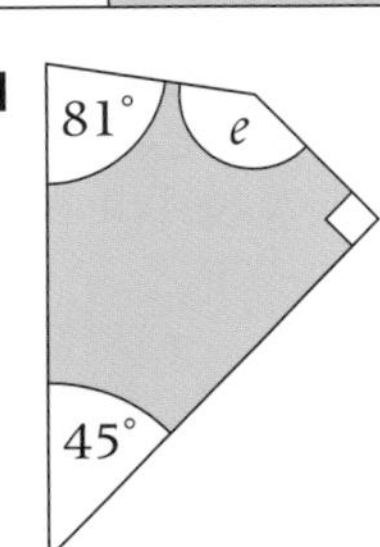

e

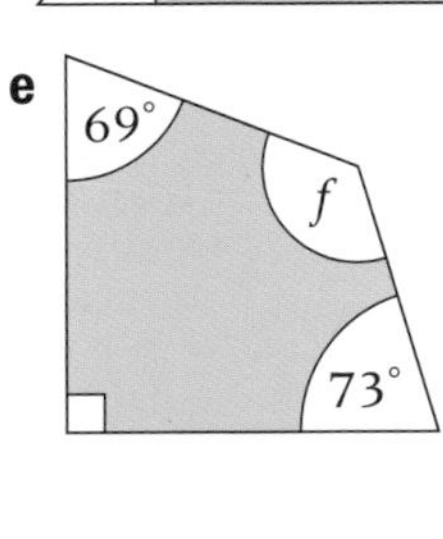

2 Find the unknown angles in each quadrilateral and state the type of quadrilateral.

a

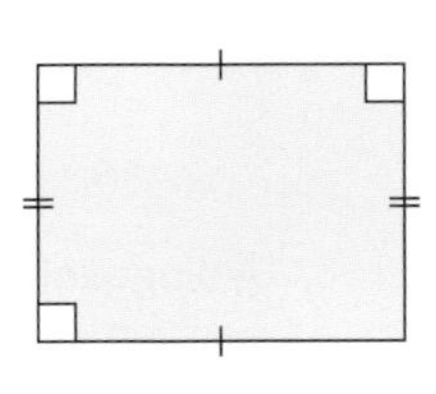

b

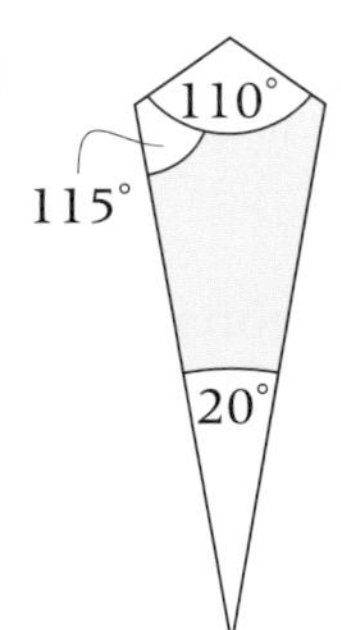

c

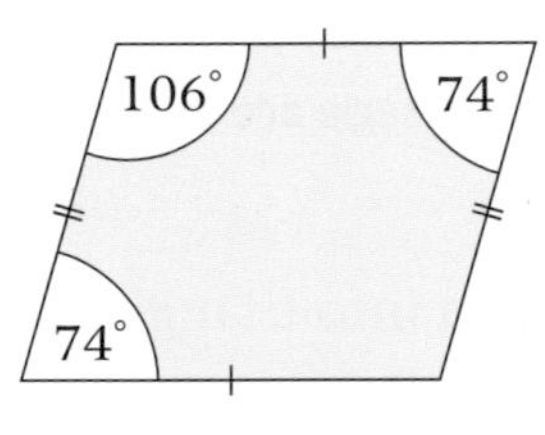

d

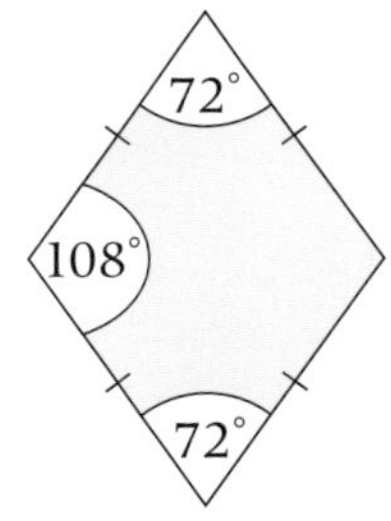

e

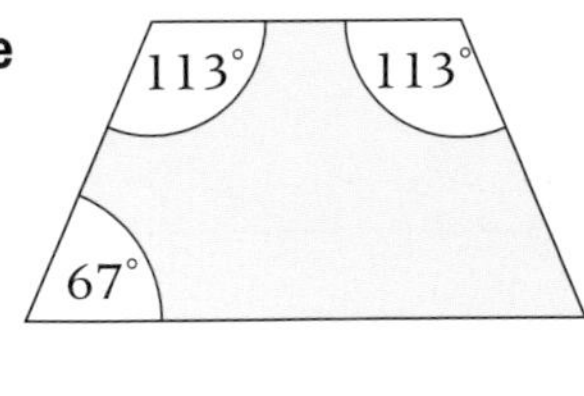

3 Calculate the value of x for each quadrilateral.

a

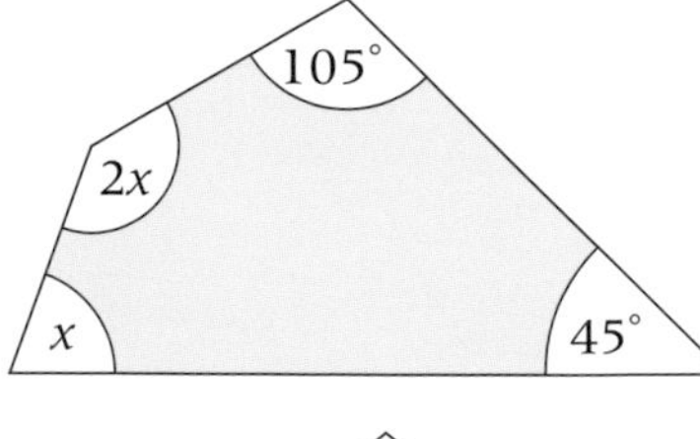

b

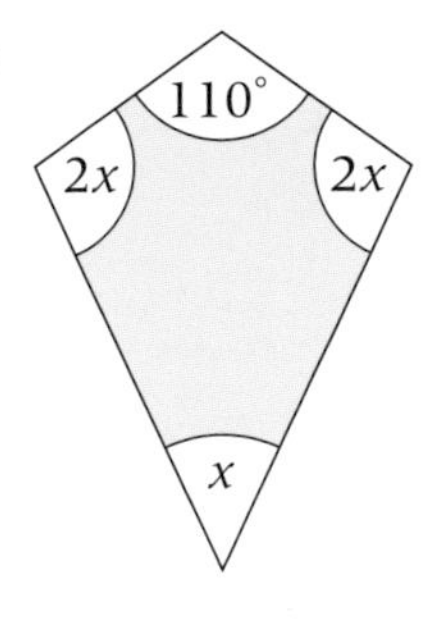

c

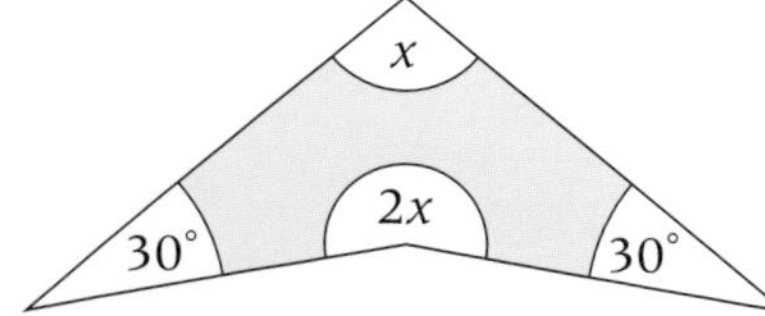

S2.3 Interior angles of a polygon

This spread will show you how to:

- Calculate and use interior angles of polygons

Keywords

Diagonal
Interior
Polygon
Regular
Vertices

A **polygon** is a 2-D shape with three or more straight sides.

- A **regular** shape has equal sides and equal angles.

A regular hexagon has 6 equal sides and 6 equal angles.

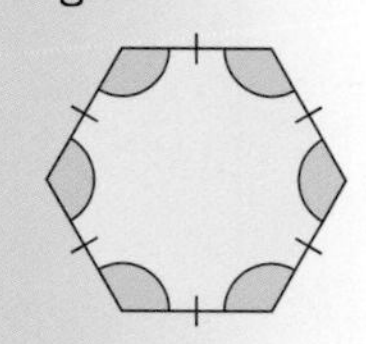

You should know the names of these polygons:

Sides	Name	Sides	Name
3	triangle	7	heptagon
4	quadrilateral	8	octagon
5	pentagon	9	nonagon
6	hexagon	10	decagon

- The angles inside a shape are called **interior** angles.

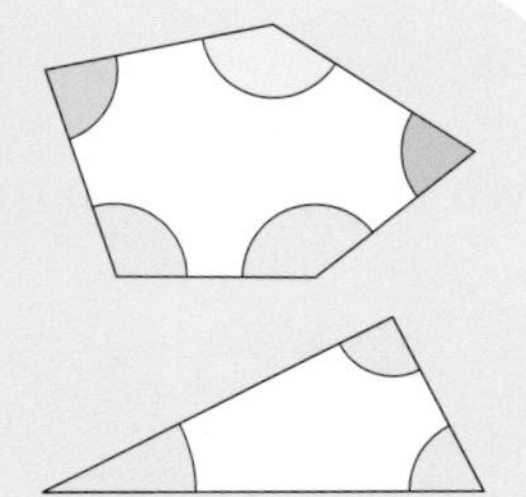

'interior' means inside.

- The interior angles in a triangle add to 180°.

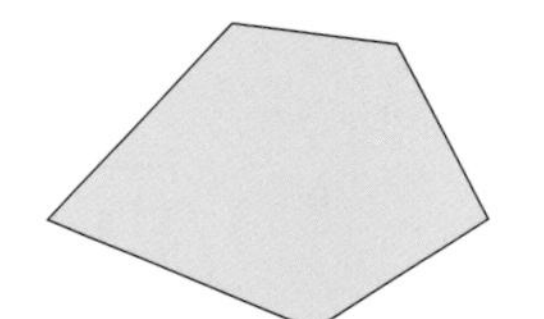

You can split a polygon into triangles by drawing diagonals from the same vertices.

A **diagonal** joins two vertices, but is not a side.

Example

Calculate the sum of the interior angles for a pentagon.

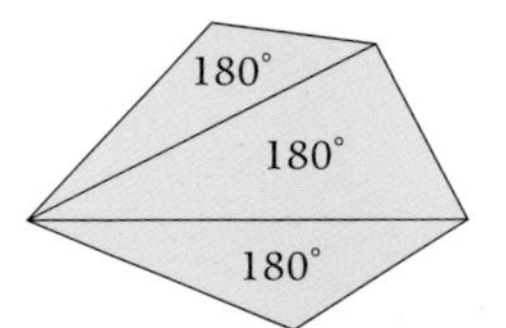

Draw in the two diagonals.
Three triangles formed: $3 \times 180° = 540°$
Sum of interior angles =
sum of all angles in each triangle = 540°

Example

Calculate the value of x in this regular hexagon.

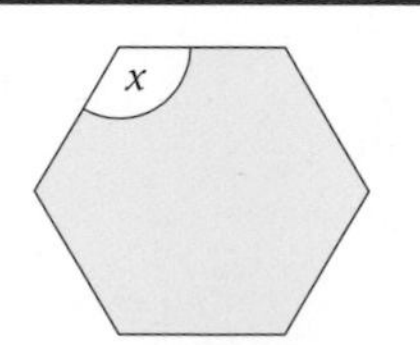

4 triangles
$4 \times 180° = 720°$
Sum of interior angles = 720°
There are 6 interior angles, so:
One interior angle, $x = 720° \div 6 = 120°$

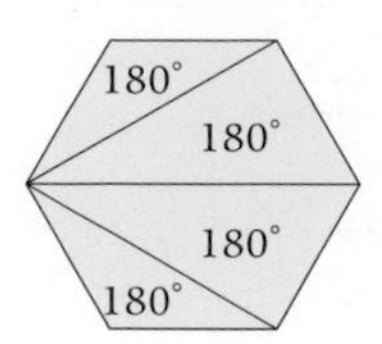

Exercise S2.3

1 a Calculate the value of one interior angle of an equilateral triangle.

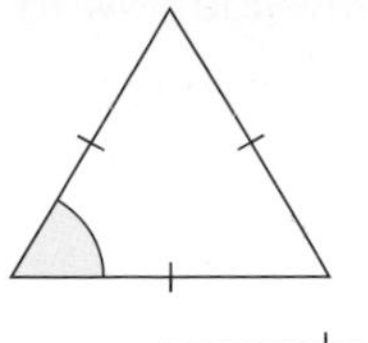

Two equilateral triangles are placed together to form a rhombus.

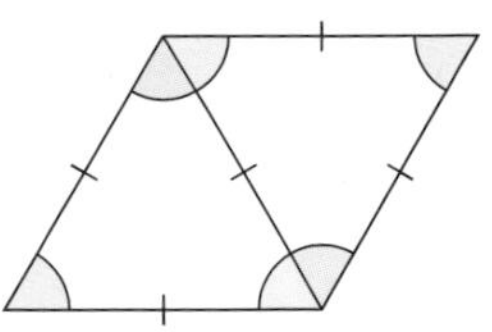

b Calculate the value of each interior angle of this rhombus.

c Calculate the sum of the interior angles of a rhombus.

2 Draw these polygons. Draw diagonals from **one** vertex.
(The quadrilateral is done for you.)
Copy and complete this table of results.

Number of sides	Number of triangles	Sum of the interior angles
3	1	180°
4	2	
5		
6		
7		
8		
9		
10		

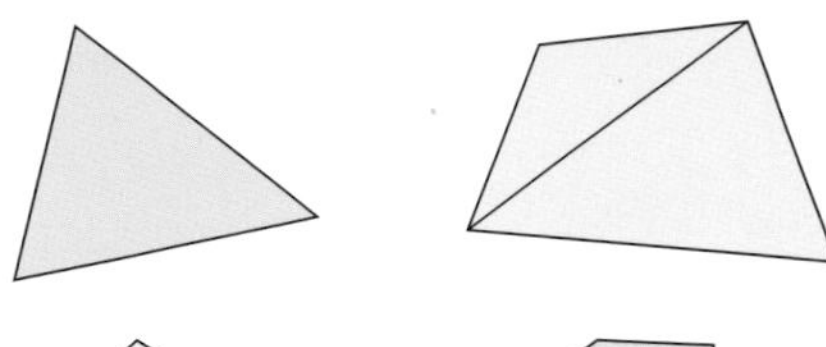

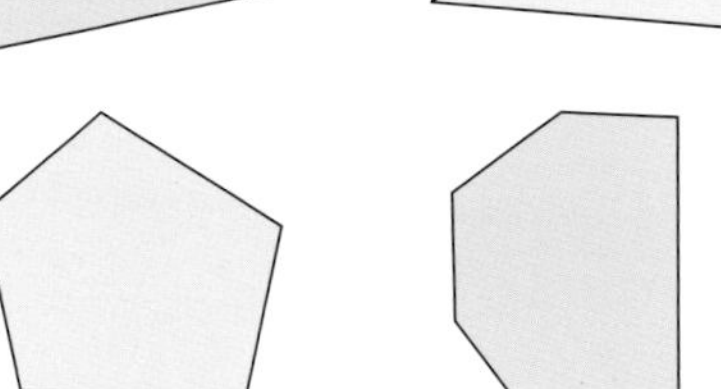

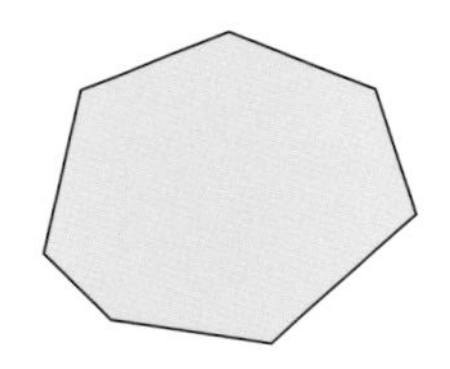

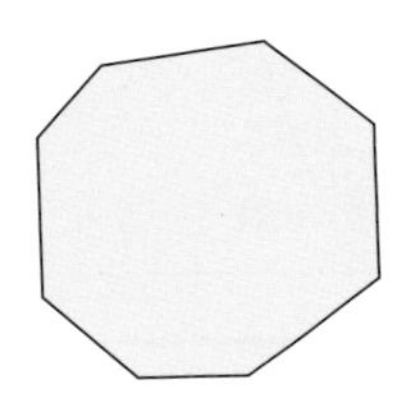

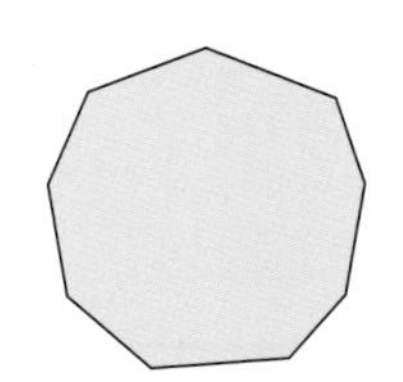

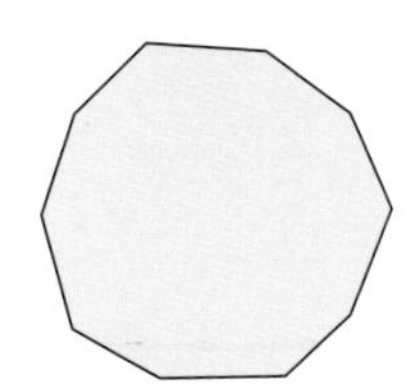

3 a Calculate the sum of the interior angles for a regular octagon.

b Calculate the value of one interior angle of a regular octagon.

c Copy and complete this table for regular polygons.

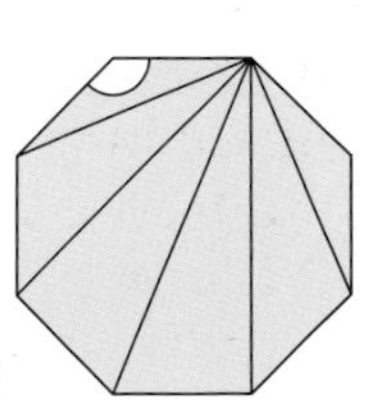

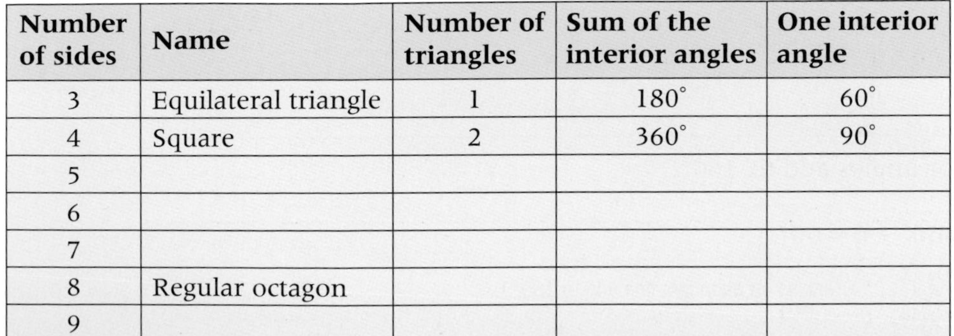

Number of sides	Name	Number of triangles	Sum of the interior angles	One interior angle
3	Equilateral triangle	1	180°	60°
4	Square	2	360°	90°
5				
6				
7				
8	Regular octagon			
9				
10				

S2.4 Exterior angles of a polygon

This spread will show you how to:

- Calculate and use interior and exterior angles of polygons

Keywords
Exterior
Interior
Polygon
Regular

The angles inside a shape are called interior angles.

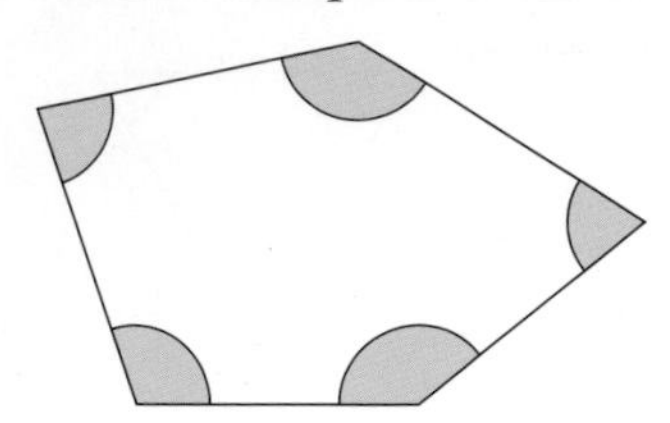

Interior = inside

You find the exterior angles of a polygon by extending each side of the shape in the same direction.

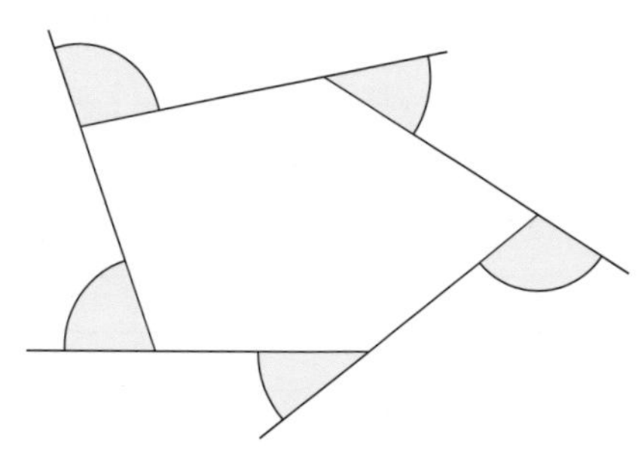

Exterior = outside

- **The exterior angles of any polygon add to 360°.**

Interior angle + exterior angle = 180°
(angles on a straight line add to 180°)

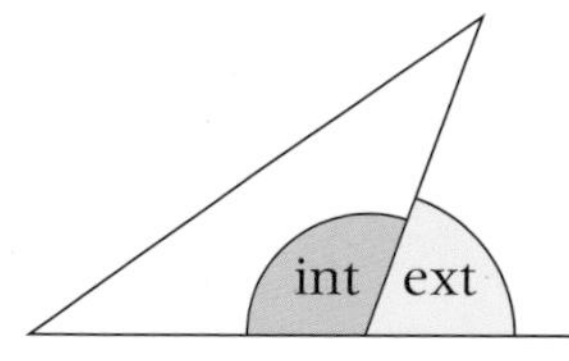

Example

Calculate the values of x and y in this regular hexagon.

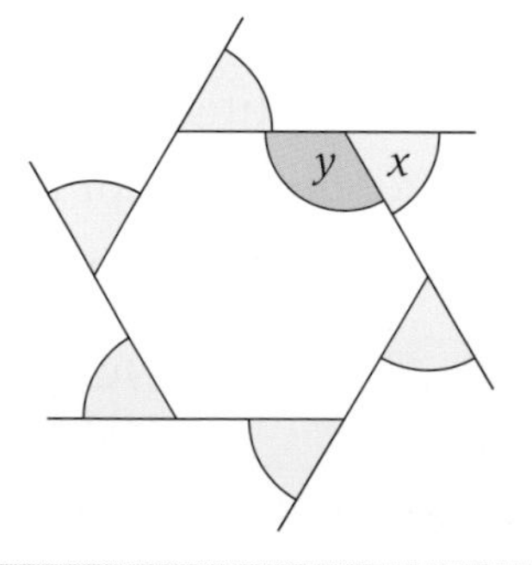

The six exterior angles add to 360°.

$x = 360° \div 6 = 60°$

$180° - 60° = 120°$ (angles on a straight line add to 180°)

$y = 120°$

Exercise S2.4

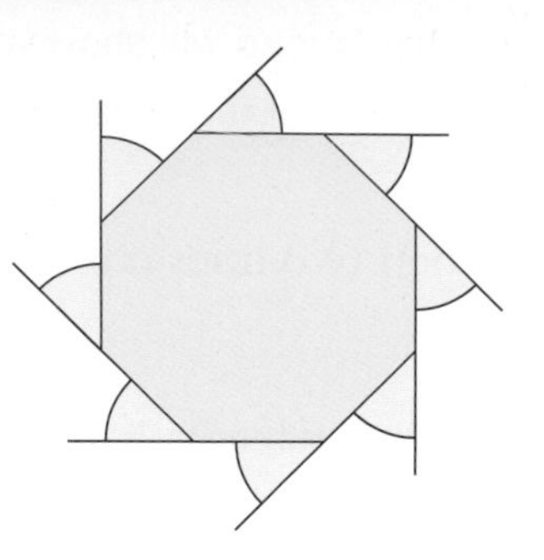

1 **a** State the total of the exterior angles of this regular octagon.

b Calculate the value of one of the exterior angles.

c Copy and complete this table of results for regular polygons.

Number of sides	Name	Sum of exterior angles	One exterior angle
3	Equilateral triangle		
4	Square		
5			
6			
7			
8	Regular octagon		
9			
10			

2 The interior angle of a regular polygon is 162°.

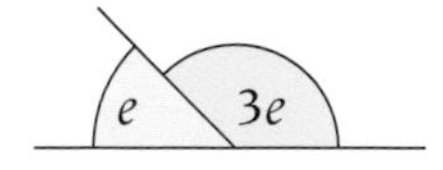

a Calculate the value of an exterior angle.

b State the sum of the exterior angles of the polygon.

c Calculate the number of exterior angles in the polygon.

d State the number of sides of the polygon.

3 A regular polygon has 15 sides.

a Calculate the value of an exterior angle.

b Calculate the value of an interior angle.

4 Calculate the size of the angles marked with letters in these polygons.

a

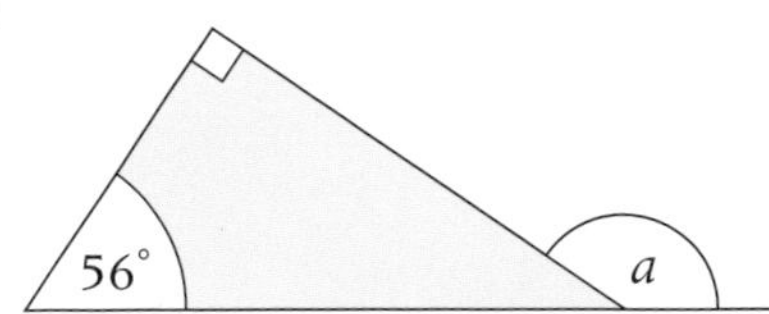

b

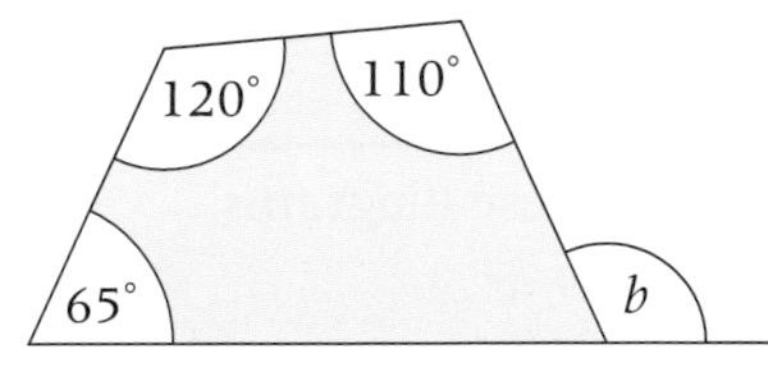

5 An interior angle of a regular polygon is three times the exterior angle.

e $3e$

a Calculate the value of each exterior angle.

b Calculate the value of each interior angle.

c Give the name of the regular polygon.

S2.5 Angles in parallel lines

This spread will show you how to:
- Use parallel lines, alternate angles and corresponding angles

Keywords
Alternate
Corresponding
Parallel
Vertically opposite

When two lines cross, four angles are formed.

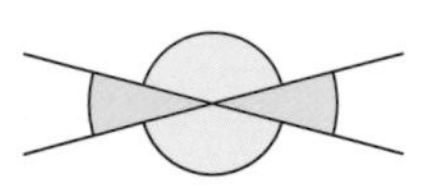

- **Vertically opposite** angles are equal.

When a line crosses two **parallel** lines, eight angles are formed.

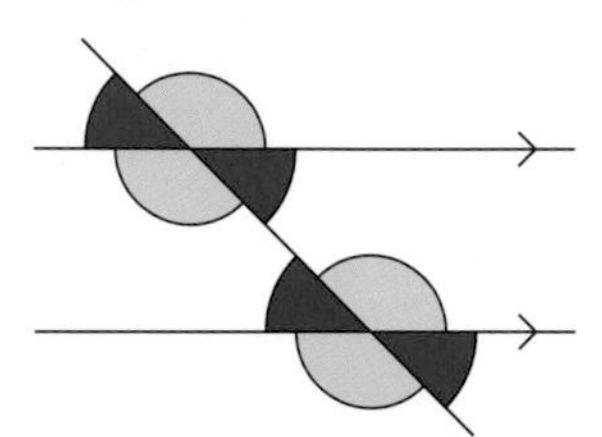

The four red **acute** angles are equal.

The four purple **obtuse** angles are equal.

Acute + obtuse = 180°

Parallel lines are always the same distance apart.

An acute angle is less than 90°.
An obtuse angle is more than 90° but less than 180°.

- **Alternate** angles are equal.

  They are called Z angles.

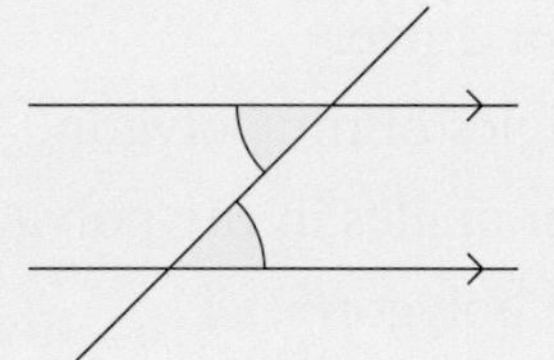

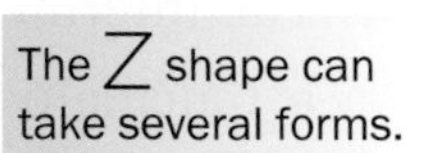

The Z shape can take several forms.

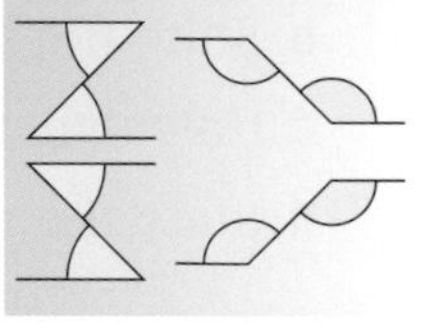

- **Corresponding** angles are equal.

  They are called F angles.

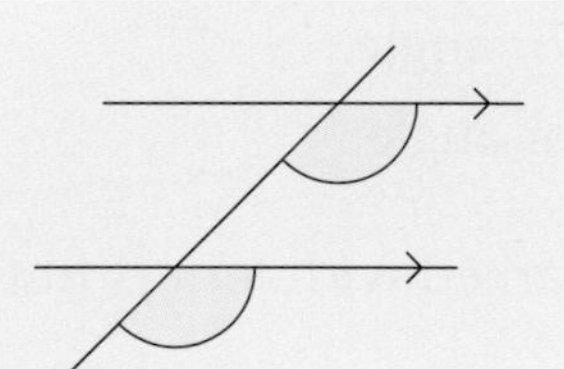

The F shape can take several forms.

Example

Find the unknown angles in these diagrams.
Give reasons for your answers.

a

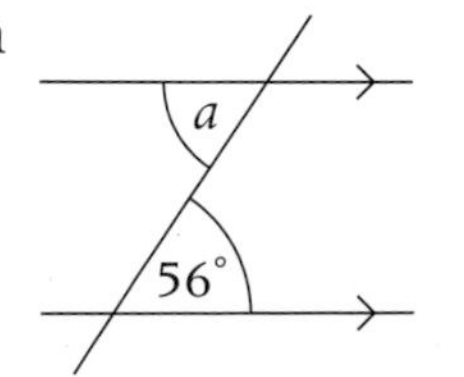

b

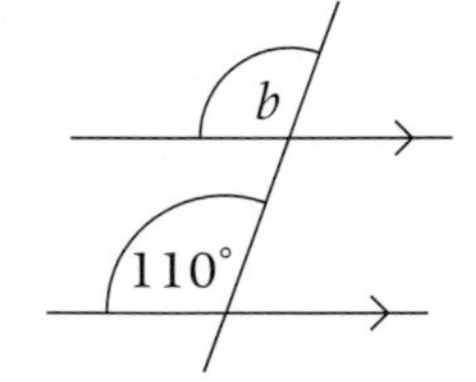

c

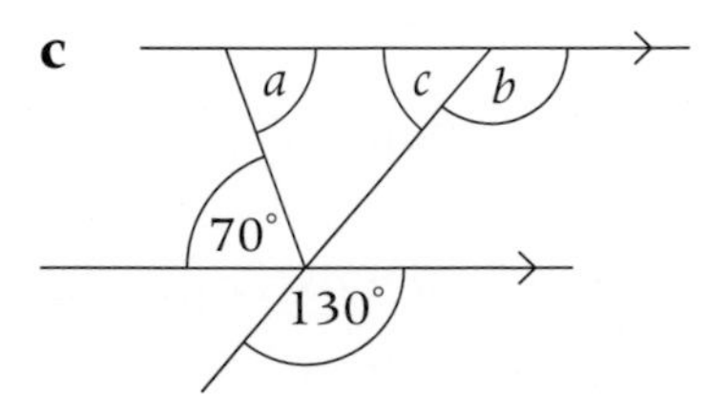

a $a = 56°$
(alternate angles)

b $b = 110°$
(corresponding angles)

c $a = 70°$ (alternate angles)
$b = 130°$ (corresponding angles)
$c = 180° - 130°$
$= 50°$ (angles on straight line add to 180°)

Exercise S2.5

1 Calculate the size of the angles marked by a letter in each diagram. Give a reason for each answer.

> The diagrams are not drawn accurately.

a

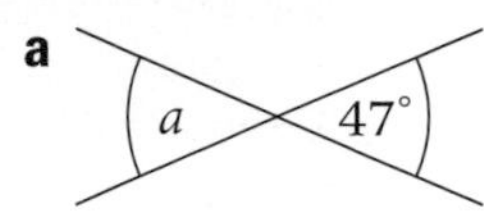

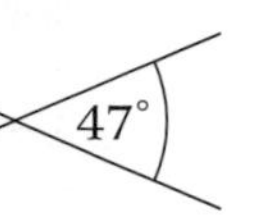

b

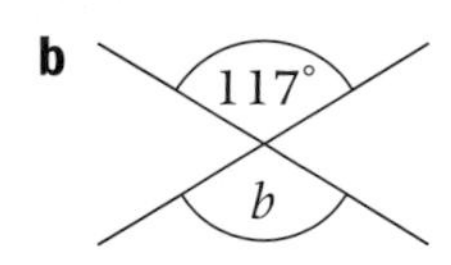

c

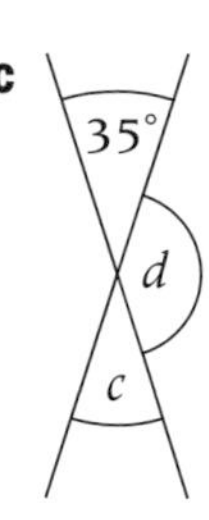

d

103° e
f d

e

g 2g
2g g

2 Find the value of each angle marked with a letter. Give a reason for each answer.

a

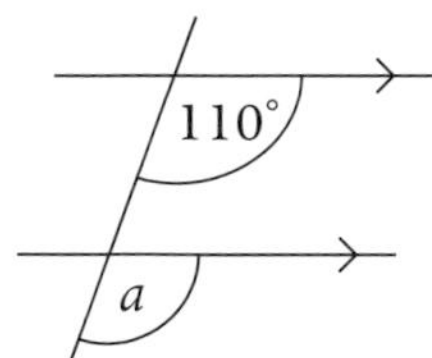

b

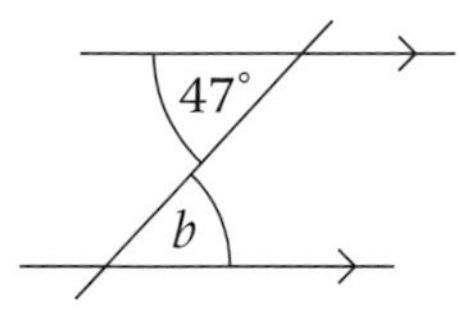

c

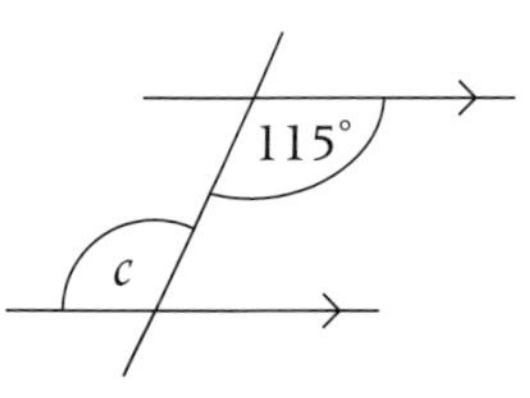

d

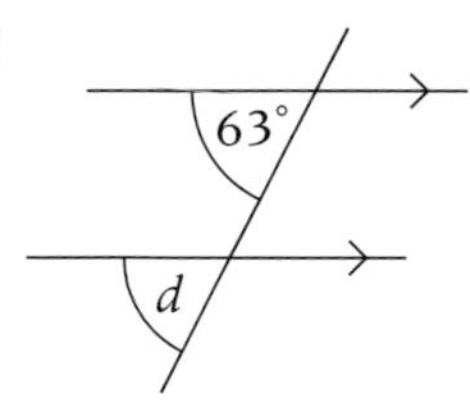

e

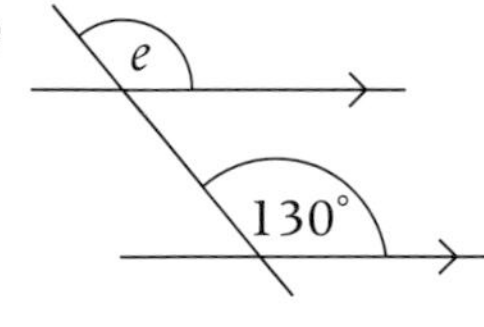

f

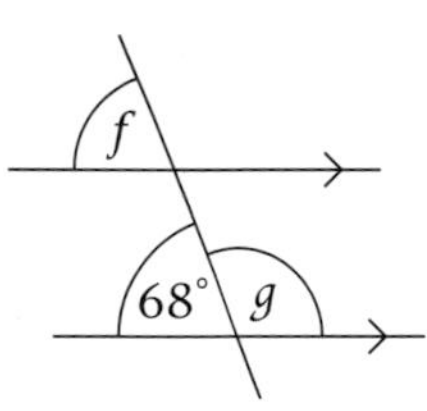

g

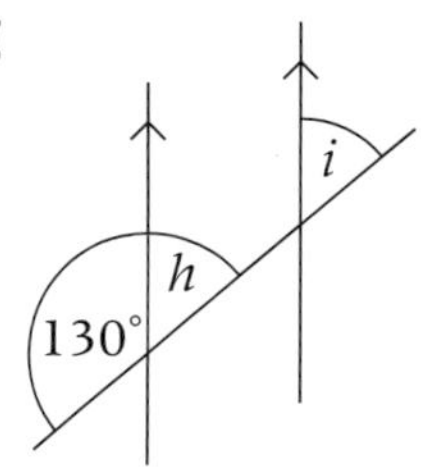

h

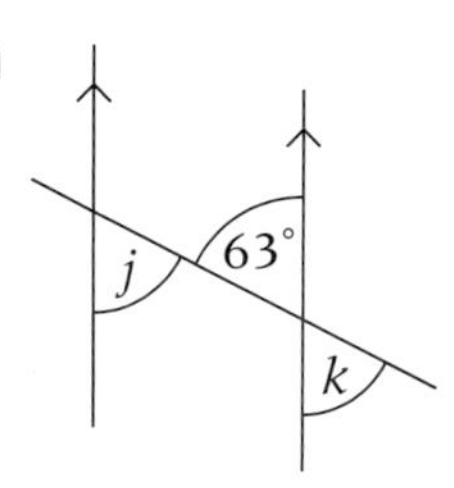

i

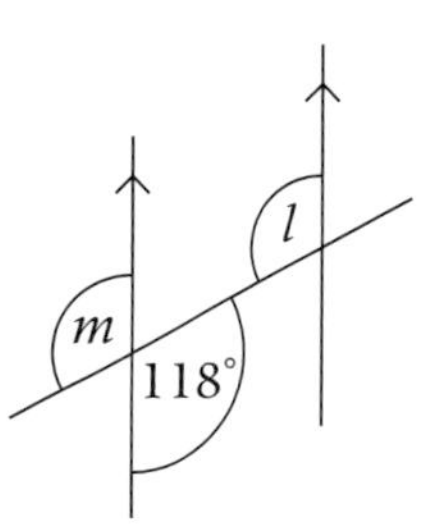

3 Find the value of each angle marked with a letter. Give a reason for each angle.

a

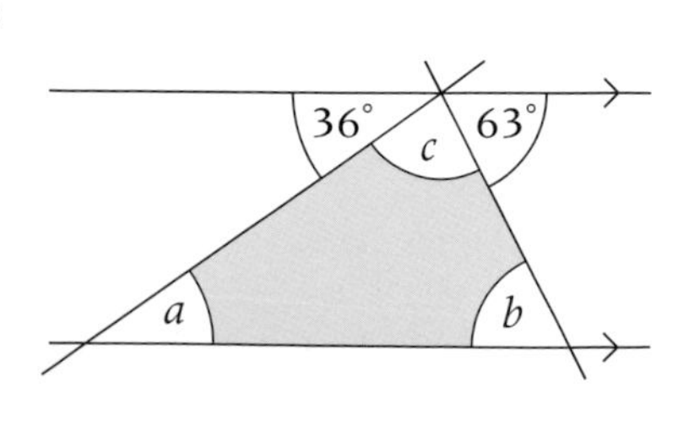

b

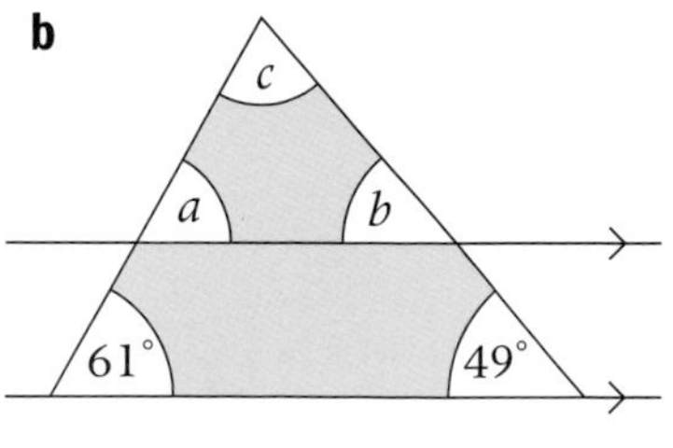

c

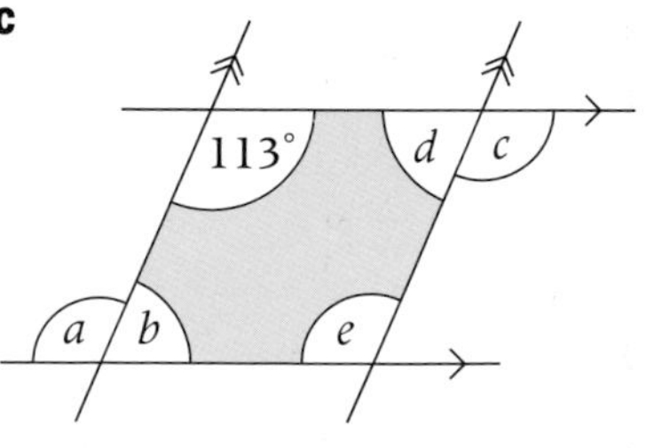

S2.6 Using parallel lines

This spread will show you how to:

- Understand and use angle properties

Keywords

Alternate angles
Corresponding angles
Exterior angle
Interior angle
Parallel
Parallelogram
Proof

When a line crosses **parallel** lines,

- **Alternate angles** are equal
- **Corresponding angles** are equal

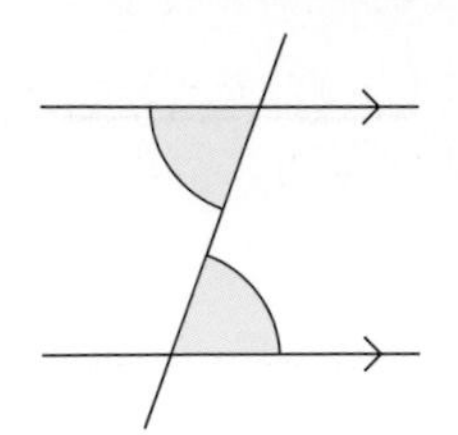

These are Z angles

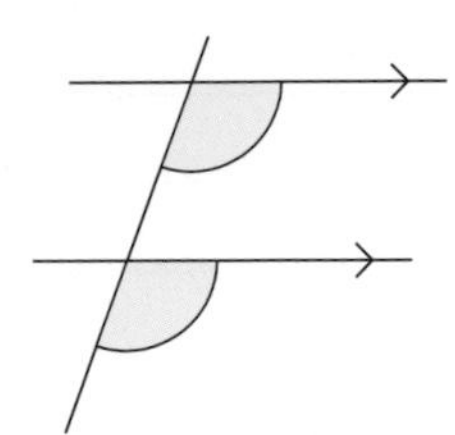

These are F angles

You can use these parallel line angle properties for three **proofs**.

Proof 1

The **exterior angle** of a triangle is equal to the sum of the **interior angles** at the other two vertices.

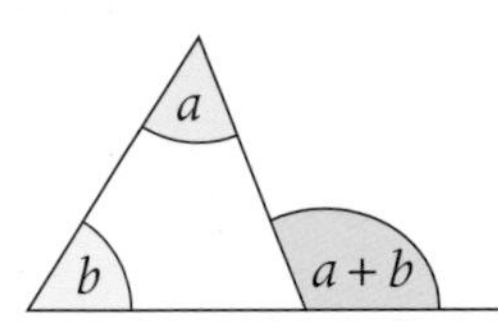

The exterior angle is formed by extending a side.

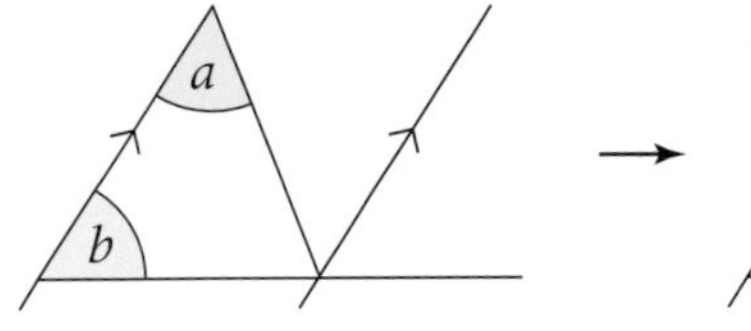

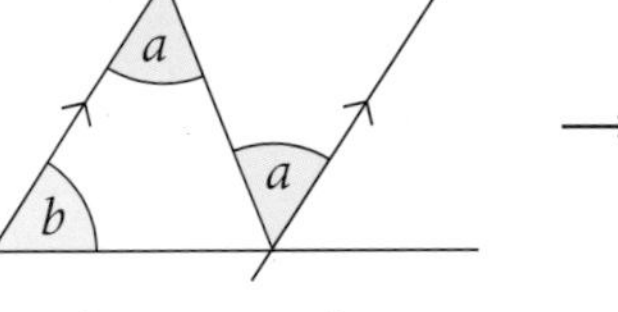

alternate angles

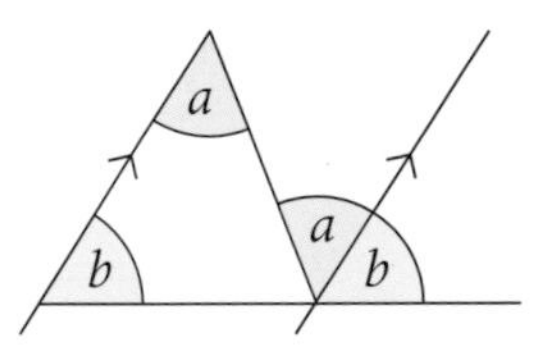

corresponding angles

Proof 2

The sum of the interior angles of a triangle is 180°.

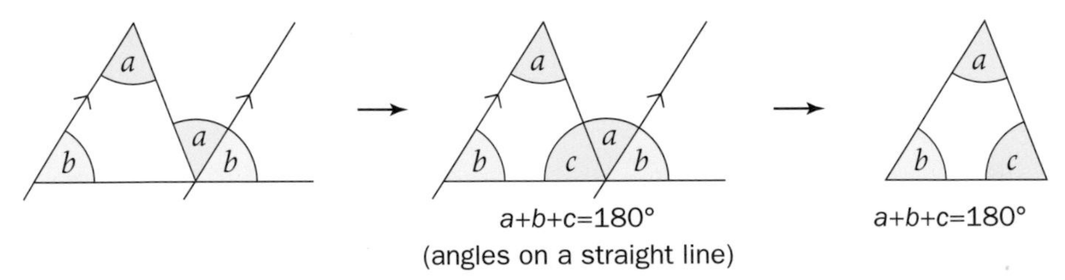

$a+b+c=180°$
(angles on a straight line)

$a+b+c=180°$

Proof 3

The opposite angles of a **parallelogram** are equal.

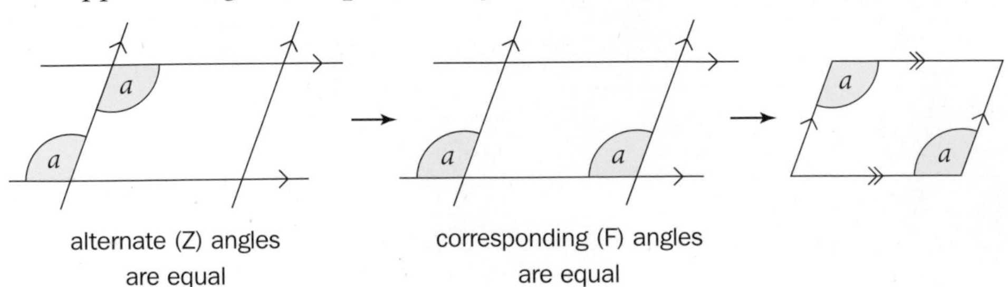

alternate (Z) angles are equal

corresponding (F) angles are equal

Exercise S2.6

1 Find the value of the angles marked by a letter. Give a reason for each answer.

a

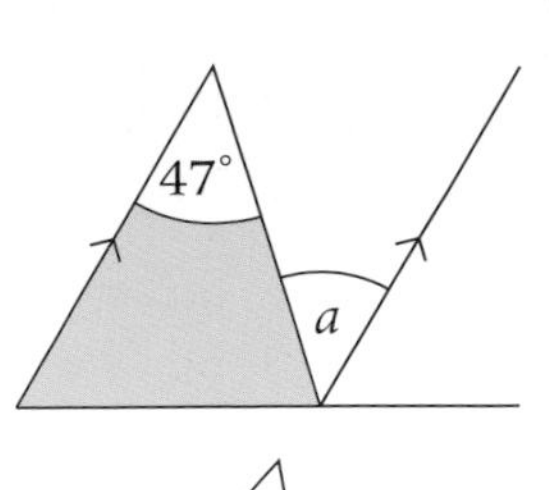

b

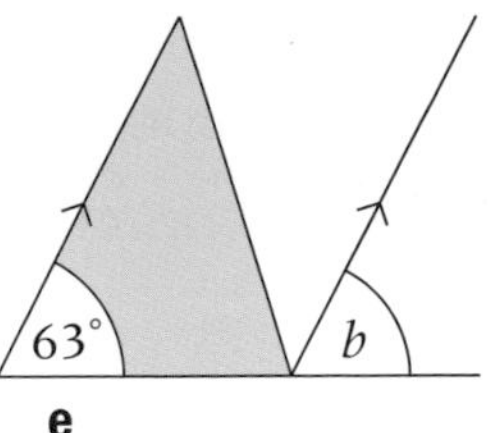

c

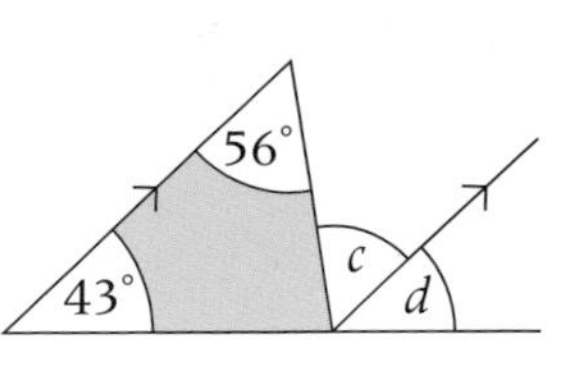

d

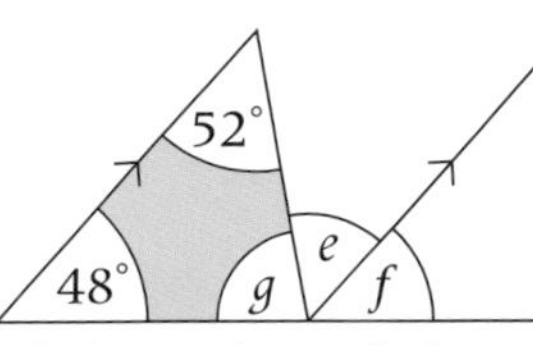

e

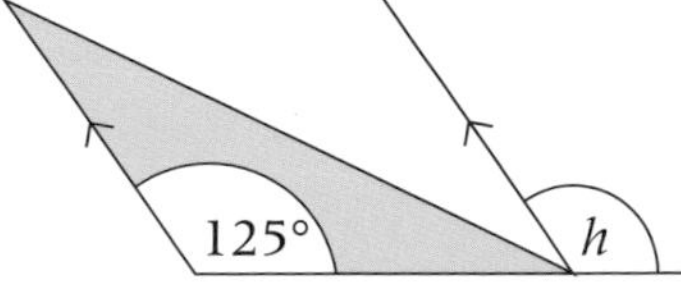

2 Find the value of the angles marked by a letter.

a

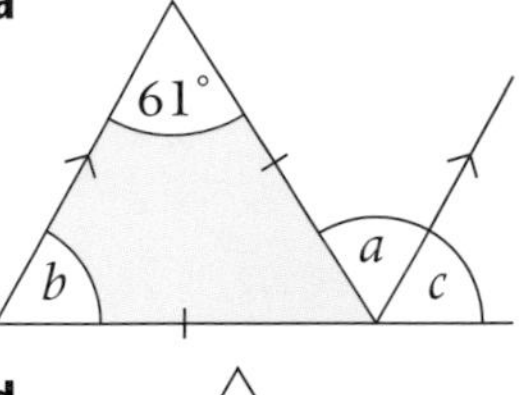

b

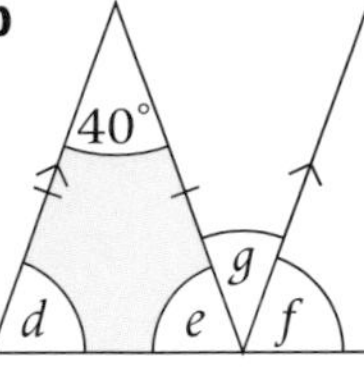

c

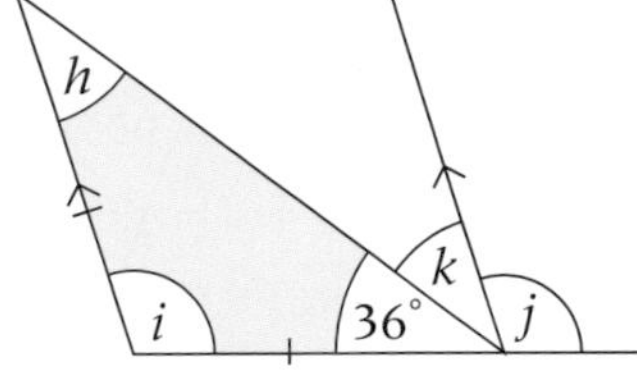

d

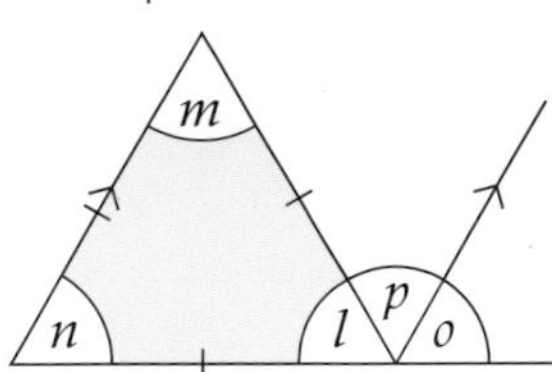

e

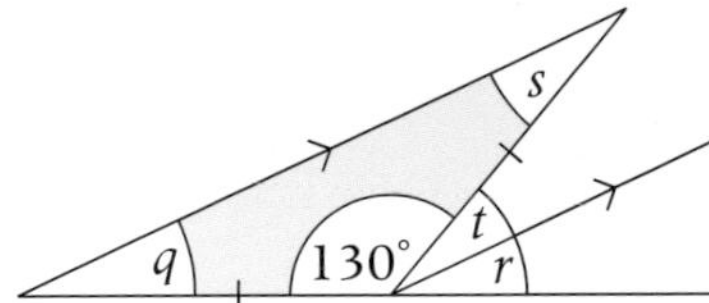

3 Find the value of the angles marked by a letter.

e
d b a
c
108°
112°

4 Find the value of the angles marked by a letter.

a

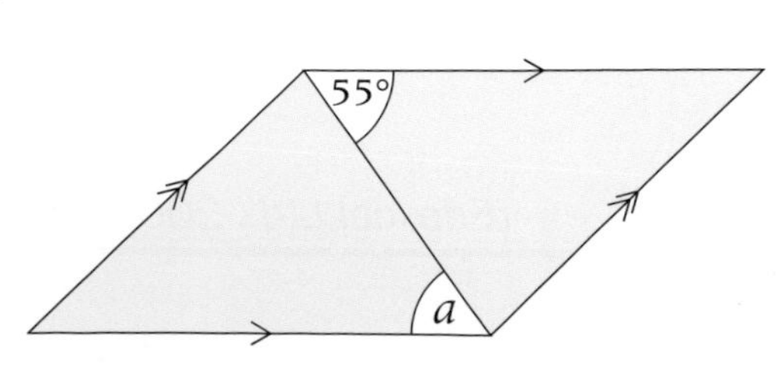

b

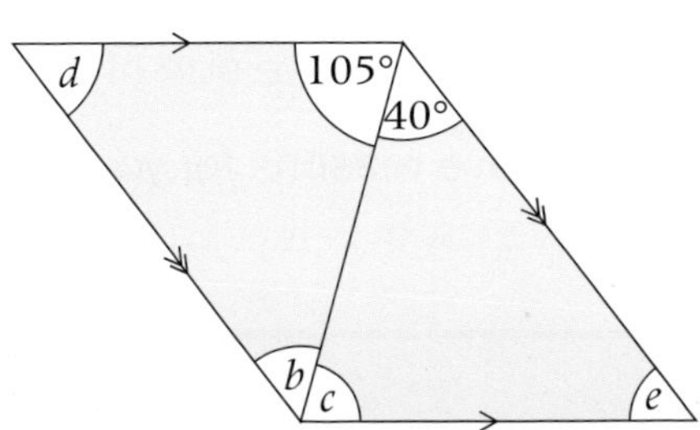

S2 Exam review

Key objectives

- Use parallel lines, alternate angles and corresponding angles
- Calculate and use the sums of the interior and exterior angles of polygons

1

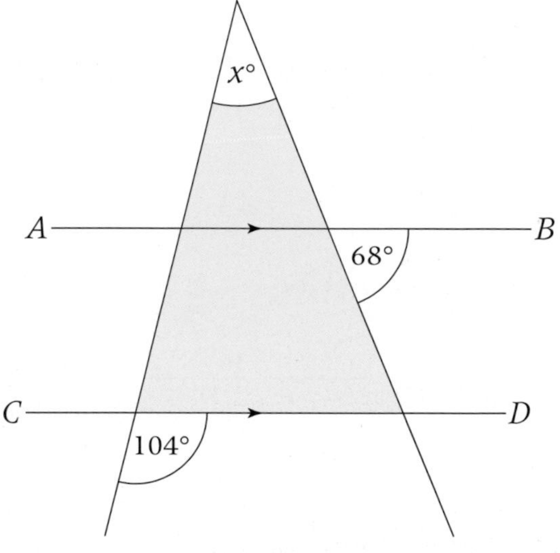

Diagram **not** to scale.

In the diagram *AB* is parallel to *CD*.
Find the value of x. (2)

(OCR, 2003)

2

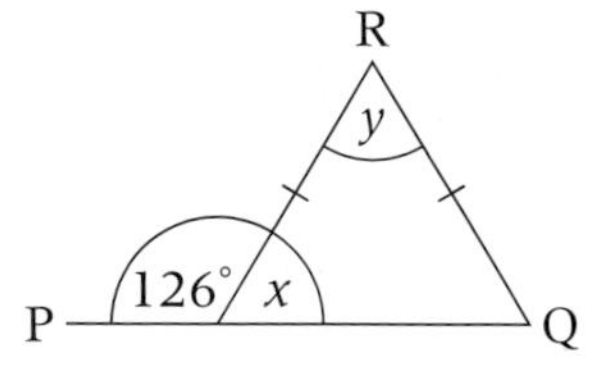

PQ is a straight line.

a Work out the size of the angle marked x. (1)

b i Work out the size of the angle marked y. (3)

ii Give reasons for your answer.

(Edexcel Ltd., 2003)

S3 Transformations

This unit will show you how to

- Understand, recognise and describe reflections, rotations and translations
- Transform triangles and other 2-D shapes by reflection, rotation, translation and a combination of these transformations
- Recognise, visualise and construct enlargements of objects
- Understand that enlargements are specified by a centre and positive scale sector
- Recognise reflection symmetry of 2-D and 3-D shapes and rotational symmetry of 2-D shapes

Before you start ...

You should be able to answer these questions.

1 Give the coordinates of

a A **b** B

c C **d** D

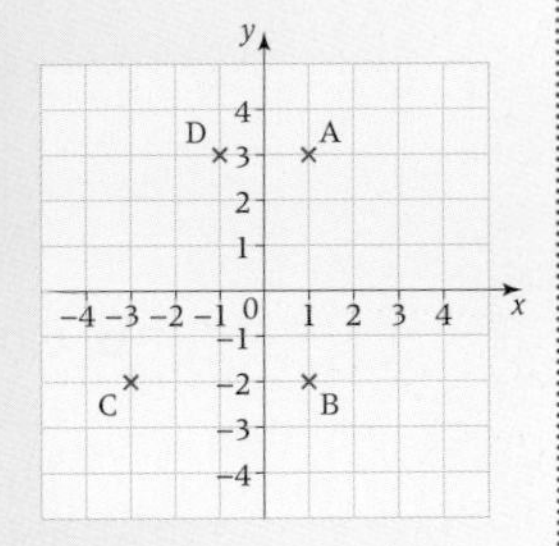

2 State the direction of the turn.

a **b**

3 Give the equation of each straight line.

a

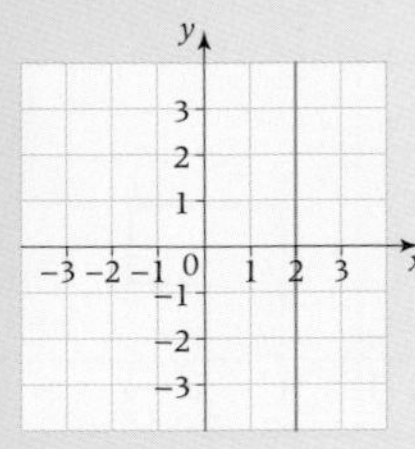

b

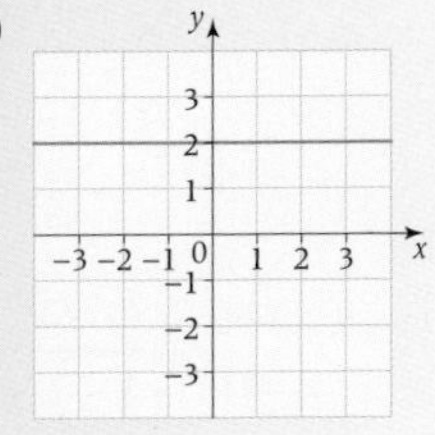

c

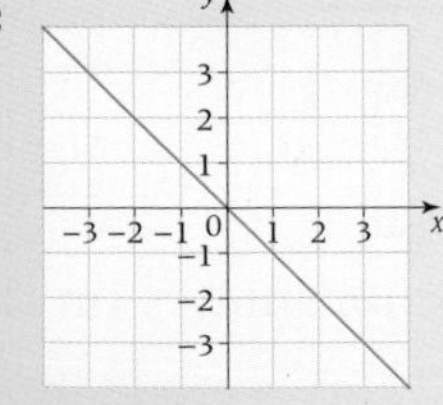

d

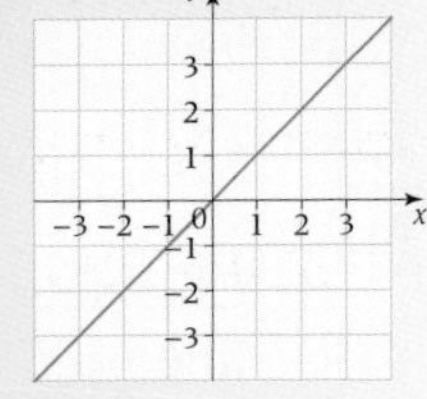

4 State the sum of the three angles a, b, c.

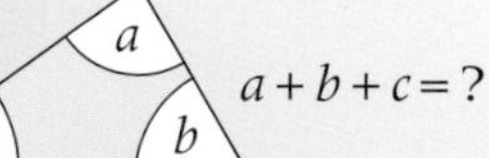

S3.1 Reflections

This spread will show you how to:

- Understand, recognise and describe reflections

Keywords
Equation
Equidistant
Flip
Mirror line
Reflection

A reflection **flips** a shape over.

- **You describe a reflection using the mirror line or reflection line.**

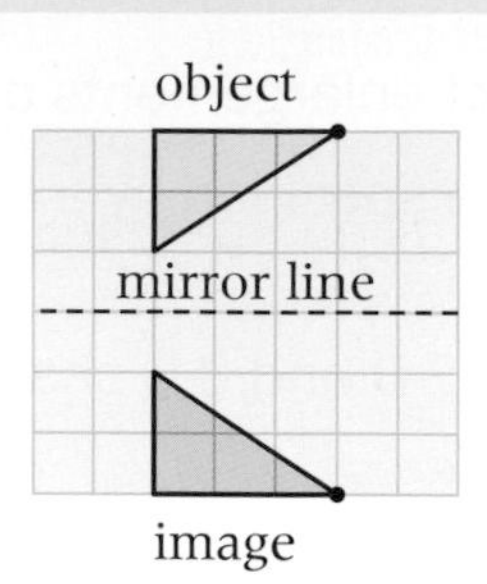

Choose a point on the object to find the corresponding point on the image.

Corresponding points are **equidistant** from the mirror line.

Equidistant means the same distance.

Example

a Draw a mirror line so that shape B is a reflection of shape A.
b Give the **equation** of the mirror line.

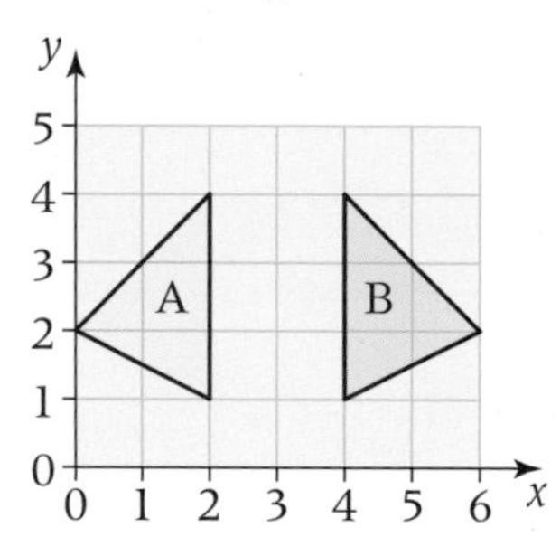

a

b Choose some coordinates on the line: (3, 0) (3, 1) (3, 2).
The x-coordinates are all 3.
The line is $x = 3$.

Example

a Draw the reflection of the shaded shape in the mirror line.
b Give the equation of the mirror line.

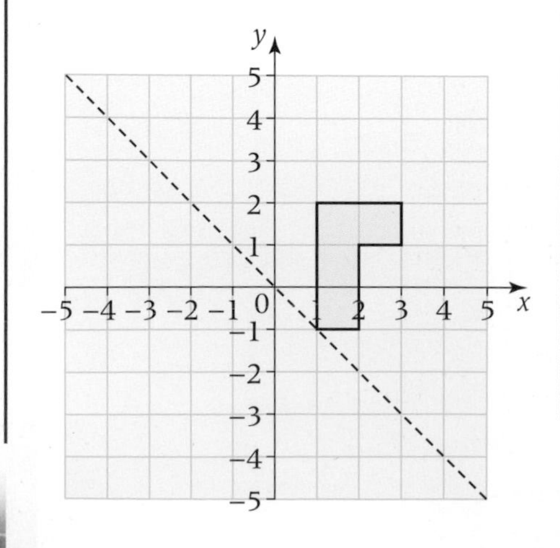

a

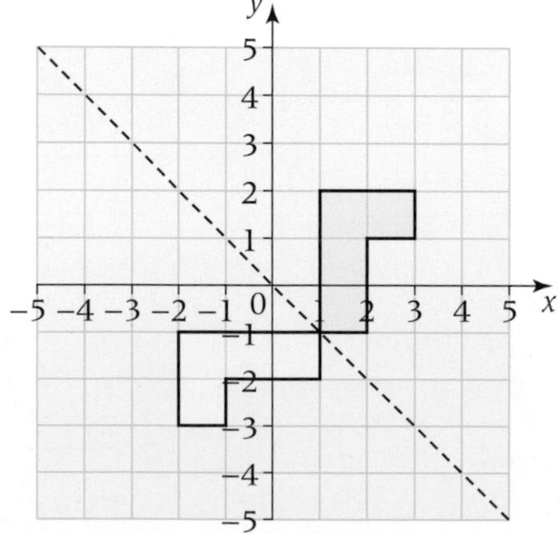

b Choose some coordinates on the line:
(−2, 2), (3, −3), (5, −5)
The line is $y = -x$

Exercise S3.1

1 Copy and complete the diagrams to show the reflections of the triangles in the mirror line $x = 2$.

a

b

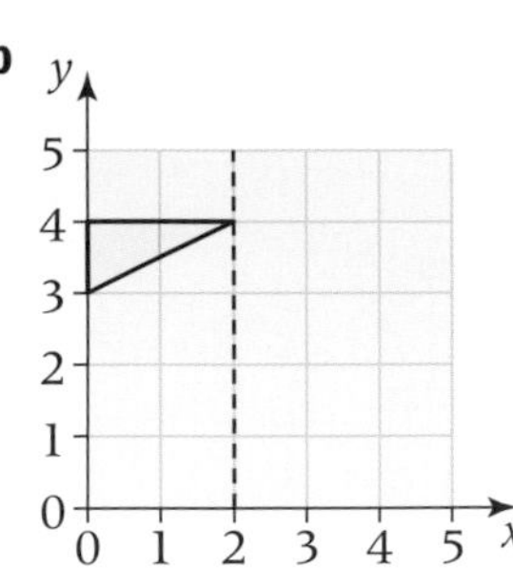

c

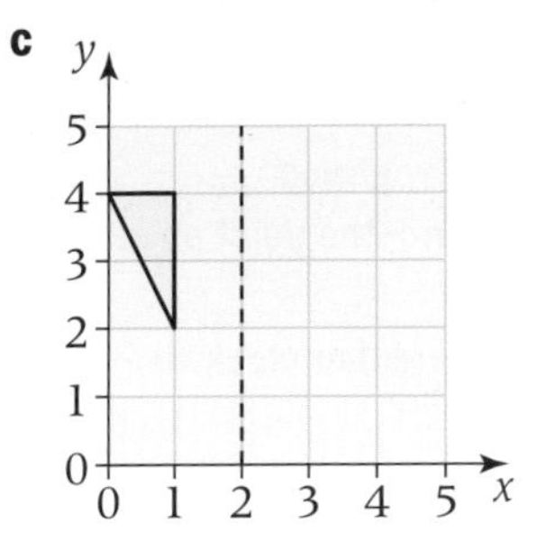

2 Give the equation of the mirror line for each reflection.

a

b

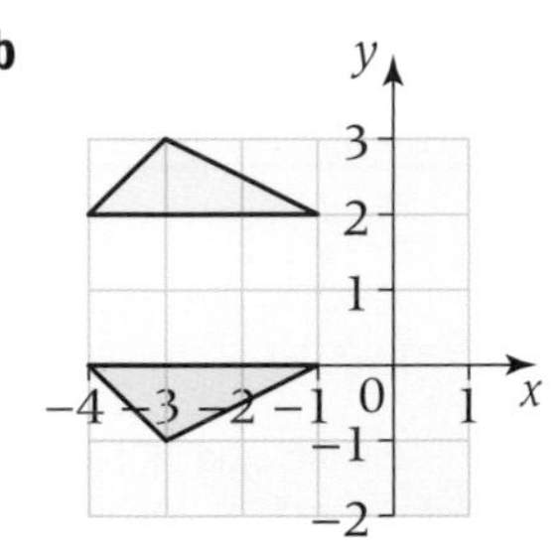

c

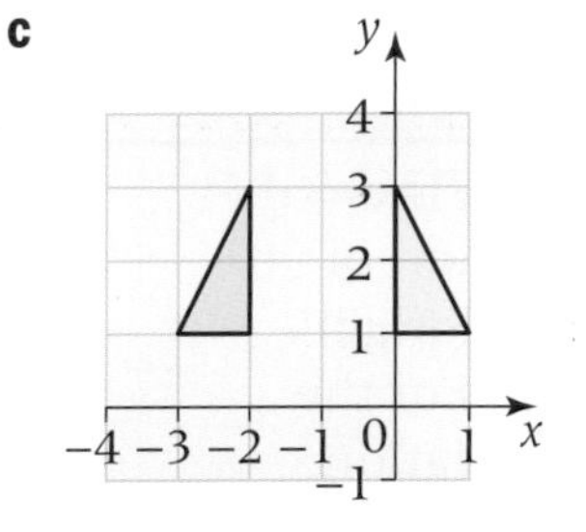

3 **a** Copy the diagram.

b Reflect the triangle in the mirror line.

c Give the equation of the mirror line.

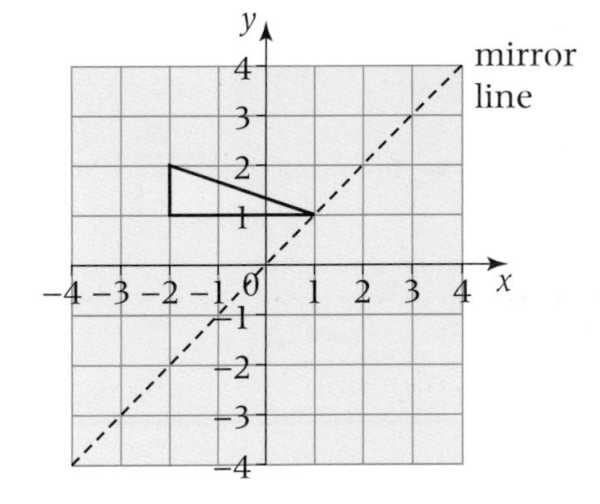

4 **a** Copy the diagram.

b Plot the points (0, 1) (3, 1) (3, 2) to form a triangle.

c Reflect the triangle in the mirror line.

d Give the coordinates of the reflected points.

e Give the equation of the mirror line.

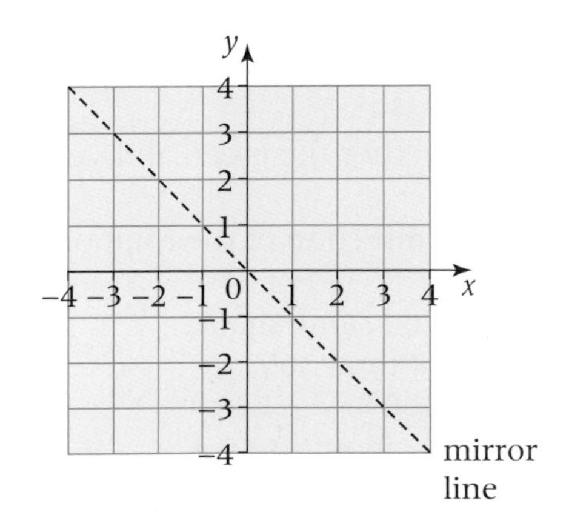

S3.2 Rotations

This spread will show you how to:

- Transform triangles and other 2-D shapes by rotation

Keywords
Anticlockwise
Centre of rotation
Clockwise
Congruent
Rotation

A **rotation** turns a shape about a fixed point.

- You describe a rotation by giving:
  - the **centre of rotation** – the point about which it turns
  - the angle of turn
  - the direction of turn – either **clockwise** or **anticlockwise**.

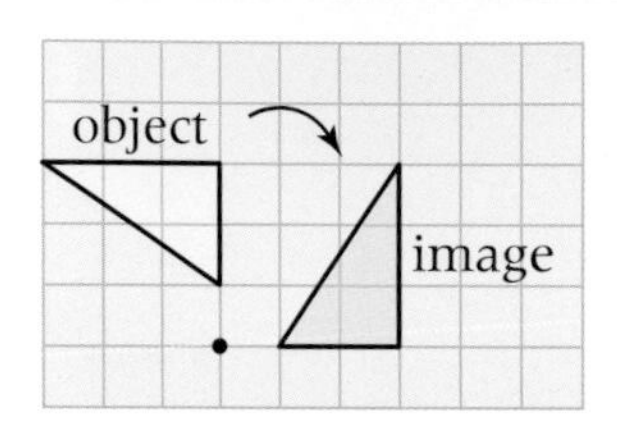

The dot is the centre of rotation.
The turn is 90°.
The direction is clockwise.

The two shapes are **congruent**. They are exactly the same size and the same shape.

Example

a Draw the position of the red triangle after a rotation of 90° clockwise about the origin.
b Give the coordinates of the point A **after** the rotation.

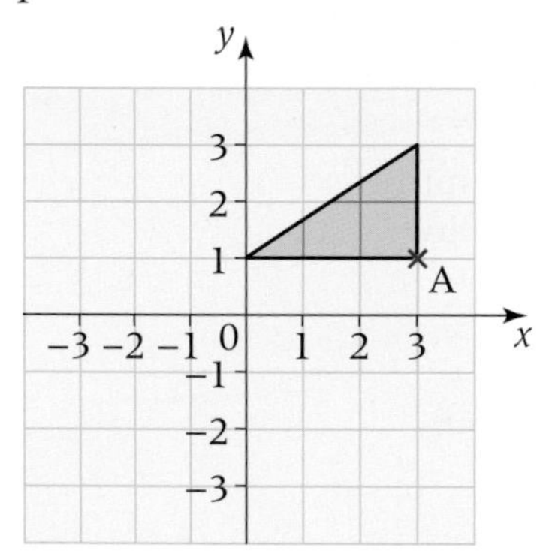

a

centre of rotation

A

b (1, −3)

The origin is the point (0, 0).

Use tracing paper to find the position of the blue triangle.

Example

A regular pentagon is divided into five isosceles triangles. The centre of the pentagon is marked with a dot (•). The green triangle is rotated about the dot onto the yellow triangle.

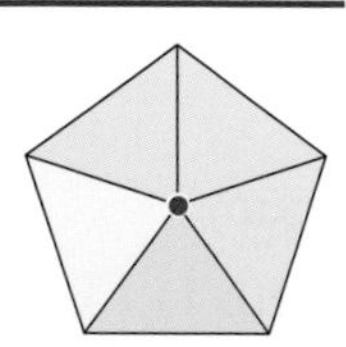

a State, with a reason, whether the green and yellow triangles are congruent.
b Calculate the angle and direction of the rotation.

a Congruent – same size and same shape.
b The five angles at the dot total 360°.
One angle at the dot is 360° ÷ 5 = 72°.
Rotation is 72° clockwise about the dot.

Exercise S3.2

1 State the angle and direction of turn for each of these rotations about the dot (●), green shape to blue shape:

2 The diagram shows the pattern made by repeated 90° rotations of a right-angled triangle about the dot (●).
Draw a similar pattern using repeated 45° rotations about the dot (●).

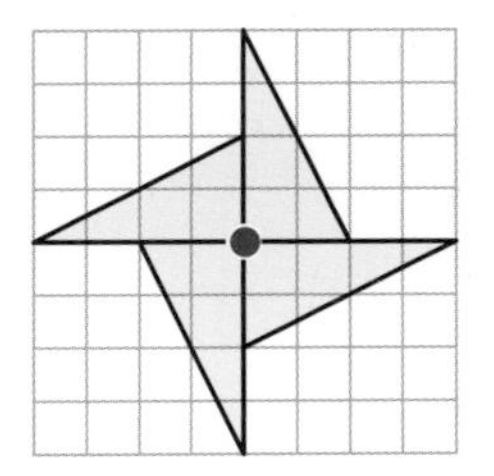

3 Copy this grid.

a Plot and join the points O (0, 0), A (2, 1), B (3, 0) and C (2, −1).

b Give the mathematical name of this shape.

c Rotate the shape through 90° anticlockwise about the origin.

d Give the coordinates of the point A after the rotation.

e Are the two shapes congruent?

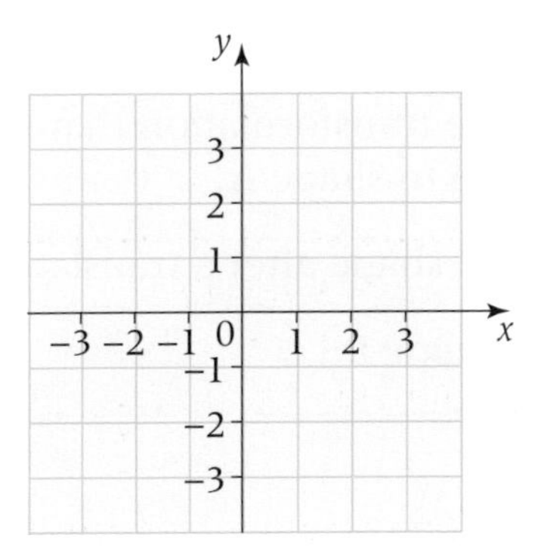

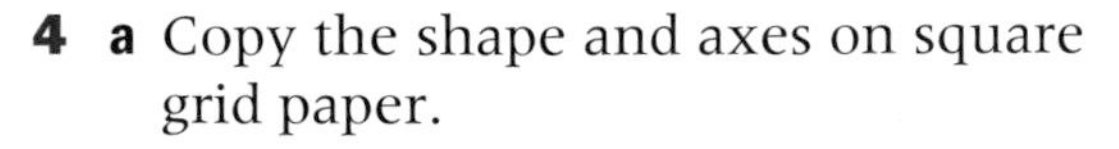

4 **a** Copy the shape and axes on square grid paper.

b Rotate the triangle by 90° anticlockwise about the point (1, 2).

c Give the coordinates of the vertices of the triangle after the rotation.

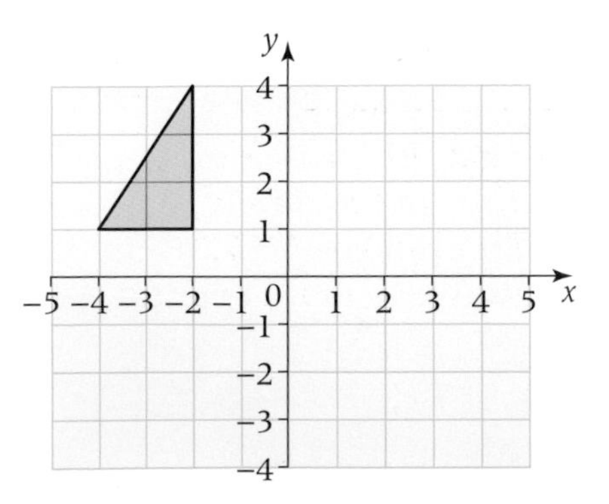

S3.3 Translations

This spread will show you how to:

- Understand, recognise and describe translations

Keywords
Congruent
Slide
Translation
Vector

A **translation** is a **sliding** movement.

To move the green shape to the blue shape, you translate the object 1 unit right and 3 units down.

You can write this translation as a **vector**, $\begin{pmatrix} 1 \\ -3 \end{pmatrix}$.

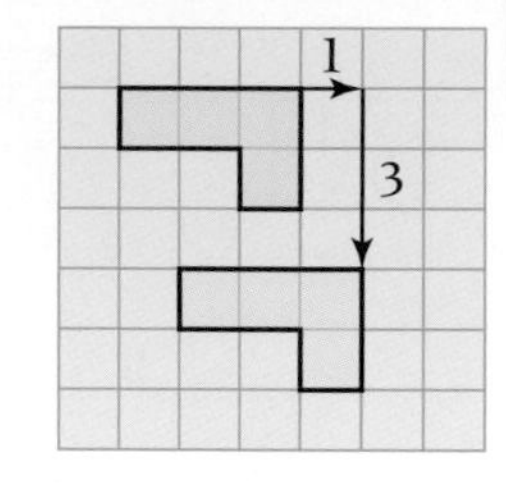

- To describe a translation you give
  - the distance moved right or left, then — right = positive, left = negative
  - the distance moved up or down. — up = positive, down = negative
- You can use a **vector** to describe the translation.

The two shapes are **congruent**.

Choose a point on the object to locate the corresponding point on the image.

This is a translation of $\begin{pmatrix} 3 \\ 1 \end{pmatrix}$.

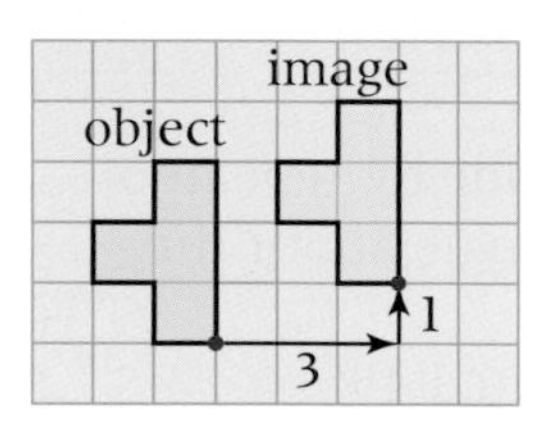

Congruent shapes are the same size and shape.

Example

a Give the mathematical name of the shaded shape.
b Describe fully the transformation that moves the shaded shape to shape A.
c Draw the shaded shape after a translation of $\begin{pmatrix} 3 \\ -1 \end{pmatrix}$.

Label the new shape B.

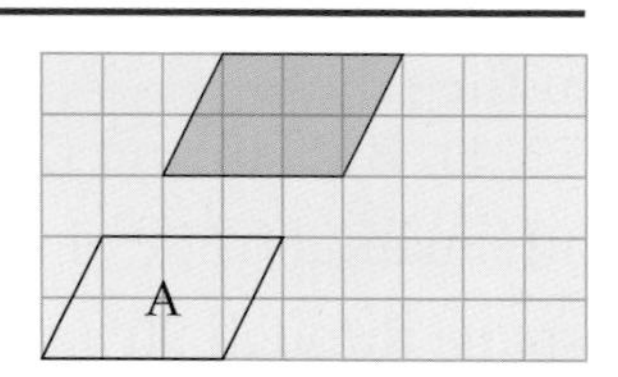

---

a Parallelogram

b Translation of $\begin{pmatrix} -2 \\ -3 \end{pmatrix}$.

c

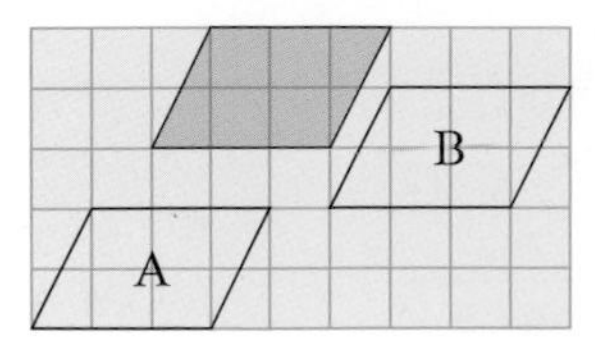

Exercise S3.3

1 Which shapes are translations of the green shape?

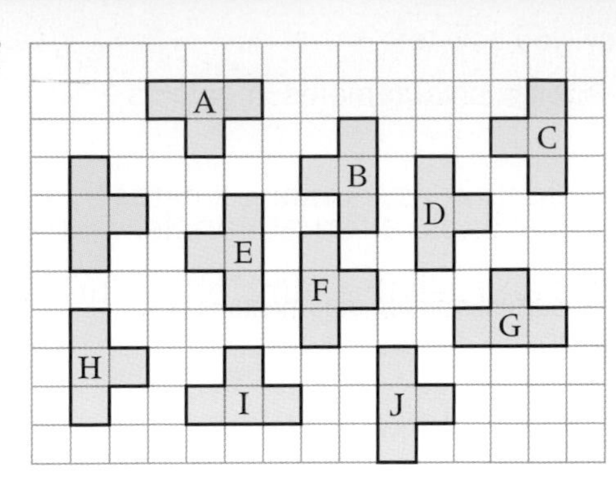

2 Describe the translation that moves the green triangle to the other triangles.

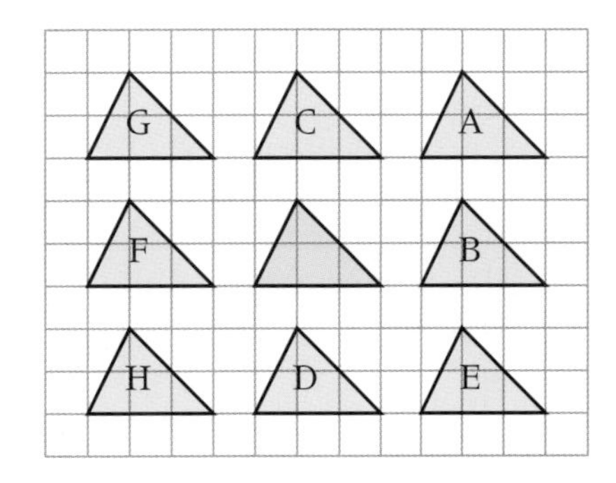

3 Describe these translations.

a A to B **b** D to B

c B to D **d** A to C

e C to D **f** D to A

g C to A **h** B to A

i B to C **j** A to D

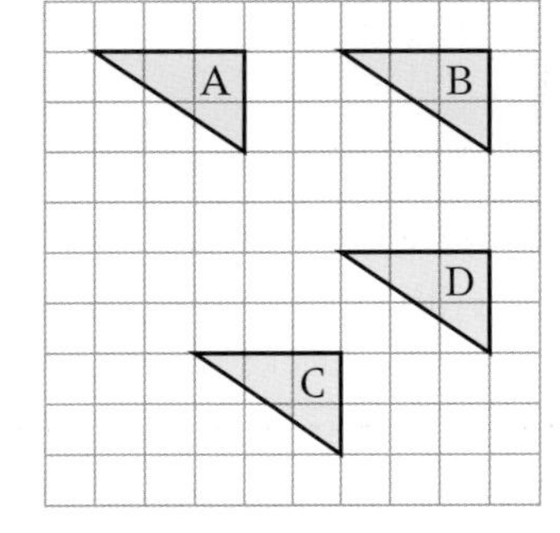

4 **a** State the coordinates of point A.

b What is the mathematical name of the shape?

c Draw the shape after a translation of $\begin{pmatrix} 5 \\ -2 \end{pmatrix}$.

d State whether the two shapes are congruent.

e Give the coordinates of the point A after the translation.

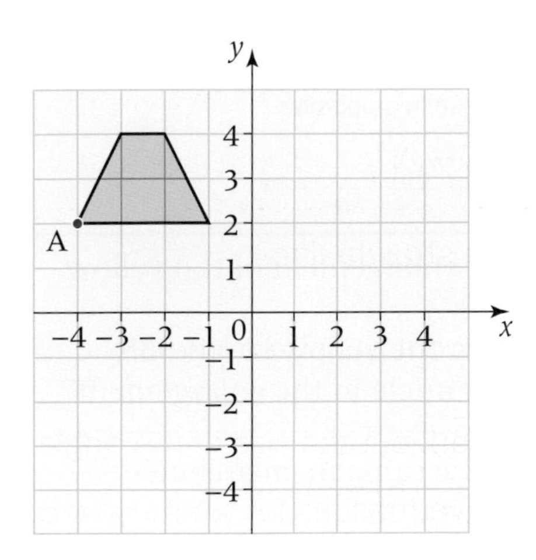

S3.4 Enlargements

This spread will show you how to:

- Recognise and visualise enlargements of objects

Keywords
Enlargement
Multiplier
Scale factor
Similar

To enlarge a shape, multiply corresponding lengths by the same scale factor.

- **The scale factor is the multiplier in the enlargement.**

The green trapezium is an enlargement of the yellow trapezium.

Corresponding lengths are multiplied by 2, $1 \times 2 = 2$, $2 \times 2 = 4$.

The scale factor of this enlargement is 2.

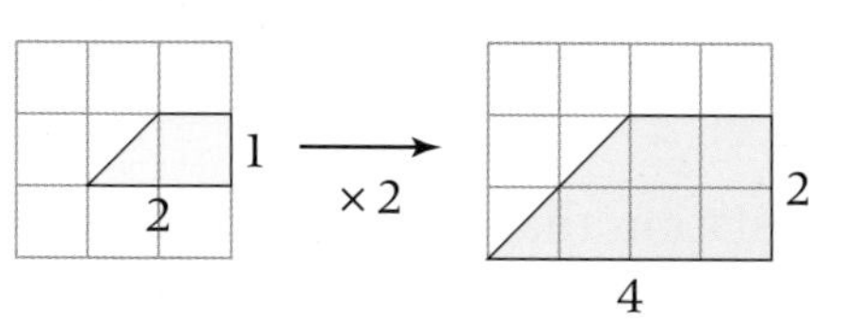

The two trapeziums are **similar** – same shape but different size.

Example

The green shape is an enlargement of the yellow shape.
Calculate the scale factor for each enlargement.

a

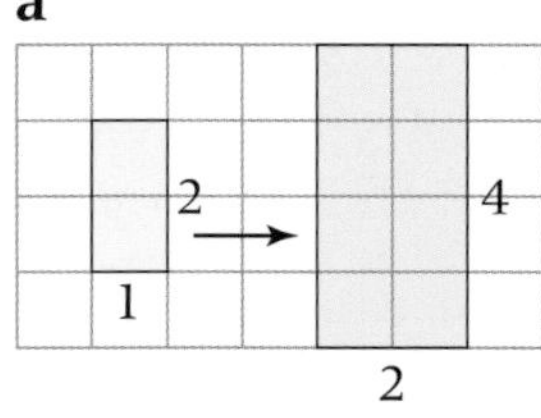

b

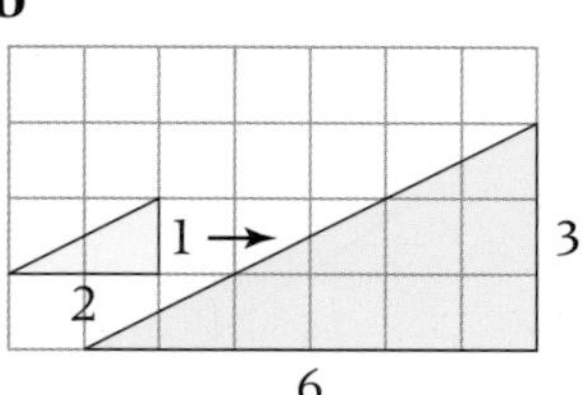

c

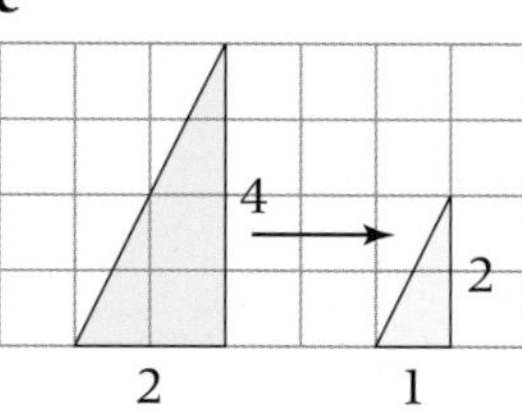

a Scale factor $= 4 \div 2 = 2$
Check: $1 \times 2 = 2$

b Scale factor $= 6 \div 2 = 3$
Check: $1 \times 3 = 3$

c Scale factor $= 2 \div 4 = \frac{1}{2}$
Check: $2 \times \frac{1}{2} = 1$

The scale factor is less than 1 as the shape actually reduces during the enlargement.

- **In an enlargement**
  - **the angles stay the same**
  - **the lengths increase in proportion.**

Example

The green kite is an enlargement of the yellow kite by scale factor 2.
The smallest angle in the yellow kite is 53°.
What is the smallest angle in the enlargement?

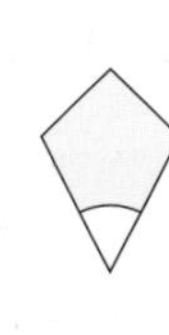

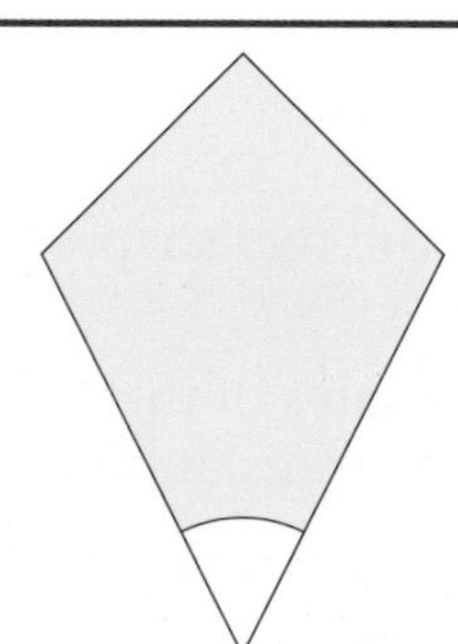

53°, as angles stay the same in enlargements.

Each length is multiplied by 2.

Exercise S3.4

1 Calculate the scale factor of these enlargements.

a

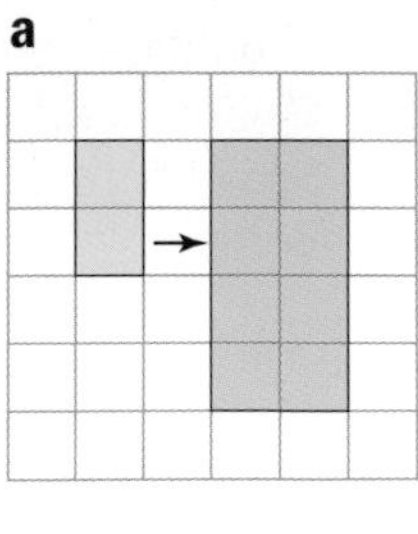

b

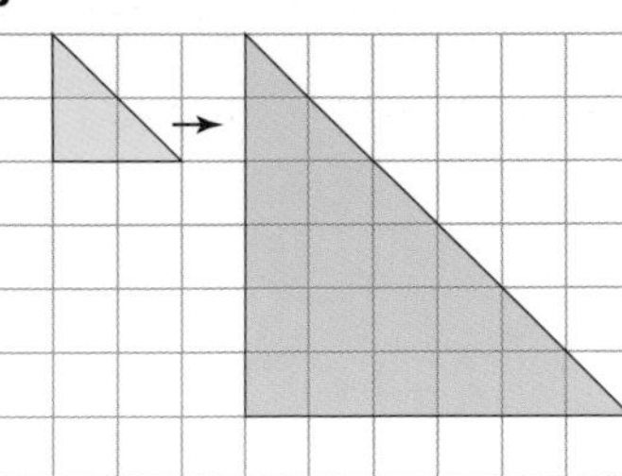

c

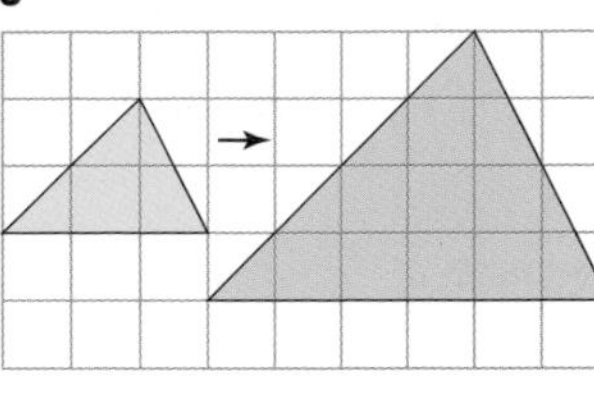

d

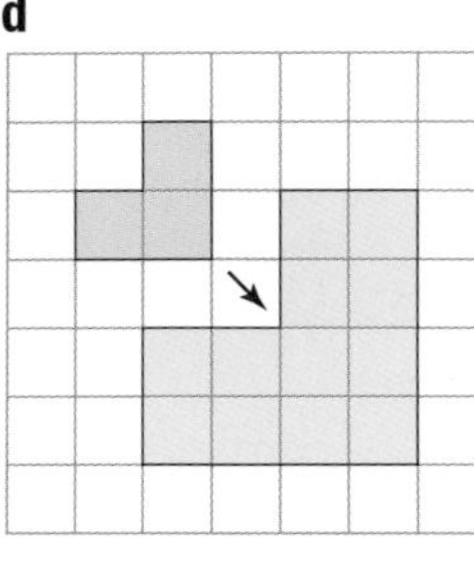

e

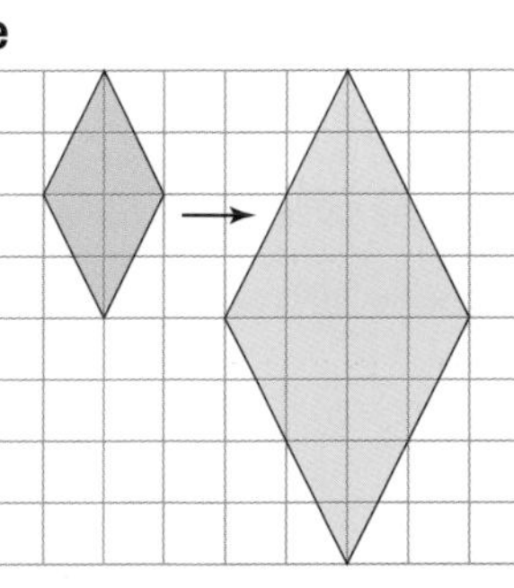

f

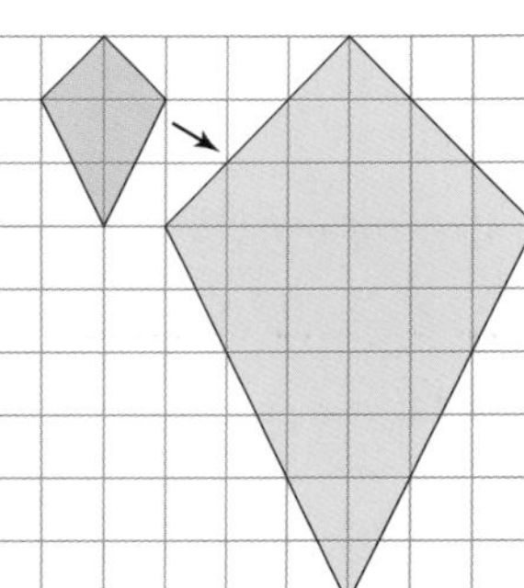

g

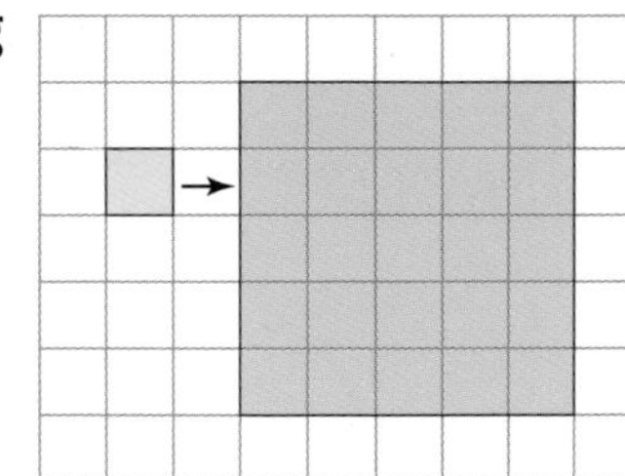

h

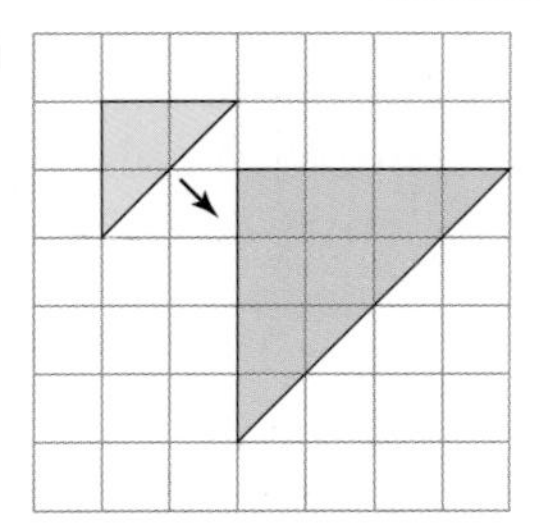

2 **a** Decide if these rectangles are enlargements of the yellow rectangle. If so, calculate the scale factor.

b List the rectangles that are similar to the yellow rectangle.

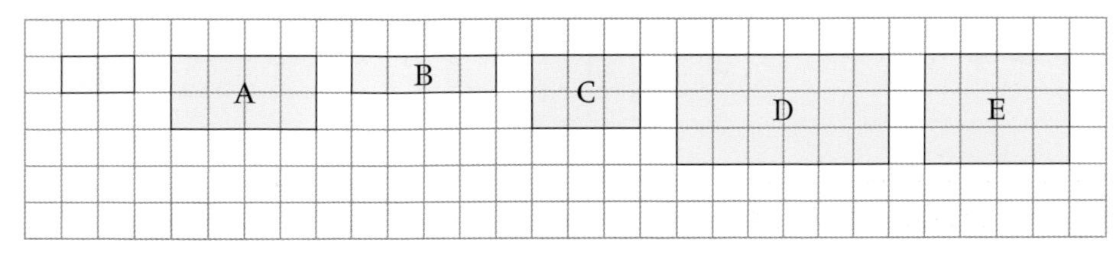

3 **a** Decide if these triangles are enlargements of the yellow triangle. If so, calculate the scale factor.

b List the triangles that are similar to the yellow triangle.

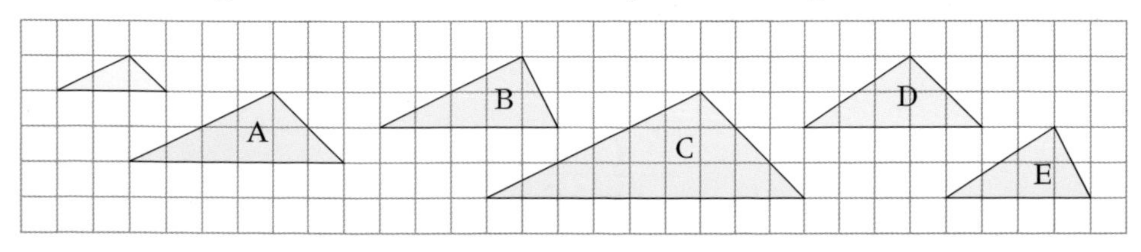

S3.5 More enlargements

This spread will show you how to:

- Understand that enlargements are specified by a centre and positive scale factor

Keywords
Centre of enlargement
Enlargement
Scale factor
Vertices

- In an **enlargement**
  - the angles stay the same
  - the lengths increase in proportion.

The position of an enlargement is fixed by the **centre of enlargement**.

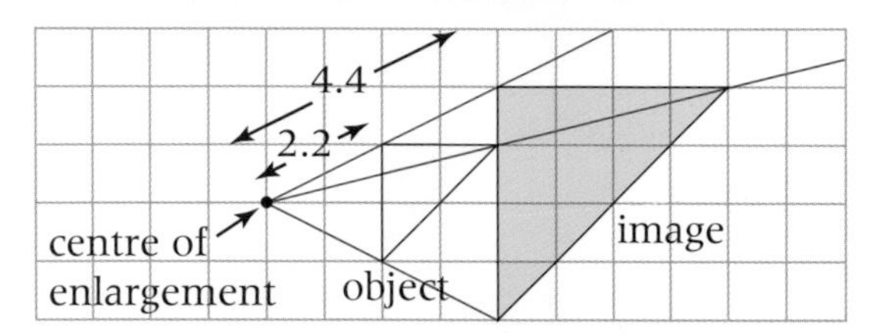

You multiply the distance from the centre to the object by the **scale factor**. This gives the distance to the image along the same extended line.

The scale factor of the enlargement is 2.

The red lines start from the centre and pass through corresponding **vertices** of the two shapes.

- To describe an enlargement, you give
  - the scale factor
  - the centre of enlargement.

Example

Find the centre of enlargement and calculate the scale factor of the enlargement from A to B.

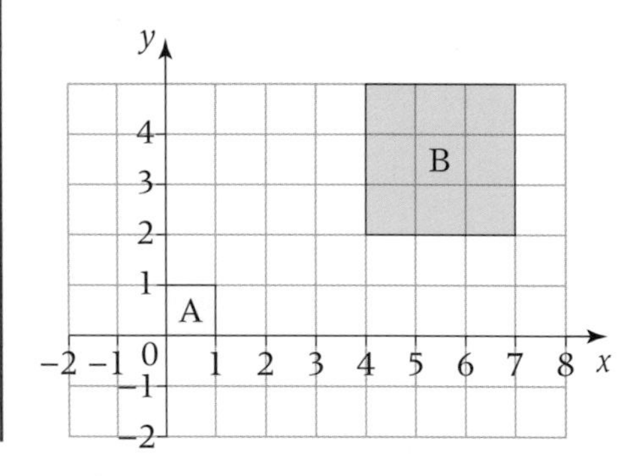

Draw the red lines to find the centre of enlargement.
Centre of enlargement is (−2, −1)
Scale factor = 3 ÷ 1 = 3

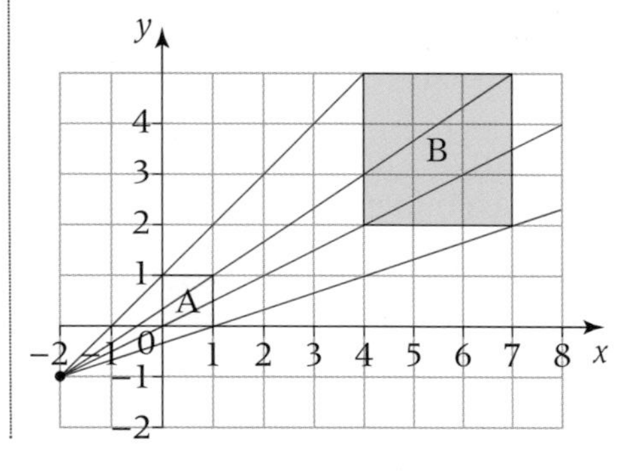

Example

Draw the enlargement of the yellow shape, using scale factor 2 and P as the centre of enlargement.

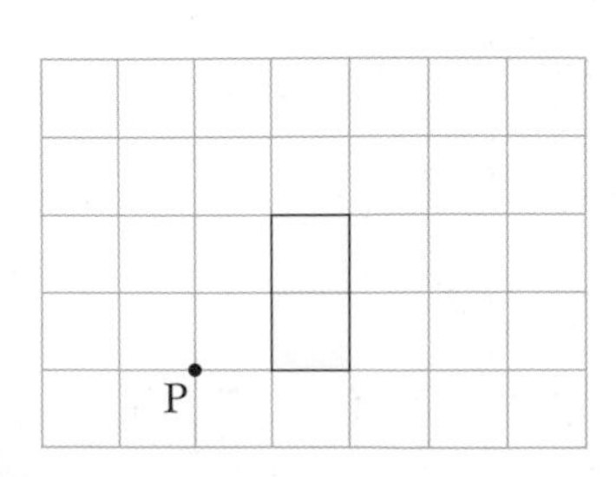

Draw lines from P to each vertex. Multiply the distances from the centre by 2.

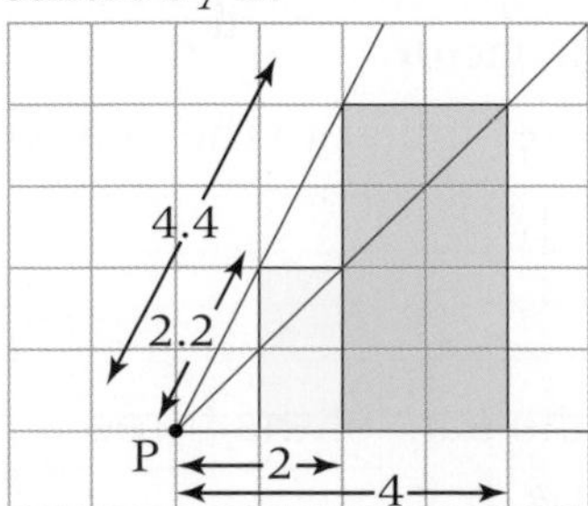

$2 \times 2 = 4$
$2.2 \times 2 = 4.4$

Exercise S3.5

1 Copy each diagram on square grid paper. Find the centre of enlargement and calculate the scale factor for these enlargements.

a

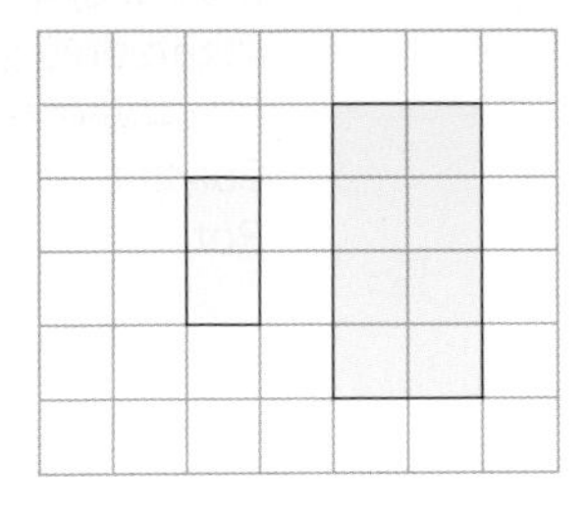

b

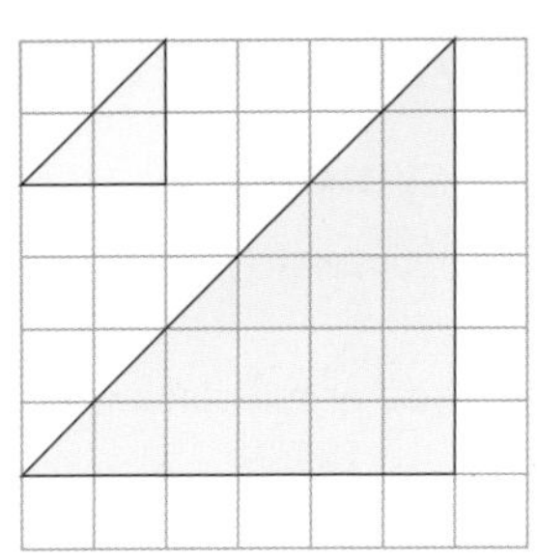

c

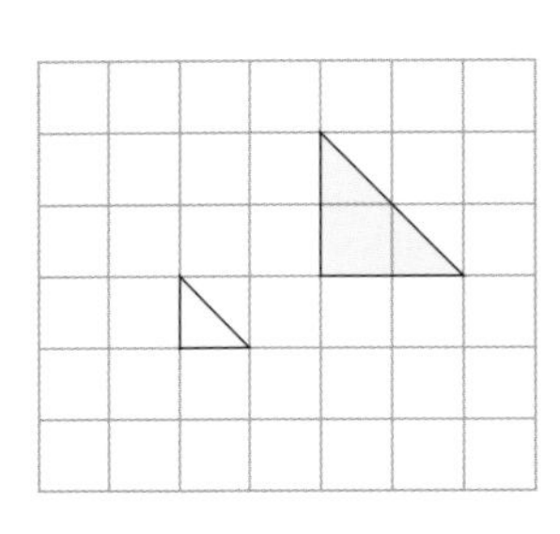

d

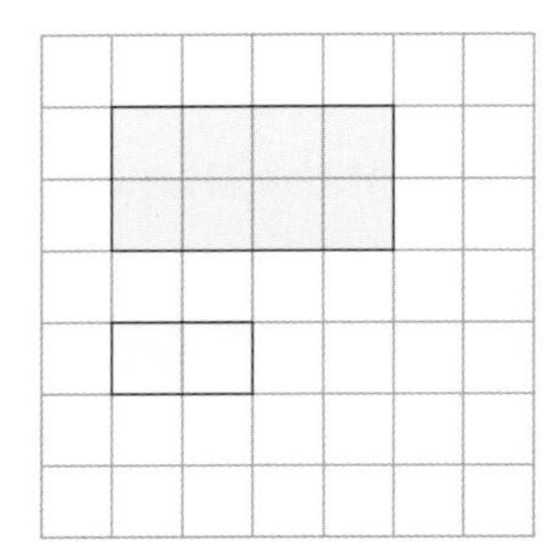

2 Copy each diagram on square grid paper.
Enlarge each shape by the given scale factor using the given centre of enlargement.

a

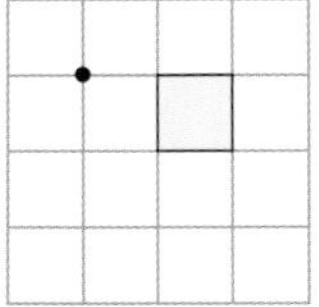

Scale factor 3

b

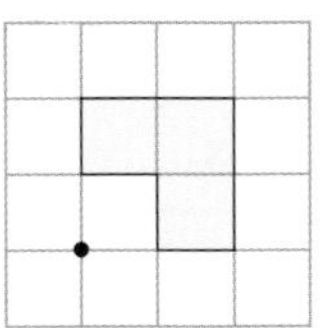

Scale factor 2

c

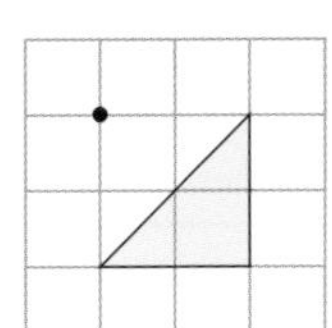

Scale factor 2

d

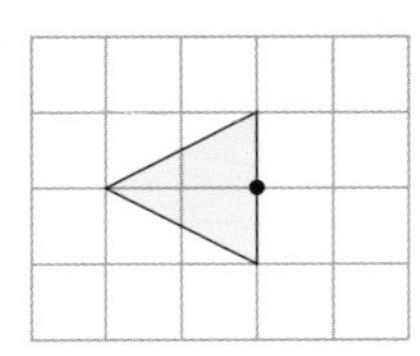

Scale factor 3

e

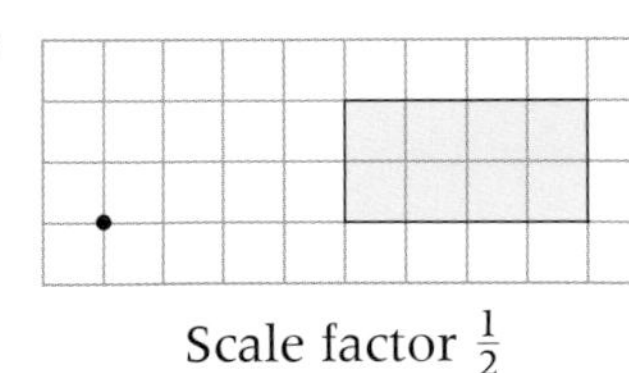

Scale factor $\frac{1}{2}$

f

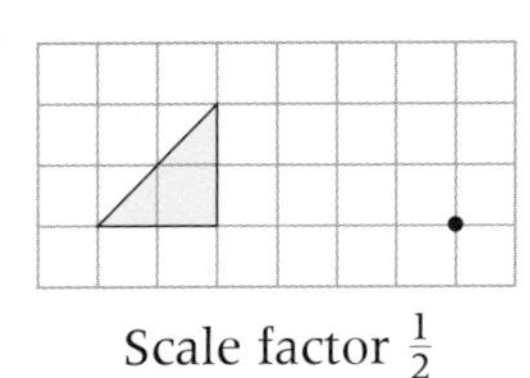

Scale factor $\frac{1}{2}$

3 The vertices of a triangle are (1, 4), (2, 1) and (1, 1). Enlarge the triangle by scale factor 2, with (0, 0) as the centre of enlargement.

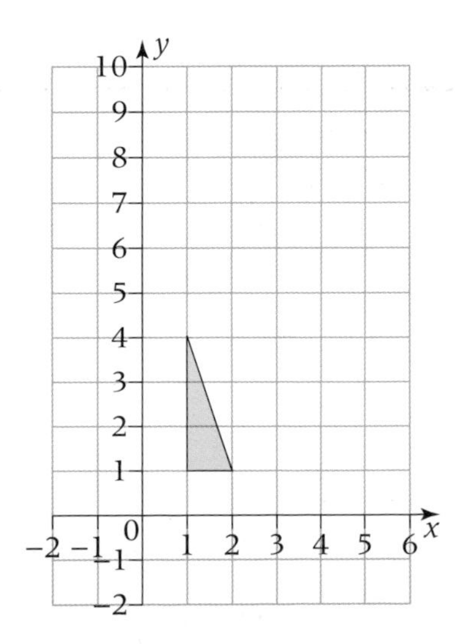

S3.6 Symmetry

This spread will show you how to:

- Recognise reflection symmetry of 2-D and 3-D shapes and rotational symmetry of 2-D shapes

Keywords

Cross-section
Line of symmetry
Plane of symmetry
Polygon
Reflection symmetry
Regular
Rotational symmetry

You can describe shapes by their **symmetry**.

A shape has **reflection symmetry** if the shape divides into two identical halves.

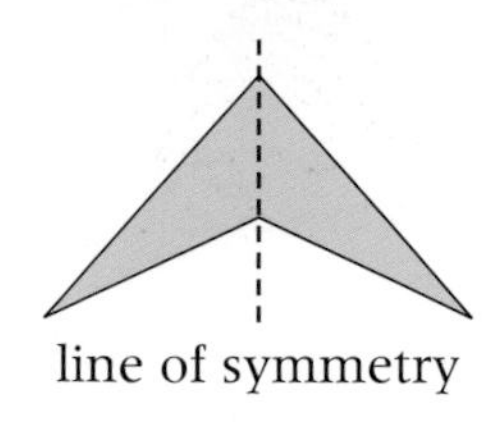

The **line of symmetry** divides the shape into identical halves.

A shape has **rotational symmetry** if the shape looks like itself more than once in a full turn.

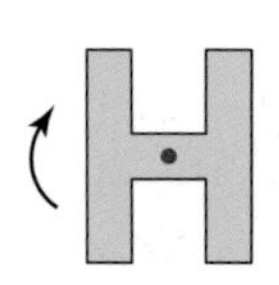 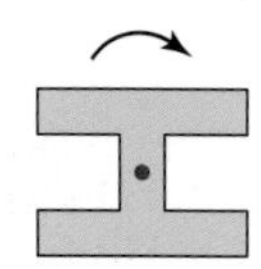 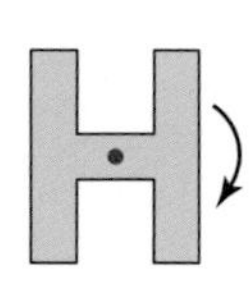 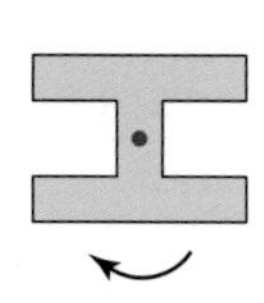 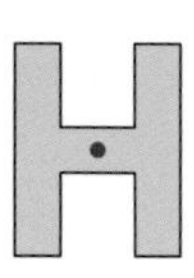

The order of rotational symmetry is 2.

- **The order of rotational symmetry is the number of times a shape looks exactly like itself in a complete turn.**

Example

a Add one extra square so that the shaded shape has 2 lines of symmetry.
b Draw the two lines of symmetry.
c State the order of rotational symmetry of the final shape.

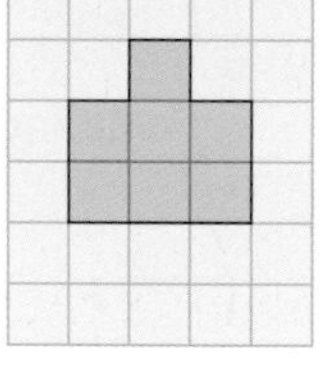

a, b

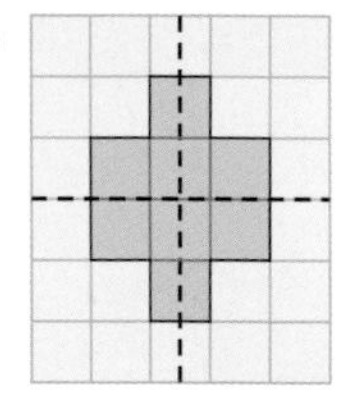

c Rotational symmetry of order 2

- **A plane of symmetry divides a 3-D shape into two identical halves.**

A cuboid has 3 planes of symmetry.

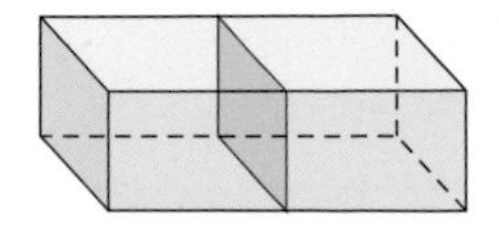 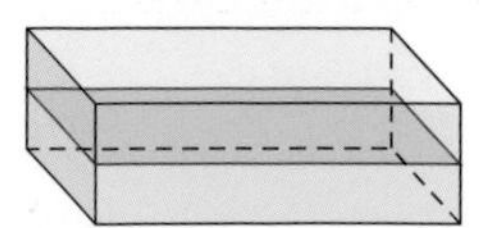 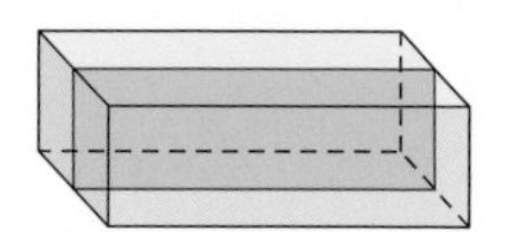

A **cross-section** is a plane through a 3-D shape (in red).

Exercise S3.6

1 **a** Write the 26 letters of the alphabet, in upper case.
b Draw any lines of symmetry on each letter.

2 State the order of rotational symmetry of these 2-D shapes.

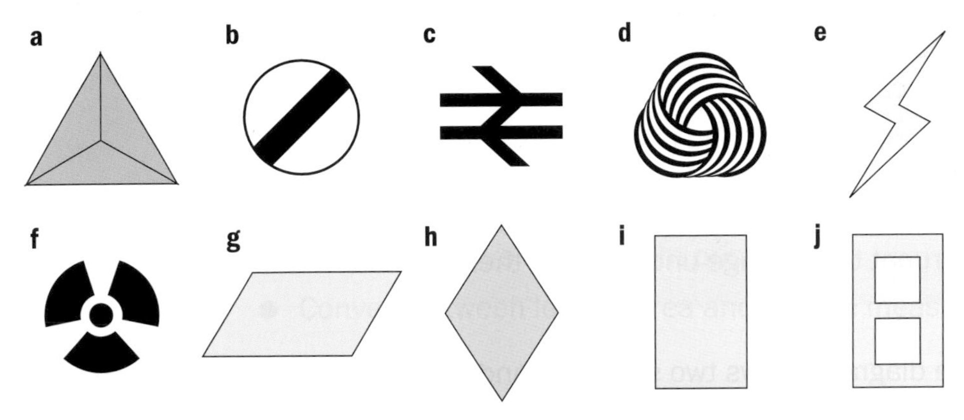

3 Copy these regular polygons.

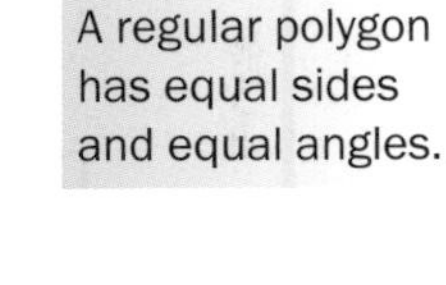

a Draw the lines of symmetry for each shape.
b State the order of rotational symmetry for each shape.

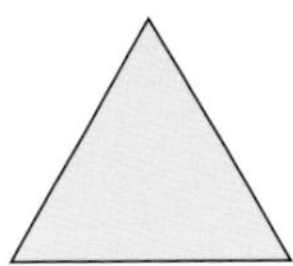

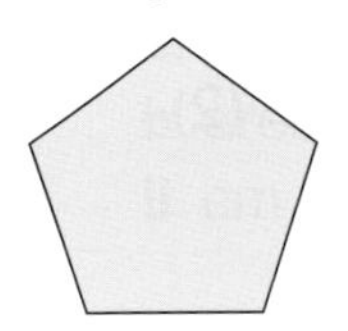
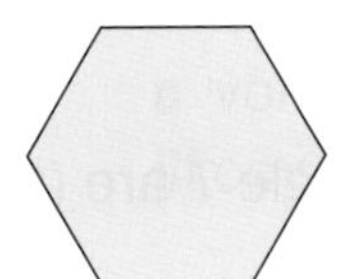
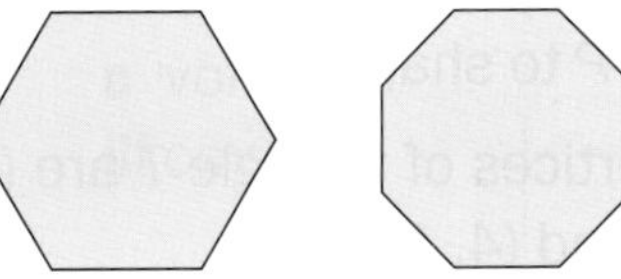

4 Make two copies of this grid.

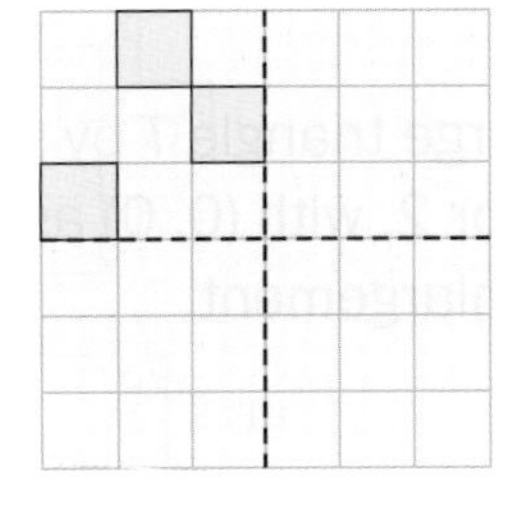

a On one copy, shade in squares so that there are two lines of symmetry.
b On the other copy, shade in squares so that there is rotational symmetry of order 4.

5 Draw copies of these 3-D shapes.
On each diagram draw one plane of symmetry.
Write the number of planes of symmetry for each shape.

a

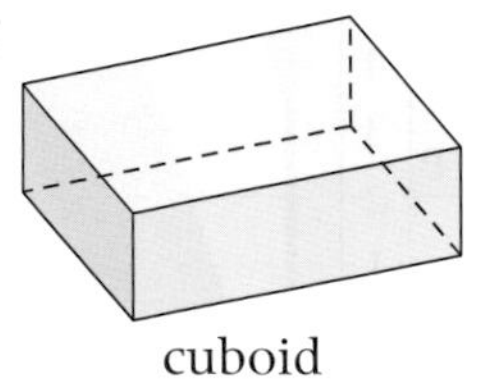

cuboid

b

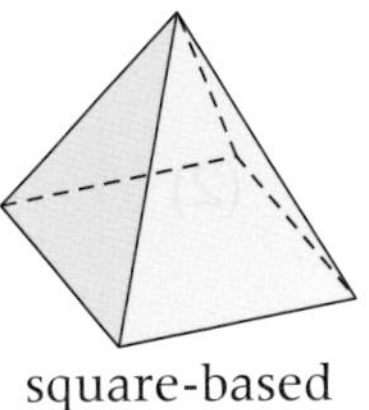

square-based pyramid

c

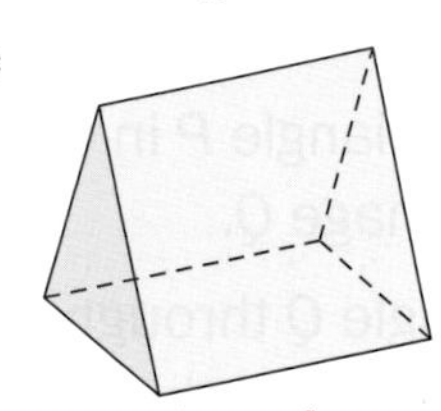

isosceles triangular prism

S4.1 Properties of triangles and quadrilaterals

This spread will show you how to:

- Use angle properties of equilateral, isosceles and right-angled triangles
- Recall the geometric properties of quadrilaterals

Keywords
Diagonal
Equilateral
Isosceles
Line symmetry
Parallel
Quadrilateral
Regular
Right-angled triangle
Rotational symmetry
Scalene
Triangle

You need to know the properties of these triangles.

Equilateral	Isosceles	Right-angled	Scalene
3 equal angles 3 equal sides 3 lines of symmetry Rotational symmetry of order 3	2 equal angles 2 equal sides 1 lines of symmetry Rotational symmetry of order 1	One 90° angle marked ⊾ No lines of symmetry Rotational symmetry of order 1	No equal angles No equal sides No lines of symmetry Rotational symmetry of order 1

The equilateral triangle is a **regular** shape as it has equal sides and equal angles. All its interior angles are 60°.

You need to know the properties of these quadrilaterals.

Square	Rectangle	Rhombus	Parallelogram
4 right angles 4 equal sides 2 sets parallel sides 4 lines of symmetry Rotational symmetry of order 4	4 right angles 2 sets equal sides 2 sets parallel sides 2 lines of symmetry Rotational symmetry of order 2	2 pairs equal angles 4 equal sides 2 sets parallel sides 2 lines of symmetry Rotational symmetry of order 2	2 pairs equal angles 2 sets equal sides 2 sets parallel sides No lines of symmetry Rotational symmetry of order 2

The square is the only regular quadrilateral

Trapezium	Isosceles trapezium	Kite	Arrowhead
1 set of parallel sides No lines of symmetry Rotational symmetry of order 1	2 sets equal angles 1 set equal sides 1 set parallel sides 1 line of symmetry Rotational symmetry of order 1	1 pair equal angles 2 sets equal sides No parallel sides 1 line of symmetry Rotational symmetry of order 1	1 pair equal angles 2 sets equal sides No parallel sides 1 reflex angle 1 line of symmetry Rotational symmetry of order 1

Exercise S4.1

1 The points A(−1, −2) and B(3, −2) are shown. Give the coordinates of a point C, so that triangle ABC

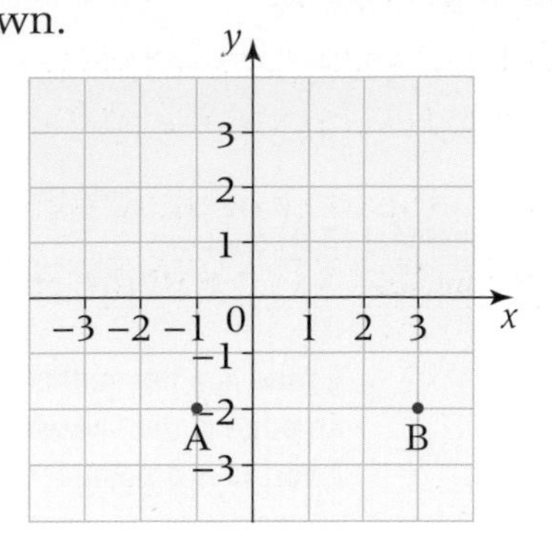

a is isosceles

b is right-angled but scalene

c is right-angled and isosceles

d is scalene

e is equilateral (only an approximate value of y is possible)

f has an area of 4 cm^2.

2 The **diagonals** of a rectangle

- are equal in length
- bisect each other
- are not perpendicular.

Copy and complete the table of results for the diagonals of these shapes.

Shape	Equal in length	Bisect each other	Perpendicular
Rectangle	✓	✓	✗
Kite			
Isosceles trapezium			
Square			
Parallelogram			
Rhombus			
Ordinary trapezium			

3 Plot the points (−2, −1), (0, −1) and (1, 2) on a copy of this grid. These points are three vertices (corners) of a parallelogram.

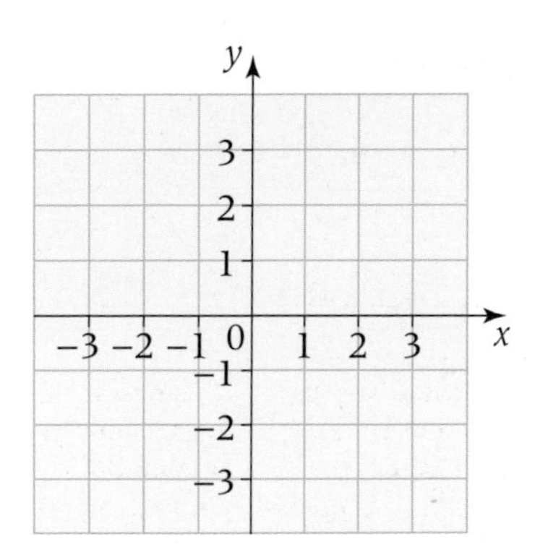

a Write the coordinates of the fourth point.

b Draw the parallelogram.

c Calculate the area of the parallelogram.

S4.2 3-D shapes

This spread will show you how to:

- Investigate 3-D shapes made from cuboids, using 2-D representations of 3-D shapes

Keywords

Cube
Cuboid
Edge
Face
Net
Prism
Pyramid
Solid
Three-dimensional (3-D)
Vertex

- A **solid** is a **three-dimensional (3-D)** shape.

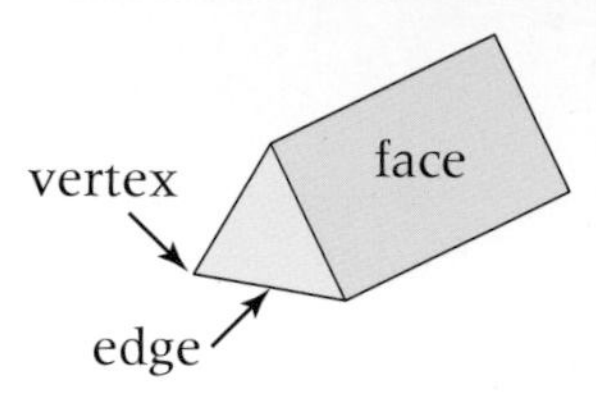

A **face** is a flat surface of a solid.
An **edge** is the line where two faces meet.
A **vertex** is a point at which three or more edges meet.

You need to know the names of these 3-D shapes.

The plural of vertex is vertices.

A **cube** has 6 square faces
12 equal edges
8 vertices

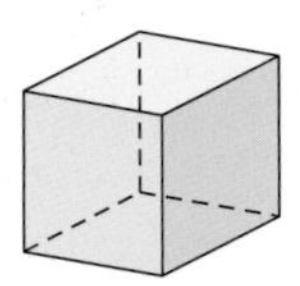

A **cuboid** has 6 rectangular faces
12 edges
8 vertices

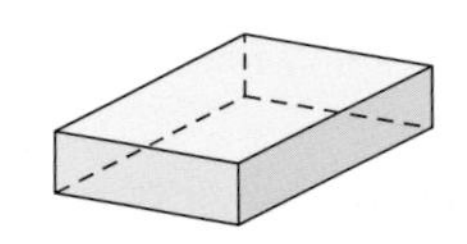

A **prism** has a constant cross-section.

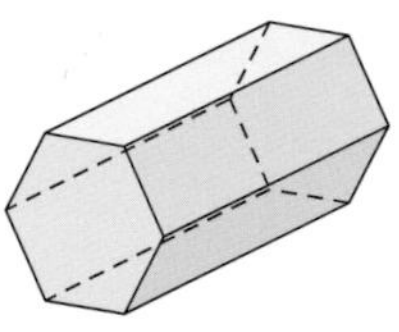

hexagonal prism

A **pyramid** has faces that taper to a common point.

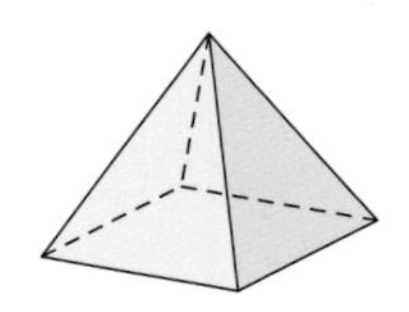

square-based pyramid

You name a prism by the shape of its cross-section.

You name a pyramid by the shape of its base.

- A **net** is a 2-D shape that can be folded to form a 3-D shape.

Example

The nets of four solids are shown. Name the solid that can be made from each net.

a

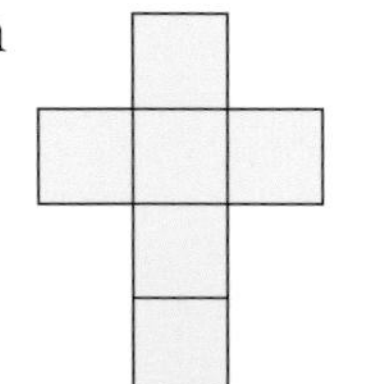

b

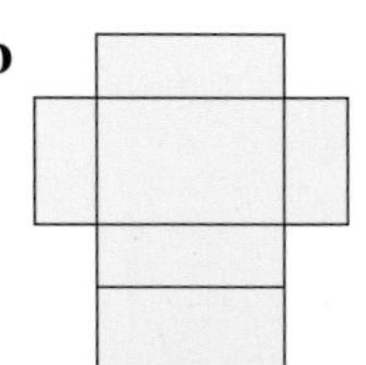

c

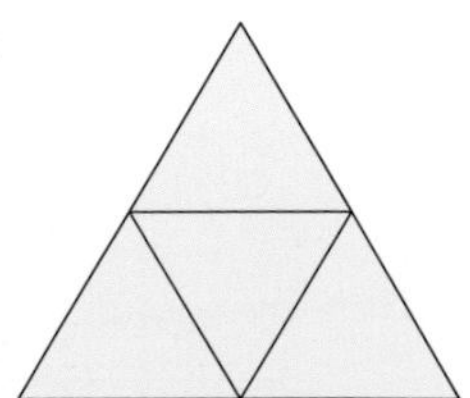

d

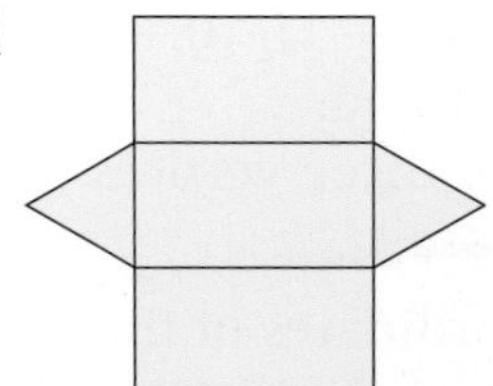

a cube
b cuboid
c tetrahedron
d triangular prism

A tetrahedron has 4 faces – all equilateral triangles.

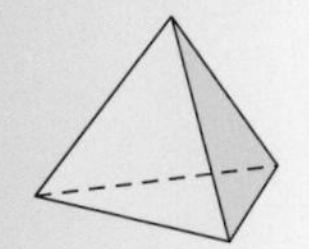

Exercise S4.2

1 Give the mathematical name of each of these solids.

a 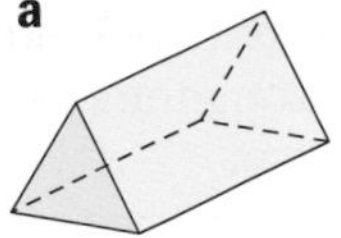

b 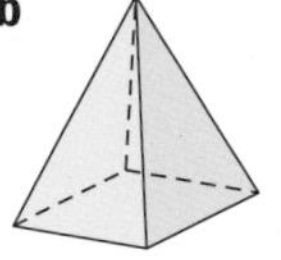

c

d 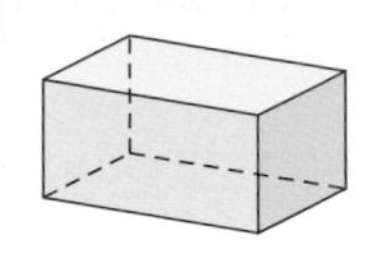

e 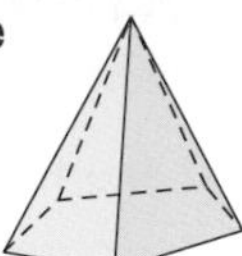

f 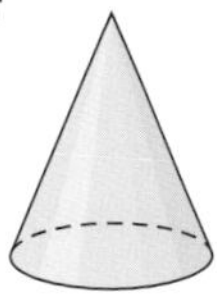

g

h 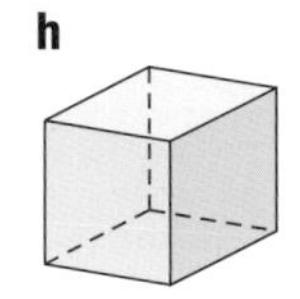

i 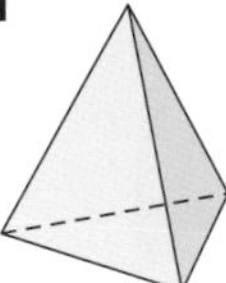

j 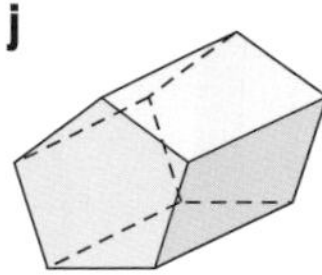

2 Draw

a a prism with a square cross-section

b a pyramid with a hexagonal base

c a tetrahedron.

DID YOU KNOW?

In science, a glass triangular prism is used to separate white light into the rainbow colours.

3 This solid consist of eight triangles.
It is called an octahedron.
Write

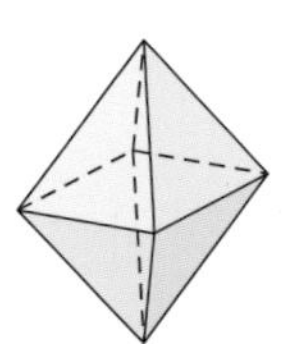

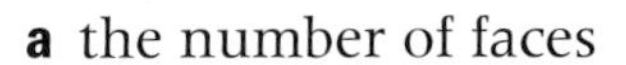

a the number of faces

b the number of edges

c the number of vertices of this solid.

4 **a** Copy and complete this table.

Name of solid	Number of faces (f)	Number of edges (e)	Number of vertices (v)
Cuboid	6	12	8
Triangular prism			
Square-based pyramid			
Tetrahedron			
Pentagonal prism			
Square-based prism			
Cube			
Hexagonal pyramid			
Octagonal prism			
Pentagonal pyramid			

b Write a relationship between f, e and v.

5 Sketch six different nets of a cube.

S4.3 Plans and elevations

This spread will show you how to:

- Analyse 3-D shapes through plans and elevations

Keywords
3-D
Front elevation
Isometric grid
Plan
Side elevation

You can look at this car from different directions.

from above, ... from the front, ... and from the side.

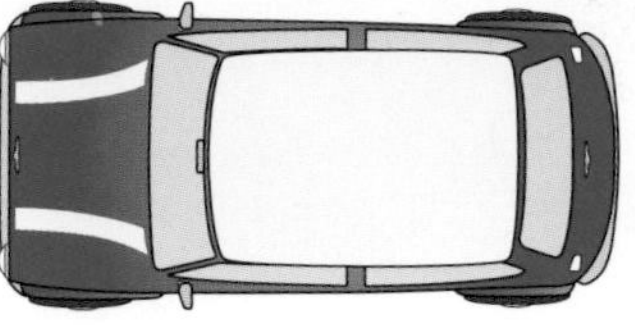

Plan

Front elevation

Side elevation

The plan is the 'birds-eye view'.

Example

This solid is made from 8 cubes.
Draw

a the plan
b the side elevation
c the front elevation

on square grid paper.

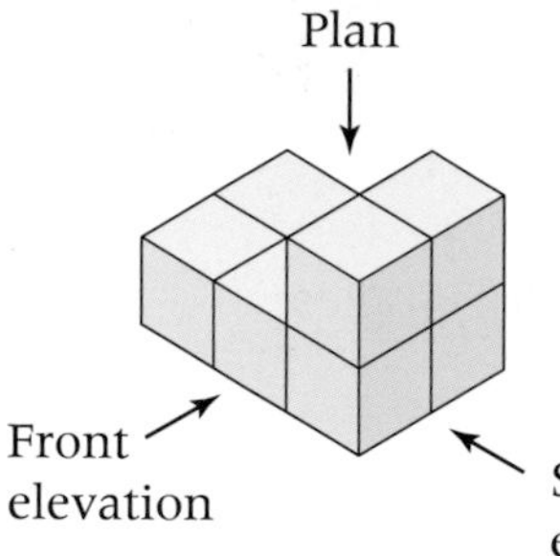

a b c

Plan Side elevation Front elevation

Notice the extra bold line in the plan, when the level of the cubes alters.

Example

The plan and front elevation of a prism are shown.

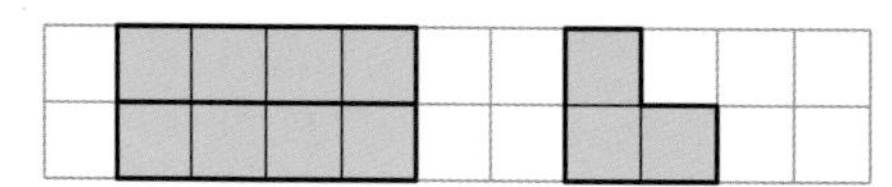

Plan Front elevation

a Draw a 3-D sketch of the prism.
b Draw the side elevation on square grid paper.

a

b

Side elevation

Exercise S4.3

1 On square grid paper, draw the plan (P), the front elevation (F) and the side elevation (S) for each solid.

a

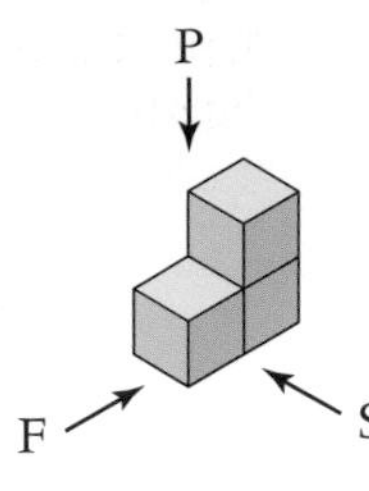

b

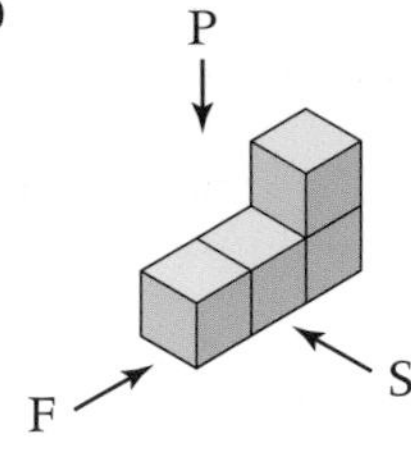

c

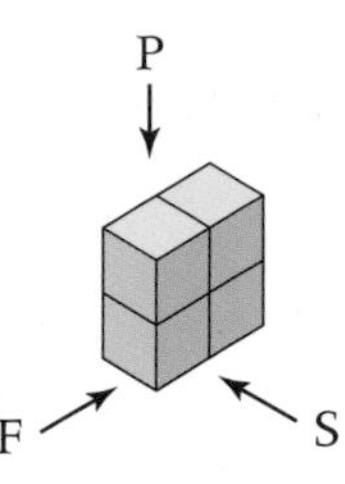

d

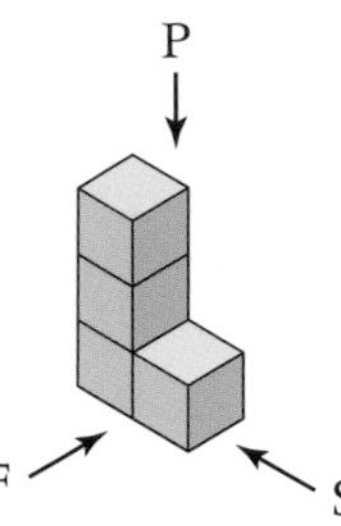

e

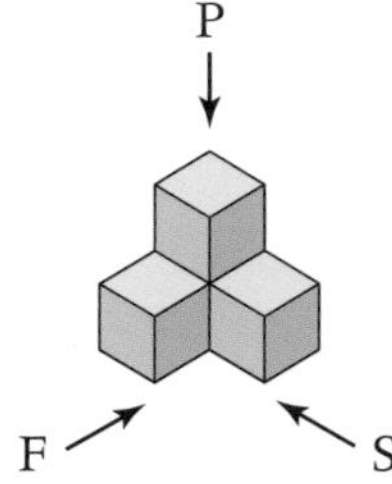

f

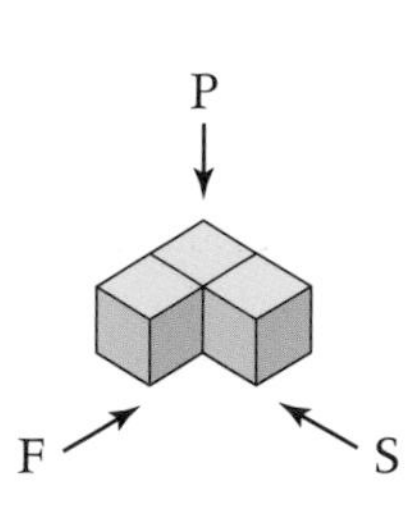

2 The plan, front elevation and side elevation are given for these solids made from cubes. Draw a 3-D sketch of each solid and state the number of cubes needed to make it.

a

plan

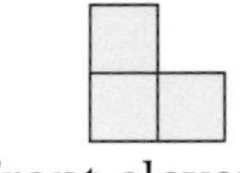
front elevation

side elevation

b

plan

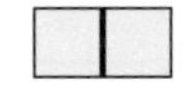
front elevation

side elevation

c

plan

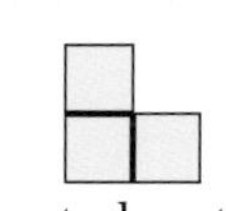
front elevation

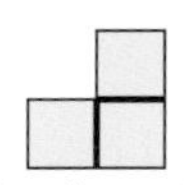
side elevation

3 Sketch the plan (P), the front elevation (F) and the side elevation (S) for each solid.

a

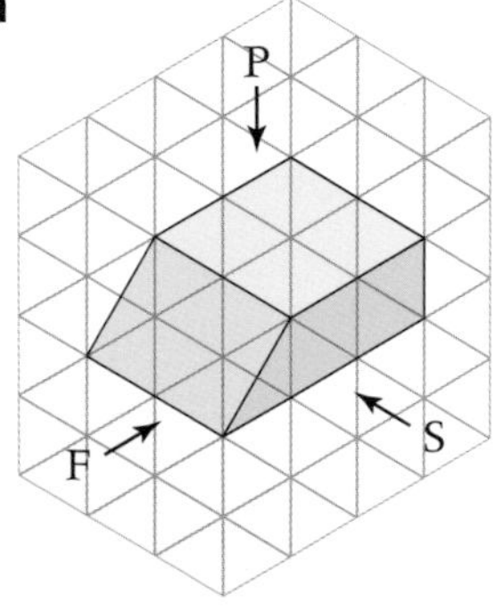

b

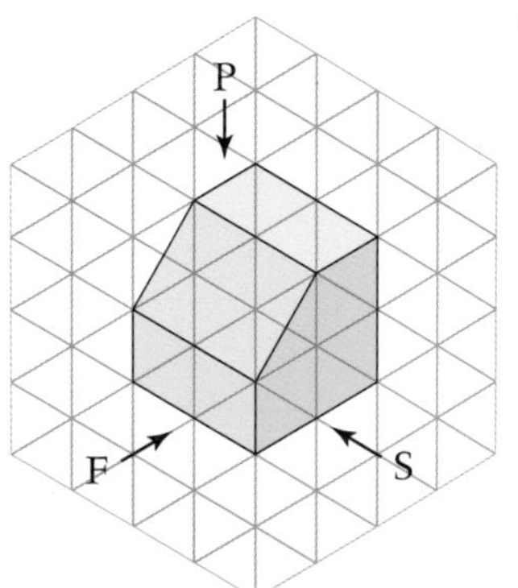

c

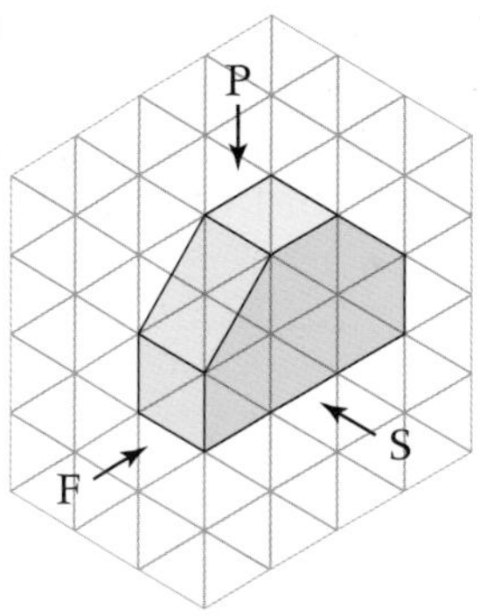

These solids are drawn on an **isometric grid**.

S4.4 3-D coordinates

This spread will show you how to:

- Understand and use coordinates in one, two and three dimensions

Keywords
Coordinates
1 dimension (1-D)
2 dimensions (2-D)
3 dimensions (3-D)

In **1 dimension (1-D)**, you only need **one** number to show a point.

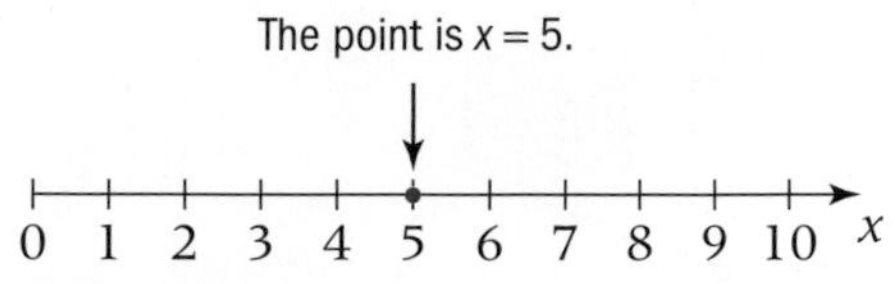

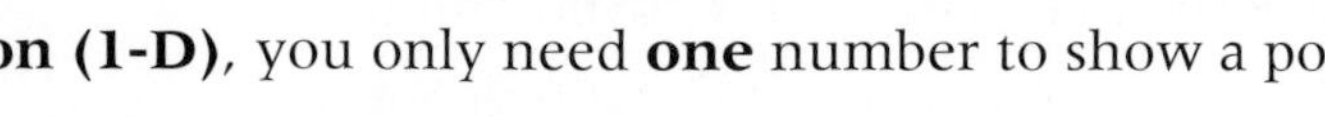

This is an ordinary number line.

In **2 dimensions (2-D)**, you need **two** numbers to show a point.

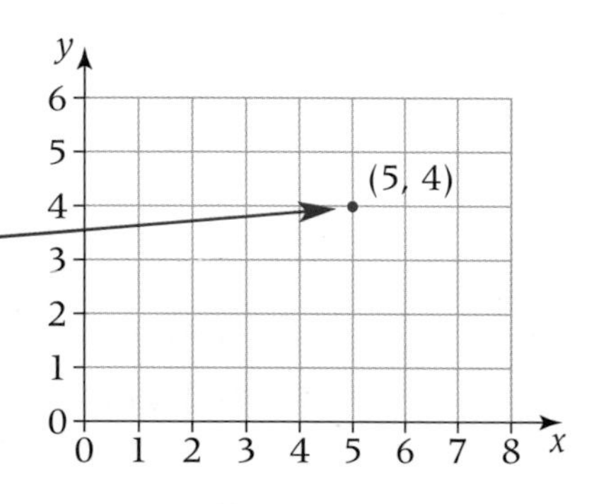

The x-axis and the y-axis are at right angles.

In **3 dimensions (3-D)**, you need **three** numbers to show a point.

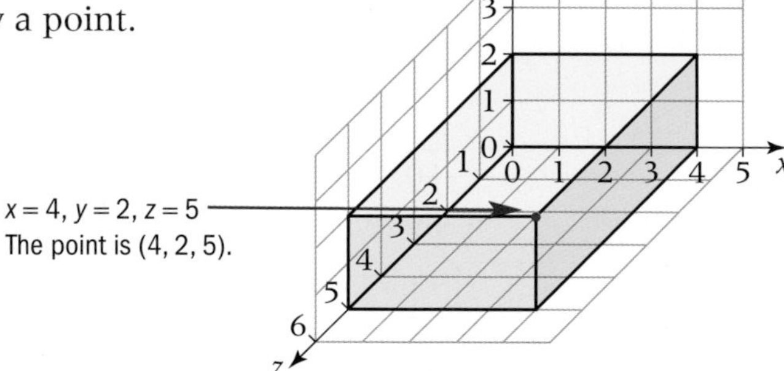

The x-axis, the y-axis and the z-axis are at right angles to each other.

Example

A 2 by 3 by 4 cuboid is placed on the axes as shown.

a Give the values of p, q and r.

b Complete the **coordinates** of A, B, C and D.

A(_, _, _)
B(3, 0, 4)
C(_, _, _)
D(_, _, _)

c Calculate the volume of the cuboid.

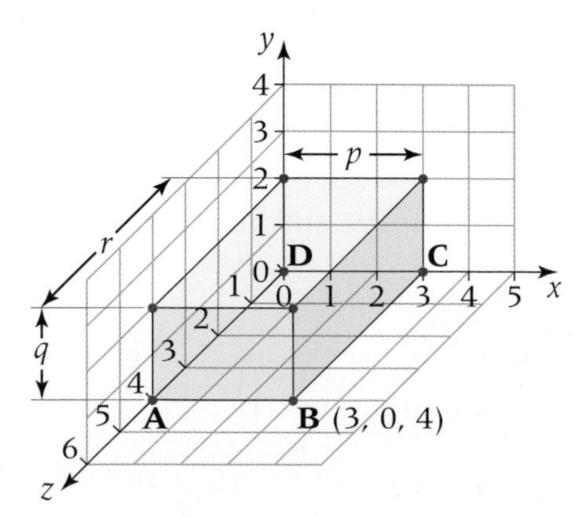

a $p = 3$, $q = 2$, $r = 4$

b A(0, 0, 4), B(3, 0, 4), C(3, 0, 0), D(0, 0, 0)

c Volume $= 4 \times 3 \times 2 = 24$ cubic units

Exercise S4.4

1 For each diagram, write the coordinates of the point P.

a

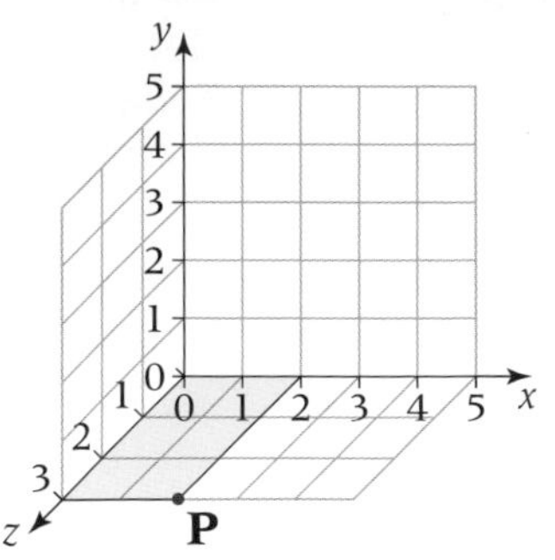

b

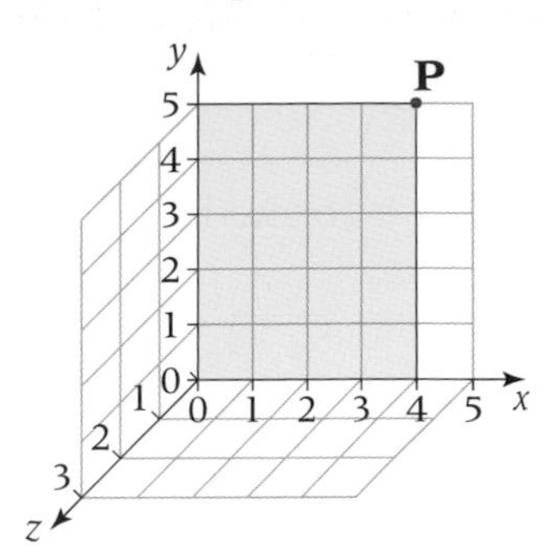

c

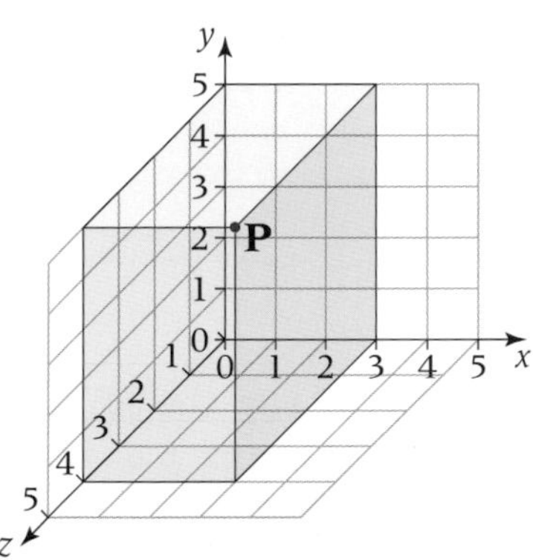

d

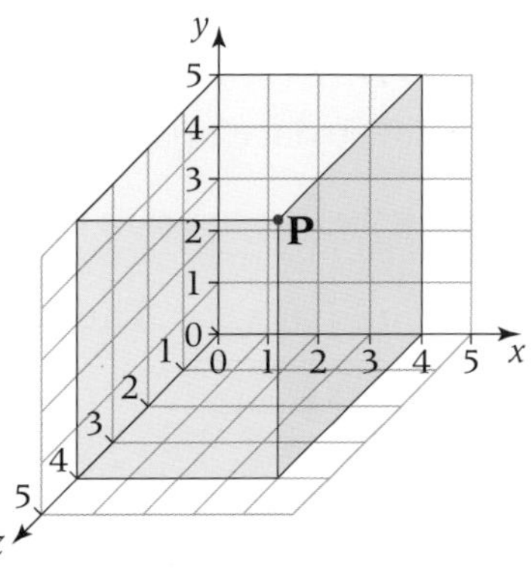

2 Draw diagrams to illustrate these points.

a (0, 2, 0) **b** (0, 3, 1) **c** (2, 0, 3)

d (5, 3, 4) **e** (4, 4, 4)

3 A 3 cm by 3 cm by 3 cm cube is placed on the axes as shown.
Give the coordinates of these points.

a A **b** B

c C **d** D

e E **f** F

g G **h** H

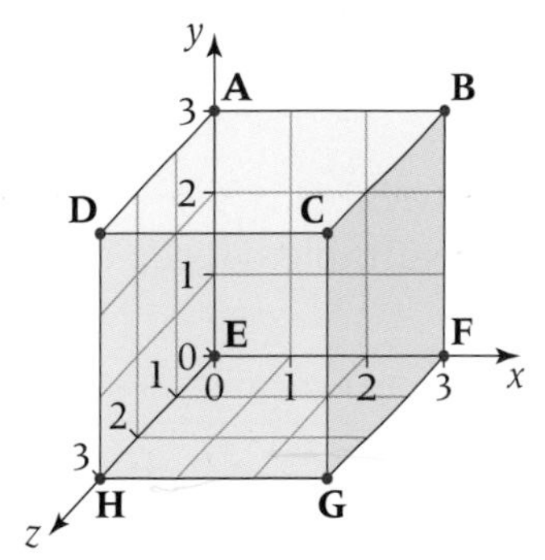

4 A 2 by 4 by 5 cuboid is placed on the axes as shown.

a Give the values of p, q and r.

b Give the coordinates of A to H.

c Calculate the volume of the cuboid.

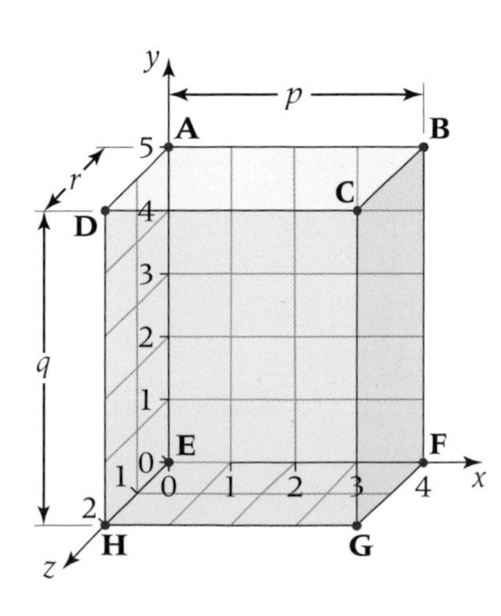

S4.5 Surface area

This spread will show you how to:

- Find the surface area of simple shapes using the area formulae for triangles and rectangles

Keywords
Faces
Net
Surface area

When you unfold a shape its **net** is formed.

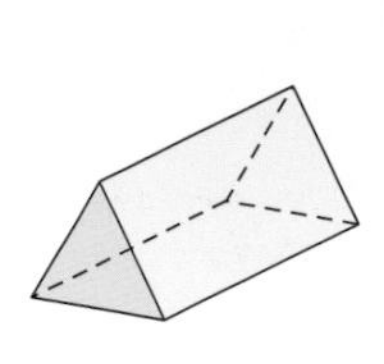

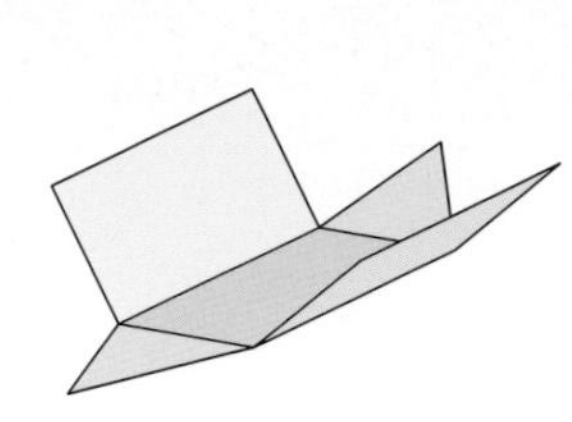

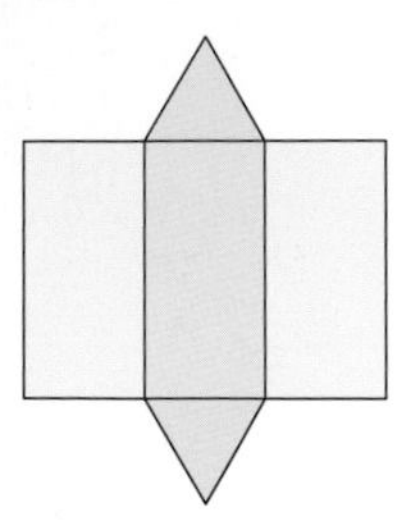

The area of the net is called the surface area.

- **The surface area of a 3-D shape is the total area of its faces.**

A cuboid has 6 faces. They are in pairs.

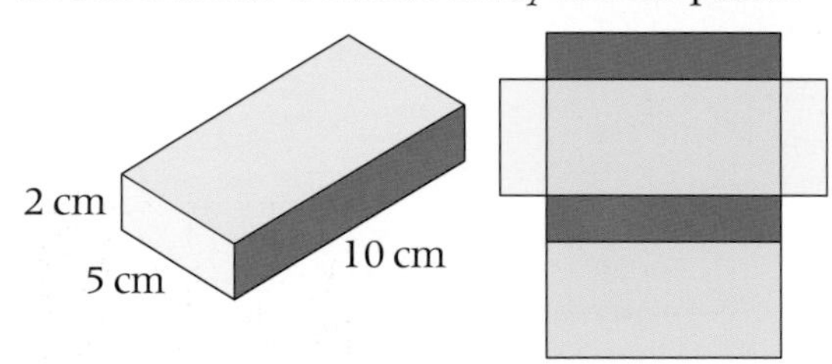

The surface area is

$2 \times \text{red} = 2 \times 2 \times 10 = 40\ \text{cm}^2$

$2 \times \text{green} = 2 \times 10 \times 5 = 100\ \text{cm}^2$

$2 \times \text{yellow} = 2 \times 5 \times 2 = 20\ \text{cm}^2$

$\text{surface area} = 160\ \text{cm}^2$

Units of area are cm^2.

Example

A tin of tomato soup is shown. The diameter of the circle is 8 cm and the height is 10 cm. The label fits exactly round the tin.

Calculate the area of the label.
Give your answer to a suitable degree of accuracy.

$\pi = 3.14...$

If the label is unfurled, the length is the same as the circumference.

The label is a rectangle.

Circumference of a circle $= \pi \times d$

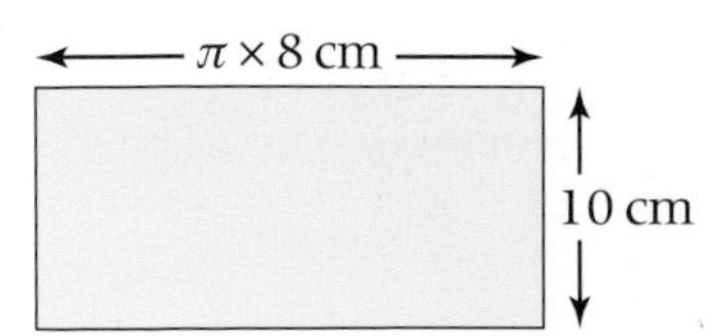

$$\begin{aligned}\text{Area of label} &= \pi \times 8 \times 10 \\ &= 251.327\,41 \\ &= 251\ \text{cm}^2 \text{ (to nearest whole number)}\end{aligned}$$

Exercise S4.5

1 A 4 cm by 6 cm by 8 cm cuboid is shown.
Calculate

a the area of the red rectangle

b the area of the orange rectangle

c the area of the green rectangle

d the surface area of the cuboid.

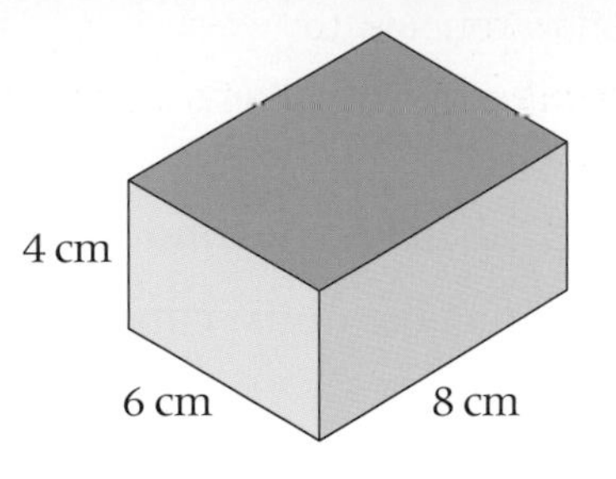

2 Calculate the surface area of these 3-D shapes.

State the units of your answer.

a

4 cm
8 cm
3 cm

b

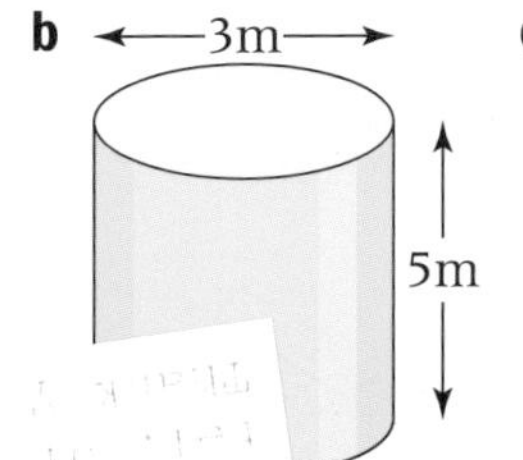

c

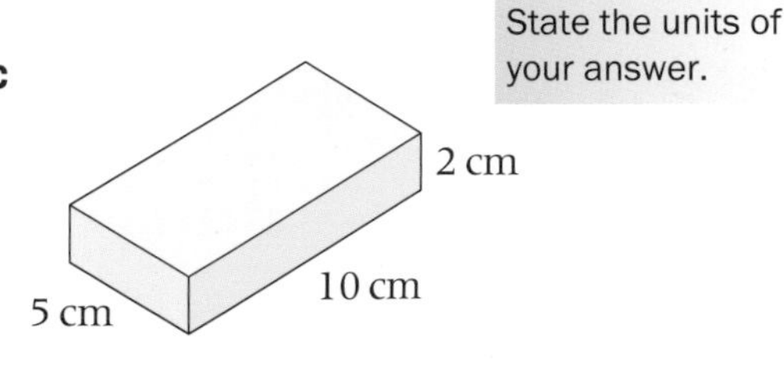

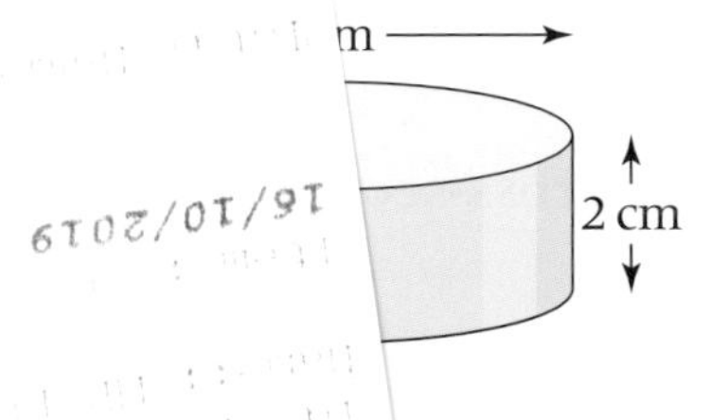

re shown.

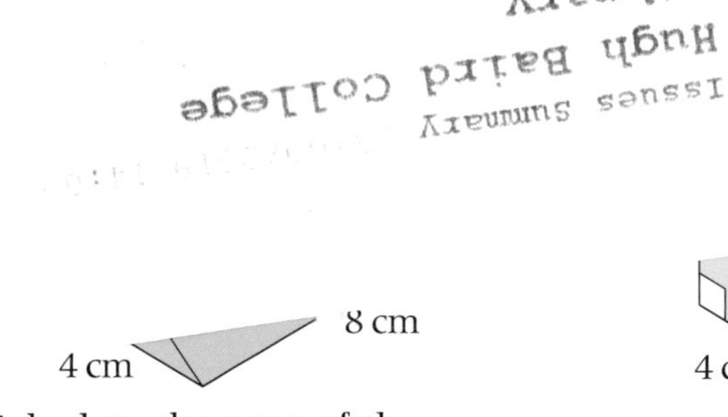

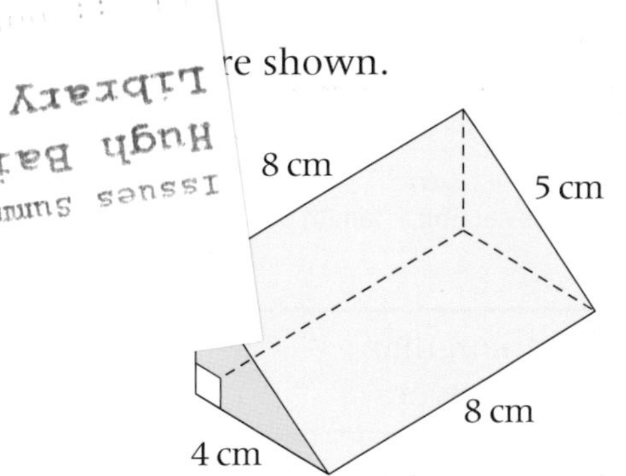

Calculate the area of the

a red rectangle **b** grey rectangle **c** green rectangle

d orange triangle **e** the surface area of the triangular prism.

DID YOU KNOW?

The Great Pyramid of Giza was 145.75 m tall with four 229 m long bases when it was built. That's a surface area of 66 753.5 m^2!

4 For each of these cuboids, calculate
i the surface area
ii the length, width and height.

Use trial and improvement for part **ii**

a

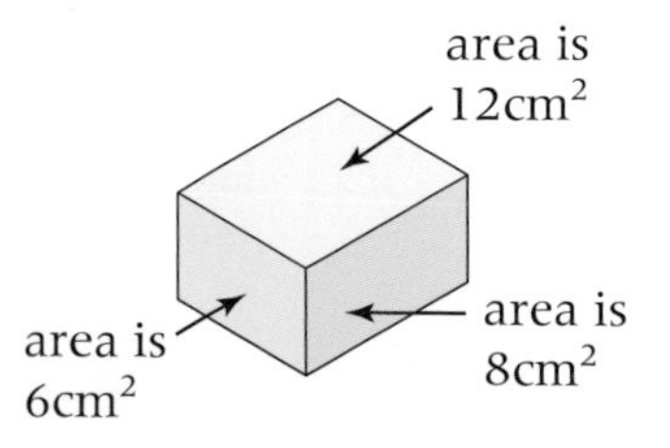

b

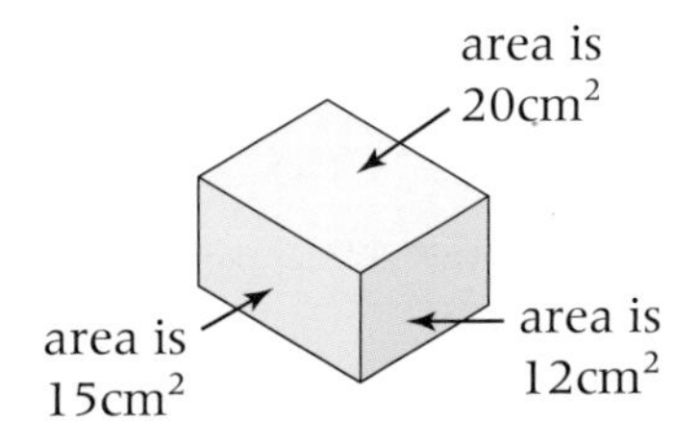

S4.6 Volume of a prism

This spread will show you how to:

- Calculate the volume of right prisms

Keywords
Cross-section
Cubic centimetre (cm^3)
Cubic metre (m^3)
Prism
Volume

The **volume** of a 3-D shape is the amount of space it takes up.

Volume is measured in cubic units: cubic centimetres (cm^3), cubic metres (m^3).

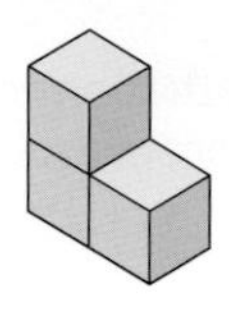

Volume = $3cm^3$

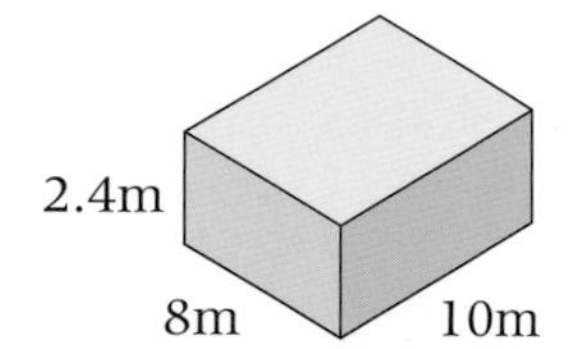

Volume = $192m^3$

The ^{3} in cm^3 shows there are 3 dimensions in a cube: length, width and height.

- **Volume of a cuboid = length × width × height**

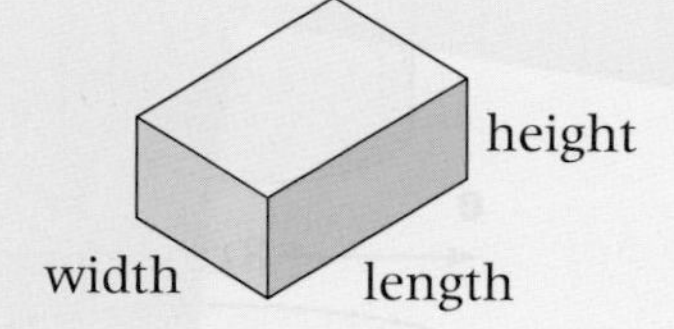

A **prism** is a 3-D shape with the same **cross-section** throughout its length.

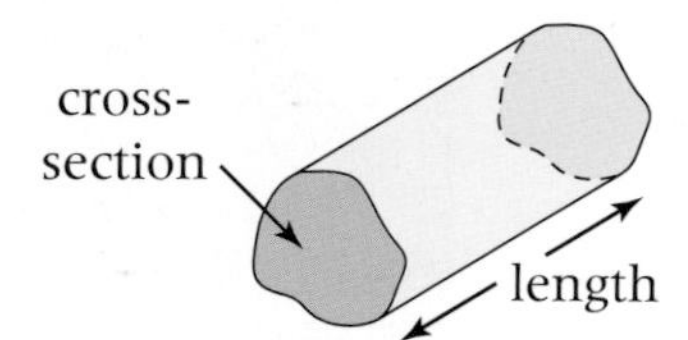

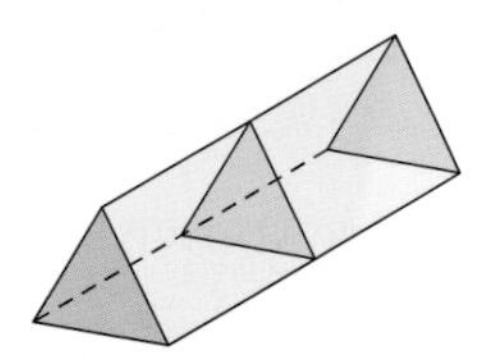

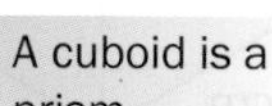

A cuboid is a prism.

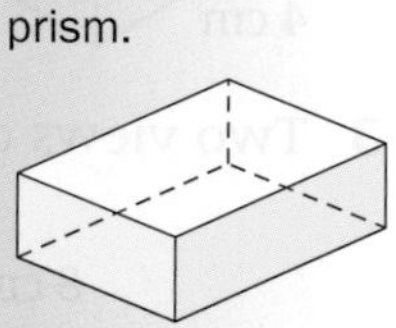

- **Volume of a prism = area of cross-section × length**

Example

Calculate the volume of this triangular prism. State the units of your answer.

2 m
4 m
8 m

Area of triangle $= \frac{1}{2} \times 4 \times 2 = 4\ m^2$

Volume of prism $=$ area of triangle $\times 8$

$= 4 \times 8 = 32\ m^3$

- **Volume of cylinder = area of circle × height**

A cylinder is a prism with a circular cross-section.

Example

Calculate the volume of this cylinder.

10 cm
4 cm

Area of circle $= \pi \times r^2$

$= \pi \times 5^2$

$= 78.539\ 816 \qquad (\pi = 3.14)$

Volume of cylinder $= 78.539\ 816 \times 4$

$= 314.159\ 265$

$= 314\ cm^3$ (to nearest whole number)

Do not round intermediate workings.

Exercise S4.6

Give the units of your answers.

1 Calculate the volume of these cuboids.

a

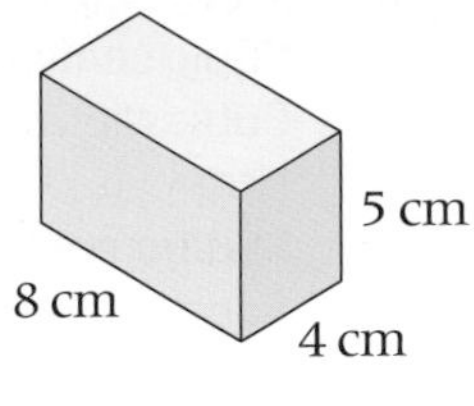

b

9 m
2 m
6 m

c

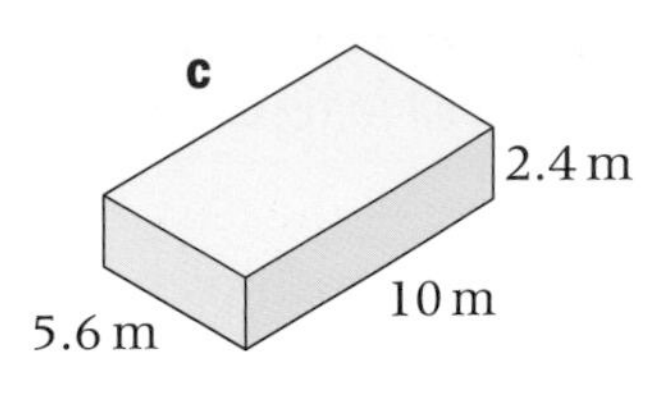

2 Calculate the volume of these cuboids.

a

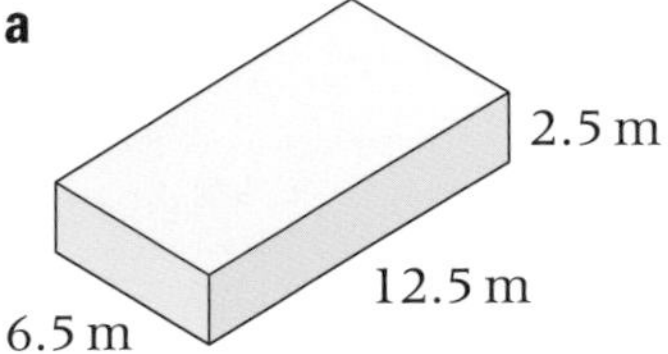

b

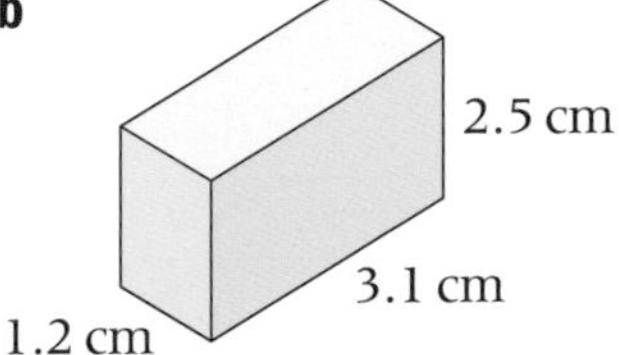

c

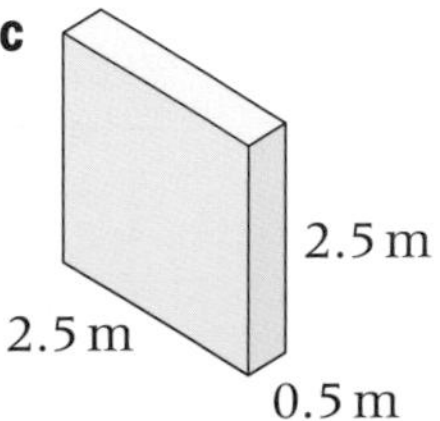

3 Calculate the area of cross-section and the volume for each prism.

a

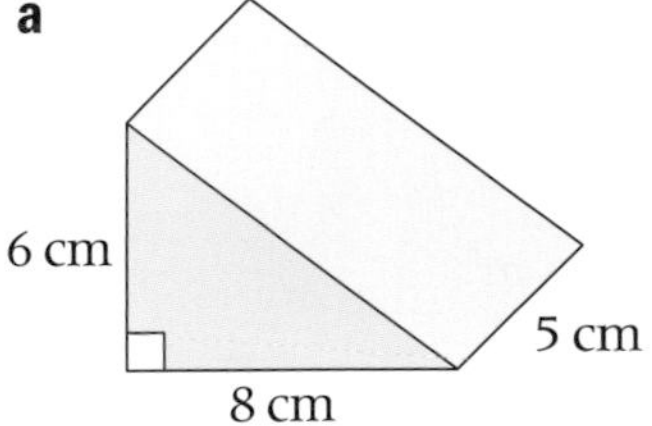

b

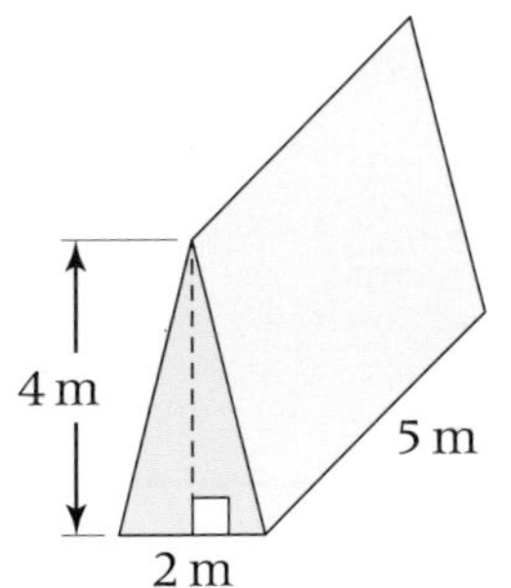

c

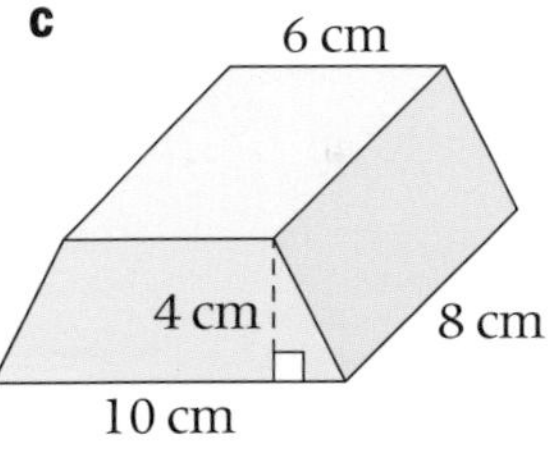

4 Calculate the volumes of these cylinders.

a

4 cm
8 cm

b

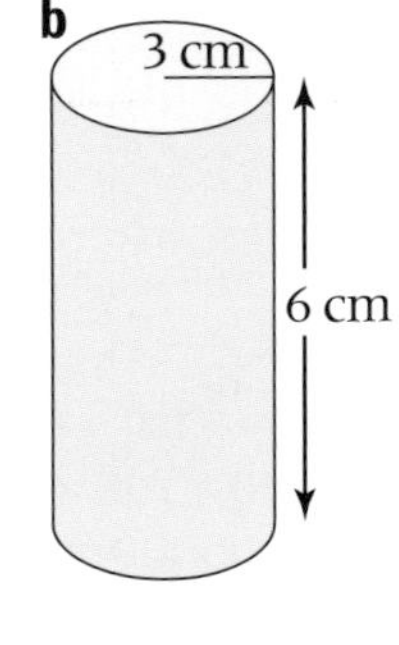

c

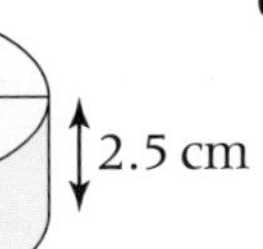

2.5 cm

d

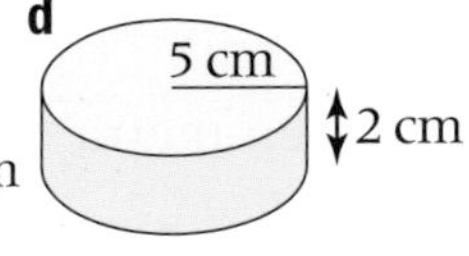

In questions **4** and **5**, give your answers to a suitable degree of accuracy.

5 Guttering can have either a cross-section of a semicircle or an isosceles trapezium.
For each type of guttering calculate
i the area of cross-section
ii the volume of a one metre length.

$\pi = 3.14$

a

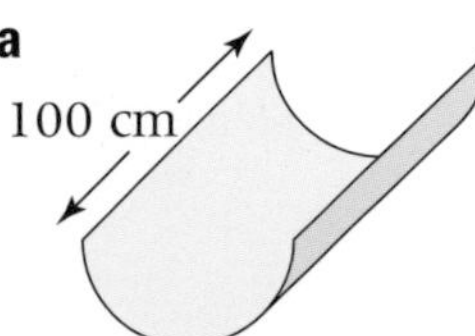

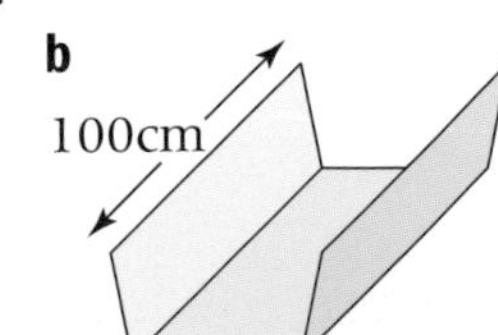

b

100cm

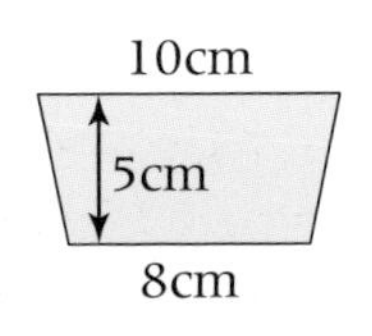

S4.7 2-D and 3-D measures

This spread will show you how to:

- Convert between length measures, area measures including cm^2 and m^2, and volume measures including cm^3 and m^3

Keywords

Area
Cubic centimetre
Cubic metre
Litre
Square centimetre
Square metre
Volume

You should know these relationships between the metric units of length:

- 1 cm = 10 mm
- 1 m = 100 cm
- 1 km = 1000 m

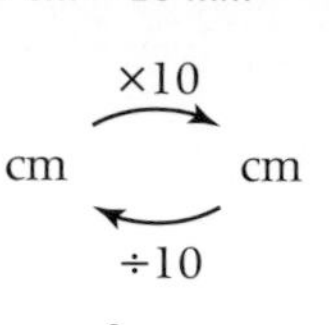

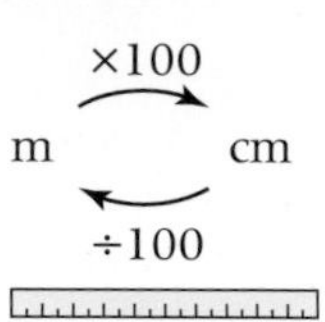

×1000
km m
÷1000

0 10
mm
cm
0 1

←100 cm→
←1 m→

The relationships between the metric units of area are:

- $1\ cm^2 = 10 \times 10\ mm^2$
  $= 100\ mm^2$
- $1\ m^2 = 100 \times 100\ cm^2$
  $= 10\ 000\ cm^2$
- $1\ km^2 = 1000 \times 1000\ m^2$
  $= 1\ 000\ 000\ m^2$

$\times 10^2$: $cm^2 \rightarrow mm^2$; $\div 10^2$: $mm^2 \rightarrow cm^2$

$\times 100^2$: $m^2 \rightarrow cm^2$; $\div 100^2$: $cm^2 \rightarrow m^2$

$\times 1000^2$: $km^2 \rightarrow m^2$; $\div 1000^2$: $m^2 \rightarrow km^2$

Example

Change $5\ m^2$ to cm^2.

$5\ m^2 = 5 \times 10\ 000\ cm^2$
$= 50\ 000\ cm^2$

You expect a larger number, so multiply.

The relationship between metric units of **volume** are

- $1\ cm^3 = 10 \times 10 \times 10\ mm^3$
  $= 1000\ mm^3$
- $1\ m^3 = 100 \times 100 \times 100\ cm^3$
  $= 1\ 000\ 000\ cm^3$

$\times 10^3$: $cm^3 \rightarrow mm^3$; $\div 10^3$: $mm^3 \rightarrow cm^3$

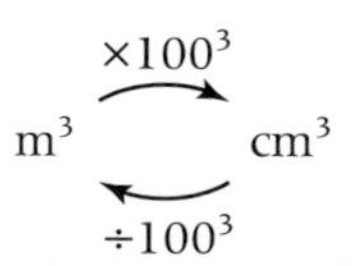

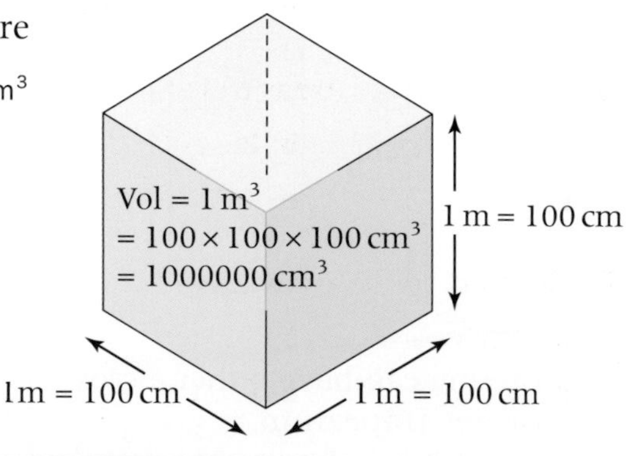

Example

Change $6\ 000\ 000\ cm^3$ to cubic metres (m^3).

$6\ 000\ 000\ cm^3 = 6\ 000\ 000 \div 1\ 000\ 000\ m^3$
$= 6\ m^3$

You expect a smaller number, so divide.

Exercise S4.7

1 Convert these metric measurements of length.

a 180 cm to mm **b** 45 mm to cm **c** 350 cm to m
d 2000 m to km **e** 3500 m to km **f** 4500 mm to m
g 85 cm to m **h** 2500 mm to cm **i** 2500 mm to m
j 800 m to km

2 Here are two identical rectangles, A and B.

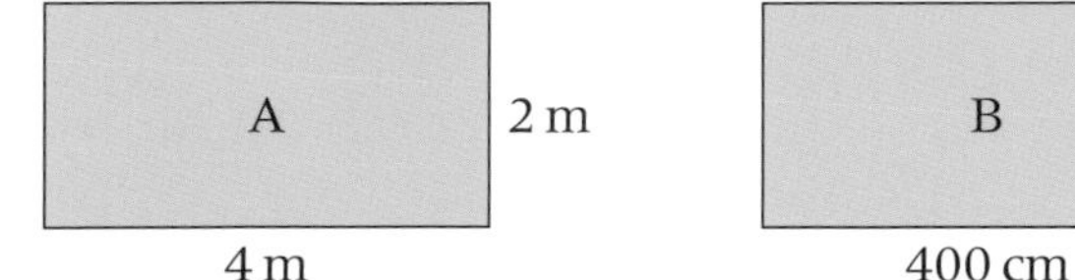

a Calculate the area of rectangle A in m^2.

b Calculate the area of rectangle B in cm^2.

3 **a** Calculate the area of this rectangle in m^2.

b Convert your answer to cm^2.

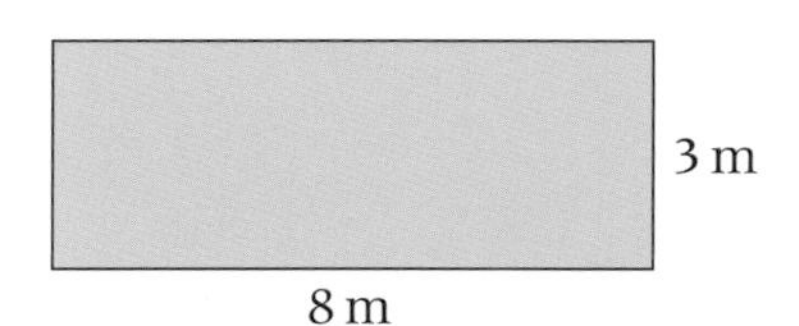

4 Convert these areas to mm^2.

a 4 cm^2 **b** 7.3 cm^2 **c** 10.9 cm^2
d 2.5 cm^2 **e** 400 cm^2

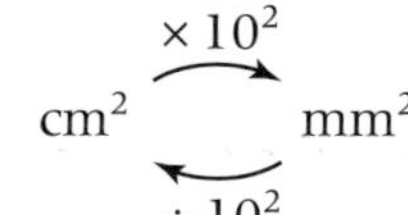

$10^2 = 100$

5 Convert these areas to cm^2.

a 600 mm^2 **b** 1200 mm^2 **c** 850 mm^2
d 6500 mm^2 **e** 10 000 mm^2

6 Convert these areas to m^2.

a 40 000 cm^2 **b** 85 000 cm^2 **c** 1 000 000 cm^2
d 125 000 cm^2 **e** 5000 cm^2

$m^2 \xrightarrow{\times 100^2} cm^2$, $cm^2 \xrightarrow{\div 100^2} m^2$

$100^2 = 10\ 000$

7 Convert these areas to cm^2.

a 5 m^2 **b** 10 m^2 **c** 6.5 m^2
d 7.75 m^2 **e** 0.6 m^2

8 Convert these areas to km^2.

a 4 000 000 m^2 **b** 18 000 000 m^2
c 500 000 m^2 **d** 1 500 000 m^2

$km^2 \xrightarrow{\times 1000^2} m^2$, $m^2 \xrightarrow{\div 1000^2} km^2$

$1000^2 = 1$ million

9 Convert these volumes to litres.

a 1 m^3 **b** 6 m^3 **c** 7.5 m^3

1 m^3 = 1000 litres

S4.8 Dimensions

This spread will show you how to:

- Understand the difference between formulae for perimeter, area and volume, considering dimensions

Keywords
Area
Dimensions
Formula
Length
Volume

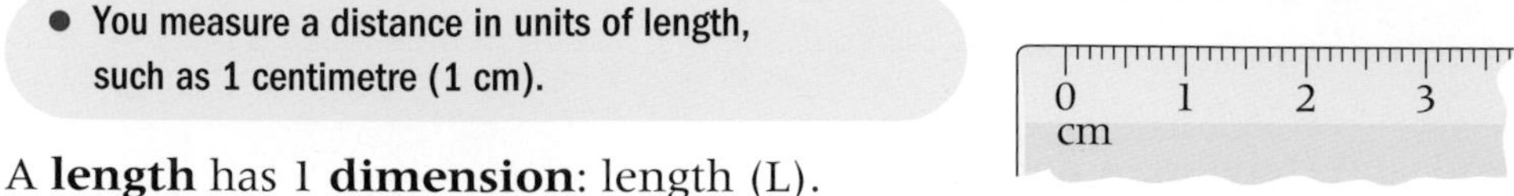

- **You measure a distance in units of length, such as 1 centimetre (1 cm).**

A **length** has 1 **dimension**: length (L).

- **You measure area in squares, such as 1 square centimetre (1 cm^2).**

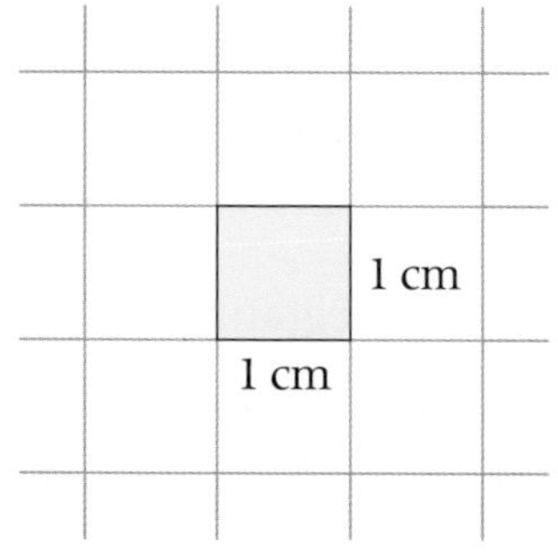

Area has 2 dimensions: length × length (L^2).

- **You measure volume in cubes, such as 1 cubic centimetre (1 cm^3).**

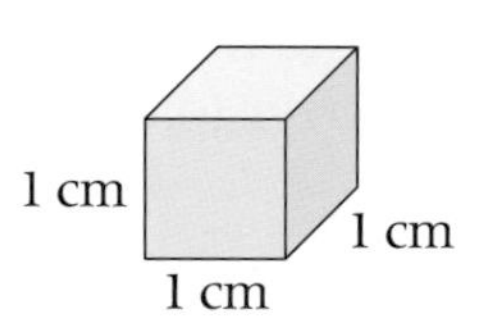

Volume has 3 dimensions:
length × length × length (L^3).

Numbers such as 2 and π have no dimensions.

You can use dimensions to check **formulae**.

For a rectangle, with sides of length a and b,

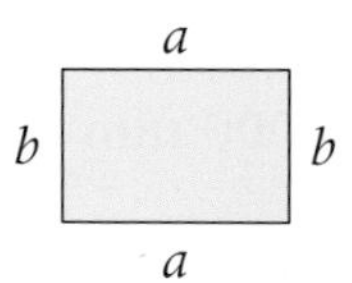

Perimeter $= a + b + a + b$
$= 2a + 2b$

Dimensions: number × length + number × length = L + L = 2L = length

Area $= a \times b$

Dimensions: length × length = L^2 = area

Example

David cannot remember the formula for the volume of a cylinder. He thinks it might be

$2\pi r$ or $\pi(r + h)$ or πr^2 or $\pi r^2 h$

How can he work out which one it might be?

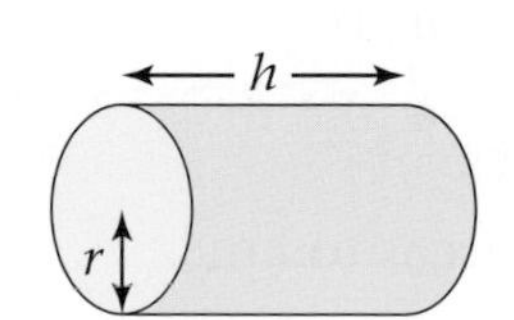

$2\pi r$	number × number × length = L = a length	NO
$\pi(r + h)$	number × (length + length) = L = a length	NO
πr^2	number × length × length = L^2 = an area	NO
$\pi r^2 h$	number × length × length × length = L^3 = a volume	YES

π is just a number and has no dimensions.

Exercise S4.8

1 Choose one of length (L), area (L^2), volume (L^3) or none of these (N) for the dimensions of each of these.

a The surface of a cube
b The space inside a gym
c The distance from London to Paris
d 6
e The amount of liquid in a mug
f The mass of a guinea pig
g The height of a mountain
h 90 litres
i The surface covered by a lawn
j 45°
k The perimeter of an airport
l 3 metres
m Your weight
n £10
o 80 kilograms
p The surface covered by a wall
q 4 m^2
r π
s The amount of space inside a car
t 9 m^3

2 State whether these expressions represent length (L), area (L^2) or volume (L^3).

a length × width
b length + width + length + width
c length × width × height
d base length × height
e $\frac{1}{2}$ × base × height
f diameter
g π × diameter
h π × radius × radius

3 One of these formulae gives the volume of a sphere. Which one?

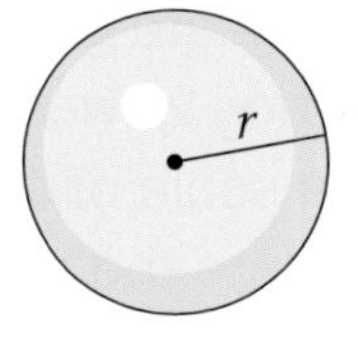

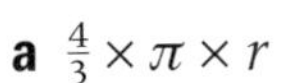
a $\frac{4}{3} \times \pi \times r$
b $\frac{4}{3} \times \pi \times r^2$
c $\frac{4}{3} \times \pi \times r^3$

4 One of these formulae gives the surface area of a cylinder. Which one?

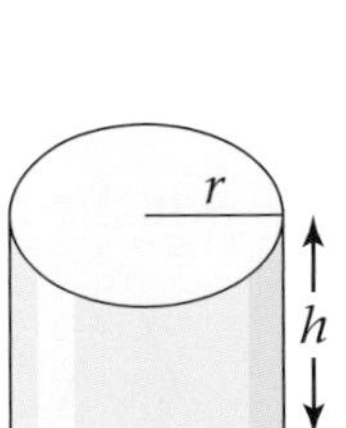

a $2\pi r + 2\pi h$
b $2\pi r^2 h$
c $2\pi r^2 + 2\pi rh$

5 One of these formulae gives the volume of a cone. Which one?

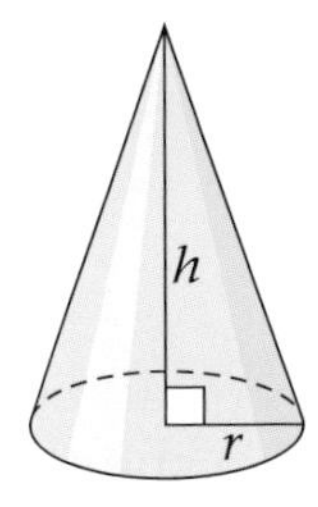

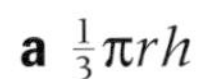
a $\frac{1}{3}\pi rh$
b $\frac{1}{3}\pi r^2 h$
c $\frac{1}{3}r + \frac{1}{3}h$

S4

Exam review

Key objectives

- Use angle properties of equilateral, isosceles and right-angled triangles
- Classify quadrilaterals by their geometric properties
- Use 2-D representations of 3-D shapes and analyse 3-D shapes through 2-D projections and cross-sections, including plan and elevation
- Understand that one coordinate identifies a point on a number line, two coordinates identify a point in a plane and that three coordinates identify a point in space, using the terms '1-D, 2-D and 3-D'
- Find the surface area of sample shapes using the area formulae for triangles and rectangles
- Calculate the volume of right prisms

1 A kitchen waste bin is a prism. The cross-section is a rectangle 30 cm wide and 50 cm high, topped by a semicircle of radius 15 cm. The bin has a square base.

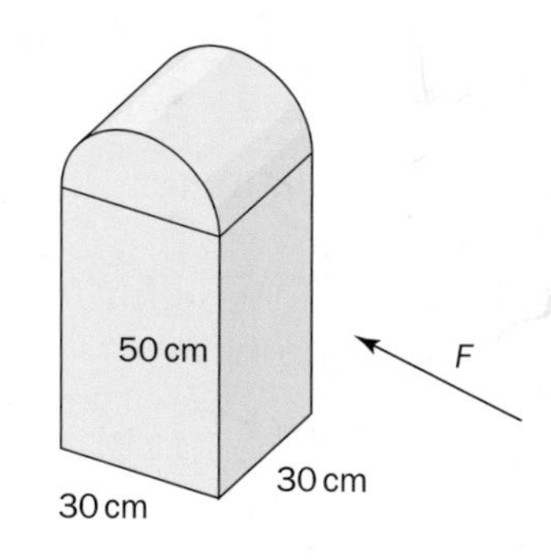

a On a square grid, draw the plan and the front elevation viewed from *F*.
Use a scale of 1 cm to 10 cm. (3)

b Calculate the total volume of the waste bin. (5)

(OCR, 2003)

2 **a** Write down the mathematical name of these 3-D shapes:

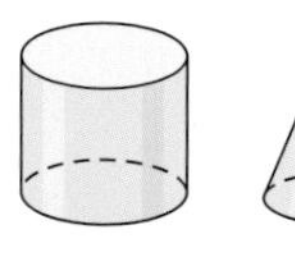

b Here are nets of two different 3-D shapes: (2)

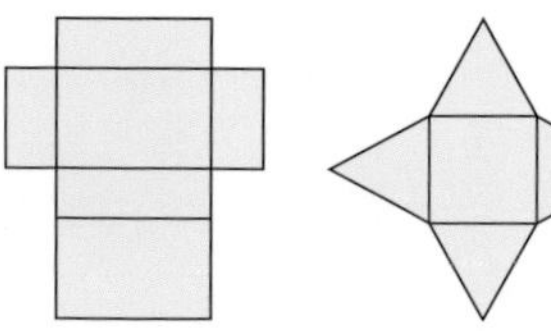

Write down the mathematical name of each of these 3-D shapes. (2)

(Edexcel Ltd., 2004)

S5 Constructions and loci

This unit will show you how to

- Understand angle measure using the associated language
- Construct triangles using a straight edge, protractor and compasses
- Use a straight edge and compasses to construct the perpendicular from a point to a line and the perpendicular from a point on the line
- Use straight edge and compasses to do standard constructions, including the bisector of an angle
- Find loci, both by reasoning and by using diagrams

Before you start ...

You should be able to answer these questions.

1 Using a protractor, measure these angles.

a **b** **c**

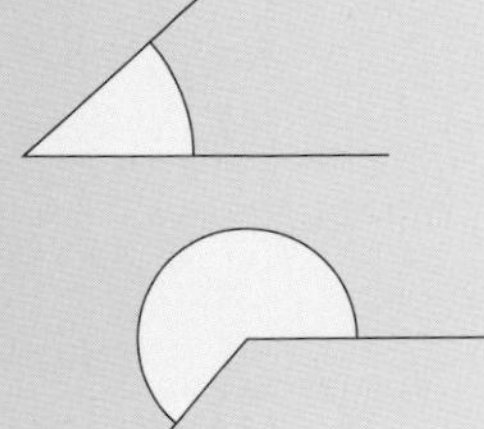

2 State the value of the marked angle.

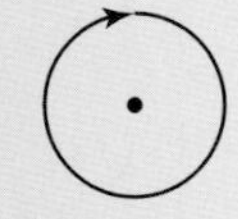

3 Using compasses, draw a circle with a diameter of 4.6 cm.

4 Calculate the area of this triangle, stating the units of your answer.

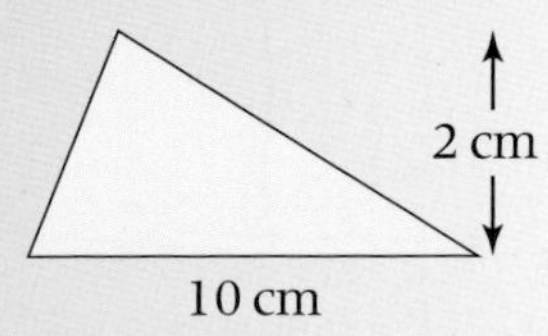

5 Measure this line

---

a in millimetres **b** in centimetres.

S5.1 Bearings

This spread will show you how to:

- Understand angle measure using the associated language

Keywords
Bearing
Direction
Scale
Three-figure bearing

North, East, South or West are not enough to give an accurate direction on most occasions.

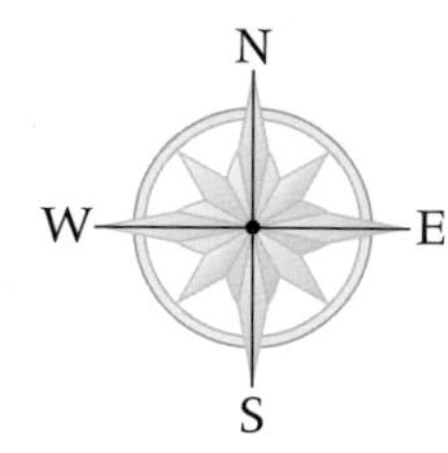

- **A bearing is an angle measured clockwise from North.**

To give a **direction** accurately, you need to find an angle measured on a 360° **scale**.

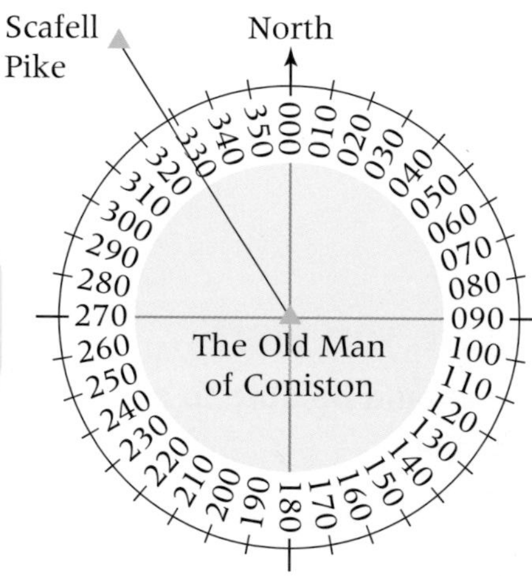

000° = North.

- **To specify a direction with a bearing:**
  - **measure from North**
  - **measure clockwise**
  - **use three figures.**

The bearing of Scafell Pike from the Old Man of Coniston is 328°

Example

A boat is sinking.

The bearing of the boat from Dawlish Warren is 168°.

The bearing of the boat from Holcombe is 085°.

a Mark the position of the boat on the map.

b How far is the boat from Dawlish?

c What is the bearing of the boat from Dawlish?

---

a See map.

b 2 cm represents 1 km.
4.6 cm represents 2.3 km.
The boat is 2.3 km from Dawlish.

c 125°

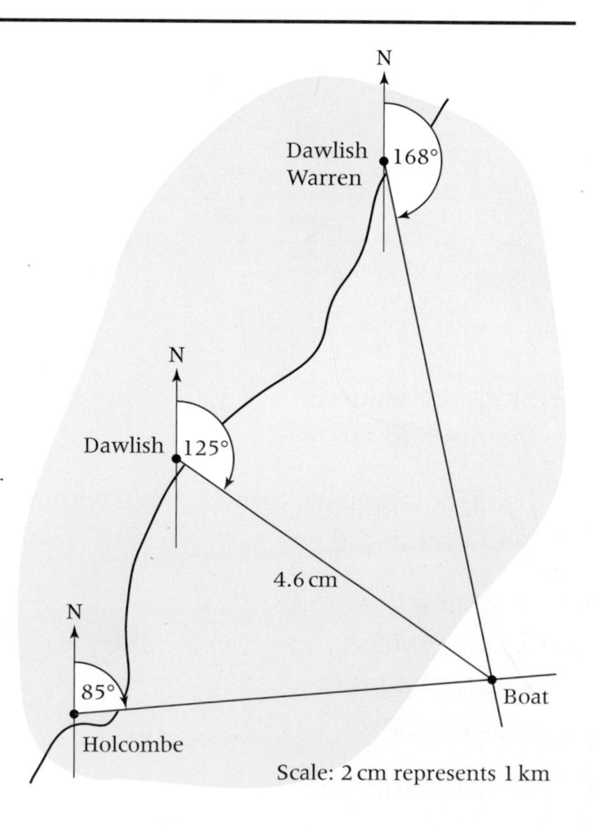

From Dawlish Warren means centre the protactor **at** Dawlish Warren.

Do not rub out the construction lines.

Exercise S5.1

1 For each question, put a cross on your page.
Plot the points and join them to form a quadrilateral.
Name the shape, then measure and calculate the perimeter.

a

Bearing from the cross	060°	120°	240°	300°
Distance from the cross	5 cm	5 cm	5 cm	5 cm

b

Bearing from the cross	000°	090°	180°	270°
Distance from the cross	2.5 cm	5 cm	2.5 cm	5 cm

c

Bearing from the cross	035°	145°	215°	325°
Distance from the cross	5 cm	5 cm	5 cm	5 cm

2 Measure and write the bearing of

a Leeds from Manchester

b Sheffield from Leeds

c Manchester from Leeds

d Manchester from Sheffield

e Leeds from Sheffield.

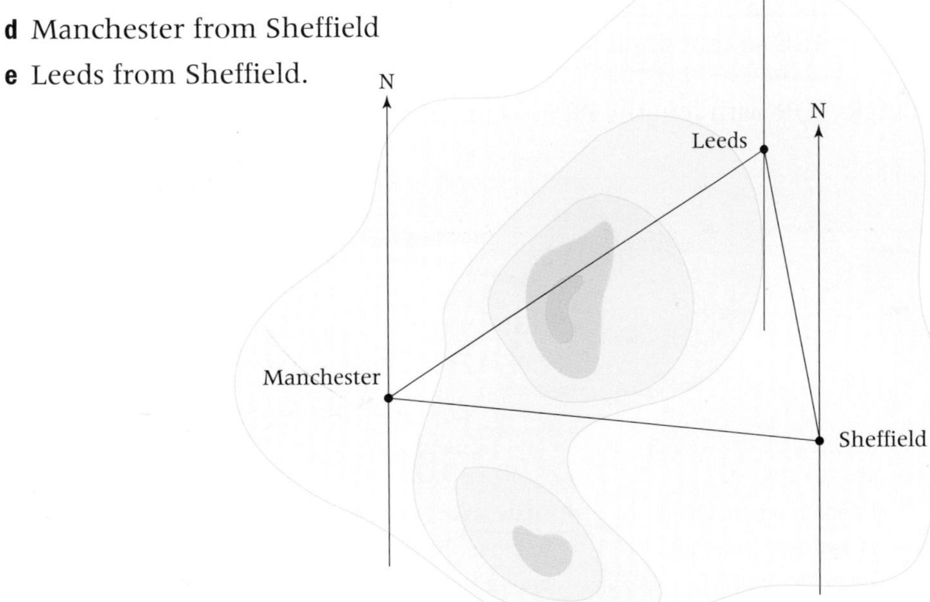

3 Copy the diagram.

The distance from Truro to Falmouth is 14 km.
The bearing of St. Mawes from Falmouth is 080°.
The bearing of St. Mawes from Truro is 170°.

a Mark the position of St. Mawes on your diagram.

b Calculate the distance from Falmouth to St. Mawes.

c Calculate the distance from Truro to St. Mawes.

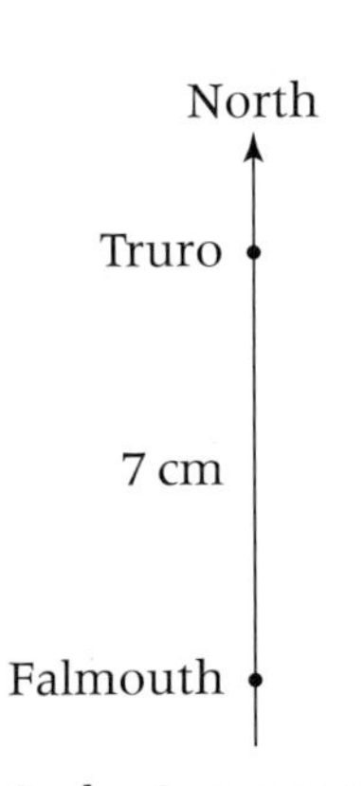

Scale: 1cm represents 2 km

S5.2 Constructing triangles

This spread will show you how to:

- Construct triangles using a straight edge, protractor and compasses

Keywords
Arc
Base
Compasses
Construct
Construction lines
Hypotenuse
Protractor
Straight edge

You can **construct** a triangle when you know

Two sides and the angle between them (SAS) or Two angles and a side (ASA) or Right angle, the hypotenuse and a side (RHS) or Three sides (SSS)

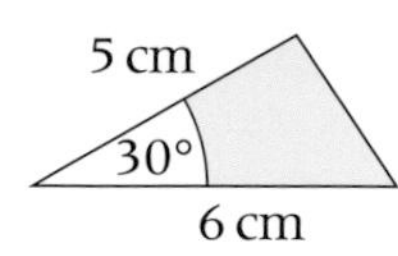

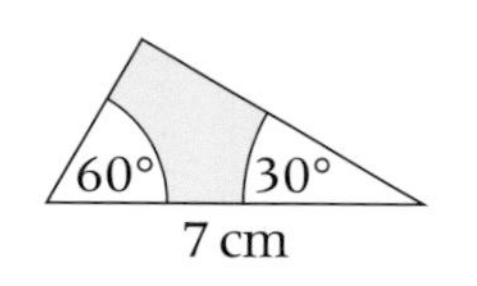

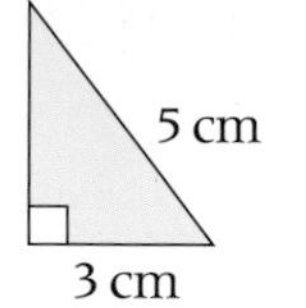

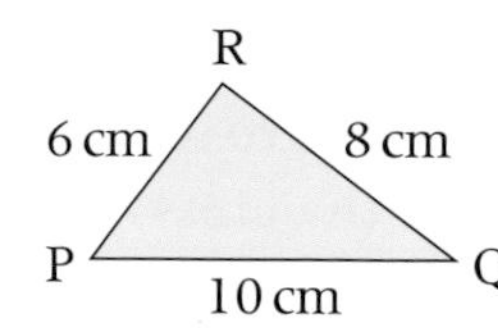

You will need a ruler and a protractor for SAS, ASA and RHS triangles.

You will need a ruler and compasses for SSS triangles.

The longest side of a **right-angled** triangle is called the **hypotenuse**.

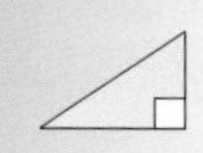

Example

a Construct the triangle ABC so that angle C = 90°, AB = 6 cm and BC = 3 cm.

b Construct the triangle PQR with lengths PR = 6 cm, QR = 8 cm and PQ = 10 cm.

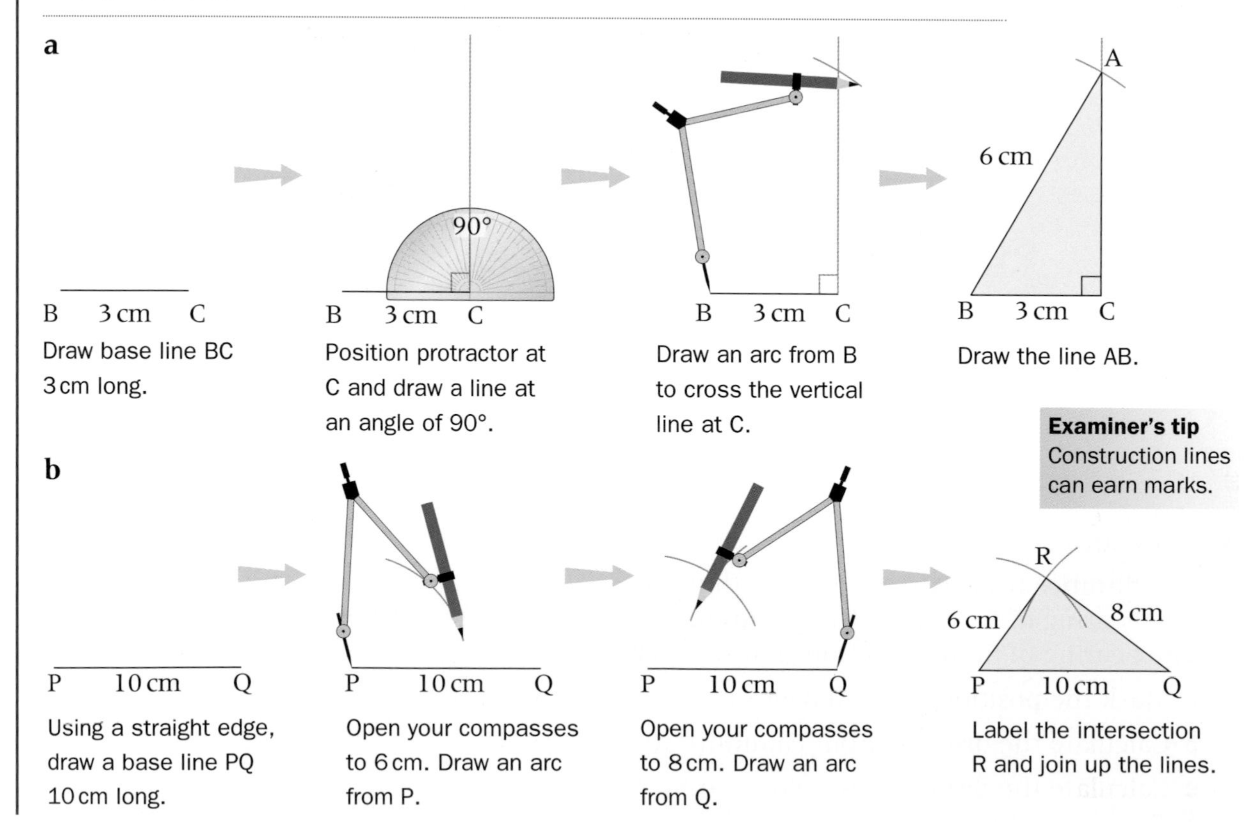

Examiner's tip
Construction lines can earn marks.

Exercise S5.2

> Leave your **construction lines** on your drawing to show your method.

1 Make accurate drawings of these triangles (SAS).
Measure the unknown length in each triangle.

a

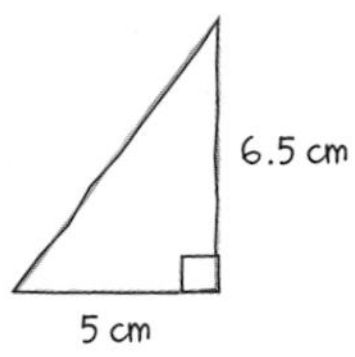

b

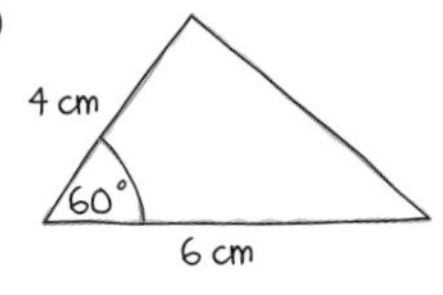

c

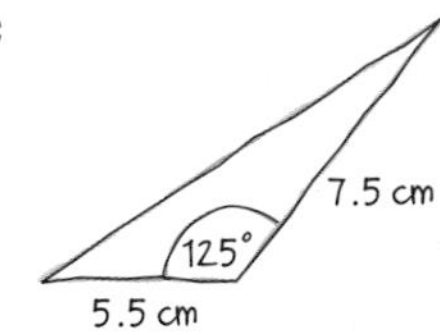

2 Make accurate drawings of these triangles (ASA).
Measure the two unknown lengths in each triangle.

a

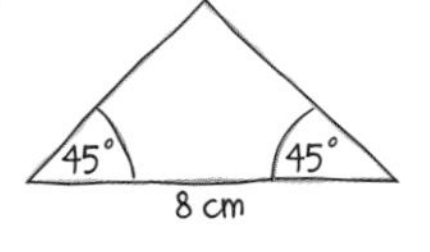

b

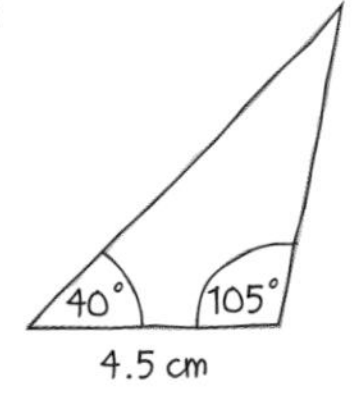

c

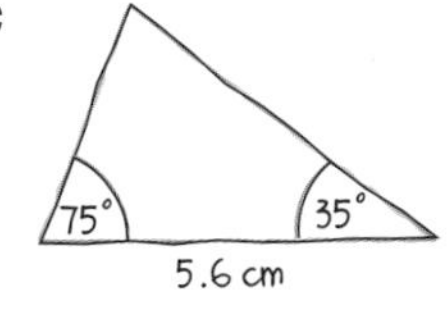

3 Make accurate drawings of these triangles (SSS).
Measure the marked angle in each triangle.

a

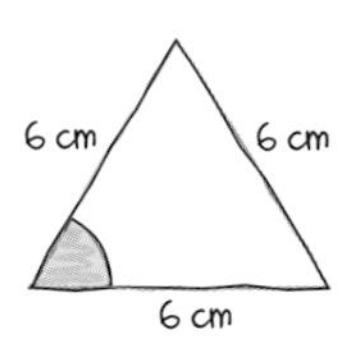

b

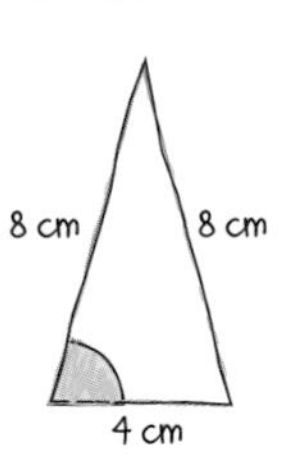

c

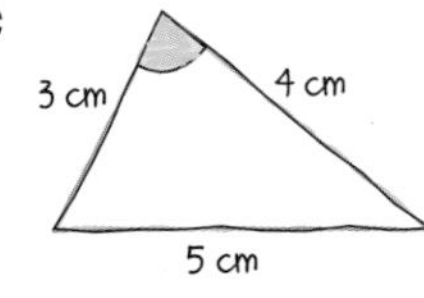

4 Make accurate drawings of these triangles (RHS).
Measure the unknown length and the marked angle in each triangle.

a

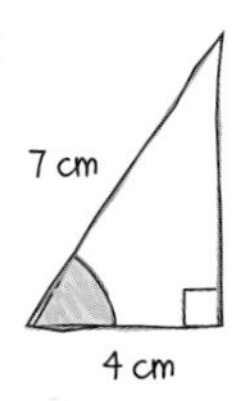

b

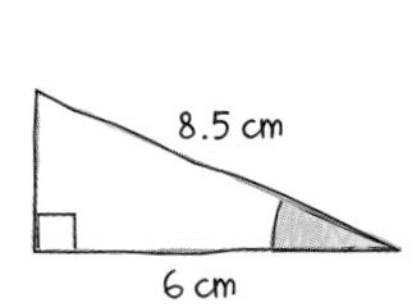

c

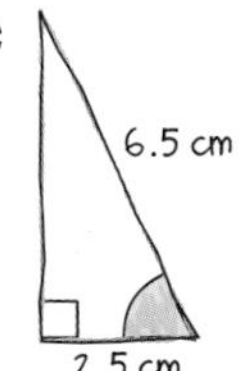

5 Make accurate drawings of these triangles (SSA).
Measure the unknown length in each triangle.

a

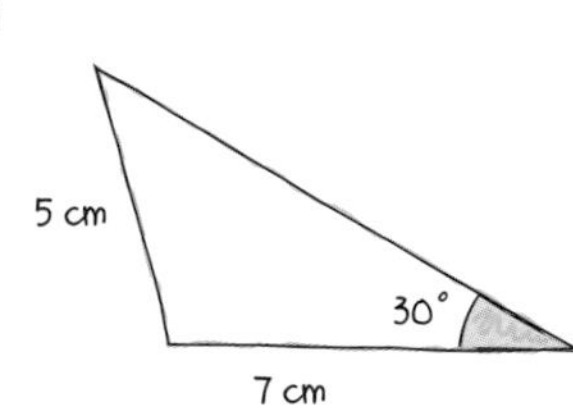

b

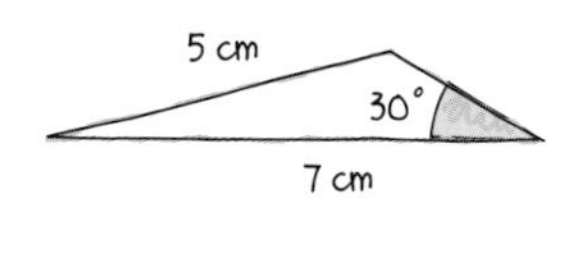

c How many different triangles ABC can you draw so that
AB = 5.5 cm, BC = 8 cm, and angle C = 40°?

S5.3 Perpendicular lines

This spread will show you how to:

- Use a straight edge and compasses to construct the perpendicular from a point to a line and the perpendicular from a point on the line

Keywords
Arc
Compasses
Construct
Midpoint
Perpendicular
Perpendicular bisector

The shortest distance from a point to a line is the **perpendicular** distance.

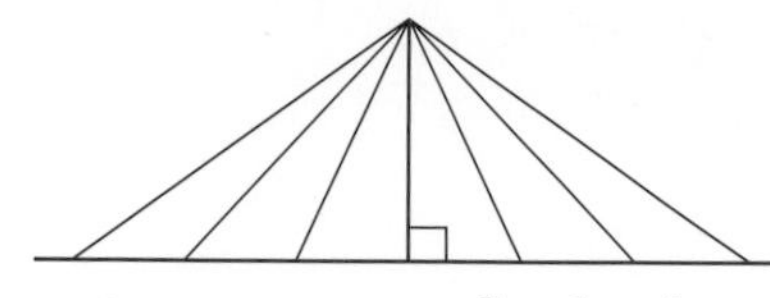

The perpendicular meets the line at right angles.

You construct a perpendicular from a point P to a line like this.

Construct implies 'use compasses'.

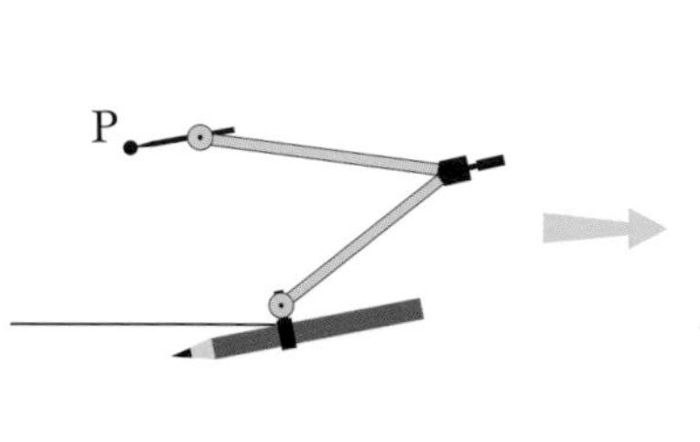

Open your compasses so that the distance is longer than the distance from the point to the line.

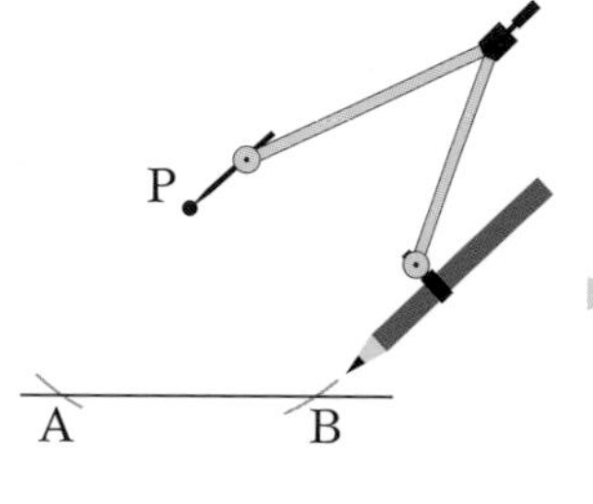

Construct two arcs from the point to the line.

Keep the compasses the same width and construct an arc from A and from B to meet at C. Join P to point C.

You construct the perpendicular from a point P on the line like this.

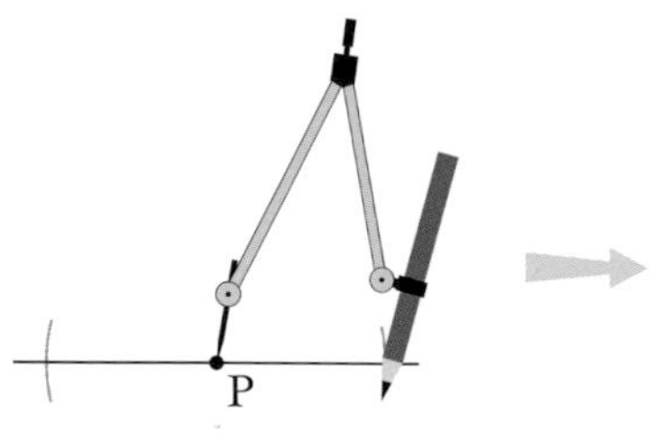

Construct two arcs on the line equidistant from point P.

Keep the compasses the same width apart. Construct arcs above and below the line from point A and from point B.

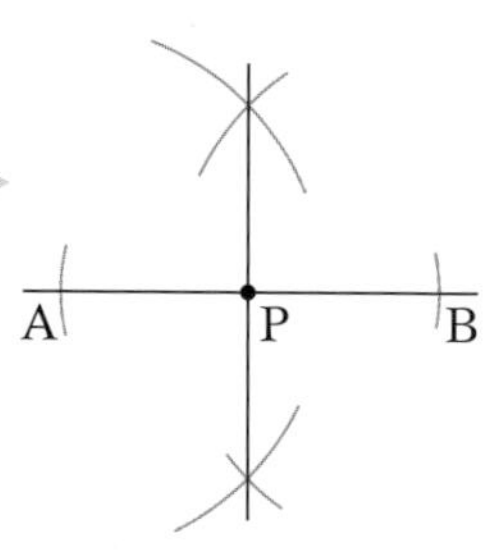

Draw the perpendicular bisector of AB.

P is the **midpoint** of AB.

- **A perpendicular bisector divides a straight line into two equal parts at right angles.**

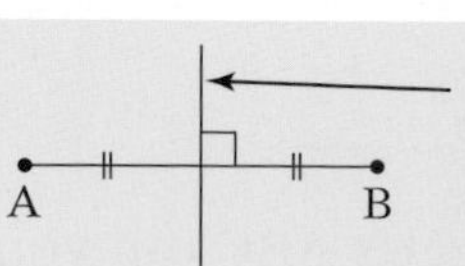

This line is the perpendicular bisector of AB.

Exercise S5.3

For all constructions use a pencil and do not rub out your construction lines.

1 **a** Draw a line AB, so that AB = 8 cm.

b Using compasses, construct the perpendicular bisector of AB.

c Label the midpoint of AB as M.

d Measure the length AM.

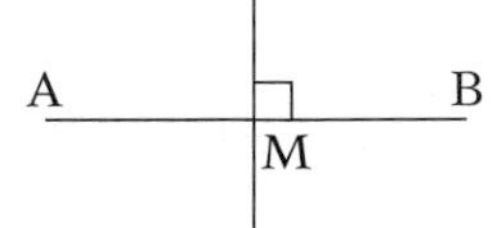

2 **a** Draw a line of length 64 mm.

b Construct the perpendicular bisector of the line.

c Check by measuring that the perpendicular bisector passes through the midpoint of the line.

3 **a** Draw a line AB, so that AB = 10 cm.

b Mark the point P, so that AP = 7 cm.

c Construct the perpendicular to AB that passes through the point P.

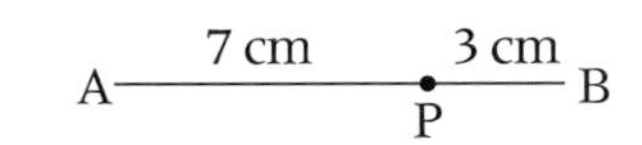

4 **a** Draw a line AB, with a point P above the line.

b Using compasses, construct the perpendicular to AB that passes through the point P.

c Measure the angle between the line AB and the perpendicular from P.

5 Construct these rhombuses using compasses and ruler.

a

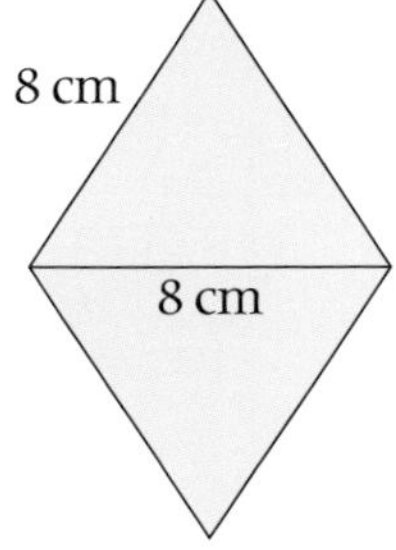

b

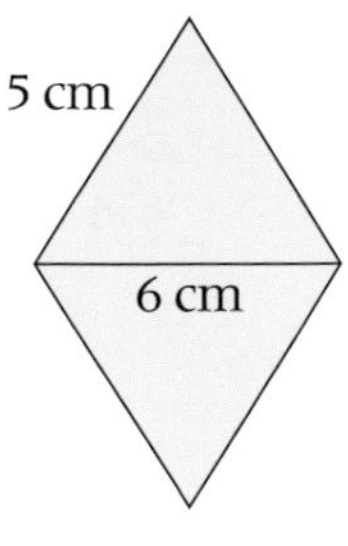

c

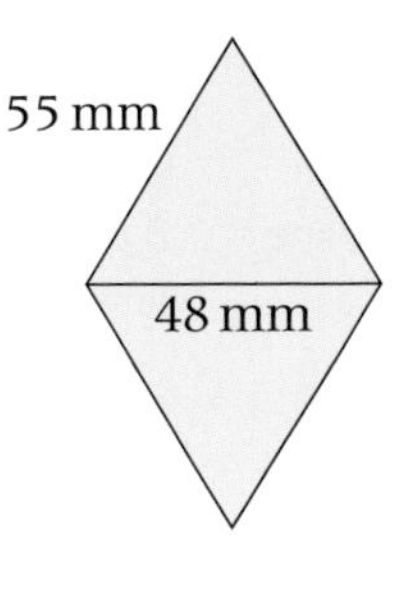

Draw the vertical diagonals on your diagrams.
For each rhombus, check that each diagonal is a perpendicular bisector of the other.

S5.4 Angle bisectors

This spread will show you how to:

- Use straight edge and compasses to do standard constructions, including the bisector of an angle

Keywords
Angle bisector
Arc
Bisect
Compasses
Construct

- **To bisect an angle, you cut the angle exactly in half.**

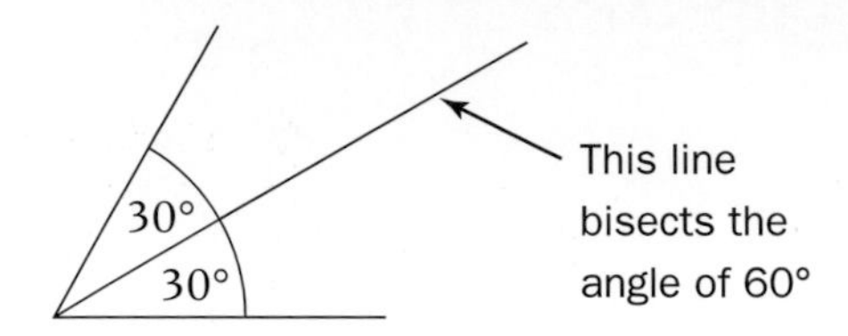

You use **compasses** to construct an **angle bisector**.

Use compasses to draw equal arcs on each arm.

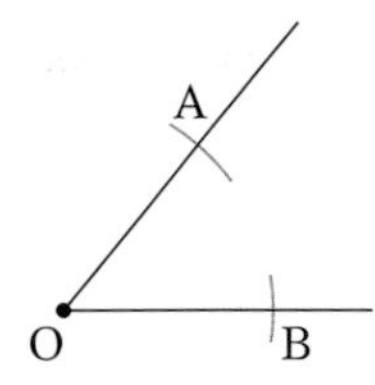

Draw equal arcs from these arcs that intersect at C.

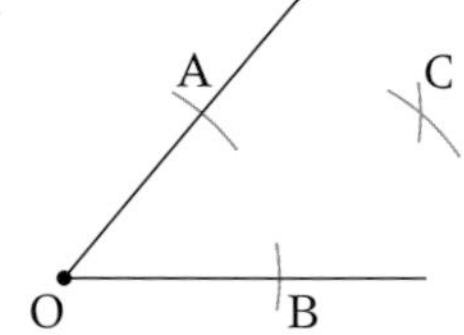

Join O to C, the vertex of the angle

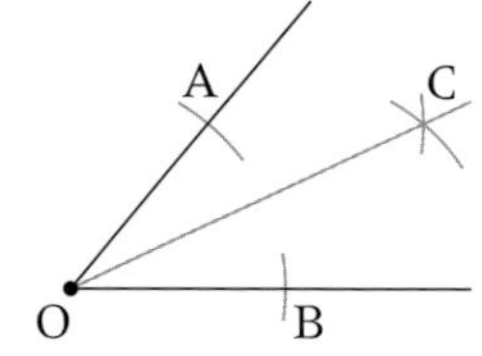

OACB is a rhombus.

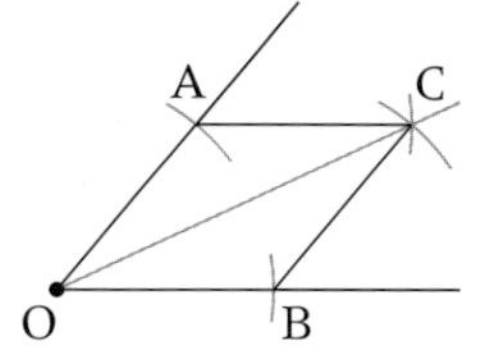

Angle AOC = angle BOC

Example

Using compasses, construct

a an angle of 60°
b an angle of 30°.

---

a

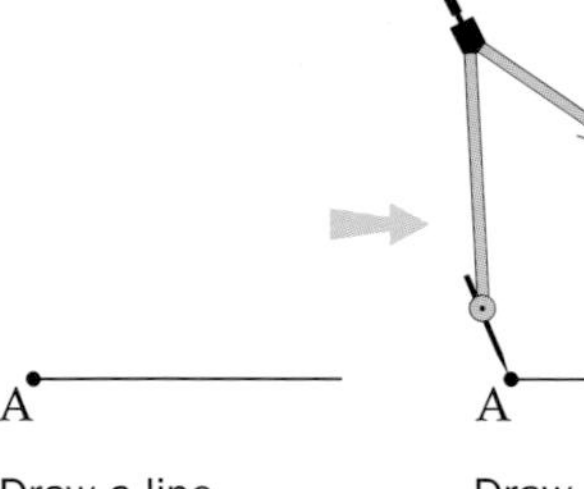

Draw a line.

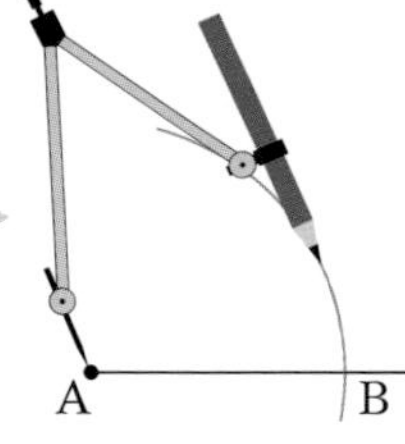

Draw an arc from A crossing the line at B.

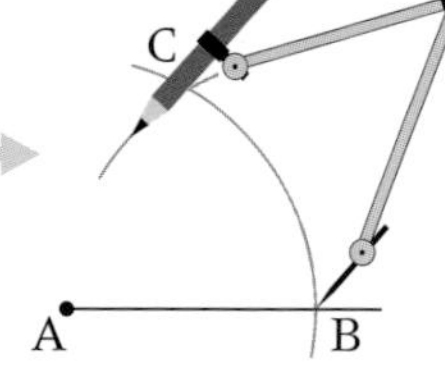

Draw an arc from B crossing the first arc at C.

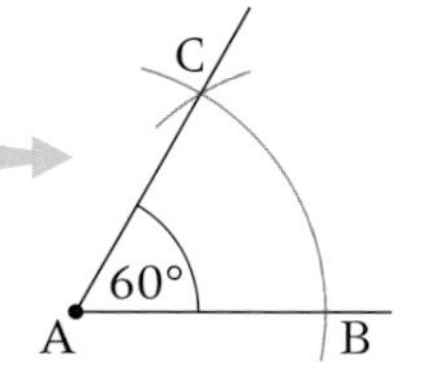

Draw a line from A to C.

b

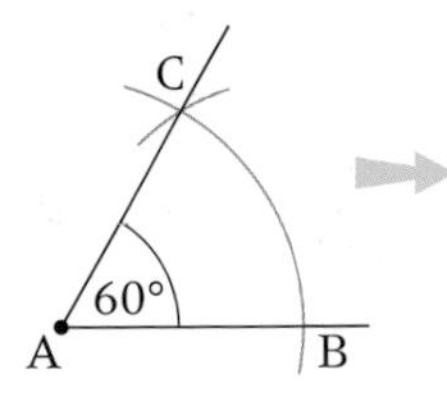

Use the angle of 60°.

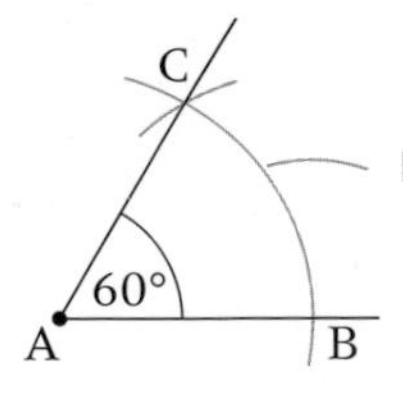

Draw an arc from B.

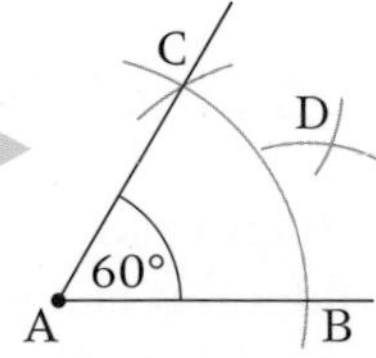

Draw an arc from C crossing the previous arc at D.

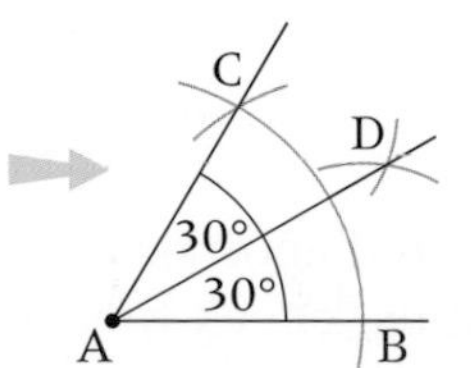

Draw a line from A to D.

Exercise S5.4

1 Use a protractor to draw these angles.

a 70° **b** 110° **c** 90° **d** 130° **e** 50°

Using compasses, construct the angle bisectors for each angle.
Use a protractor to check each angle bisector.

2 **a** Using compasses, construct an angle of 60°.

b Use a protractor to check the angle.

3 Construct a triangle that is similar to this equilateral triangle.

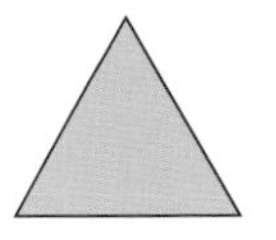

Similar shapes are the same in shape but different in size.

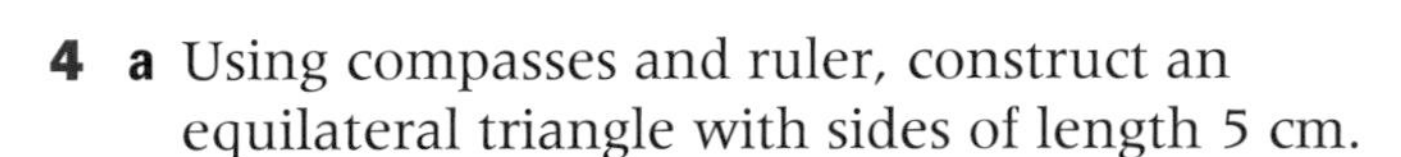

4 **a** Using compasses and ruler, construct an equilateral triangle with sides of length 5 cm.

b Use a protractor to check the angles.

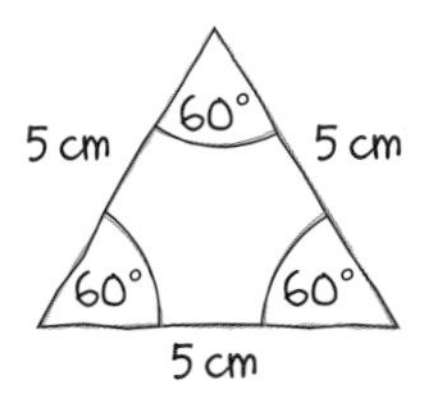

5 Using compasses, construct an angle of

a 30° **b** 120°.

Use a protractor to check your answers.

Bisect a 60° angle.

6 **a** Use a protractor to draw two perpendicular lines as shown.

b Using compasses, construct the bisectors of the right angles.

c Label the eight lines N, NE, E, SE, S, SW, W, NW.

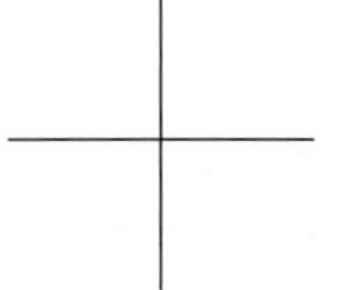

7 **a** Draw any two intersecting lines.

b Construct the angle bisectors for the acute angles.

c Construct the angle bisectors for the obtuse angles.

d Copy and complete this sentence:

The bisector of the acute angles is ______ to the bisector of the obtuse angles.

e Explain why the sentence is true for any two intersecting lines.

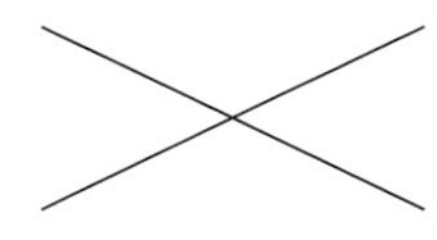

8 **a** Using compasses, construct the triangle PQR.

b Construct the angle bisectors for angle P, angle Q and angle R.

c Label the point of intersection of the angle bisectors as O.

d Draw a circle, centre O, that just touches the lines PQ, QR and PR.

e State the radius of this circle.

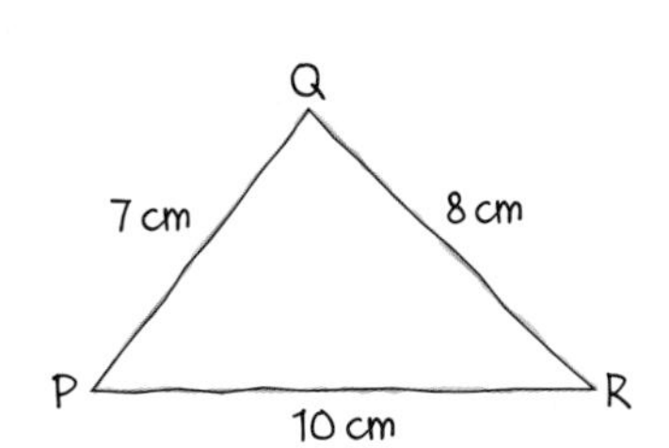

Loci

This spread will show you how to:

- Find loci, both by reasoning and by using diagrams

Keywords
Arc
Compasses
Construct
Equidistant
Loci
Locus
Perpendicular bisector

Loci is the plural of locus.

- **The locus of an object is its path.**
- **A locus is a set of points that move according to a rule.**

The red counters are all the same distance from the blue counter.

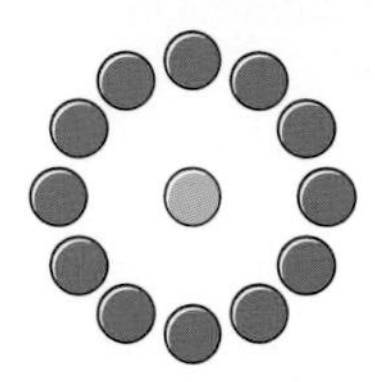

The locus is a circle.

The red counters are the same distance from the two blue lines.

The locus is a straight line.

The red counters are the same distance from the two blue lines.

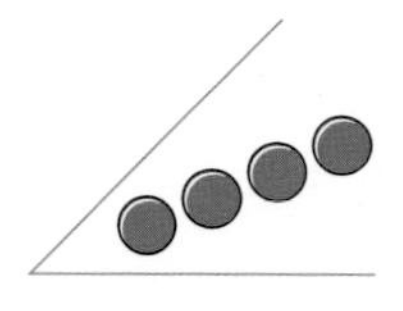

The locus is the angle bisector of the angle between the two blue lines.

The red counters are the same distance from the two blue counters.

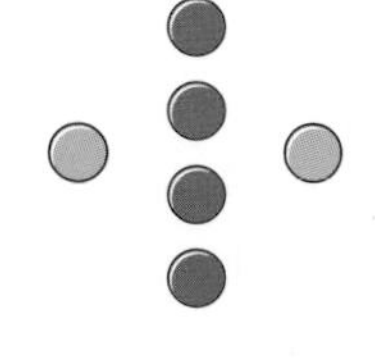

The locus is the perpendicular bisector of the line joining the two blue counters.

Example

Some treasure is positioned 2 m from C, but **equidistant** from A and B.
Using ruler and compasses only, mark the possible positions of the treasure.

×C

A× ×B

Draw a circle, centre C, radius 2 m.
Join A and B.
Construct the perpendicular bisector of AB.
Mark where the line crosses the circle.

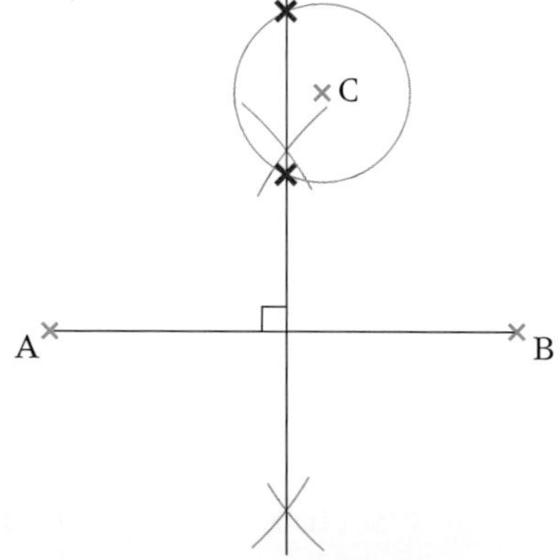

Example

A bull is tethered to a stake by a 5m chain in the middle of a field. Use a scale of 1cm represents 1m to show the extent the bull can move.

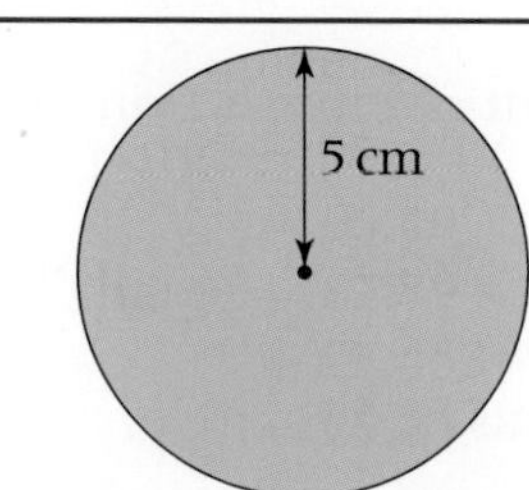

Exercise S5.5

1 **a** Using a protractor, draw and label an angle of 50°.

b Draw the locus of the points that are the same distance from AB and BC.

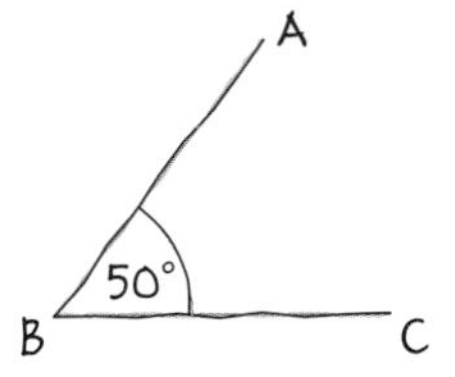

2 Draw a point and label it O. Construct the locus of the point that are 3 cm from the point O.

• O

3 Draw two parallel lines.
Draw the locus of the points that are equidistant from the two lines.

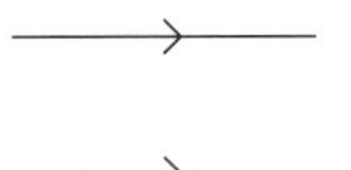

4 **a** Draw a line AB so that AB = 8 cm.

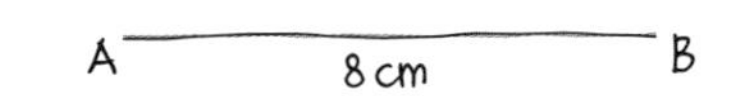

b Construct the locus of the points that are equidistant from A and B.

c On your diagram, indicate the region that has points that are nearer to A than B.

5 Copy this diagram.
The diagram represents a triangular field using a scale of 1 cm represents 10 cm.
Use constructions to find the position of the hidden treasure that

- is 20m from the point A.
- lies along the perpendicular bisector of BC.

A

5 cm 5 cm

1 cm represents 10 m

B 4 cm C

6 A 3 metre length of rope is used to restrain a dog. The rope is fixed to the corner of a building. Draw a diagram to show the extent to which the dog can move.

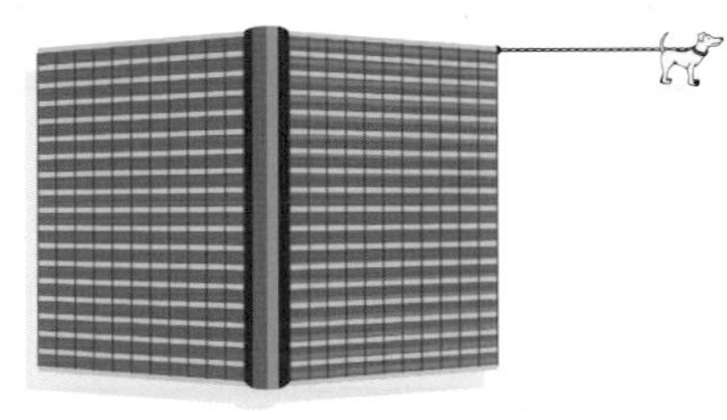

1 cm represents 1 m

S6.3 Relationships of scale

This spread will show you how to:

- Understand the effects of enlargement on the angles, perimeter, area and volume of shapes and solids

Keywords
Area
Enlargement
Multiplier
Perimeter
Scale factor
Volume

In this **enlargement**, corresponding lengths are multiplied by 2, so the **scale factor** is 2.

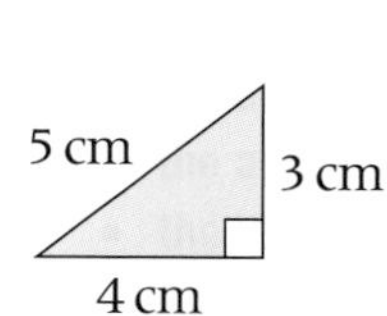

Perimeter = 12 cm
Area = 6 cm^2

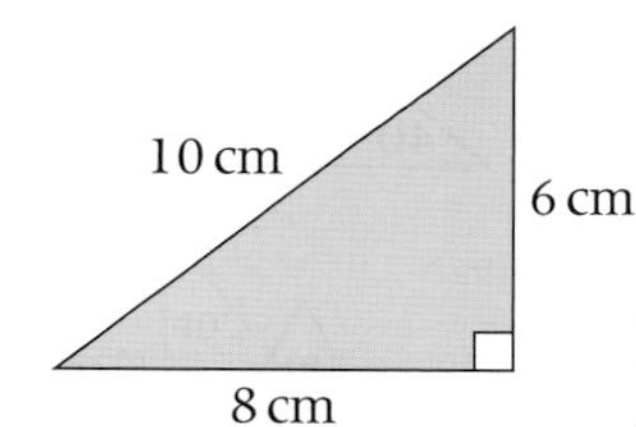

Perimeter = 24 cm
Area = 24 cm^2

Area of a triangle $= \frac{1}{2} \times$ base $\times$ height.

If scale factor for length is 2 then scale factor for **area** is $2 \times 2 = 4$.

Scale factor is 2 but area is ×4.

In this enlargement, corresponding lengths are multiplied by 2, so the scale factor is 2.

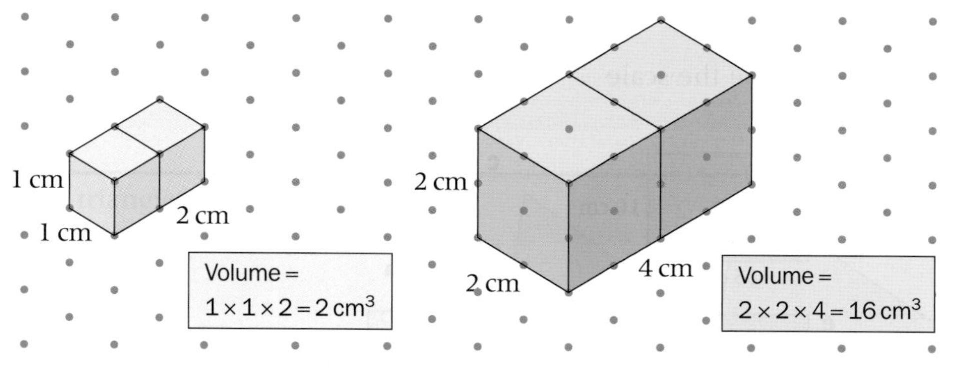

Volume of a cuboid = length × width × height.

If scale factor for length is 2 then scale factor for **volume** is $2 \times 2 \times 2 = 8$.

Scale factor is 2 but volume is ×8.

- **If scale factor for length is L then the multiplier for area is $L \times L = L^2$ and the multiplier for volume is $L \times L \times L = L^3$.**

Example

A 2 cm by 3 cm by 4 cm cuboid is shown. Calculate

a the area of the base rectangle

b the volume of the cuboid.

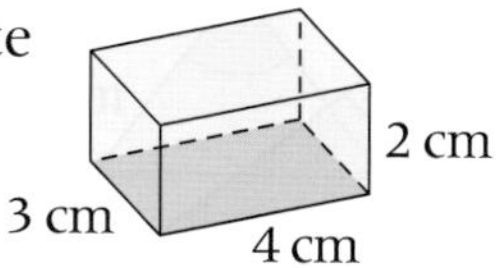

The cuboid is enlarged by scale factor 3.

c Calculate the area of the new base rectangle.

a Area = $3 \times 4 = 12$ cm^2

b Volume = $2 \times 3 \times 4 = 24$ cm^3

c Scale factor is 3, so new lengths are
$3 \times 3 = 9$ cm, $4 \times 3 = 12$ cm, $2 \times 3 = 6$ cm
Area = $9 \times 12 = 108$ cm^2

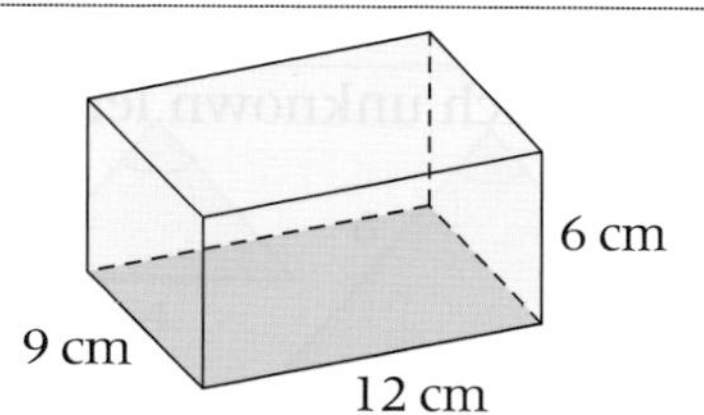

Scale factor is 3 but area is ×9.

Note that enlarged volume = $6 \times 9 \times 12$
= 648 cm^3

Scale factor is 3 but volume is ×27.

Exercise S6.3

1 **a** Calculate the perimeter of this rectangle. State the units of your answer.

b Find the area of the rectangle. State the units of your answer.

The rectangle is enlarged by scale factor 5.

c Calculate the length and width of the enlarged rectangle.

d Calculate the perimeter of the enlarged rectangle.

e Calculate the area of the enlarged rectangle.

f Copy and complete this sentence:

For an enlargement scale factor 5, the perimeter increases by multiplying by ___ and the area increases by multiplying by ___.

2 **a** Calculate the volume of this cuboid. State the units of your answer.

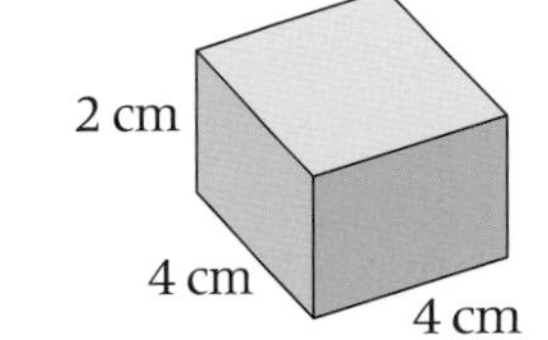

The cuboid is enlarged by scale factor 3.

b Calculate the dimensions of the new cuboid.

c Calculate the volume of the new cuboid.

d Copy and complete this sentence:
For an enlargement scale factor 3, the volume increases by multiplying by ___.

3 **a** Calculate the area of this triangle. State the units of your answer.

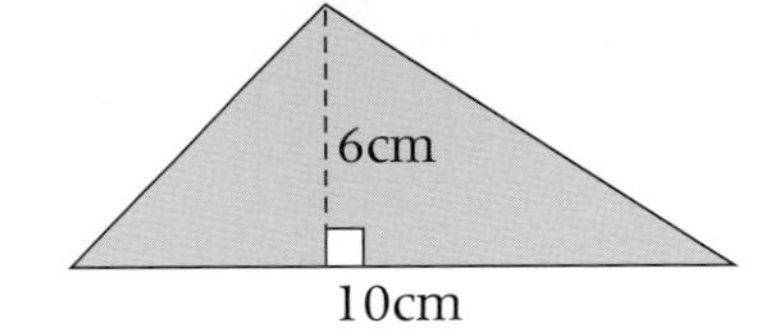

The triangle is enlarged by scale factor 2.

b Calculate the area of the enlarged triangle.

4 **a** Calculate the volume of this cuboid. State the units of your answer.

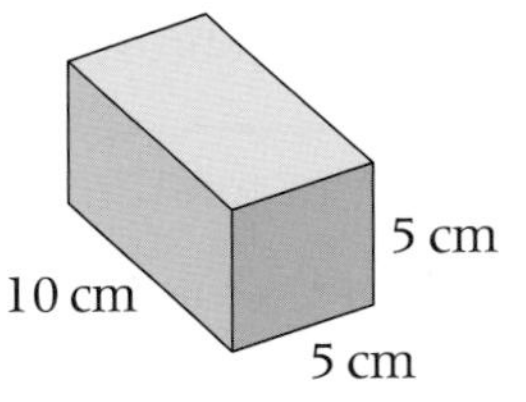

The cuboid is enlarged by scale factor 2.

b Calculate the volume of the enlarged cuboid.

5 Copy and complete this table for enlargements.

Scale factor	Multiplier for length	Multiplier for area	Multiplier for volume
2	2	4	8
3	3	9	27
4			
5			
6			
7			

6 A shape has a perimeter of 14 cm and an area of 10 cm^2. Calculate the perimeter and area of the shape after an enlargement of scale factor 4.

S6.4 Maps and scale drawings

This spread will show you how to:

- Use and interpret maps and scale drawings

Keywords
Enlarged
Reduced
Scale drawing
Scale factor

In **scale drawings**, lines and shapes are **reduced** or **enlarged**.

Corresponding lengths are multiplied by the same **scale factor**.
You can write the scale factor as a ratio.

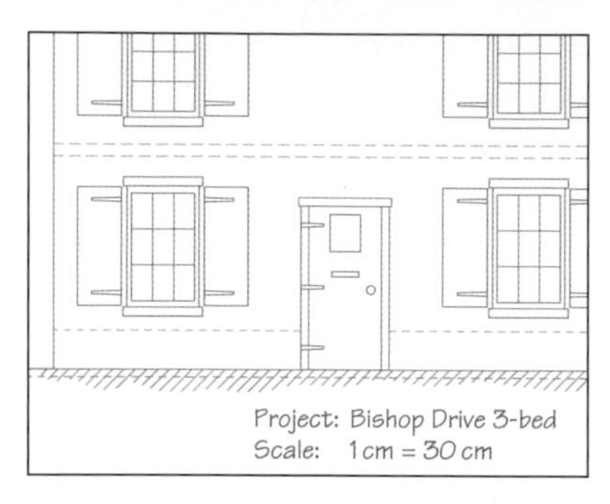

1 cm on the architect's plan = 30 cm of the actual house.

You can write this scale factor as 1 cm represents 30 cm
or 1 : 30.

Maps are scale drawings.

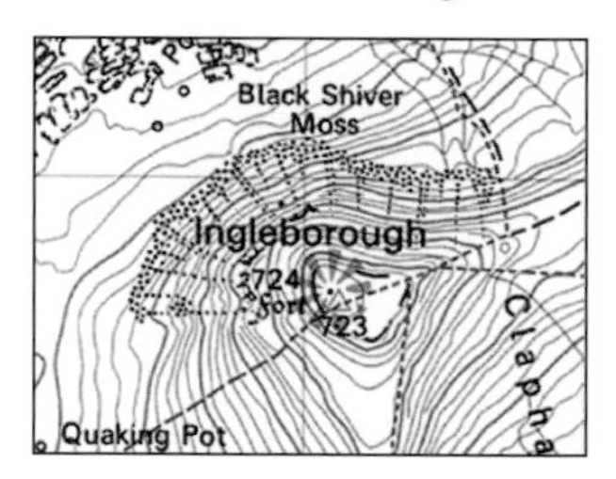

This is an enlargement of scale factor 50 000.

The map scale can be written 1 : 50 000.

You write the scale as 1 cm represents 50 000 cm
or 1 cm represents 500 m.

Example

In this scale drawing 1 cm represents 3 m.

3 cm

7 cm

a Calculate the height of the building.
b Calculate the length of the building.
c Calculate the area of the front.
State your units.
d There are 26 windows in the scale drawing.
How many windows are there on the front of the real house?

---

a 1 cm represents 3 m
3 cm represents $3 \times 3 = 9$ m
Height of building = 9 m

b 7 cm represents $7 \times 3 = 21$ m
Length of building = 21 m

c Area of front = $9 \text{ m} \times 21 \text{ m} = 189 \text{ m}^2$

d 26 windows

The number of windows is the same on the scale drawing as on the real house.

Exercise S6.4

1 The scale on a drawing is 1 cm represents 10 cm.
Calculate the distance represented by

0 10 20 cm

a 4 cm **b** 10 cm **c** 0.5 cm

d 6.5 cm **e** 12.5 cm

2 This is an accurate scale drawing of a Dalek.
For the real Dalek, calculate

a the height

b the width of the base.

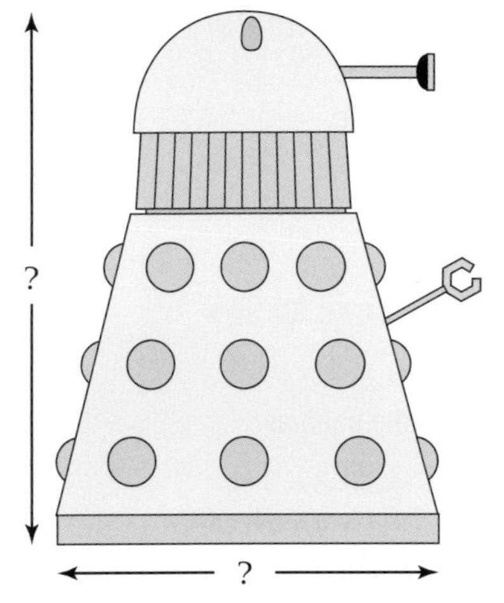

Scale: 1 cm represents 20 cm

3 On the plan of a house, a door measures 3 cm by 8 cm.
If the plan scale is 1 cm represents 25 cm, calculate the dimensions of the real door.

4 This is a scale drawing of a volleyball court.

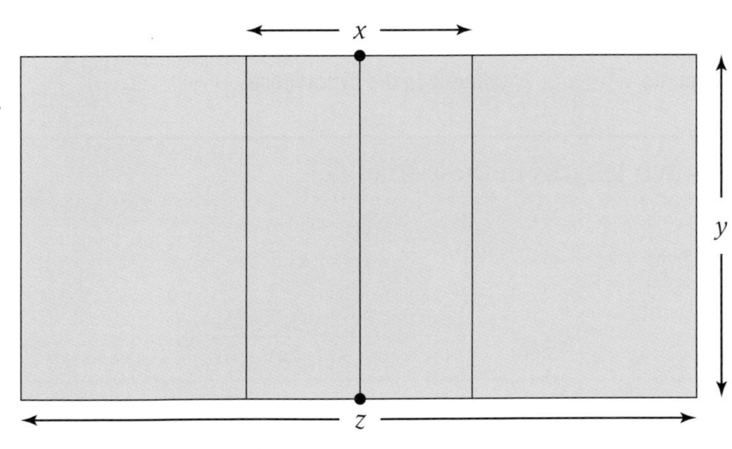

Scale 1 : 200
or 1 cm
represents 2 m

Calculate

a the actual distances marked x, y and z

b the area of the court. State the units of your answer.

5 A map has a scale of 1 : 50 000 or 1 cm represents 50 000 cm.
Calculate in metres the actual distance represented on the map by

a 2 cm **b** 8 cm **c** 10 cm

d 0.5 cm **e** 14.5 cm.

S6.5 Pythagoras' theorem

This spread will show you how to:

- Understand, recall and use Pythagoras' theorem

Keywords
Hypotenuse
Pythagoras' theorem
Right-angled triangle
Square
Square root

The longest side of a **right-angled triangle** is called the **hypotenuse**.
The hypotenuse is always opposite the right angle.

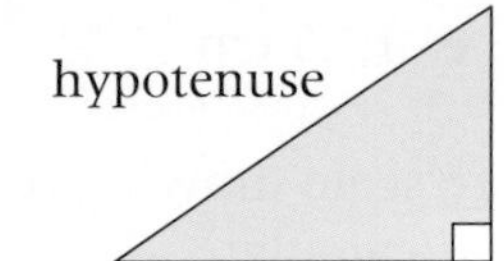

This is a right-angled triangle.

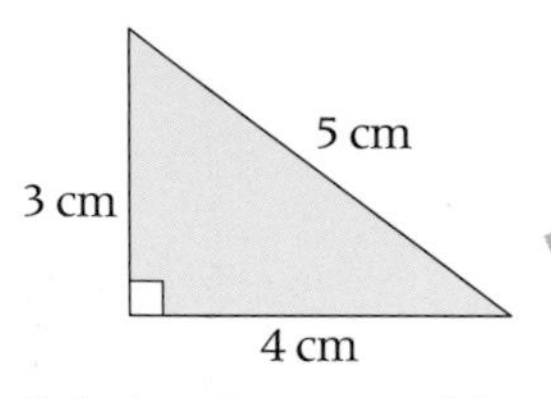

Draw the **squares** on each side.

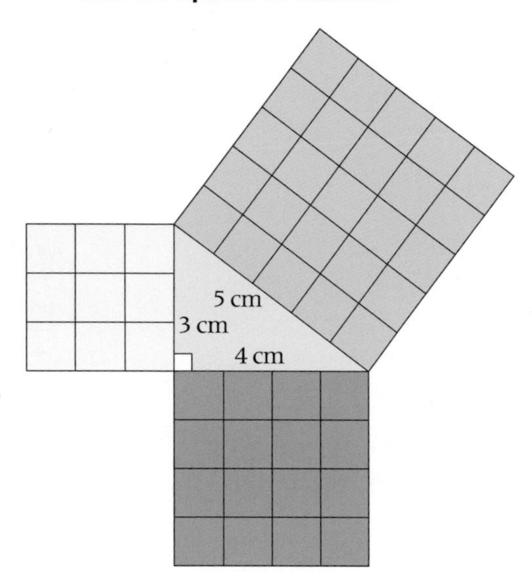

Calculate the areas of the squares.

Area of yellow square $= 3 \times 3 = 9\ \text{cm}^2$
Area of red square $= 4 \times 4 = 16\ \text{cm}^2$
Area of orange square $= 5 \times 5 = 25\ \text{cm}^2$

Area of orange square = area of yellow square + area of red square.

This is **Pythagoras' theorem**.

- **In a right-angled triangle, $c^2 = a^2 + b^2$ where c is the hypotenuse.**

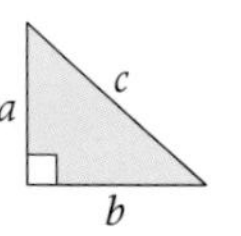

Example

Calculate the unknown lengths in these triangles.

a

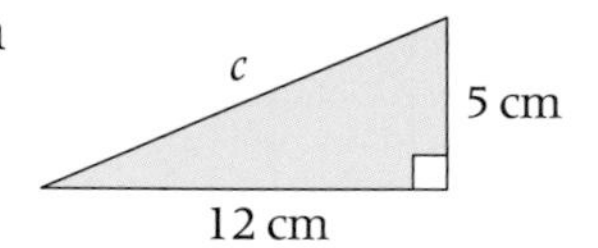

b

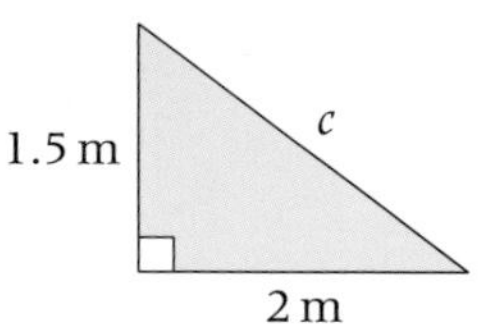

a Label the sides.

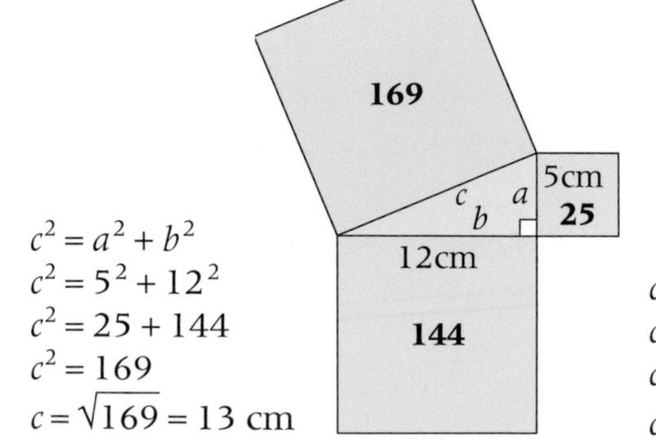

$c^2 = a^2 + b^2$
$c^2 = 5^2 + 12^2$
$c^2 = 25 + 144$
$c^2 = 169$
$c = \sqrt{169} = 13$ cm

b Label the sides.

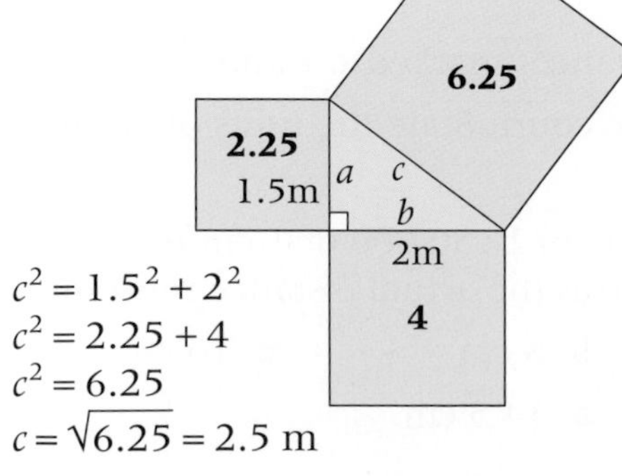

$c^2 = 1.5^2 + 2^2$
$c^2 = 2.25 + 4$
$c^2 = 6.25$
$c = \sqrt{6.25} = 2.5$ m

$\sqrt{\ }$ means **square root**.

$\sqrt{169} = 13$ because $13 \times 13 = 169$.

Exercise S6.5

1 Calculate the area of these squares. State the units of your answers.

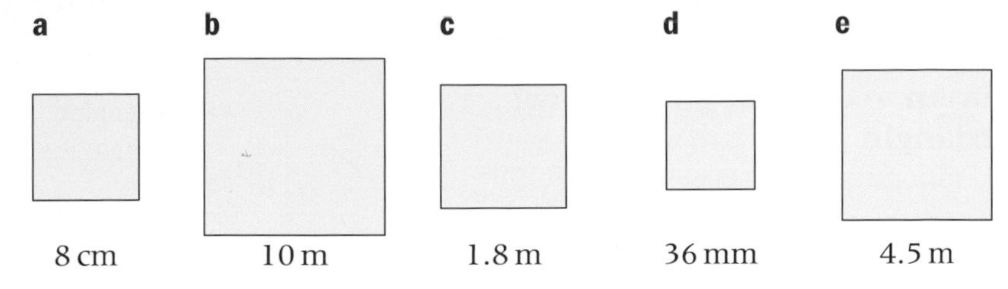

2 Calculate the length of a side of these squares.
State the units of your answers.

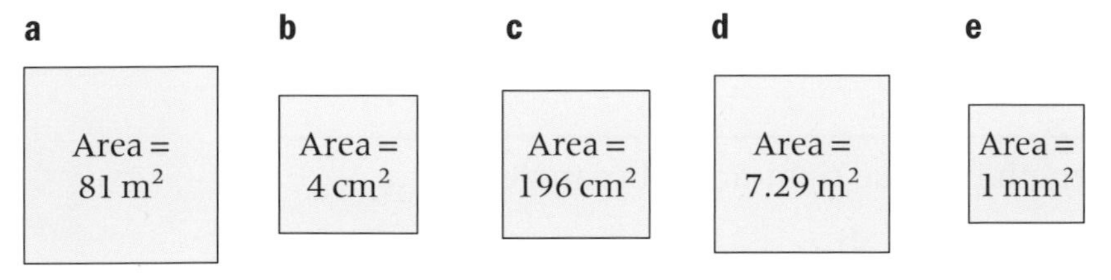

3 Calculate the unknown area for these right-angled triangles.

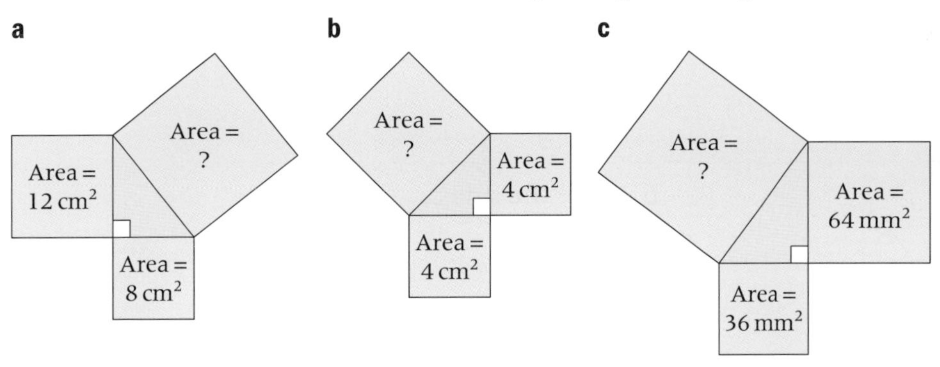

DID YOU KNOW?

Pythagoras was a Greek mathematician most famous for his theorem, who taught his students that 'all things are numbers'.

4 Calculate the length of the hypotenuse in these right-angled triangles. State the units of your answers.

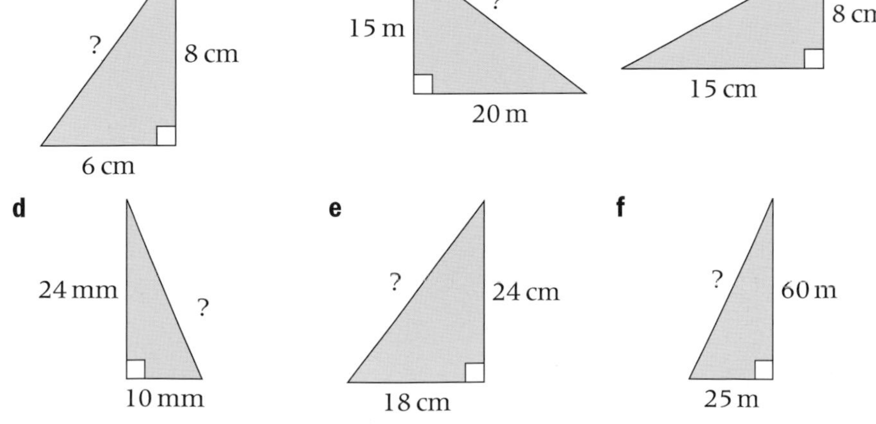

S6.6 More Pythagoras' theorem

This spread will show you how to:

- Understand, recall and use Pythagoras' theorem

Keywords
Hypotenuse
Pythagoras' theorem
Right-angled triangle
Square
Square root

You use **Pythagoras' theorem** to calculate an unknown length in a **right-angled triangle**.

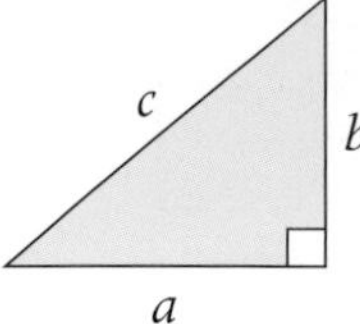

- **Pythagoras' theorem states**
  For any right-angled triangle, $c^2 = a^2 + b^2$ where c is the hypotenuse.

You can use Pythagoras' theorem to find a side given two other sides.

- You add to find the **hypotenuse**.
- You subtract to find the other sides.

The triangle must be right-angled.

Example

Calculate the unknown length in these right-angled triangles.

a

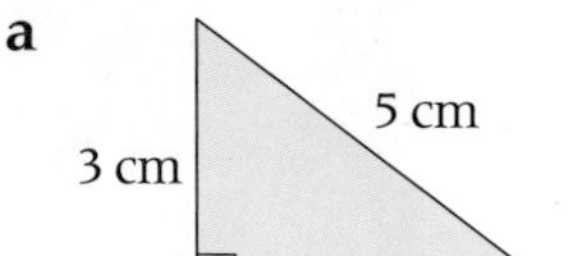

b

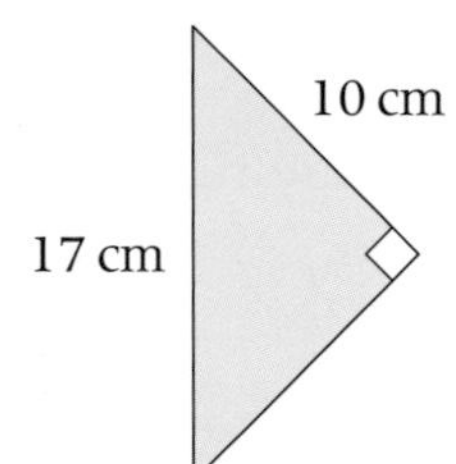

a Label the sides.

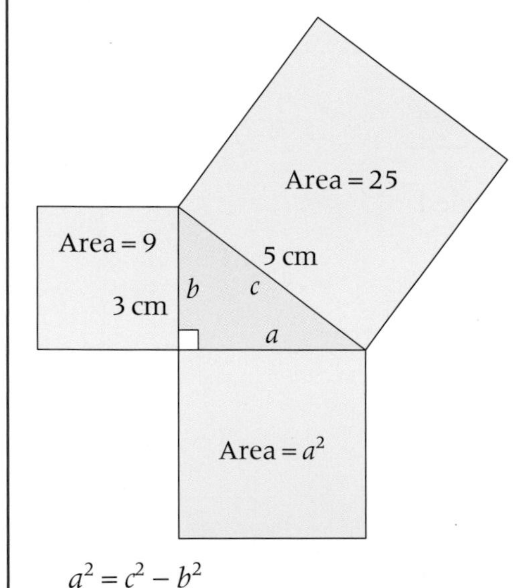

$a^2 = c^2 - b^2$
$a^2 = 5^2 - 3^2$
$= 25 - 9$
$a^2 = 16$
$a = \sqrt{16} = 4$ cm

b Label the sides.

a
c
b

$b^2 = c^2 - a^2$
$b^2 = 17^2 - 10^2$
$= 189$
$b = \sqrt{189}$
$= 13.7$ cm (to 1 dp)

You don't need to draw the squares on the sides.

Exercise S6.6

1 Calculate the unknown area of the square for these right-angled triangles.

a

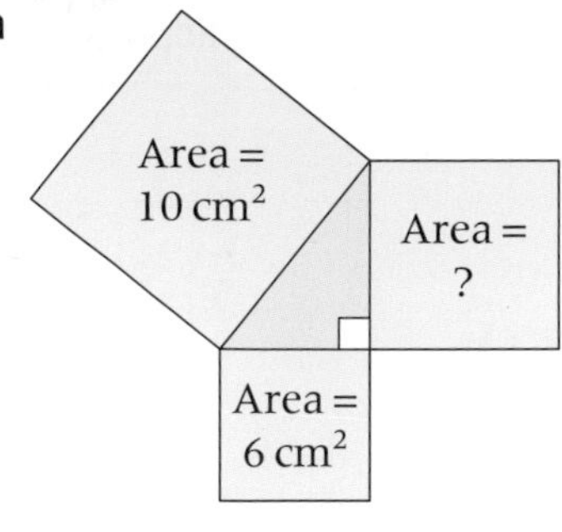

b

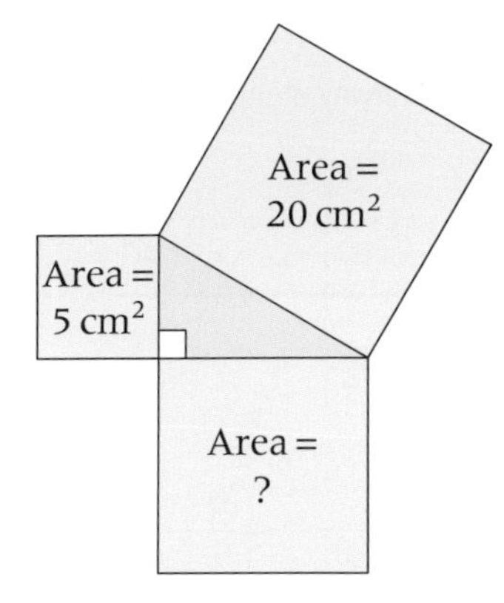

2 Calculate the unknown length in these right-angled triangles. Give the units of your answers.

a **b**

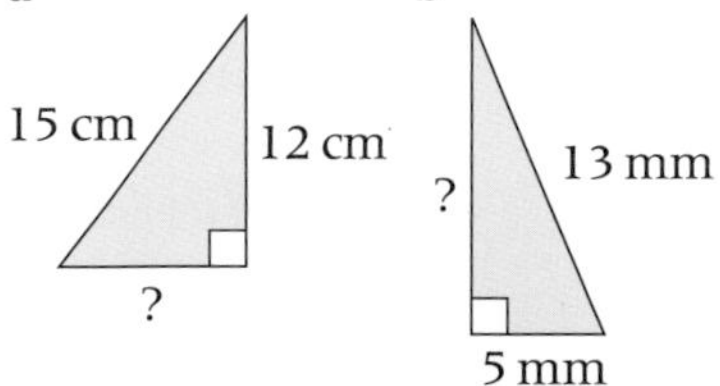

c **d**

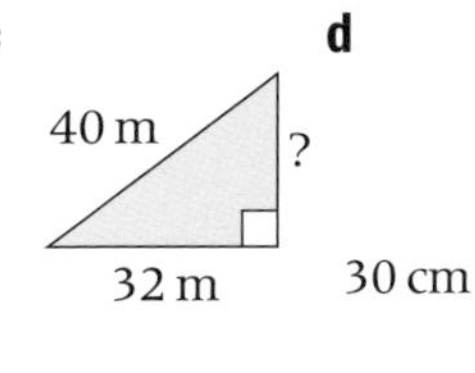

e

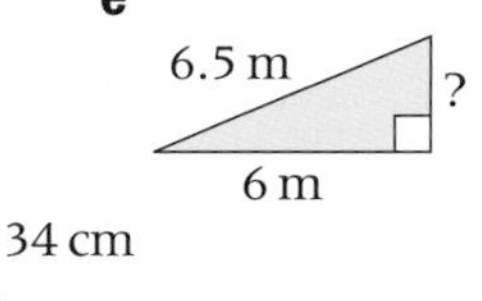

3 Calculate the unknown length in these right-angled triangles. Give your answers to a suitable degree of accuracy.

a

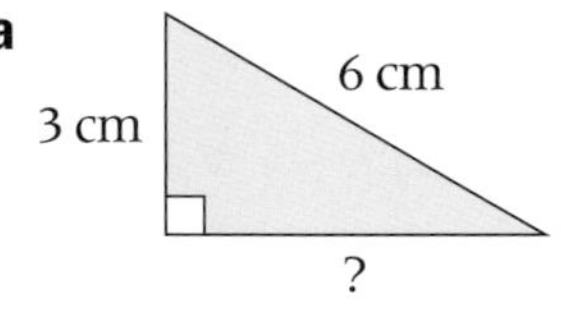

b

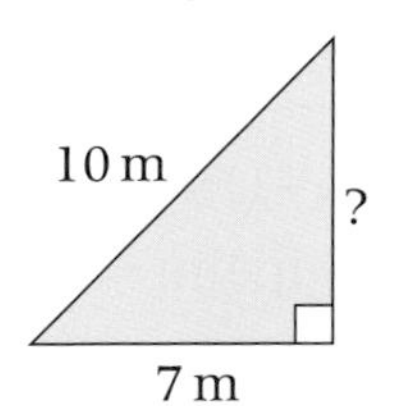

c

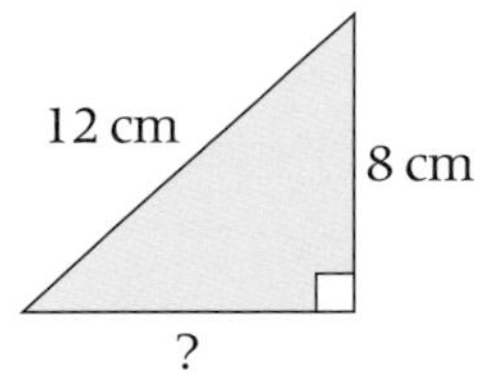

d

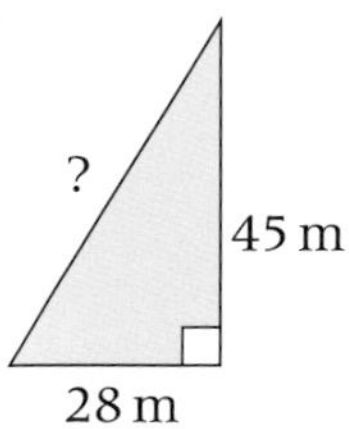

e

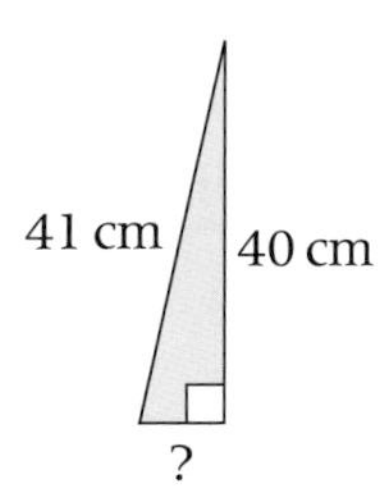

f

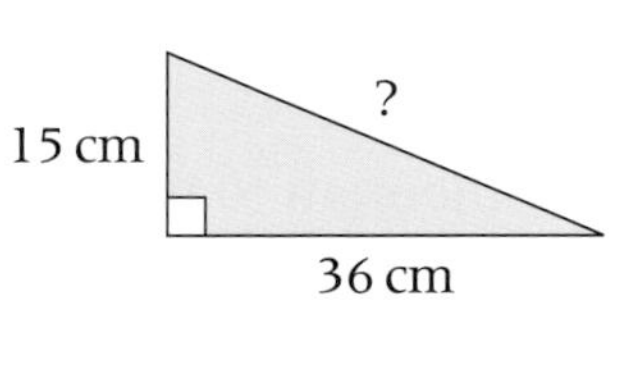

4 **a** Calculate the perpendicular height of this right-angled triangle.

b Calculate the perimeter of the triangle.

c Calculate the area of the triangle.

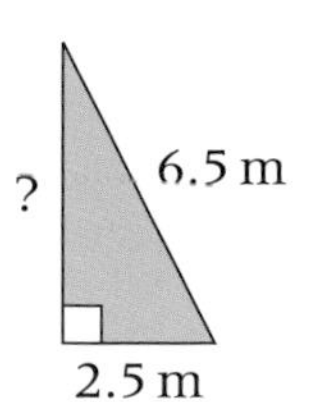

S6.7 Coordinates and Pythagoras

This spread will show you how to:

- Calculate the length and midpoint of the line AB

Keywords
Coordinates
Line segment
Midpoint

- **The midpoint M of a line AB is halfway along it.**

If A = (−1, 3)
and B = (3, 3)
then M_1 = (1, 3)

If E = (−2, 2)
and F = (0, −1)
then $M_3 = (-1, \frac{1}{2})$

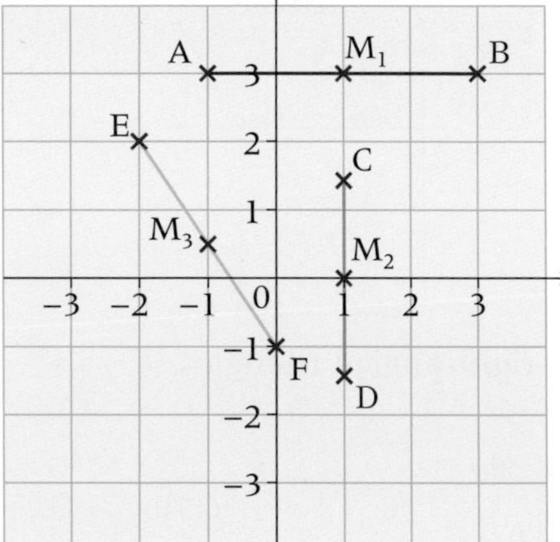

If $C = (1, 1\frac{1}{2})$
and $D = (1, -1\frac{1}{2})$
then M_2 = (1, 0)

M is the midpoint.

- **If A = (x_1, y_1) and B = (x_2, y_2) then M =** $\left(\frac{x_1 + x_2}{2}, \frac{y_1 + y_2}{2}\right)$

The midpoint of AB is the mean of the **coordinates** of points A and B.

Example

Calculate the coordinates of the midpoint between the points

a (7, 1) and (−3, 5) **b** (4, −1) and (2, −2)

a (7, 1) = (x_1, y_1) and (−3, 5) = (x_2, y_2)

Midpoint = $\left(\frac{7 + -3}{2}, \frac{1 + 5}{2}\right)$

$= \left(\frac{4}{2}, \frac{6}{2}\right)$

= (2, 3)

b (4, −1) = (x_1, y_1) and (2, −2) = (x_2, y_2)

Midpoint = $\left(\frac{4 + 2}{2}, \frac{-1 - 2}{2}\right)$

$= \left(\frac{6}{2}, \frac{-3}{2}\right)$

$= (3, -1\frac{1}{2})$

You use Pythagoras' theorem to find the length of a line on a grid.

Example

Calculate the length of the **line segment** from (2, 4) to (5, 2).

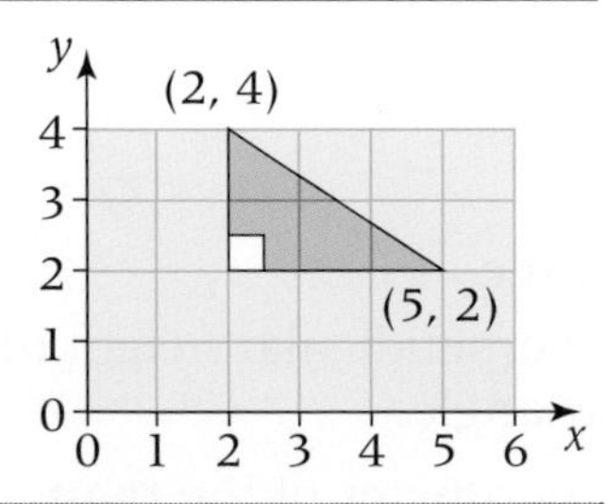

$c^2 = 2^2 + 3^2$

$c^2 = 4 + 9$

$c^2 = 13$

$c = \sqrt{13} = 3.60555 = 3.6$ units (to 1 dp)

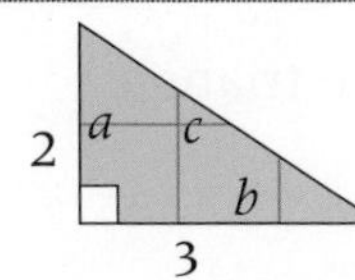

Label the sides.

Exercise S6.7

1 **a** Draw the points A(1, 3) and B(5, 1) on a copy of the grid.

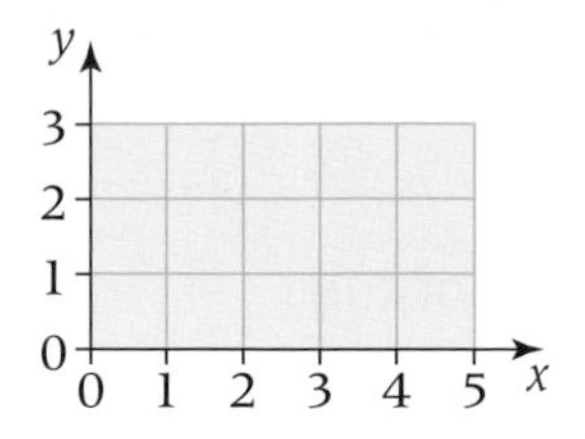

b M is the midpoint of the line AB.
Find the coordinates of the point M.

2 Calculate the coordinates of the midpoint between the points

a (3, 1) and (3, 5)

b (0, 4) and (4, 4)

c (2, −2) and (2, 4)

d (3, 1) and (7, 7)

e (−2, −1) and (6, 7)

3 The point (3, 4) is the midpoint between (x, 6) and (1, y).
Find the values of x and y.

4 Calculate the distance between these points.
Give your answers to a suitable degree of accuracy.

a (1, 2) and (4, 6)

b (2, 2) and (6, 5)

c (1, 2) and (2, 5)

d (0, 5) and (4, 1)

e (3, 6) and (6, 0)

5 **a** Plot the points A(0, 3), B(3, 6), C(6, 3) and D(3, 0) on square grid paper.

b What is the name of the shape ABCD?

c Calculate the length of AB.

d Calculate the length of AD.

e Calculate the area of the shape ABCD.

S6 Exam review

Key objectives

- Understand congruence
- Identify the scale factor of an enlargement as the ratio of the lengths of any two corresponding line segments
- Understand similarity of triangles and other plane figures and use this to make geometric inferences
- Use and interpret maps and scale drawings
- Understand and use Pythagoras' theorem

1 One end of a piece of rope 8.4 m long is tied to the top of a vertical pole *XY* and the other end is tied to the ground at the point *Z* which is at a horizontal distance of 6.6 m from the foot of the pole. Calculate the height of the pole.

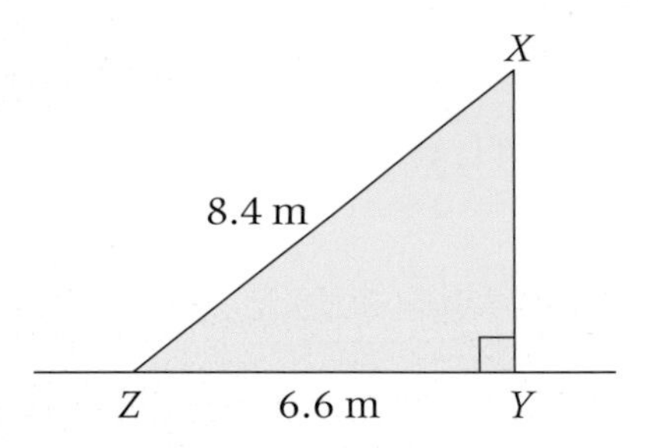

(3)

(WJEC, 2003)

2 BE is parallel to CD.
ABC and AED are straight lines.
AB = 4 cm
BC = 6 cm
BE = 5 cm
AE = 4.8 cm.

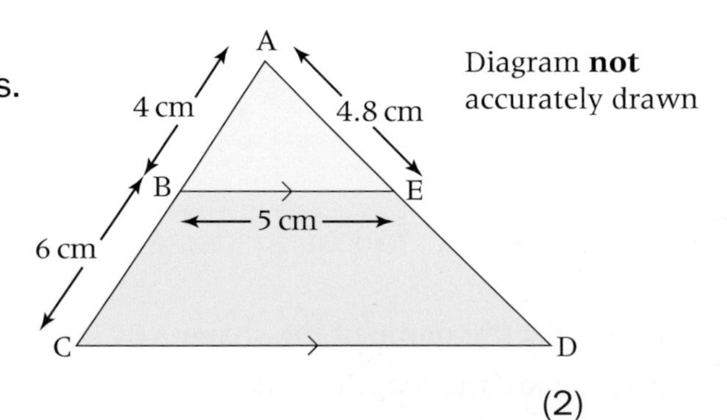

a Calculate the length of CD. (2)

b Calculate the length of ED. (2)

(Edexcel Ltd., 2004)

Collecting data

This unit will show you how to

- Recognise the difference between primary and secondary data
- Design and use data-collection sheets for discrete and grouped data
- Understand and use frequency tables
- Collect primary and secondary data using a variety of methods
- Use efficient methods of random sampling
- Discuss how data relates to a problem, identifying and minimising possible sources of bias
- Design and use two-way tables

Before you start ...

You should be able to answer these questions.

1 Put these numbers in order of size, smallest first.

a 56, 47, 48, 55, 65, 44, 61, 59.

b 1.7, 2.5, 0.8, 2.1, 1.5, 0.5.

c 31.2, 32.1, 23.1, 13.2, 13.1, 21.3.

2 Calculate.

a $36 + 45$ **b** $82 + 97$

c $36 + 45 + 24$ **d** $138 + 67$

e $325 + 116$ **f** $91 - 17$

g $148 - 13$ **h** $256 - 79$

i $345 - 126$ **j** $125 - 84$

3 This table shows the average sunrise and sunset times for six months of the year.

	Sunrise	Sunset
January	08.22	17.00
March	07.25	18.17
May	07.09	22.03
July	06.20	23.04
September	07.48	20.27
November	08.01	17.08

At what average time does the sun

a rise in September **b** set in March?

D1.1 Frequency tables

This spread will show you how to:

- Recognise the difference between primary and secondary data
- Design and use data-collection sheets for discrete and grouped data
- Understand and use frequency tables

Keywords
Data
Data-collection sheet
Frequency table
Primary data
Secondary data
Tally chart

Before this headline could be written, information, or **data**, was collected.

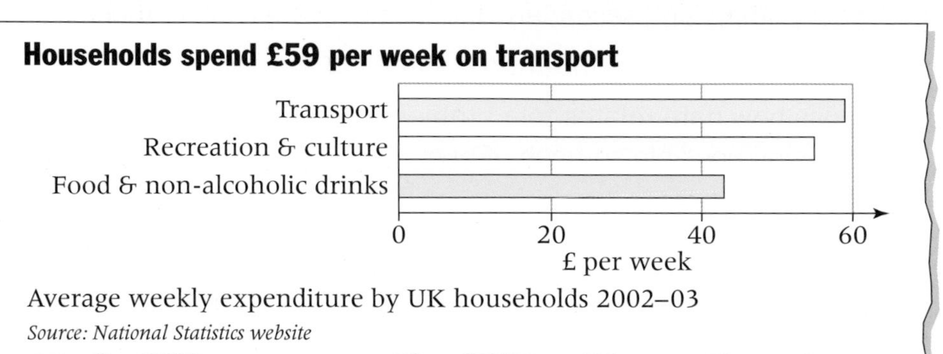

Average weekly expenditure by UK households 2002–03
Source: National Statistics website

- **Primary data is data you collect yourself.**
  You count the scores when rolling a dice.

- **Secondary data is data someone else has already collected.**
  This includes information from newspapers or the internet.
- **You can collect data using a data-collection sheet.**
  This one is a **tally chart**.

Coin face	Tally	Frequency
Head	卌 卌 III	13
Tail	卌 卌 II	12

卌 = 5

The coin was spun $13 + 12 = 25$ times.

Data can also be shown using a **frequency table**.

Example

The number of televisions in each house in my street is shown in the frequency table.

a Calculate the number of houses in my street.
b Calculate the total number of televisions in my street.

Number of TVs	Number of houses
0	1
1	5
2	12
3	9
4	1

The numbers in the table are

0, 1, 1, 1, 1, 1, 2, 2, 2, 2, 2, 2, 2, 2, 2, 2, 2, 2, 3, 3, 3, 3, 3, 3, 3, 3, 3, 4

a $1 + 5 + 12 + 9 + 1 = 28$ houses

b $0 + 5 + 24 + 27 + 4 = 60$ televisions

TVs	Houses	TVs × Houses
0	1	$0 \times 1 = 0$
1	5	$1 \times 5 = 5$
2	12	$2 \times 12 = 24$
3	9	$3 \times 9 = 27$
4	1	$4 \times 1 = 4$

Exercise D1.1

1 **a** Copy and complete the tally chart to find the frequency of the vowels a, e, i, o, u in this sentence.

b Which vowel occurs the most often?

c Which vowel occurs the least often?

d Calculate the total number of vowels in the sentence.

e Find a paragraph of writing in a newspaper and complete a similar tally chart.

Vowel	Tally	Frequency
a		
e		
i		
o		
u		

2 Rainfall, measured in millimetres, is recorded daily for the month of April.

4 2 1 0 0 1 2 2 3 5
7 8 5 3 3 2 0 0 0 1
2 3 2 4 6 7 8 8 1 2

a Copy and complete the tally chart to show this information.

b State the number of completely dry days in April.

c Calculate the total amount of rain to fall throughout April. State the units of your answer.

Rainfall (mm)	Tally	Number of days
0		
1		
2		
3		
4		
5		
6		
7		
8		

3 Nails can be bought in bags.
There are approximately 20 nails in each bag.
The numbers of nails in 40 bags are recorded.

19 20 19 21 20 20 20 21
22 19 19 19 21 20 20 21
19 21 21 22 20 19 20 21
20 20 21 21 20 20 22 21
19 19 20 20 21 20 20 20

a Copy and complete the tally chart to show this information.

b Calculate the total number of nails in all 40 bags.

Number of nails	Tally	Number of bags
19		
20		
21		
22		

4 Sophie did a survey to find the number of DVDs watched by the students in her class over a 1-week period. The results are shown in the frequency table.

a Calculate the number of students in Sophie's class.

b Calculate the total number of DVDs watched by the whole class.

Number of DVDs	Number of students
0	1
1	8
2	6
3	8
4	2
5	3

D1.2 Observation, controlled experiment and sampling

This spread will show you how to:

- Collect primary and secondary data using a variety of methods
- Use efficient methods of random sampling
- Discuss how data relates to a problem, identifying and minimising possible sources of bias

Keywords
Biased
Controlled experiment
Data logging
Observation
Random sample

- You can collect data by **observation**.
  To find the average daily rainfall, you would have to measure the rainfall every day for a period of time.
- You can collect data by **controlled experiment**.
  To see if a Head or a Tail occurs more often, you would have to spin a coin a number of times.
- You can collect data by **data logging**.
  For example, your heart rate can be measured when exercising.

You will need a data-collection sheet, whichever method you use.

An observation usually involves watching.

An experiment usually involves setting up a test.

Data is usually collected automatically when data logging.

Example

Design a suitable data-collection sheet to find the number of cars in households in your class.

Number of cars	Tally	Frequency
0		
1		
2		
3		
4 or more		

Covers all possibilities

Sometimes it is impossible to collect data from all the population and so a **random sample** is used.

- **In a random sample, each person or item must be equally likely to be chosen.**

The sample must not be **biased**.
The sample is biased if each person or item is not equally likely to be chosen.

Example

Describe a method to choose a random sample of 30 students for a year group of 120 students.

- Number each student from 1 to 120.
- Put a different number on 120 pieces of paper.
- Place the pieces of paper in a bag.
- Pick out 30 numbers from the bag.

Taking every 4th student from an alphabetical or form list is biased. Why is this so?

Exercise D1.2

1 Decide whether these data collections are an observation, a controlled experiment, or data logging.

a Spinning a spinner

b The types of drink bought from a vending machine at break

c Choosing a colour from a given set of colours

d Automatically measuring the temperature of ice as it is heated

e Measuring the 'bounce' of different rubber balls dropped from the same height

f The make of vehicles passing the college gates

g The number of passes a player makes during a game

h Automatically measuring the number of vehicles travelling on a road

i The usage of the slide, the swing and the roundabout in a children's playground

j Automatically measuring pulse rate on a jogging machine.

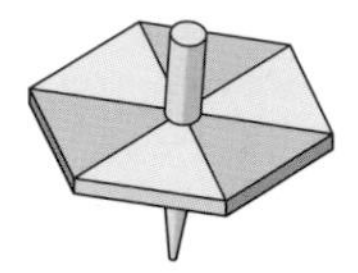

A spinner is an instrument for creating random outcomes, usually in probability experiments.

2 The number of occupants in passing cars are counted. The results are

1 2 1 4 3 1 2 1 5 4
1 1 1 2 1 3 1 4 1 1
1 2 1 1 1 2 1 1 4 2
2 2 1 2 1 3 2 1 2 1

a Is this data collection an observation, a controlled experiment or data logging?

b Construct a data-collection sheet to show this data.

c Calculate the total number of cars that passed.

d Calculate the total number of occupants of the cars that passed.

DID YOU KNOW?

In 2008 the UK's first carpool lane will open on the M1. Lone drivers will be banned, encouraging drivers to share their cars and ease congestion.

3 A dice, numbered 1 to 6, is rolled and the scores recorded.

a Is this data collection an observation, a controlled experiment or data logging?

b Construct a suitable data-collection sheet.
The scores are

1 5 6 2 1 4 2 1 2 5 6 1 3 2 2
2 1 2 1 2 2 1 6 5 4 3 2 1 1 1
5 5 6 1 2 2 1 4 3 3 2 2 1 4 6

c Complete your data-collection sheet with the data.

d Which score occurred the most often?

e Calculate the total number of rolls of the dice.

f Do you think the dice is biased? Explain your answer.

g How could you improve the reliability of your answer?

D1.3 Surveys and questionnaires

This spread will show you how to:

- Collect primary and secondary data using a variety of methods

Keywords
Data-collection sheet
Hypothesis
Questionnaire
Survey

Surveys are used to find people's opinions or to test a **hypothesis**.

A hypothesis is a predictive statement, for example 'boys are taller than girls'.

- You can collect data by using a survey with
  - a **data-collection sheet**
  - a **questionnaire**.

This data-collection sheet allows you to ask one question and collects all the data on one sheet.

Do you smoke?	Tally	Frequency
Yes	~~||||~~ ~~||||~~ |	
No	~~||||~~	

A questionnaire gives you more data, but you need one questionnaire for each person in your survey.

Male ☐ Female ☐
Age group: Under 20 ☐
20 or over ☐
Do you smoke? Yes ☐
No ☐

You must be careful what questions you ask in a survey.

Never ask a leading question, such as: The speed of cars today, it's very bad, isn't it?	• Avoid giving your opinion. What do you think of the speed of cars today? Too slow ☐ About right ☐ Too fast ☐
Never ask a vague question, such as: Do you eat cereal?	• Ask for factual responses. Did you eat cereal for breakfast today?
Never ask a question that gives too many responses, such as: What do you like to eat?	• Limit the choice of responses. Meat ☐ Fish ☐ Vegetables ☐
Never ask a complicated, wordy question.	• Ask simple, straightforward questions.
Don't forget to allow for all possible responses.	• Use 'Don't know' or 'Other'.
Never use 'occasionally', 'regularly' or 'sometimes' as they mean different things to different people.	• Ask for numerical responses. 1–10 ☐ 11–20 ☐ 21–30 ☐

Example

Give two reasons why the response section of this questionnaire is unsatisfactory.

How much money are you carrying?
£0.01–£1 ☐ £1–£2 ☐ Over £2 ☐

- No response is possible for £0.
- £1 is in two categories.

Exercise D1.3

1 Lauren wants to find out how often the students in her class eat 'Take-away' food.

a One of the questions in her questionnaire is shown. Write one criticism of this question.

Which 'Take-away' meals do you like?

b Another question is shown. Write one criticism of this question.

Have you had a 'Take-away' meal recently?
Yes ☐ No ☐

2 a Devise a question that could be used to give this frequency table.

b Calculate the number of people that completed this survey.

Fruit	Number of people
Apple	43
Banana	35
Pear	13
Other	9

3 Ross intends to survey shoppers about their shopping habits.

a One of the questions in his questionnaire is shown. Write two criticisms of the response section.

How old are you?
Under 20 ☐ 20–30 ☐ 30–40 ☐

b Another question is shown. Write one criticism of the question and one criticism of the response section.

Do you shop often?
Seldom ☐ Rarely ☐ Sometimes ☐

c Ross is going to use the questionnaire outside one shop. Write one criticism of his plan.

d Rewrite the questions in parts **a** and **b**.

4 a Devise a question that could produce this data-collection sheet.

b Calculate the total number of people that completed the survey.

c Write one criticism of the choice of categories for the channel.

Channel	Tally	Frequency
BBC1	𝍸 II	7
BBC2	III	3
ITV	𝍸 I	6
Channel 4	II	2
Five	I	1
Other	𝍸 𝍸 I	11

5 One question in a questionnaire about reading habits is 'Do you read a newspaper regularly?'

a Write one criticism of this question.

b Rewrite the question, including a response section.

c Another question is:

Write two criticisms of the response section.

When was the last time you bought a book?
1 year ago ☐ 2 years ago ☐ 3 or more years ago ☐

D1.4 Grouped data

This spread will show you how to:

- Design and use data-collection sheets for discrete and grouped data

Keywords
Class intervals
Continuous
Discrete
Group
Grouped frequency table

Some surveys produce data with many different values.

- You can **group** data into **class intervals** to avoid too many categories.

Example

The exam marks of a particular class in Clarendon College are shown:

35	47	63	25	31	8	19	55	47	14
24	36	56	61	15	43	22	50	66	10
36	45	18	20	53	31	40	60	44	47

Complete the **grouped frequency table**.

Mark	Tally	Frequency
1–20		
21–40		
41–60		
61–80		

→

Mark	Tally	Frequency
1–20	𝍸 II	7
21–40	𝍸 IIII	9
41–60	𝍸 𝍸 I	11
61–80	III	3

𝍸 = 5

Check that the frequencies add to 30.

- **Discrete** data can only take exact values (usually collected by counting), for example the number of students in each class in a school.

- **Continuous** data can take any value (collected by measuring), for example the heights of the students in your class.
  Continuous data cannot be measured exactly.

- You can write class intervals for discrete data like this:

People
1–10
11–20
21–30

11–20 means between 11 and 20 including 11 and 20.

- You can write class intervals for continuous data like this:

Weight (w)
$50 \leqslant w < 55$
$55 \leqslant w < 60$
$60 \leqslant w < 65$

The first interval is between 50 and 55, including 50 but not including 55.

Exercise D1.4

1 Decide whether this data is discrete or continuous.

a The number of taxis waiting at a station

b The number of drinks in a machine

c The heights of people

d The number of magazines in a shop

e The weights of animals

f The number of people in an office

g The handspans of people

h The speed of cars

i Dress sizes

j Shoe sizes.

2 **a** Copy and complete the frequency table using these test marks.

8 14 21 4 15 22 25 24 15 11
10 17 24 20 13 16 12 9 3 14
20 10 16 15 7 23 23 14 15 16
8 2 9 19 12 10 10 20 13 13
15 17 11 14 19 20 23 23 24 5

Test mark	Tally	Frequency
1–5		
6–10		
11–15		
16–20		
21–25		

b Calculate the number of people who took the test.

3 **a** Copy and complete the frequency table using these weights of people, given to the nearest kilogram.

48 63 73 55 59 61 70 63 58 67
46 45 57 58 63 71 60 47 49 51
53 61 68 65 70 60 52 59 50 49
48 47 63 61 58 71 53 51 60 70

Weight (kg)	Tally	Number of people
$45 \leqslant w < 50$		
$50 \leqslant w < 55$		
$55 \leqslant w < 60$		
$60 \leqslant w < 65$		
$65 \leqslant w < 70$		
$70 \leqslant w < 75$		

b Calculate the number of people who were weighed.

4 Every student in Emma's class threw a ball as far as possible. The lengths of the throws, to the nearest metre, are shown.

8 15 18 11 10 20 24 11 12
21 22 9 7 12 14 18 6 19
20 22 21 17 13 15 20

a Draw and complete a frequency table, using class intervals $0 \leqslant l < 5$, $5 \leqslant l < 10$, $10 \leqslant l < 15$, ...

b How many students are in Emma's class?

5 The masses of parcels, in kilograms, are shown.

1.8 1.9 2.3 2.0 0.7 1.9 3.4 1.8
2.1 1.2 3.2 3.1 1.5 1.7 2.9 2.7
3.7 0.9 2.5 3.3 0.2 2.7 2.8 3.5
1.9 1.8 0.3 0.8 2.5 1.7 0.5 1.0
0.5 2.4 3.0 3.6 1.3 3.3 3.2 1.9

a Draw and complete a frequency table, using class intervals $0 < m \leqslant 1.0$, $1.0 < m \leqslant 2.0$, ...

b Which class interval has the greatest number of parcels?

c Calculate the total number of parcels that were weighed.

D1.5 Two-way tables

This spread will show you how to:

- Design and use two-way tables

Keywords
Column
Frequency table
Row
Two-way table

- You can organise data in a table, such as a **frequency table**, using **rows** and **columns**.

Method of travel	Number of students
Car	14
Bus	11
Walk	5

11 students travelled by bus.

- A **two-way table** shows more detail and links two types of information, for example, method of travel and gender.

	Boys	Girls
Car	4	10
Bus	6	5
Walk	2	3

5 girls travelled by bus.

Example

Design a data-collection sheet, in the form of a two-way table, that could be used to survey the audience at a pantomime.
The survey must distinguish between children and adults.

	Child	Adult
Male		
Female		

or

	Male	Female
Child		
Adult		

Example

Katie and Marcus counted their music collection.
Complete the two-way table.

Type	Katie	Marcus	Totals
CD	6		
MP3		10	
Totals	15		28

Calculate the missing values

$15 - 6 = 9$
$9 + 10 = 19$
$28 - 19 = 9$
$9 - 6 = 3$
$3 + 10 = 13$

Type	Katie	Marcus	Totals
CD	6	3	9
MP3	9	10	19
Totals	15	13	28

Check: $15 + 13 = 28$

Exercise D1.5

1 The two-way table shows the numbers of cars and vans that are crushed or saved for spare parts.

	Crushed	Spare parts
Cars	15	24
Vans	7	12

a State the number of

i cars that are crushed

ii vans that are saved for spare parts.

b Calculate the total number of

i cars

ii crushed vehicles.

c Calculate the total number of vehicles shown in the table.

2 The 32 members of a sports club play either football ⚽ or badminton 🏸. There are 15 women in the club, with 5 men and 8 women playing badminton.

a Copy and complete the two-way table.

	⚽	🏸
Men		
Women		

b How many men play football?

3 One hundred mobile phones are surveyed for colour (either black or silver) and for the payment method (either pay as you go or contract).

a Devise a two-way table that would show this information.

b Choose four suitable numbers for your table.

4 A vending machine only sells tea and coffee.
Janice is carrying out a survey about the use of sugar in drinks.
Devise a two-way table that Janice could use to show this information.

5 A car salesman sells vehicles that are either saloons or hatchbacks and are bought part-exchange or cash.
Devise a suitable two-way table to summarise his sales.

6 A computer repair centre does a stock check.
There are 25 PCs and 35 laptops. 9 of the PCs are broken and 17 of the laptops are not broken.
a Devise a two-way table that could show this information.
b What is the total number of broken computers?

D1 Exam review

Key objectives

- Collect data from a variety of suitable sources, including experiments and surveys, and primary and secondary sources
- Design an experiment or survey
- Design and use two-way tables
- Discuss how data relate to a problem, identify possible sources of bias and plan to minimise it

1 The table shows the number of bunches of flowers bought by the customers of a florist's shop on the day before Mother's Day:

Number of bunches of flowers bought	Number of customers
1	24
2	10
3	5

a Calculate the total number of bunches of flowers bought from the shop that day. (2)

b Explain why the total number of bunches of flowers bought from the shop that day may not be representative of the normal daily sales. (1)

(AQA, 2003)

2 The manager of a school canteen has made some changes. She wants to find out what students think of these changes. She uses this question on a questionnaire:

What do you think of the changes in the canteen?

Excellent ☐ Very good ☐ Good ☐

a Write down what is wrong about this question. (1)

This is another question on the questionnaire:

How much money do you normally spend in the canteen?

A lot ☐ Not much ☐

b i Write down one thing that is wrong with this question. (1)

ii Design a better question for the canteen manager to use. You should include some response boxes. (2)

(Edexcel Ltd., 2004)

Displaying data

This unit will show you how to

- Draw and produce relevant diagrams and charts to display data
- Recognise the difference between discrete and continuous data, using appropriate diagrams and charts
- Draw and use frequency diagrams for continuous data
- Draw and use stem-and-leaf diagrams
- Draw and use line graphs for time series, recognising data trends
- Draw and use scatter graphs, and compare two data sets
- Understand correlation and use lines of best fit

Before you start ...

You should be able to answer these questions.

1 Calculate the value of the angle x.

a

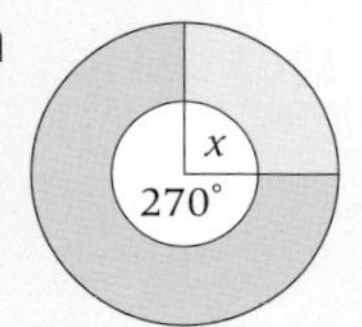

b

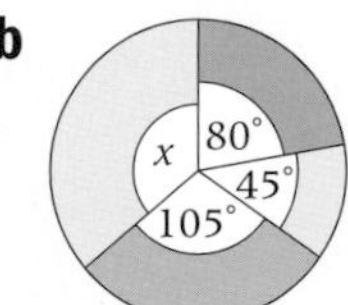

2 Calculate

a $360 \div 3$ **b** $360 \div 8$ **c** $360 \div 6$
d $360 \div 5$ **e** $360 \div 60$ **f** $360 \div 18$
g $360 \div 12$ **h** $360 \div 20$ **i** $360 \div 36$

3 Match the axis with its direction.

a x-axis **b** y-axis
i vertical **ii** horizontal

4 Put these numbers in order of size, smallest first.

a 746, 751, 665, 714, 661, 756, 741.
b 1.8, 0.9, 1.5, 2.1, 2.2, 0.5.
c 45.9, 44.6, 49.5, 46.4, 45.6.

5 Give the coordinates of

a point A
b point B

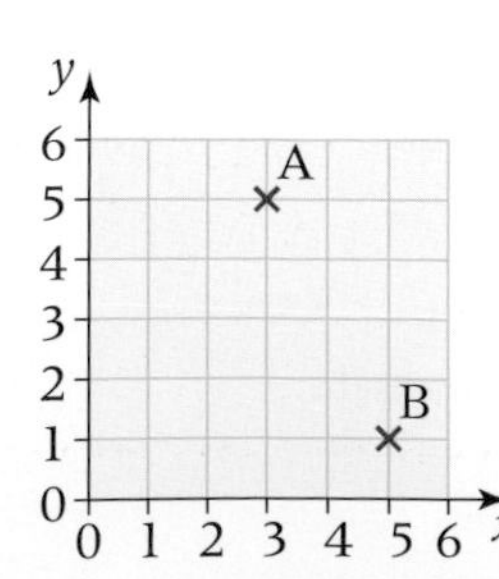

D2.1 Diagrams and charts

This spread will show you how to:

- Draw and produce relevant diagrams and charts to display data

Keywords
Bar chart
Bar-line chart
Category
Pictogram
Pie chart
Sector

You can use a variety of diagrams and charts to display data.

Pictograms use symbols to represent the size of each category.

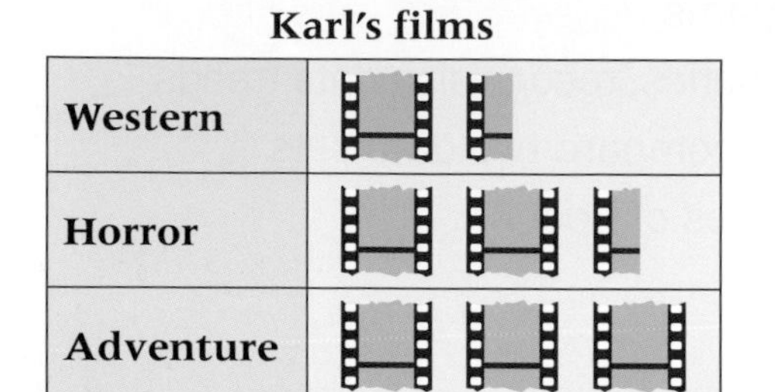

represents 2 films

Bar charts use bars to represent frequencies.

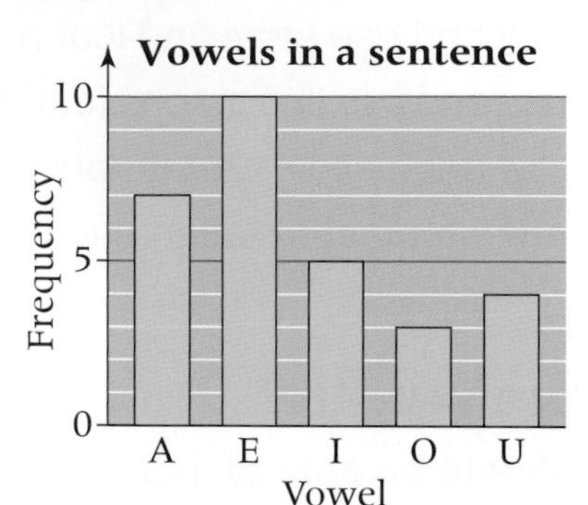

Notice the gaps between the bars.

Bar-line charts use vertical lines to represent numerical data.

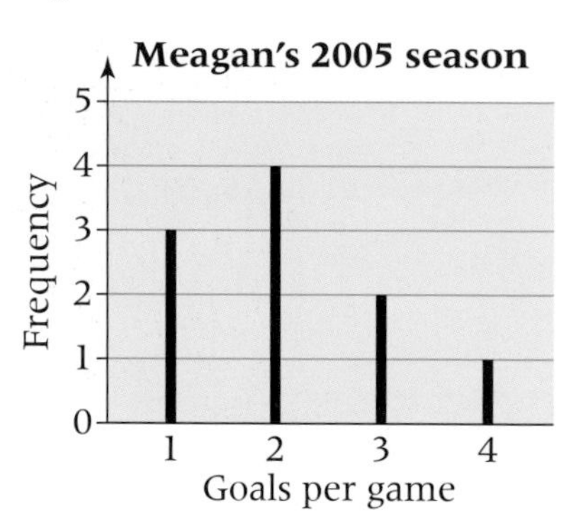

Pie charts use sectors of a circle to represent the size of each category.

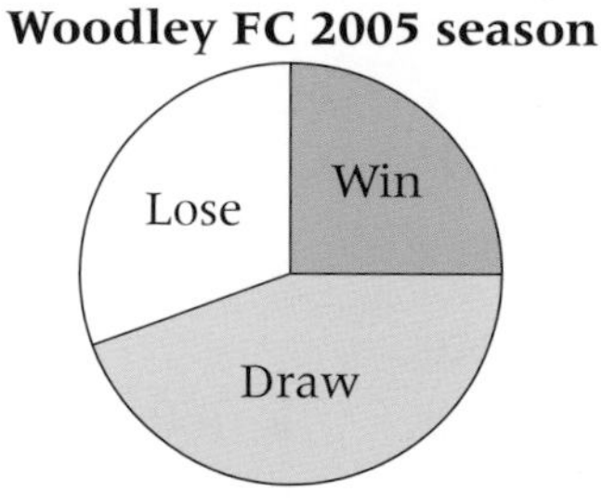

The size of each **sector** is proportional to the corresponding frequency.

Example

240 people are asked to name their favourite fruit. The results are shown.

Fruit	Apple	Banana	Orange	Other
Number of people	50	80	72	38

Draw a pie chart to illustrate the information.

Calculate the angle for one person:

$360° \div 240 = 1.5°$

Calculate the angles for each category:

Coldplay	Muse	Oasis	Arct
500	350	300	

Measure, colour and label the sectors.

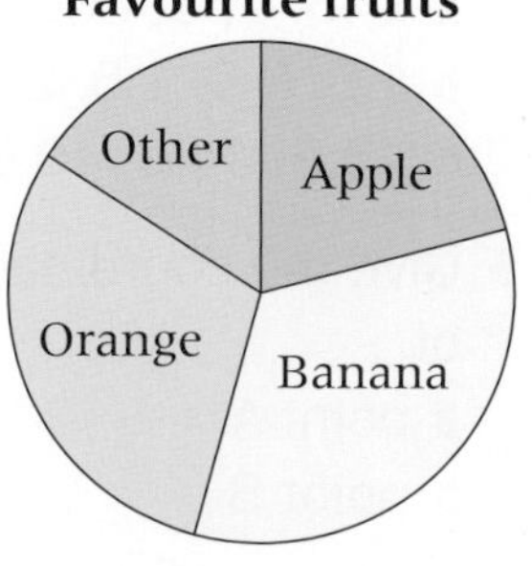

Check by adding the angles of the sectors.

Exercise D2.1

1 The pictogram shows the results for the 'Best Live Band' survey.

Coldplay	🎸🎸🎸🎸🎸
Muse	🎸🎸🎸½
Oasis	
Arctic Monkeys	
Radiohead	

🎸 represents 100 votes

The votes were as follows

Coldplay	Muse	Oasis	Arctic Monkeys	Radiohead
500	350	300	250	150

a Calculate the total number of votes made.

b Copy and complete the pictogram.

2 A football team plays 36 matches in the season. The results are

Win	Draw	Lose
15	8	?

a Calculate the number of matches that were lost.

b Calculate the angle one match represents in a pie chart.

c Calculate the angle of each category in the pie chart.

d Draw a pie chart to show the information.

DID YOU KNOW?

Florence Nightingale was a celebrated statistician as well as a nurse. She used pie charts and diagrams to show the need for improved hygiene conditions.

3 The weather record for 60 days is shown in the frequency table.

a Calculate the angle one day represents in a pie chart.

b Calculate the angle of each category in the pie chart.

c Draw a pie chart to show the data.

Weather	Number of days
Sunny	15
Cloudy	18
Rainy	14
Snowy	3
Windy	10

4 A fishing catch consisted of 480 fish. The frequency table shows the amount of each type.

a Calculate each angle of a pie chart to illustrate this information.

b Draw the pie chart.

Fish	Frequency
Cod	120
Plaice	100
Haddock	96
Sardines	116
Mackerel	48

D2.2 Grouped frequency diagrams

This spread will show you how to:

- Recognise the difference between discrete and continuous data, using appropriate diagrams and charts

Keywords
Bar chart
Continuous
Discrete
Frequency polygon
Grouped data
Histogram

- **Discrete** data can only take exact values (usually collected by counting).
- **Continuous** data can take any value (collected by measuring).

You must be careful if the data is **grouped**.

You can use a **bar chart** to display grouped discrete data.

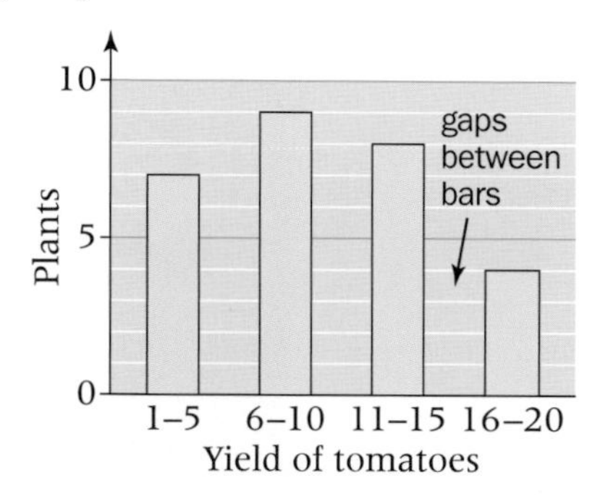

You can use a **histogram** to display grouped continuous data.

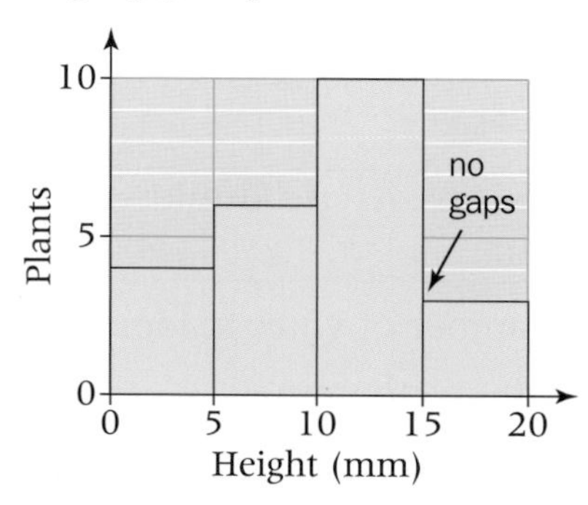

- You can use a **frequency polygon** to display grouped **continuous** data.

Frequency polygons use straight lines drawn from the top centre of each bar.

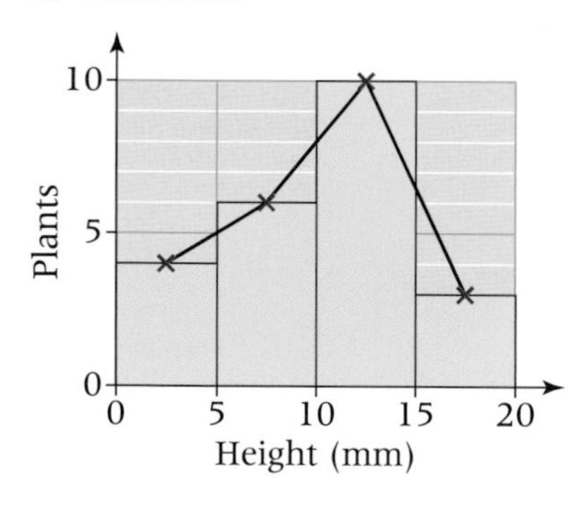

→

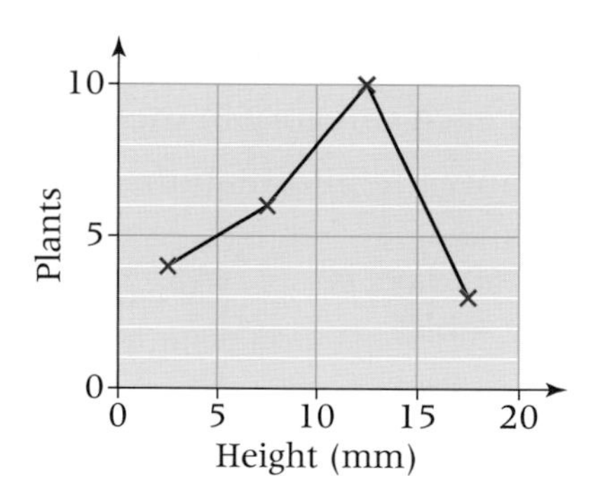

Plot the points at the midpoints of the class intervals.

Example

The times taken, in seconds, to run 100 m are shown in the table.

Time (seconds)	Number of people
$0 < t \leqslant 10$	0
$10 < t \leqslant 20$	4
$20 < t \leqslant 30$	6
$30 < t \leqslant 40$	3

Draw a frequency diagram to illustrate this information.

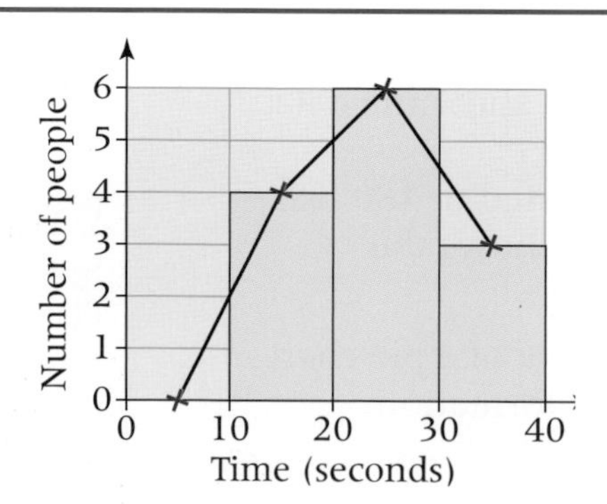

Don't draw both diagrams. Either the histogram or the frequency polygon answers this question.

Exercise D2.2

1 **a** Copy and complete the frequency table using these heights of people.

153	134	155	142	140	163	150	135
170	156	171	161	141	153	144	163
140	160	172	157	136	160	134	154
176	154	173	179	160	152	170	148
151	165	138	143	147	144	156	139

Height (cm)	Tally	Number of people
$130 < h \leqslant 140$		
$140 < h \leqslant 150$		
$150 < h \leqslant 160$		
$160 < h \leqslant 170$		
$170 < h \leqslant 180$		

b Copy and complete the histogram on graph paper.

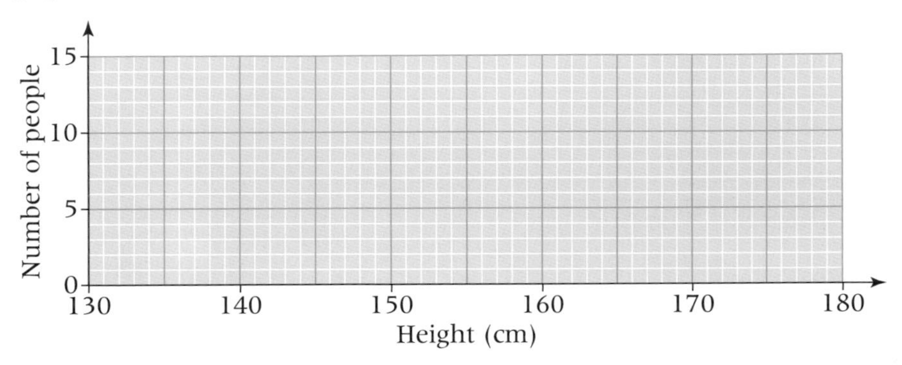

c Draw a frequency polygon on the same axes and graph paper.

2 The depth, in centimetres, of a reservoir is measured daily throughout April.

Draw a histogram to show the depths.

Depth (cm)	Number of days
$0<d\leqslant5$	1
$5<d\leqslant10$	5
$10<d\leqslant15$	14
$15<d\leqslant20$	8
$20<d\leqslant25$	2

3 The exam marks of 40 students are shown in the frequency table.

Draw a bar chart to show the exam marks.

Exam mark (%)	Number of students
1 to 20	4
21 to 40	9
41 to 60	13
61 to 80	11
81 to 100	3

4 The times taken for 50 runners of the London Marathon are shown in the frequency table.

a Draw a histogram to show the times.

b Draw the frequency polygon on the same diagram.

Time (hours)	Number of runners
$0 < t \leqslant 1$	0
$1 < t \leqslant 2$	1
$2 < t \leqslant 3$	23
$3 < t \leqslant 4$	18
$4 < t \leqslant 5$	8

D2.3 Stem-and-leaf diagrams

This spread will show you how to:

- Draw and use stem-and-leaf diagrams

Keywords
Ordered
Stem-and-leaf diagram

- You can use a **stem-and-leaf diagram** to display numerical data.

A stem-and-leaf diagram shows

- the shape of the distribution
- each individual value of the data.

This stem-and-leaf diagram is **ordered**, as the data is in numerical order.

stem	leaf
13	5
12	0 6 9
11	2 2 6 8
10	3 7

This means 135.

This means 120.

Key: | 11 | 2 | means 112

Always give a key.

Example

The weights, in kilograms, of 10 parcels are shown:

4.2 3.7 3.5 2.8 1.5
0.9 2.4 1.4 1.0 2.4

Show this data in an ordered stem-and-leaf diagram.

First choose the stem.
Go up in ones.

0	
1	
2	
3	
4	

You can order the data before you draw the diagram if you want.

Then write in the leaves.

0	9
1	5 4 0
2	8 4 4
3	7 5
4	2

order

Finally, order the leaves.

0	9
1	0 4 5
2	4 4 8
3	5 7
4	2

Key: | 1 | 5 | means 1.5kg

Examiner's tip: The key is essential to gain full marks.

Exercise D2.3

1 The times, in seconds, for a sample of 30 students to guess 2 minutes are

112	110	108	131	125	130
120	121	117	135	116	110
140	108	142	126	125	136
119	126	137	108	144	119
120	134	117	111	121	138

Copy and complete the ordered stem-and-leaf diagram.

10	
11	
12	
13	
14	

Key:

12	3

means 123 seconds

2 The times taken, in seconds, for 25 athletes to run 400 metres are given.

44.3	44.4	43.3	43.2	44.0
45.2	45.0	44.5	45.6	43.9
46.5	46.3	46.0	46.5	44.7
46.9	44.1	43.8	45.0	46.9
43.0	46.1	45.1	43.8	45.9

Draw an ordered stem-and-leaf diagram, using stems of 43.0, 44.0, 45.0 and 46.0. Remember to give the key.

3 **a** Use the temperatures in °C to draw an ordered stem-and-leaf diagram. Choose suitable stems.

	°C	°F
Athens	20	68
Berlin	31	88
Brussels	25	77
Cairo	34	93
Cape Town	17	63
Corfu	26	79
Dublin	14	57
Edinburgh	14	57
Faro	24	75
Guernsey	14	57

	°C	°F
Hong Kong	32	90
London	19	66
Los Angeles	20	68
Malaga	23	73
Malta	26	79
Miami	29	84
Moscow	22	72
Nairobi	20	68
New York	21	70
Oslo	10	50

	°C	°F
Paris	26	79
Perth	27	81
Rome	26	79
Sydney	19	66
Tel Aviv	28	82
Tenerife	23	73
Tokyo	23	73
Toronto	19	66
Vancouver	22	72
Vienna	29	84

b Use the temperatures in °F to draw another ordered stem-and-leaf diagram.

D2.4 Time series graphs

This spread will show you how to:

- Draw and use line graphs for time series, recognising data trends

Keywords
Horizontal
Line graph
Time series graph
Trend

You can use a **line graph** to show how data changes as time passes.

The data can be discrete or continuous.

The temperature of a liquid is measured every minute.

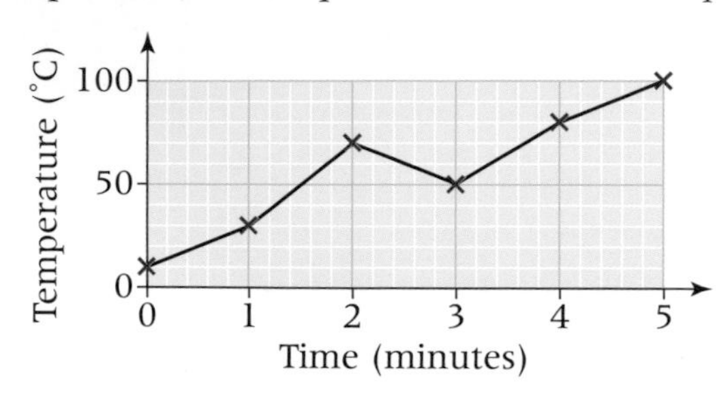

Time is always the **horizontal** axis.

This is an example of a **time series graph**.

Time could be seconds, minutes, hours, days, weeks, months or years.

- **A time series graph shows**
  - **how the data changes over time, or the trend**
  - **each individual value of the data.**

Example

The table shows the average monthly rainfall, in centimetres, in Sheffield over the last 30 years.

Month	Jan	Feb	Mar	Apr	May	Jun	Jul	Aug	Sep	Oct	Nov	Dec
Rainfall (cm)	8.7	6.3	6.8	6.3	5.6	6.7	5.1	6.4	6.4	7.4	7.8	9.2

Draw a line graph to show this information.

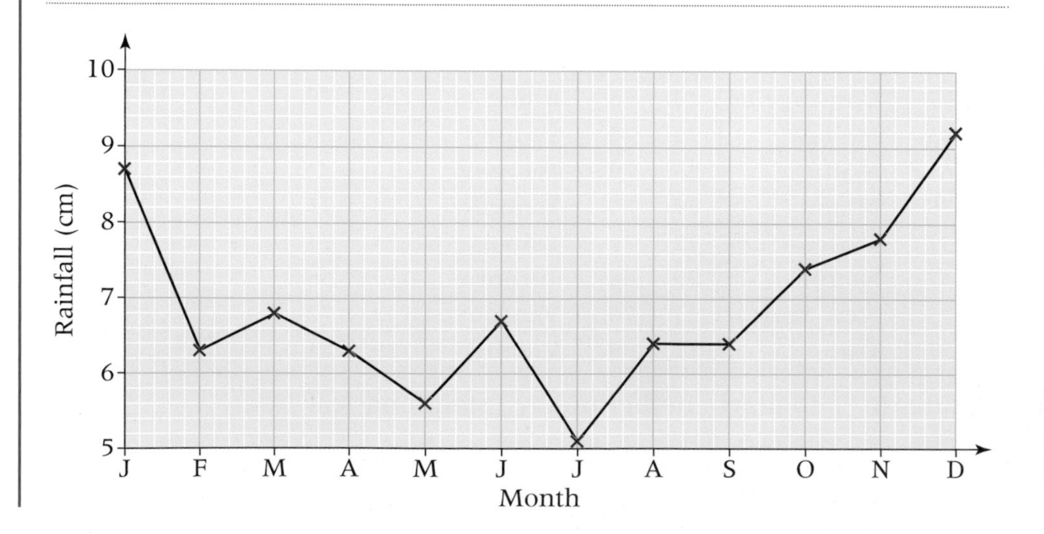

Each division on the vertical axis is 0.2 cm.

The vertical scale doesn't have to start at zero.

Exercise D2.4

1 The numbers of DVDs rented from a shop during a week are shown.

Sunday	Monday	Tuesday	Wednesday	Thursday	Friday	Saturday
18	9	7	11	15	35	36

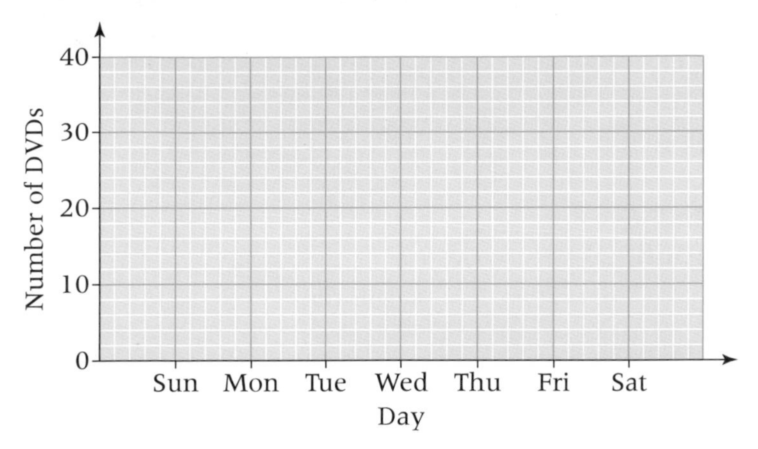

Copy and complete the line graph.

2 Every year on his birthday, Peter's weight, in kilograms, is measured.

Age	2	3	4	5	6	7	8	9	10	11	12	13
Weight (kg)	14	16	18	20	22	25	28	31	34	38.5	40	45

Draw a line graph to show the weights.

3 The hours of sunshine each month are shown in the table.

Jan	Feb	Mar	Apr	May	Jun	Jul	Aug	Sep	Oct	Nov	Dec
43	57	105	131	185	176	194	183	131	87	53	35

Draw a line graph to show the hours of sunshine.

4 The daily viewing figures, in millions, for a reality TV show are shown.

Day	Sat	Sun	Mon	Tue	Wed	Thu	Fri
Viewers (in millions)	3.2	3.8	4.3	4.5	3.1	5.2	7.1

Draw a line graph to show the viewing figures.

5 The men's world record times for running 100 m are shown.

Draw a line graph to show the world record times.

Year	Athlete	Time (sec)
1968	Jim Hines (USA)	9.95
1983	Calvin Smith (USA)	9.93
1988	Carl Lewis (USA)	9.92
1991	Leroy Burrell (USA)	9.90
1991	Carl Lewis (USA)	9.86
1994	Leroy Burrell (USA)	9.85
1996	Donovan Bailey (Can)	9.84
1999	Maurice Greene (USA)	9.79
2002	Tim Montgomery (USA)	9.78
2005	Asafa Powell (Jam)	9.77

Ben Johnson's world records of 9.83 secs in 1987 and 9.79 secs in 1988 were both annulled after he tested positive for drugs at the Seoul Olympics.

Be careful with the horizontal axis (don't miss out the 1970s!)

D2.5 Scatter graphs

This spread will show you how to:

- Draw and use scatter graphs, and compare two data sets
- Understand correlation and use lines of best fit

Keywords
Correlation
Line of best fit
Relationship
Scatter graph
Variables

You can use a **scatter graph** to compare two sets of data, for example, height and weight.

- The data is collected in pairs and plotted as coordinates.
- If the points lie roughly in a straight line, there is a **relationship** or **correlation** between the two **variables**.

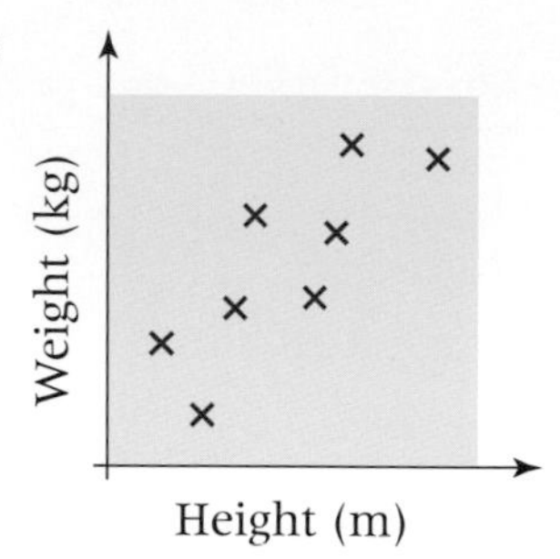

The data can be discrete or continuous.

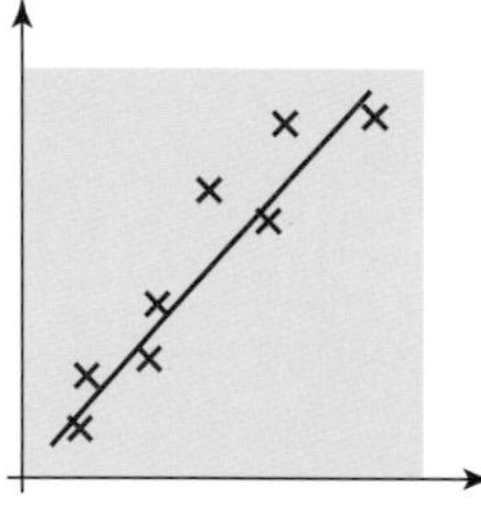
Positive correlation

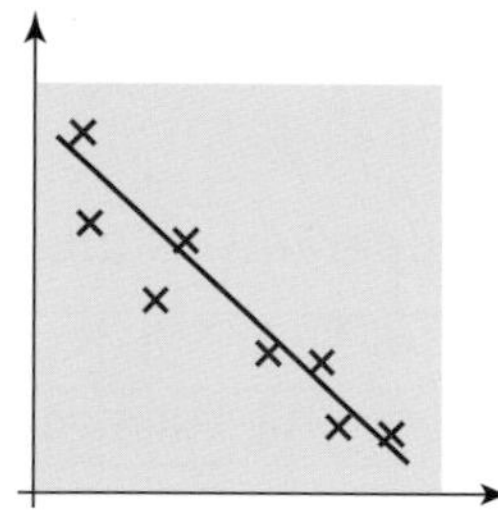
Negative correlation

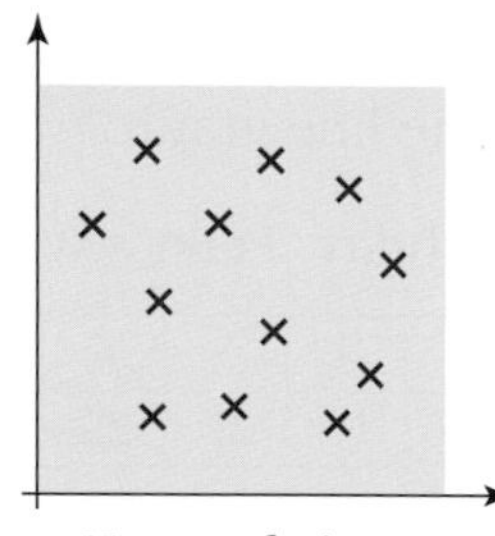
No correlation

Plotted points are not joined on a scatter diagram.

The straight line is the **line of best fit**.

- The correlation is strong if the points are close to the line of best fit.

Example

The exam results (%) for Paper 1 and Paper 2 for 10 students are shown.

Paper 1	56	72	50	24	44	80	68	48	60	36
Paper 2	44	64	40	20	36	64	56	36	50	24

a Draw a scatter graph and a line of best fit.

b Describe the relationship between the Paper 1 results and Paper 2 results.

---

a Plot the exam marks as coordinates. The line of best fit should be close to all the points, with approximately the same number of crosses on either side of the line.

b Students who did well on Paper 1 did well on Paper 2. Students who did not do well on Paper 1 did not do well on Paper 2 either.

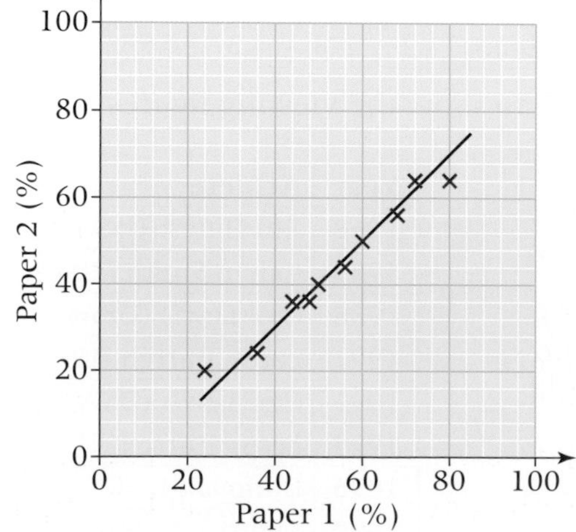

The line of best fit does not have to pass through (0, 0).

Exercise D2.5

1 Describe the type of correlation for each scatter graph.

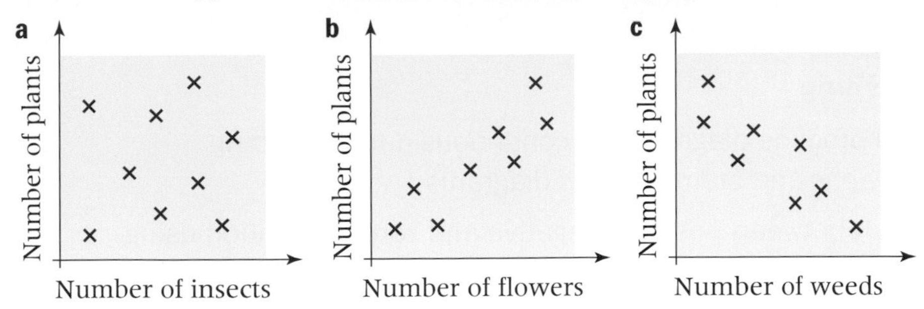

2 The table shows the amount of water used to water plants and the daily maximum temperature.

Water (litres)	25	26	31	24	45	40	5	13	18	28
Maximum temperature (°C)	24	21	25	19	30	28	15	18	20	27

a Copy and complete the scatter graph for this information.

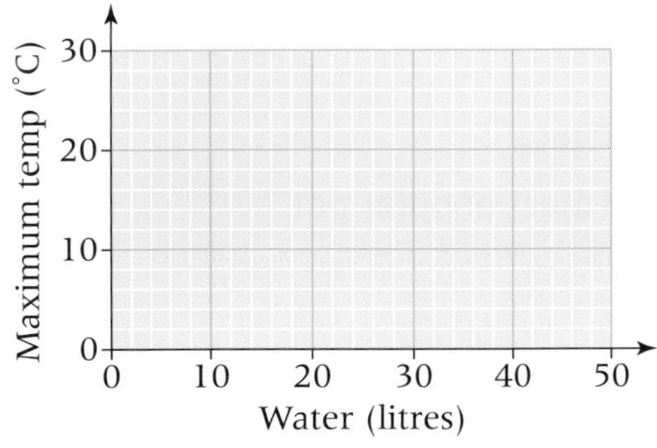

b State the type of correlation shown in the scatter graph.

c Copy and complete these sentences:

i As the temperature increases, the amount of water used _____.

ii As the temperature decreases, the amount of water used _____.

3 The times taken, in minutes, to run a mile and the shoe sizes of ten athletes are shown in the table.

Shoe size	10	$7\frac{1}{2}$	5	9	6	$8\frac{1}{2}$	$7\frac{1}{2}$	$6\frac{1}{2}$	8	7
Time (mins)	9	8	8	7	5	13	15	12	5	6

a Draw a scatter graph to show this information.

Use 2 cm to represent 1 shoe size on the horizontal axis.
Use 2 cm to represent 5 minutes on the vertical axis.

b State the type of correlation shown in the scatter graph.

c Describe, in words, any relationship that the graph shows.

D2 Exam review

Key objectives

- Draw and produce diagrams for continuous data, including scatter graphs and stem-and-leaf diagrams
- Distinguish between positive, negative and zero correlation using lines of best fit

1 Iqbal asked 120 school students, "What is your favourite pet?"
His results are shown in the table.

Pet	Number of student
Dog	40
Cat	55
Rabbit	15
Horse	10

a Represent this information in a pie chart. (4)

b Iqbal said, "In my town, cats are the most popular pet." (1)
Explain why his statement may not be correct.

(OCR, 2003)

2 Here are the times, in minutes, taken to change some tyres:

5 10 15 12 8 7 20 35 24 15
20 33 15 25 10 8 10 20 16 10

Draw a stem-and-leaf diagram to show these times. (3)

(Edexcel Ltd., 2003)

Probability

This unit will show you how to

- List all outcomes for single events and identify different mutually exclusive outcomes
- Understand and use the probability scale
- Understand two-way tables and use them to calculate probabilities
- Understand and use estimates or measures of probability and relative frequency
- Compare experimental data and theoretical probabilities
- Know that the sum of the probabilities of all the outcomes is 1
- List all of the outcomes for two successive events in a systematic way

Before you start ...

You should be able to answer these questions.

1 Choose a number from the rectangle that is

4 7 3 1 9
10
5 2 8 6

a prime **b** square
c triangular **d** a multiple of 3
e a factor of 8

2 Put these fractions in simplest form.
a $\frac{6}{9}$ **b** $\frac{5}{10}$ **c** $\frac{15}{20}$ **d** $\frac{2}{8}$ **e** $\frac{20}{20}$

3 Copy and complete these calculations.
a $\frac{3}{4} + \frac{1}{4} = ?$ **b** $\frac{7}{10} + \frac{3}{10} = ?$
c $1 - \frac{7}{10} = ?$ **d** $1 - \frac{3}{5} = ?$

4 Copy and complete these calculations.
a $1 - 0.1 = ?$ **b** $1 - 0.6 = ?$
c $1 - 0.15 = ?$

5 Calculate
a $\frac{1}{3} \times 21$ **b** $\frac{2}{3} \times 21$
c 0.2×80 **d** 0.4×50
e 0.3×60

6 Convert these fractions into decimals.
a $\frac{1}{100}$ **b** $\frac{25}{100}$ **c** $\frac{35}{100}$ **d** $\frac{5}{100}$ **e** $\frac{36}{100}$

D3.1 Probability

This spread will show you how to:

- List all outcomes for single events and identify different mutually exclusive outcomes

Keywords
Equally likely
Event
Outcome
Probability
Systematically
Trial

- **A trial is an activity.**
  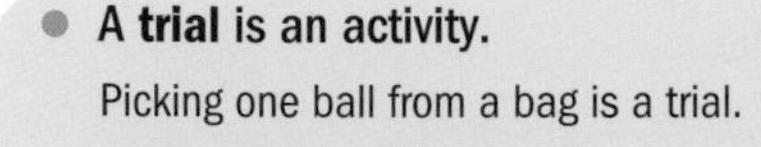
  Picking one ball from a bag is a trial.
- **An outcome is one possible result of a trial.**
  When picking a ball from the bag, all possible outcomes are red, red, red, red, red, green, green, green, green, green.
- **An event is a set of one or more outcomes of a trial.**
  'Pick a red ball' is an event.

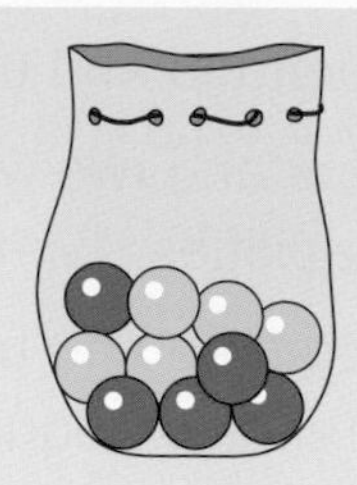

Each outcome is **equally likely** as the balls are identical in size and shape.

Probability measures how likely it is that an event will happen.

- **You can calculate probability using the formula:**

$$\text{Probability of an event occurring} = \frac{\text{Number of outcomes that satisfy the event}}{\text{Total number of possible outcomes of the trial}}$$

All probabilities have a value between 0 and 1.
The probability of an event can be written as P(event).

0 means impossible.
1 means certain.

Example

An ordinary dice is rolled.

a List the possible outcomes.
b Calculate the probability of choosing
  i a 5
  ii an even number.

---

a ⚀ ⚁ ⚂ ⚃ ⚄ ⚅
or 1, 2, 3, 4, 5, 6.

Draw the 6 outcomes in order or **systematically**.

b i There is one 5.
There are 6 possible equally likely outcomes.
$P(5) = \frac{1}{6}$

ii There are 3 even numbers.
There are 6 possible outcomes.
$P(\text{even}) = \frac{3}{6} = \frac{1}{2}$

Exercise D3.1

1 Copy these tables and put each event in the appropriate table.

The event is impossible
Probability is 0

The event is certain
Probability is 1

a Picking a blue ball from a bag of blue balls.

b Picking a red ball from a bag of blue balls.

c Next month will have 7 days.

d Next year will have 12 months.

e You will roll an 8 on an ordinary six-sided dice.

2 List all the possible outcomes for these trials.

a spinning a coin

b rolling a tetrahedron dice (4 faces, numbered 1, 2, 3, 4)

c picking a letter from C H A N G E

d spinning the spinner shown on the right

e picking a day from all the days of the week.

3 Answer each of these questions for each bag A to E.

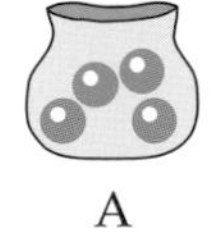

A

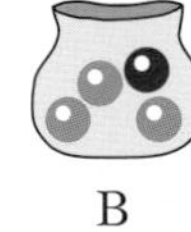

B

C

D

E

a List the 4 possible outcomes, when a ball is taken from the bag.

b Calculate the probability of taking out a blue ball.

c Calculate the probability of taking out a red ball.

d Which colour ball is the most likely to be taken out?

e Which colour ball is the least likely to be taken out?

4 A raffle has only one prize. 250 tickets are sold.
Calculate the probability of winning the prize if you buy

a one ticket **b** ten tickets.

5 There are 48 boys and 72 girls on a diploma programme at Clarendon College.
One student is selected at random.
Calculate the probability that the student is

a a boy **b** a girl.

D3.2 Probability scale

This spread will show you how to:

- Understand and use the probability scale

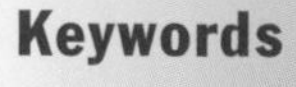

Keywords

Chance
Mutually exclusive
Outcome
Probability
Probability scale

The **chance** of an event occurring is measured by the **probability**.

All probabilities have a value between 0 and 1 and can be marked on a **probability scale**.

You can use fractions or decimals on the probability scale.

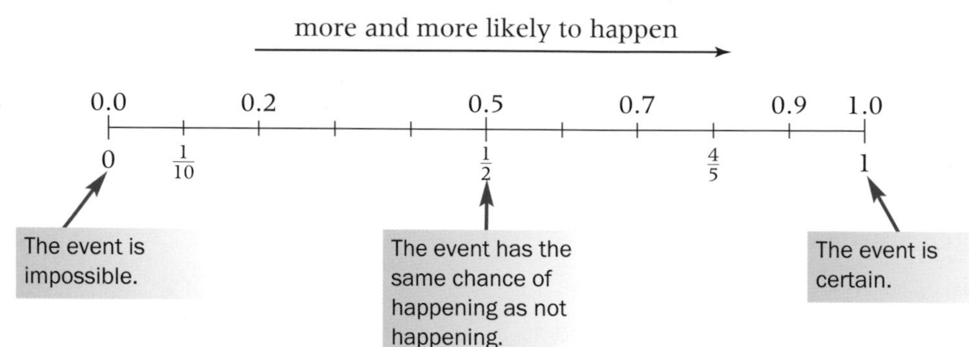

An equilateral triangle is made into a spinner as shown.
The possible **outcomes** are Orange, Orange and Green.

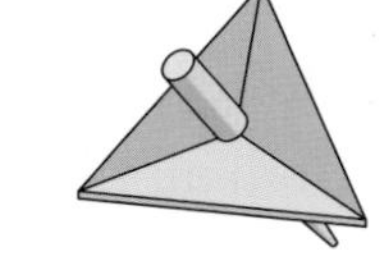

These outcomes are **mutually exclusive** because if you get one outcome you cannot get the other one.

Probability of spinning orange = $\frac{2}{3}$

Probability of **not** spinning orange = $\frac{1}{3}$

- **The probabilities of mutually exclusive outcomes add up to 1.**
- **Probability of an outcome not happening = 1 – probability of the outcome happening**

$\frac{2}{3} + \frac{1}{3} = 1$

$\frac{1}{3} = 1 - \frac{2}{3}$

Example

The probability of picking a coloured counter from the tin is shown.

Colour	Red	Orange	Yellow
Probability	0.4	0.25	

a Calculate the probability of picking a yellow counter.
b Which colour counter is the least likely to be picked?

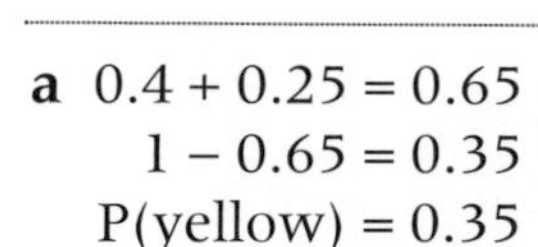

a $0.4 + 0.25 = 0.65$
$1 - 0.65 = 0.35$
$P(\text{yellow}) = 0.35$

b Orange, as orange has the smallest probability.

Exercise D3.2

1 Draw a 10 cm line. Put a mark at every centimetre.
Label the marks 0.0, 0.1, 0.2, ..., 0.9, 1.0 as shown.

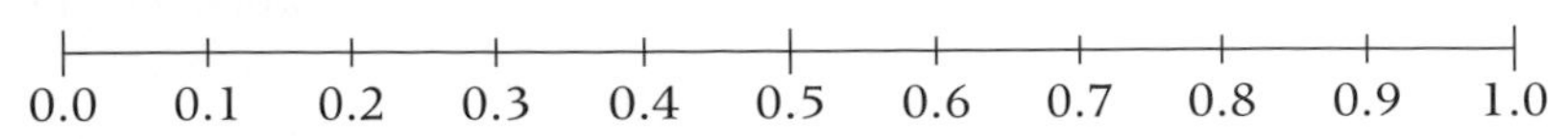

On your probability scale, mark the position of

a an impossible event

b an even chance event

c a certain event.

2 A bag contains 1 yellow, 3 green and 6 red balls.
One ball is taken out at random.

a Calculate the probability that the ball is

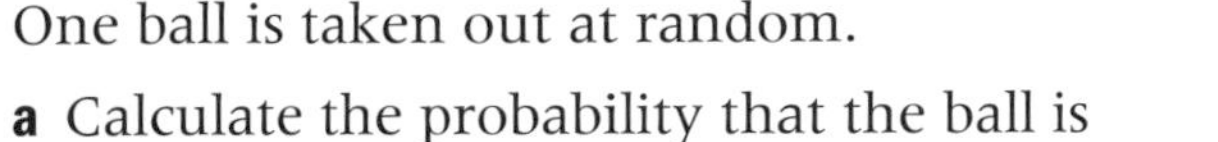

i yellow **ii** green **iii** red.

b Draw a probability scale as in question 1.
On your scale, mark the positions of P(yellow), P(green) and P(red).

c Which colour is the least likely to be picked?

d Which colour is the most likely to be picked?

e Add the answers P(yellow), P(green) and P(red).

3 A spinner is made from a regular octagon.
There are 6 pink and 2 green triangles.

a Calculate the probability that the arrow spins to

i a pink triangle

ii a green triangle.

b Draw a probability scale as in question 1.
On your scale, mark the positions of P(pink) and P(green).

c Which colour is the least likely?

d Which colour is the most likely?

e Add the answers P(pink) and P(green).

4 The probability of winning a raffle is 0.01.
What is the probability of not winning the raffle?

5 A bag contains red, yellow and green counters.
One counter is taken out at random.
The probability that the counter is a particular colour is shown in the table.

Colour	Red	Yellow	Green
Probability	0.3	0.1	

Calculate the probability of taking a green counter out of the bag.

D3.3 Two-way tables

This spread will show you how to:

- Understand two-way tables and use them to calculate probabilities

Keywords
Random
Two-way table

- A **two-way table** links two types of information.

23 children and adults attend a birthday party.
The two-way table shows this information.

	Male	Female
Adult	0	4
Child	7	12

$7 + 12 = 19$ children at the party.

There were no adult males.

$4 + 12 = 16$ females at the party.

You can use two-way tables to calculate probabilities.

Example

The shoe sizes of fifty students are shown in the two-way table.

	Size 6	Size 7	Size 8	Size 9	Size 10
Boy	3	8	9	8	2
Girl	8	9	2	1	0

One of the students is selected at random.
Calculate the probability that the student is

a a boy

b a girl

c a girl who wears size 7 shoes

d a student who wears size 8 shoes.

---

a $3 + 8 + 9 + 8 + 2 = 30$ boys

$\text{P(boy)} = \frac{30}{50} = \frac{3}{5}$

b $\text{P(girl)} = 1 - \frac{3}{5} = \frac{2}{5}$

The outcomes 'a boy' or 'a girl' are mutually exclusive.

c $\text{P(girl with size 7 shoes)} = \frac{9}{50}$

d $9 + 2 = 11$

$\text{P(size 8)} = \frac{11}{50}$

Exercise D3.3

1 The numbers of passengers in 50 cars are recorded. The results are shown in the two-way table. Calculate the probability that a car has

Number of passengers	Number of cars
0	20
1	15
2	10
3 or more	5

a 0 passengers **b** 1 passenger

c 2 passengers **d** 3 or more passengers.

2 A tray contains brown and white eggs. Some of the eggs are cracked. The two-way table shows the number of eggs in each category.

	Cracked	Not cracked
Brown	2	6
White	4	12

a Calculate the number of eggs that are

i on the tray **ii** brown **iii** white

iv cracked **v** not cracked.

One egg is selected at random.

b Calculate the probability that the egg is

i brown and cracked **ii** white and not cracked

iii a brown egg **iv** a white egg

v a cracked egg **vi** an uncracked egg.

3 A game uses a set of tiles made from squares. Each tile holds either a cross (×) or a circle (○), and is either red or green.

The number of tiles in each category is shown in the two-way table.

	Cross(×)	Circle(○)
Green	20	10
Red	4	16

a Calculate the number of tiles in the complete set.

One tile is taken at random.

b Calculate the probability that the tile is

i green with a cross

ii red with a circle.

c Copy and complete the two-way table showing the probabilities of choosing a tile.

d Add all four probabilities in your two-way table. Explain your answer.

	Cross(×)	Circle(○)
Green		$\frac{1}{5}$
Red	$\frac{2}{25}$	

D3.4 Expected frequency

This spread will show you how to:

- Understand and use estimates or measures of probability

Keywords
Expect
Expected frequency
Trial

If you know the probability of an event, you can calculate how many times you **expect** the outcome to happen.

Example

A dice is rolled 60 times.
Calculate the number of

a fours

b even numbers you would expect.

a $P(4) = \frac{1}{6}$ or 1 four for every 6 rolls
or 10 fours for every 60 rolls.

b $P(\text{even}) = \frac{1}{2}$ $\frac{1}{2}$ of $60 = 30$ times should be even.

- **The expected frequency is the number of times you expect an outcome to happen.**
- **Expected frequency = probability × number of trials.**

Each roll of the dice is called a **trial**.

Example

Red, yellow and green balls are put in a bag.
The probability of taking out each colour of ball is given in the table.

Red	Yellow	Green
0.1	0.3	

a Calculate the probability of taking out a green ball.

b Which colour is the most likely to be taken out?

c There are 60 balls in the bag.
How many of them are red, yellow and green?

a $0.1 + 0.3 = 0.4$
$1 - 0.4 = 0.6$
$P(\text{green}) = 0.6$

b Green is the most likely as 0.6 is the biggest probability.

c Red: $0.1 \times 60 = 6$ red balls
Yellow: $0.3 \times 60 = 18$ yellow balls
Green: $0.6 \times 60 = 36$ green balls
Check: $6 + 18 + 36 = 60$ balls

The outcomes like: 'a red', 'a yellow' or 'a green' are mutually exclusive.

Exercise D3.4

1 A crisp manufacturer claims that '3 out of every 4 people prefer Potayto Crisps'.
If 100 people are asked, how many would you expect to prefer Potayto Crisps?

2 The probability of a drawing pin landing point up when dropped, is 0.8.
If 100 drawing pins are dropped, how many of them would you expect to land point up?

3 Research has shown that when you take a dog for a daily walk, the probability that you will have a conversation with someone is $\frac{3}{10}$.
If you walked a dog daily throughout September, how many times would you expect to have a conversation with someone?

'30 days has September, April June and November.'

4 A spinner is made from a circle divided into 12 equal sectors, coloured green, yellow and pink.
If the spinner is spun 60 times, how many times would you expect the colour to be

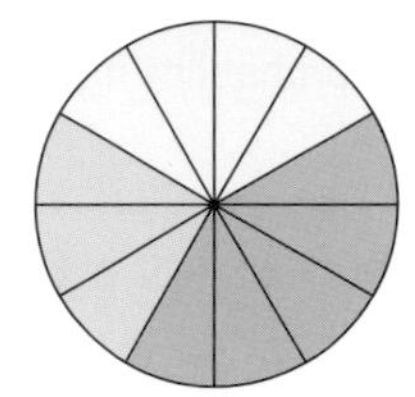

a green

b yellow

c pink?

5 The probability of rolling each number for this biased dice is shown in the table.

Number	1	2	3	4	5	6
Probability	0.1	0.15	0.3	0.2	0.1	

a Calculate the probability of rolling a 6.

b Calculate the expected frequencies for each number, if the dice is rolled

i 100 times

ii 500 times

iii 1000 times.

6 In a 'roll a penny' game, a penny is rolled into the box.
To win you need to land the penny off any black lines.
It costs 1p to have a go and winners are given 10p back.
The probability of landing the penny off the black lines is $\frac{1}{20}$.

If 200 pennies are rolled into the box, how much profit would you expect the game to make?

D3.5 Relative frequency

This spread will show you how to:

- Understand and use estimates or measures of probability and relative frequency
- Compare experimental data and theoretical probabilities

Keywords

Biased
Equally likely
Estimate
Experiment
Fair
Relative frequency
Trial

- **A dice is fair if each number is equally likely to be rolled.**
  You would normally expect a dice to be fair, as each face is identical.

- **A spinner is biased if each colour is NOT equally likely to happen.**
  This spinner is not fair as the size and shape of each colour are not identical.

It is not always possible to calculate the theoretical probability.

- **You can estimate the probability from experiments.**

Example

Sam knows the probability of a Head when spinning a coin should be $\frac{1}{2}$ (or 0.5).
She thinks the coin is biased and so she spins the coin 50 times. The results are shown in the frequency table.

	Frequency
Head	35
Tail	15

a Estimate the probability of getting a Head when spinning the coin.
b Do you think the coin is biased? Explain your answer.

a Sam got a Head on 35 out of 50 occasions.
Estimated probability of getting a Head = $\frac{35}{50} = \frac{7}{10} = 0.7$
b The coin could be biased as 0.7 is significantly larger than 0.5. However, Sam needs to spin the coin a lot more times before she can make the decision.

- **Estimated probability is called the relative frequency. This becomes more reliable as you increase the number of trials.**

Each spin of the coin is called a **trial**.

In the example, if Sam calculated the relative frequency of getting a Head after each spin, she could graph the results.

Highest relative frequency is 1.

Lowest relative frequency is 0.

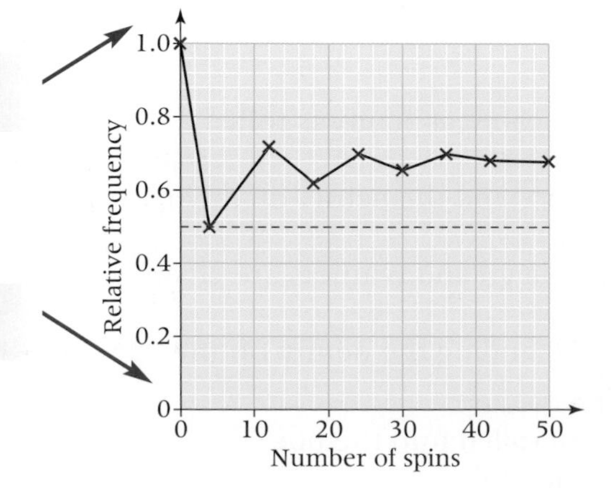

P(Head) = $\frac{1}{2}$ = 0.5
This is the theoretical probability of spinning a Head.

Exercise D3.5

1 The colours of 50 cars are recorded. The colours are shown

Blue	Red	Other	Silver	Blue	Red	Silver	Silver	Blue
Other	Red	Silver	Silver	Blue	Red	Red	Other	Silver
Silver	Red	Red	Silver	Silver	Blue	Blue	Silver	Silver
Red	Red	Other	Blue	Red	Red	Other	Red	Red
Red	Silver	Blue	Blue	Blue	Other	Silver	Other	Other
Other	Silver	Red	Red	Blue				

a Copy and complete the frequency chart to show the 50 colours.

Colour	Tally	Frequency
Blue		
Red		
Silver		
Other		

b State the most common colour.

c Give an estimate of the probability that the next car will be

i blue **ii** red **iii** silver.

2 A spinner is made from a regular pentagon. The scores are recorded

1	2	3	4	3	5	1	2
5	4	2	1	3	1	5	4
2	2	3	1	4	5	4	2
3	1	2	2	4	4	5	5
1	2	3	4	2	4	1	1

a Draw a frequency chart to show the scores.

b State the most common score.

c How many times is the spinner spun?

d Estimate the probability of scoring

i 1 **ii** 2 **iii** 3 **iv** 4 **v** 5.

e If the spinner is fair, how many times would you expect to spin a 3 from 100 spins?

3 There are 10 coloured balls in a bag.
One ball is taken out and then replaced in the bag.
The colours of the balls are shown in the frequency table.

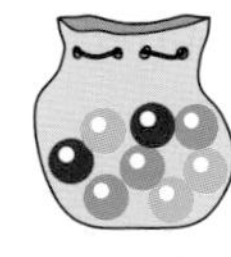

Colour	Red	Green	Blue
Frequency	9	14	27

a How many times was a ball taken out of the bag?

b Estimate the probability of taking out

i a red ball **ii** a green ball **iii** a blue ball.

c How many balls of each colour do you think are in the bag?

d How could you improve this guess?

D3.6 Mutually exclusive outcomes

This spread will show you how to:

- Identify different mutually exclusive outcomes
- Know that the sum of the probabilities of all the outcomes is 1

Keywords
Mutually exclusive
Or
Outcome

The possible **outcomes** when a dice is rolled are 1, 2, 3, 4, 5, 6.

These outcomes are **mutually exclusive** because if you get one outcome, you cannot get another one.

The probability of rolling a 4 = $\frac{1}{6}$

The probability of rolling a 5 = $\frac{1}{6}$

The probability of rolling a 4 **or** a 5 = $\frac{2}{6}$

$P(4) = \frac{1}{6}$

$P(5) = \frac{1}{6}$

Or means either of the outcomes 4 or 5.

Notice that

$P(4 \text{ or } 5) = P(4) + P(5)$

$\frac{2}{6} = \frac{1}{6} + \frac{1}{6}$

- **In general, if events A and B are mutually exclusive,**
  P(A or B) = P(A) + P(B)

Example

A spinner is made from a regular pentagon and numbered from 1 to 5. Calculate

a P(3) **b** P(3 or 4) **c** P(3 or an odd number)

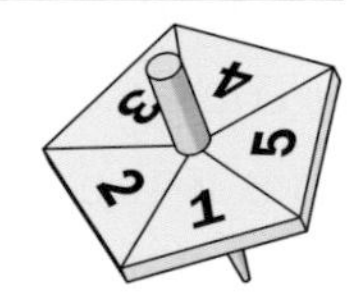

a There is one 3. There are 5 possible outcomes. $P(3) = \frac{1}{5}$

b $P(3) = \frac{1}{5}$ $\quad P(4) = \frac{1}{5}$

The outcomes are mutually exclusive and so

$$\begin{aligned} P(3 \text{ or } 4) &= P(3) + P(4) \\ &= \frac{1}{5} + \frac{1}{5} \\ &= \frac{2}{5} \end{aligned}$$

c $P(3) = \frac{1}{5}$ $\quad P(\text{odd}) = \frac{3}{5}$

The outcomes are not mutually exclusive so P(A or B) = P(A) + P(B) cannot be used.
A list of the 5 possible outcomes is

$$\begin{pmatrix} 1 \\ \text{odd} \end{pmatrix} \begin{pmatrix} 2 \\ \text{even} \end{pmatrix} \begin{pmatrix} 3 \\ \text{odd} \end{pmatrix} \begin{pmatrix} 4 \\ \text{even} \end{pmatrix} \begin{pmatrix} 5 \\ \text{odd} \end{pmatrix}$$

There are 3 outcomes that are OK.

$P(3 \text{ or an odd number}) = \frac{3}{5}$

Exercise D3.6

1 Outcomes are mutually exclusive if they cannot occur at the same time. State if these outcomes are mutually exclusive.

a spinning a Head and spinning a Tail with a coin

b rolling a 2 and rolling a 3 with a dice

c rolling a 2 and rolling an even number with a dice

d rolling a 2 and rolling an odd number with a dice

e rolling a 2 and rolling a prime number with a dice

f winning and losing a game of chess

g sunny and rainy weather

h taking out a red ball and taking out a blue ball, when taking out one ball from a bag.

2 This is a net of a tetrahedral dice. The dice is made and rolled. What is the probability of rolling

a a 3 **b** a 2 **c** a 2 or a 3?

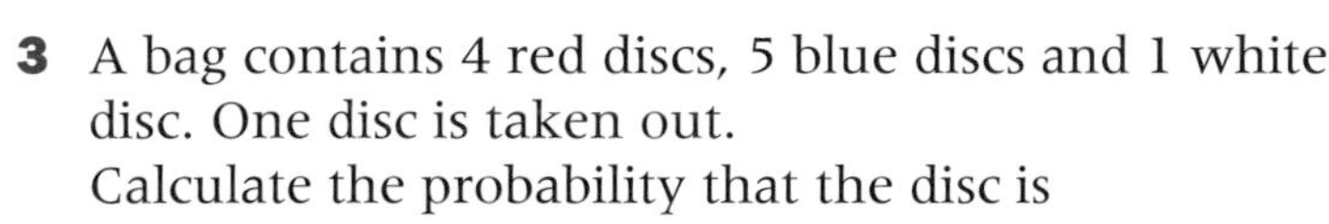

3 A bag contains 4 red discs, 5 blue discs and 1 white disc. One disc is taken out.
Calculate the probability that the disc is

a red **b** blue **c** white

d red or white **e** blue or white **f** red or blue or white

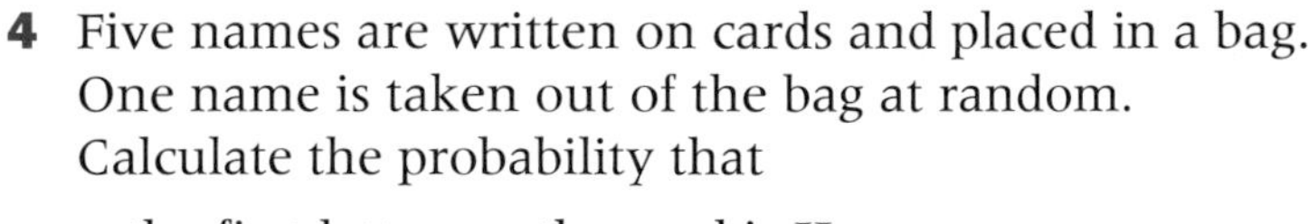

4 Five names are written on cards and placed in a bag. One name is taken out of the bag at random.
Calculate the probability that

HENRY
EDWARD
JAMES
CHARLES
GEORGE

a the first letter on the card is H

b the first letter on the card is G

c the first letter on the card is H or G

d the card has 5 letters written on it

e the card has 5 or 6 letters written on it.

5 A survey of vehicles passing the college gate is taken.

C C C C V L C C V B
C L C C C C B C C V
V V C C L C C B C C
L B C C C V L C C C
B V C V C C V C C V

C = Car
L = Lorry
B = Bus
V = Van

a Copy and complete the frequency table.

Vehicle	Tally	Frequency
Car (C)		
Lorry (L)		
Bus (B)		
Van (V)		

b Calculate the estimated probability that the next vehicle that passes will be

i a car **ii** a lorry **iii** a bus or a van.

D3.7 Two events

This spread will show you how to:

- List all of the outcomes for two events in a systematic way

Keywords
Equally likely
Sample space diagram

You can list the possible outcomes for two **successive** events using a **sample space diagram**

Successive means following on, such as 3, 4, 5.

Example

A dice is numbered 1, 2, 3, 4, 5, 6. A spinner has colours red, blue, yellow and green.
Stephanie rolls the dice and spins the spinner.

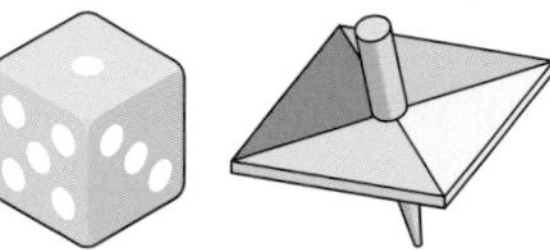

a Draw a sample space diagram to show the possible outcomes.
b What is the probability that she gets an even number and blue?

a

		Score on the dice					
		1	**2**	**3**	**4**	**5**	**6**
Colour on the spinner	**Red**	(1, R)	(2, R)	(3, R)	(4, R)	(5, R)	(6, R)
	Blue	(1, B)	(2, B)	(3, B)	(4, B)	(5, B)	(6, B)
	Yellow	(1, Y)	(2, Y)	(3, Y)	(4, Y)	(5, Y)	(6, Y)
	Green	(1, G)	(2, G)	(3, G)	(4, G)	(5, G)	(6, G)

b Even and Blue occurs 3 times. There are 24 possible outcomes.

P(Even and Blue) $= \frac{3}{24} = \frac{1}{8}$

- **The sample space diagram can only be used for equally likely outcomes.**

Example

Two spinners are numbered 1, 3, 5 and 2, 4, 6.
Each spinner is spun and the scores are added.

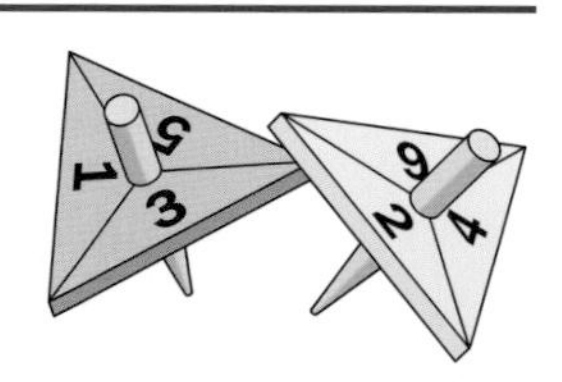

a Draw a sample space diagram to show the possible totals.
b What is the probability of getting a total of 7?

a

	1	**3**	**5**
2	3	5	7
4	5	7	9
6	7	9	11

The outcomes are equally likely as the triangles on each spinner are the same size.

b 7 occurs 3 times. There are 9 possible outcomes.

P(total of 7) $= \frac{3}{9} = \frac{1}{3}$

Exercise D3.7

1 A spinner is made from a square and coloured red (R), yellow (Y), green (G) and pink (P). Another spinner is made from a regular pentagon and labelled A, B, C, D, E. Both spinners are spun.

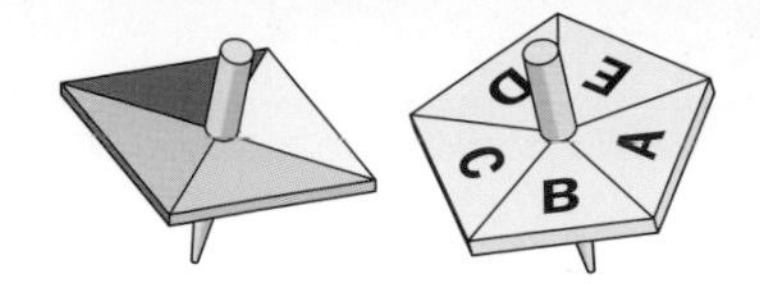

a List the 20 possible outcomes.

b Copy and complete the sample space diagram to show the outcomes.

		Colour spinner			
		R	Y	G	P
Letter spinner	A	(R, A)			
	B				
	C				
	D				
	E				

c Calculate the probability of getting Green and a C.

2 A dice is made from a tetrahedron, and numbered 1, 2, 3, 4. A coin has Heads or Tails. The dice is rolled and the coin is spun.

a Copy and complete the sample space diagram to show the possible outcomes.

		Dice			
		1	2	3	4
Coin	Heads	(1, H)			
	Tails				

b Calculate the probability of getting

i a 3 and a Head

ii an even number and a Head.

3 A 'fruit machine' has only two windows. Each window can show a Club (♣), a Diamond (♦), a Spade (♠) or a Heart (♥).

a Draw a sample space diagram to show the 16 possible outcomes.

b Calculate the probability of getting two of the same symbols.

4 Two fair dice each numbered 1 to 6 are rolled.

List all the possible outcomes.

Calculate the probability of getting a double six.

D3 Exam review

Key objectives

- Understand and use the probability scale
- Understand and use estimates or measures of probability from theoretical models, or from relative frequency
- List all outcomes for single events, and for two successive events, in a systematic way
- Identify different mutually exclusive outcomes and know that the sum of the probabilities of all these outcomes is 1

1 The probabilities of whether a student, picked at random from a school, is vegetarian or not are shown in this table.

	Boys	Girls
Vegetarian	0.08	0.2
Non-vegetarian	0.4	0.32

a What is the probability that a student chosen at random from the school is vegetarian? (1)

b There are 320 girls in the school who are vegetarian. How many students are there in the school altogether? (2)

(AQA, 2003)

2 The probability that a biased dice will land on a four is 0.2.

Pam is going to roll the dice 200 times.

Work out an estimate for the number of times the dice will land on a four. (2)

(Edexcel Ltd., 2004)

D4 Averages and range

This unit will show you how to

- Recognise the difference between discrete and continuous data
- Calculate the mean, median, mode and range for sets of data
- Use frequency tables for discrete and grouped data
- Compare distributions and make inferences
- Calculate an estimate of the mean for grouped data
- Identify the modal class and the class interval in which the median lies for grouped data

Before you start ...

You should be able to answer these questions.

1 Calculate

a 43 – 17 **b** 136 – 118

c 8.3 – 2.7

2 Order these numbers in size, smallest first.

a 6.5, 8, 4, 5.5, 7.

b 3.2, 2.3, 3.4, 4.3, 4.2, 2.4.

c 8.6, 7.5, 9.1, 7.9, 8.3.

3 List the ten numbers recorded in each frequency table.

a

Number	Frequency
5	0
6	2
7	1
8	2
9	5

b

Number	Frequency
100	2
101	1
102	2
103	0
104	5

4 Find the mid-value between

a 1 and 5 **b** 6 and 10

c 40 and 44 **d** 45 and 49

e 50 and 54

D4.1 Types of data and the range

This spread will show you how to:

- Recognise the difference between discrete and continuous data

Keywords
Continuous data
Discrete data
Range
Spread

- Numerical data can be **discrete** or **continuous**.

- **Discrete data** can only take exact values.

Shoe sizes could be 7, $7\frac{1}{2}$, 8, $8\frac{1}{2}$, 9.
There are no values between them.

The shoe size $7\frac{1}{4}$ does not exist.

- **Continuous data** can take any value within a given range.

The values of continuous data depend on the accuracy of the measurement.

Temperature can take any value between 21 °C and 22 °C.

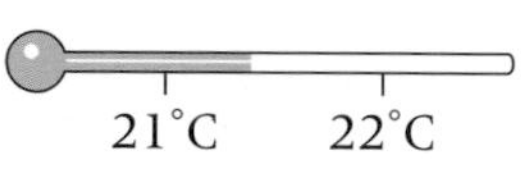

Continuous data cannot be measured exactly.

A height of 171 cm has been given to the nearest centimetre.

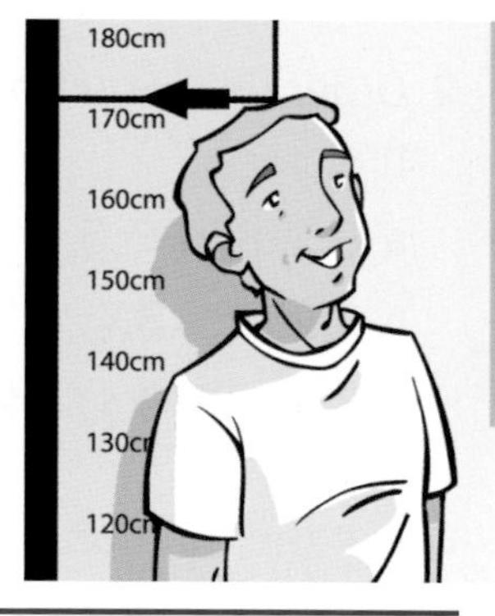

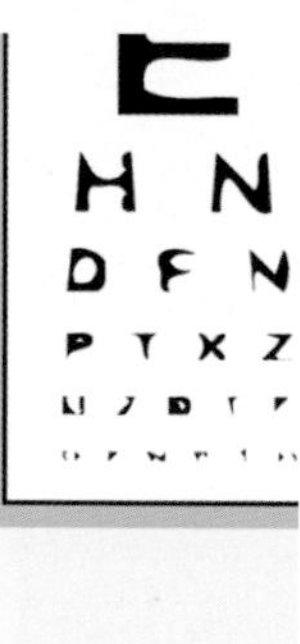

- You can measure the **spread** of a set of data by calculating the **range**.
- The **range** is the highest value minus the lowest value.

Example

There are eight tutor groups on the GCSE programme at Clarendon College. The numbers of students that were absent from each group were 3, 1, 4, 8, 4, 2, 5, 1.

a Calculate the range of absences.
b State whether the data is discrete or continuous.

a Range = highest value − lowest value
= 8 − 1
= 7

b Discrete – you cannot have $7\frac{1}{2}$ students.

Exercise D4.1

1 Decide whether each of these are discrete or continuous data.

a number of people in a room
b number of cars in a car park
c length of a piece of wood
d thickness of a piece of wood
e amount of water in a pan
f cost of buying a DVD
g time taken to walk to the shops
h weight of a piece of cheese
i number of tomatoes on a plant
j temperature in a fridge
k score on a dice
l your age

2 Calculate the range for each set of numbers.

a 0, 0, 1, 2, 2, 2, 3, 4, 4, 5, 5
b 6, 7, 8, 8, 8, 9, 9, 9, 10
c 32, 32, 33, 35, 41
d 48, 48, 48, 49, 49
e 85, 86, 87, 88, 89, 90, 91, 92
f 5, 6, 8, 4, 7, 9, 5, 6, 7
g 31, 34, 18, 25, 31, 26, 35, 27
h 17, 17, 15, 16, 21, 22, 20, 20
i $3\frac{1}{2}$, $3\frac{1}{2}$, $4\frac{1}{2}$, 4, $3\frac{1}{2}$, 5, 3, $3\frac{1}{2}$
j £3.00, £1.25, £4.70, £2.52, 85p

3 Calculate the range of these sets of numbers.

a

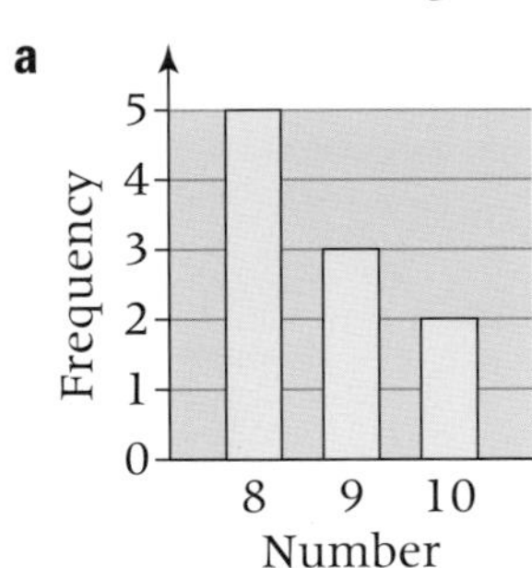

b

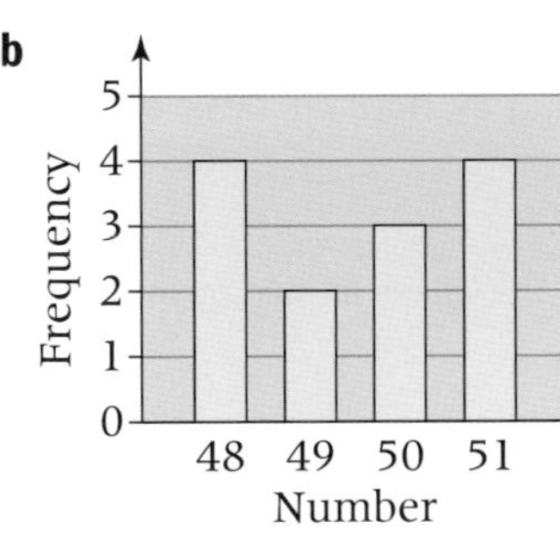

c

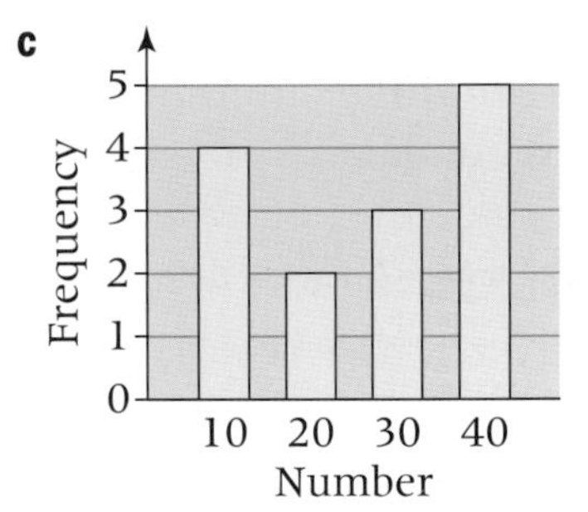

4 The weights, in kilograms, of 10 students are

46 33 53 72 56 31 60 55 48 51

Calculate the range of the weights. State the units of your answer.

5 The mean monthly temperature, in °F, is shown for Athens and Madrid.

	Jan	Feb	Mar	Apr	May	Jun	Jul	Aug	Sep	Oct	Nov	Dec
Athens	50	50	54	59	67	75	81	81	75	67	59	53
Madrid	42	45	49	53	60	69	76	76	69	58	49	44

a Calculate the range of temperatures for Athens.
b Calculate the range of temperatures for Madrid.

6 The range of these numbers is 17.
Find two possible values for the unknown number.

D4.2 Averages

This spread will show you how to:

- Calculate the mean, median and mode for sets of data

Keywords

Average
Mean
Median
Modal value
Mode
Representative value

The **average** 17-year-old boy in the UK is 174 cm tall.
This does not suggest that every boy's height is 174 cm, but that 174 cm is used to represent the height of all the 17-year-old boys in the UK.

- **You can represent a set of data with one number, called the average.**

There are three different ways to find a typical or **representative value** for a set of data.

- **The mean of a set of data is the total of all the values divided by the number of values.**
- **The mode is the value that occurs most often.**
- **The median is the middle value when the data is arranged in order.**

The mode is sometimes called the **modal value**.

Example

Calculate the mean, mode and median of

8, 3, 8, 7, 5

Mean = $(8 + 3 + 8 + 7 + 5) \div 5$
$= 31 \div 5$
$= 6.2$

Mode = 8   as 8 occurs most often

For the median, first arrange the numbers in numerical order:

3   5   7   8   8

↑ middle

Median = 7

To calculate the median of 7, 7, 8, 9, 10, 14: The middle numbers are 8 and 9. Median = $(8 + 9) \div 2 = 8.5$

Exercise D4.2

1 **a** Calculate the mean of these five numbers.

3, 3, 8, 1, 5

b Copy the bar chart to illustrate the five numbers.

c Mark the mean on your diagram with a horizontal line.

d Show how the rectangles above the mean can be moved to give five bars with equal heights.

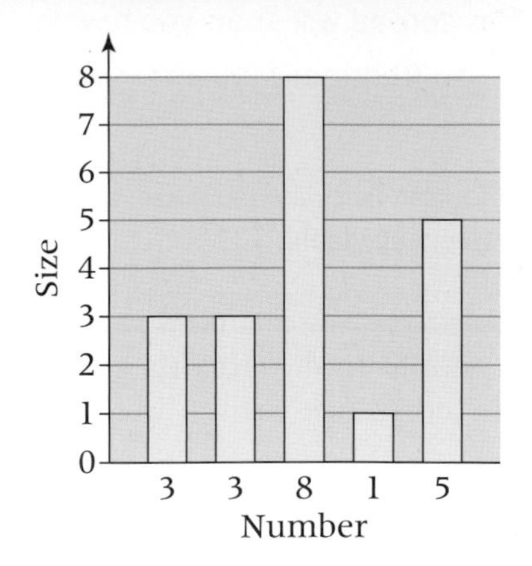

2 These 40 vehicles are recorded as they pass.

C C C T B B C C B C
T B B B C C C C C B
B C C C C T C B B C
C C B C C C C T C C

B = Bus
T = Tram
C = Car

a Copy and complete the frequency table for the vehicles.

Vehicle	Tally	Frequency
Bus (B)		
Tram (T)		
Car (C)		

b State the modal vehicle.

3 Five men were asked to count the loose change in their pockets. The results were

£2.47 £1.12 38p £2.15 85p

a Arrange the amounts in order, smallest first.

b Find the median amount of money. State the units of your answer.

4 The seven-day weather forecast for a warm week in August is shown.

Thu	Fri	Sat	Sun	Mon	Tue	Wed

80°F	82°F	85°F	80°F	80°F	81°F	75°F

Calculate

a the mean temperature **b** the modal temperature

c the median temperature.

5 Six numbers are arranged in order. The median of these numbers is 5.5.

3.2	4.5	?	6.0	7.6	8.5

a Calculate the unknown number.

b Calculate the mean of the six numbers.

D4.3 Charts and tables

This spread will show you how to:

- Use frequency tables for discrete and grouped data

Keywords
Frequency table
Mean
Median
Mode
Range

- **You can calculate the mean, mode, median and range for discrete data from a frequency table.**

Example

Ten people took part in a golf competition. Their scores are shown in the frequency table.

Calculate the mean, mode, median and range of the scores.

Score	Frequency
67	1
68	4
69	3
70	1
71	1

4 people scored 68.

1 person scored 71.

The results can be written in numerical order.

67, 68, 68, 68, 68, 69, 69, 69, 70, 71

Mean = 687 ÷ 10 = 68.7
Median = (68 + 69) ÷ 2 = 68.5
Mode = 68 (occurs the most)
Range = 71 − 67 = 4

Alternatively, you can calculate the mean, mode, median and range directly from the frequency table without rewriting the numbers.

Score	Frequency	Score × Frequency
67	1	67
68	4	272
69	3	207
70	1	70
71	1	71
	10	687

68 + 68 + 68 + 68 or 68 × 4.

The total of all the scores of the 10 golfers.

Mean = 687 ÷ 10 = 68.7
Median = (5th value + 6th value) ÷ 2 = (68 + 69) ÷ 2 = 68.5
Mode = 68 as has the highest frequency
Range = 71 − 67 = 4

The mode is 68, not 4.

Exercise D4.3

1 The numbers of flowers on eight rose plants are shown in the frequency table.

Number of flowers	Tally	Frequency
3	IIII	4
4	II	2
5	II	2

a List the eight numbers in order of size, smallest first.

b Calculate the mean, mode, median and range of the eight numbers.

2 The number of days that 25 students were present at school in a week are shown in the frequency table.

Number of days	Tally	Frequency
0		0
1	IIII	4
2	~~IIII~~ I	6
3	II	2
4	~~IIII~~	5
5	~~IIII~~ III	8

a List the 25 numbers in order of size, smallest first.

b How many students were present for 5 days of the week?

c Calculate the mean, mode, median and range of the 25 numbers.

3 Twenty people decide to buy some raffle tickets.
Some of the people buy more than one ticket.
The table gives the information.

Number of tickets	Tally	Number of people
1	~~IIII~~ I	6
2	~~IIII~~	5
3	IIII	4
4	~~IIII~~	5

Calculate the

a mean **b** mode **c** median **d** range.

4 The tetrahedron dice is rolled 50 times.
The scores are shown in the frequency table.
Calculate the

a mean **b** mode

c median **d** range.

Score	Tally	Frequency
1	~~IIII~~ ~~IIII~~	10
2	~~IIII~~ ~~IIII~~ I	11
3	~~IIII~~ ~~IIII~~ III	13
4	~~IIII~~ ~~IIII~~ ~~IIII~~ I	16

5 The test results are shown for a class of 20 students.

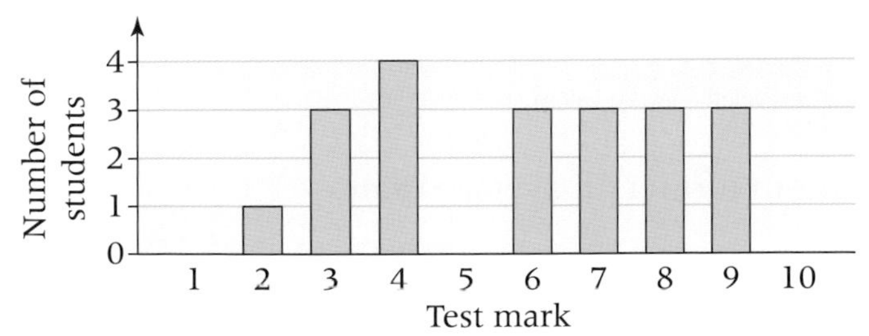

a Copy and complete the frequency table to illustrate these results.

Mark	1	2	3	4	5	6	7	8	9	10
Number of students	0	1								

b Calculate the mean, mode, median and range for the 20 marks.

D4.4 Comparing data

This spread will show you how to:

- Compare distributions and make inferences

Keywords
Compare
Mean
Median
Mode
Range
Spread

- You can **compare** sets of data using the mean, mode, median and range.

Example

A team of seven women and a team of eight men do a sponsored run for charity. The distances the men and women ran are shown.

Women

Distance (km)	Frequency
1	3
2	2
3	1
4	1
5	0

Men
3, 5, 5, 3, 4, 4, 5, 5
all distances in kilometres

a Construct a similar frequency table for the men's distances.
b By calculating the mean, median and range, compare each set of data.

a Men

Distance (km)	Tally	Frequency	Distance × Frequency
1		0	0
2		0	0
3	II	2	6
4	II	2	8
5	IIII	4	20
		8	34

b Women

Mean = (3 + 4 + 3 + 4 + 0) ÷ 7 = 14 ÷ 7 = 2
Median = 2 (the 4th distance)
Range = 4 – 1 = 3 (highest value – lowest value)

Men

Mean = 34 ÷ 8 = 4.25
Median = (5 + 5) ÷ 2 = 10 ÷ 2 = 5 (the mean of the 5th and 6th distances)
Range = 5 – 3 = 2 (highest value – lowest value)

The mean and median show the men ran further on average than the women.
The range shows that the women's distances were more **spread** out than the men's distances.

Exercise D4.4

1 The number of bottles of milk delivered to two houses is shown in the table.

	Sat	Sun	Mon	Tues	Wed	Thur	Fri
Number 45	2	0	1	1	1	1	1
Number 47	4	0	2	2	2	2	2

a Calculate the range for Number 45 and Number 47.

b Use your answers for the range to compare the number of bottles delivered to each house.

2 The number of cars at each house on Ullswater Drive are

2 4 1 0 1 2 1 2 3 2

a Copy and complete the frequency table.

Number of cars	Tally	Number of houses
0		
1		
2		
3		
4		

b Calculate the mean, mode and median number of cars for Ullswater Drive.

The mean, mode and median number of cars at each house on Ambleside Close are

Mean	Mode	Median
0.7	0	1

c Use the mean, mode and median to compare the number of cars on Ullswater Drive and Ambleside Close.

3 The number of days of rain each month in a particular year in Ireland and Spain is recorded.

	Jan	Feb	Mar	Apr	May	Jun	Jul	Aug	Sep	Oct	Nov	Dec
Ireland	27	22	27	24	23	24	25	24	26	26	26	28
Spain	11	10	10	11	10	7	2	5	6	11	11	12

a List each set of numbers in order, smallest first.

b Calculate the median days of rain for Ireland and for Spain.

c Using your answers for the median, compare the two sets of data.

d Calculate the range for each set of data.

e Using your answers for the range, compare the number of rainy days in Ireland and Spain.

D4.5 Grouped data

This spread will show you how to:

- Calculate the mean for grouped data
- Identify the modal class and the class interval in which the median lies for grouped data

Keywords

Class interval
Continuous data
Estimated mean
Grouped data
Mid-values
Modal class

The number of students in a class and their absences are shown.

Absences	Frequency
0 to 4	9
5 to 9	8
10 to 14	5
15 to 19	6
20 to 24	2

5 students had either 10, 11, 12, 13 or 14 absences.

You cannot tell the **exact** number of absences in this frequency table.

Therefore you cannot calculate the exact mean, mode or median.

- **For grouped data in a frequency table, you can calculate**
  - **the estimated mean**
  - **the modal class**
  - **the class interval in which the median lies.**

Example

The times taken for 10 people to run a race are shown.

Use the frequency table to find
the estimated mean,
the modal class
and the class interval in which the median lies.

Time (t minutes)	Mid-value	Frequency	Mid-value × Frequency
$40 < t \leqslant 50$	45	1	45
$50 < t \leqslant 60$	55	2	110
$60 < t \leqslant 70$	65	5	325
$70 < t \leqslant 80$	75	2	150
	Total	10	630

Use < Δ for **continuous data**. 40ä< *t* Δ ā0 means more than 40, but less than or equal to 50.

By using the **mid-values**, the 10 times are taken as 45, 55, 55, 65, 65, 65, 65, 65, 75, 75.

45 + 55 + 55 + 65 + 65 + 65 + 65 + 65 + 75 + 75 = 630

Estimated mean = 630 ÷ 10
= 63 minutes

Modal class = $60 < t \leqslant 70$ as this class has the highest frequency

The median is given by the times of the 5th and 6th runners, which are within the class interval $60 < t \leqslant 70$.

Exercise D4.5

1 The weights, to the nearest kilogram, of 25 men are shown.

69 82 75 66 72
73 79 70 74 68
84 63 69 88 81
73 86 71 74 67
80 86 68 71 75

a Copy and complete the frequency table.

Weight (kg)	Tally	Number of men
60 to 64		
65 to 69		
70 to 74		
75 to 79		
80 to 84		
85 to 89		

b State the modal class.

c Find the class interval in which the median lies.

2 The speeds of 10 cars in a 30 mph zone are shown in the frequency table.

Speed (mph)	Mid-value	Number of cars	Mid-value × Number of cars
21 to 25		1	
26 to 30		6	
31 to 35		2	
36 to 40		1	

a Calculate the number of cars that are breaking the speed limit.

b Copy the frequency table and calculate the mid-values for each class interval.

c Complete the last column of your table and find an estimate of the mean speed.

3 The heights, in centimetres, of some students are shown in the frequency table.

Height (cm)	Number of students
$140 < h \leqslant 150$	3
$150 < h \leqslant 160$	9
$160 < h \leqslant 170$	8
$170 < h \leqslant 180$	10

a Calculate the total number of students shown in the table.

b Find the class interval in which the median lies.

c State the modal class.

d Calculate an estimate for the mean height.

D4 Exam review

Key objectives

- Calculate mean, range and median of small data sets with discrete data
- Identify the modal class for grouped data
- Calculate an estimate of the mean for large data sets with grouped data
- Compare discrete distributions and make inferences, using the shapes of distributions and measures of average and range

1 The masses of 90 pupils were measured to the nearest kilogram. The table shows a grouped frequency distribution of the results.

Mass, m (to the nearest kg)	Number of pupils
$30 \leqslant m < 40$	3
$40 \leqslant m < 50$	24
$50 \leqslant m < 60$	30
$60 \leqslant m < 70$	22
$70 \leqslant m < 80$	11

Find an estimate for the mean mass of the pupils. (4)

(WJEC, 2003)

2 20 students scored goals for the school hockey team last month. The table gives information about the number of goals they scored:

Goals scored	Number of students
1	9
2	3
3	5
4	3

Copy the table.

a Write down the modal number of goals scored. (1)

b Work out the range of the number of goals scored. (1)

c Work out the mean number of goals scored. (3)

(Edexcel Ltd., 2004)

Interpreting diagrams and charts

This unit will show you how to

- Draw and produce, using paper and ICT, pie charts for categorical data, and diagrams for data, including line graphs for time series and frequency diagrams
- Interpret a wide range of graphs and diagrams and draw conclusions
- Compare distributions and make inferences
- Look at data to find patterns and exceptions
- Calculate the mean, median, mode and range of small sets of discrete data
- Have a basic understanding of correlation, including lines of best fit

Before you start ...

You should be able to answer these questions.

1 Find the value of each angle x.

a

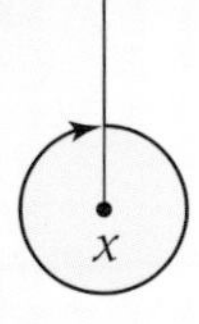

b

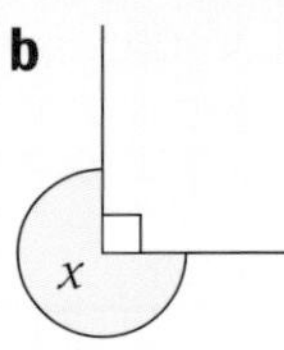

c

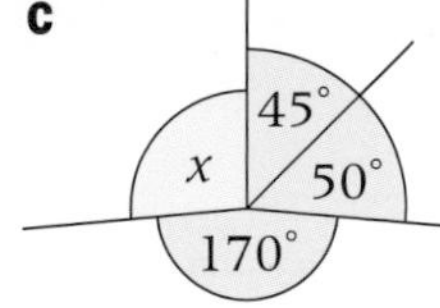

2 Calculate

a $360 \div 12$ **b** $360 \div 18$

c $360 \div 24$ **d** $360 \div 36$

e $360 \div 120$

3 Find the mean, mode, median and range of

a 4, 8, 10, 24, 24.

b 2, 2, 8, 12.

c 5, 3, 8, 16, 3.

4 Give the coordinates of

a point A

b point B

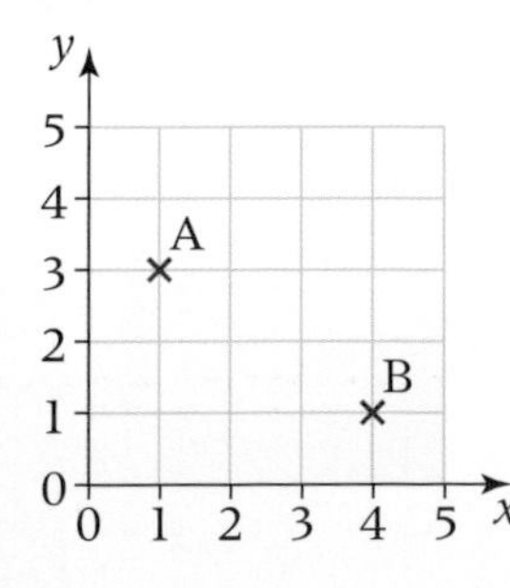

D5.1 Interpreting statistical diagrams

This spread will show you how to:

- Interpret a wide range of graphs and diagrams and draw conclusions

Keywords
Bar chart
Bar-line chart
Pictogram
Pie chart
Sector

You can interpret data from a variety of diagrams.

Pictograms use symbols to represent the size of each category.

Food	○ ○ ◖
Heat	○
Rent	○ ○ ○ ○ ◖

Key: ○ represents £10

Bar charts use horizontal or vertical bars to represent the frequencies.

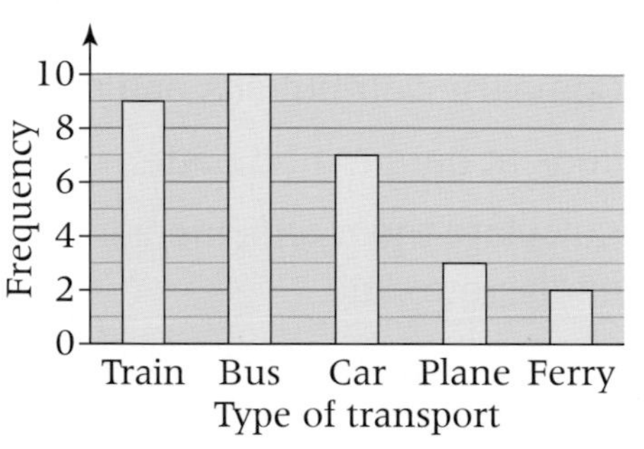

Notice the gaps between the bars.

Bar-line charts use vertical lines to represent numerical data.

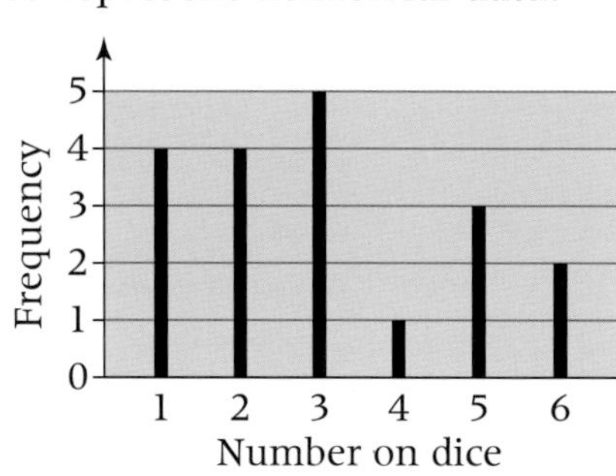

Pie charts use sectors of a circle to represent the size of each category.

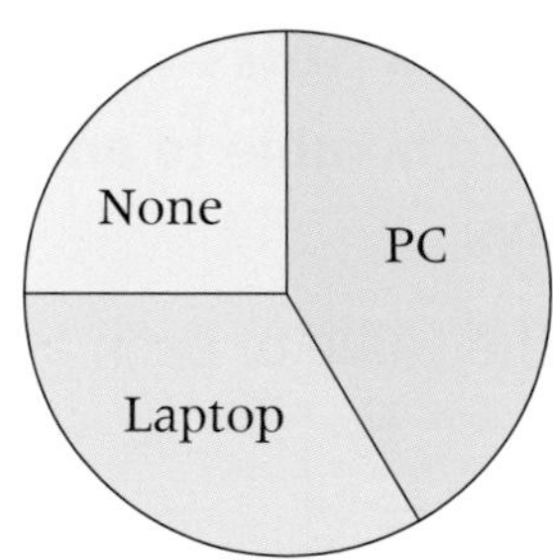

The size of the **sector** is proportional to the frequency.

Example

The pie chart shows 24 hours in Syd's life.
Syd spends 8 hours sleeping.
Calculate

a the value of x
b the fraction of time spent sleeping
c the percentage of time spent watching TV
d the number of hours for each category.

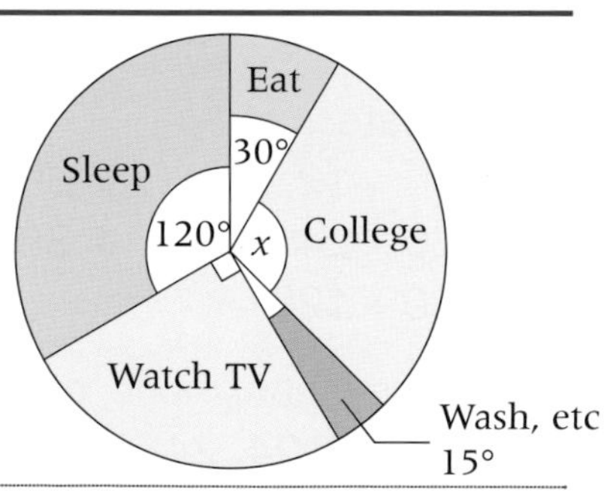

a $360° - (120° + 30° + 15° + 90°) = 105°$

b $\frac{120}{360} = \frac{1}{3}$

c $\frac{90}{360} = \frac{1}{4} = 25\%$

d 8 hours is represented by 120°
1 hour is represented by 15°

Eat	$30 \div 15 = 2$ hours
College	$105 \div 15 = 7$ hours
Wash etc	$15 \div 15 = 1$ hour
Watch TV	$90 \div 15 = 6$ hours

The angles at a point add to 360°.

Check:
$2 + 7 + 1 + 6 + 8$
$= 24$ hours

Exercise D5.1

1 A survey of the number of lorries passing through four villages each day is shown in the pictogram.

Abbey	
Batty	
Cotton	
Ditty	

Key: represents 8 lorries

a Calculate the number of lorries passing daily through

i Abbey **ii** Batty **iii** Cotton **iv** Ditty

v all four villages taken together.

b Which village should be considered for a bypass?

2 Travel insurance prices are shown on the bar chart.

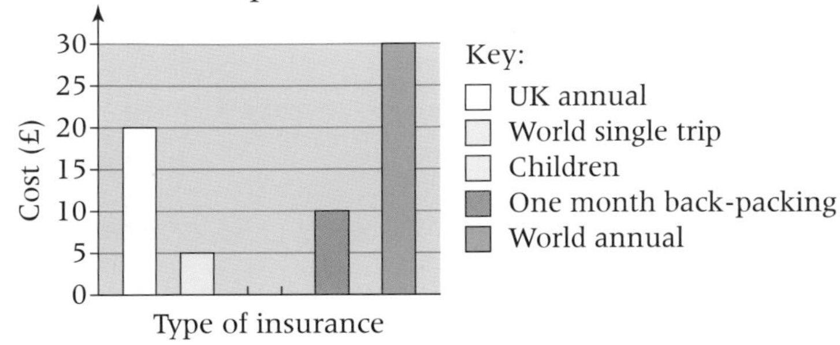

a State the cost of

i annual insurance in the UK

ii one month back-packing insurance.

b Who is offered free travel insurance?

c Suzie plans to make four trips abroad during the year.
She can either buy World Single Trip insurance each time or World Annual insurance.
Which is her cheaper option? Show your working.

3 The bar-line chart shows the number of tickets bought by 10 people.

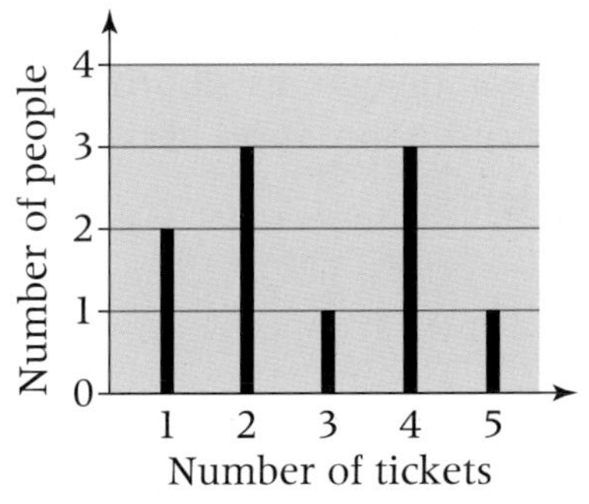

a State the number of people who bought

i 5 tickets **ii** 4 tickets.

b Calculate the total number of tickets that were bought by the 10 people.

4 The pie chart shows the pathways for 120 students.
Calculate

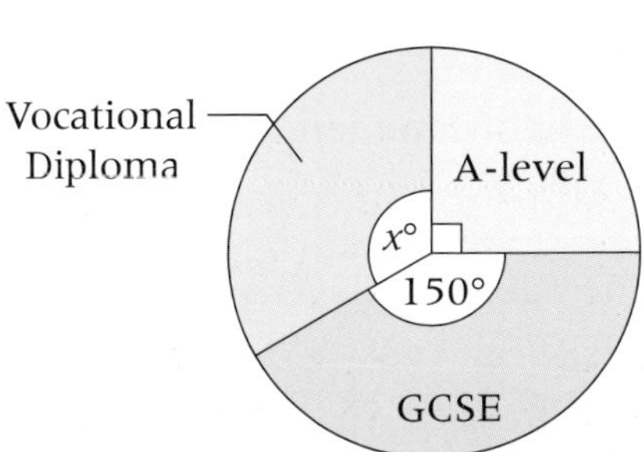

a the value of x

b the angle that represents one student

c the number of students in

i A-level **ii** GCSE **iii** Diplomas

D5.2 Interpreting more statistical diagrams

This spread will show you how to:

- Interpret a wide range of graphs and diagrams and draw conclusions
- Compare distributions and make inferences
- Look at data to find patterns and exceptions

Keywords

Comparative bar chart
Frequency polygon
Grouped
Histogram
Modal

You can interpret **grouped** continuous data from a **histogram** and a **frequency polygon**.

The lengths of eight pieces of string are shown in the histogram.

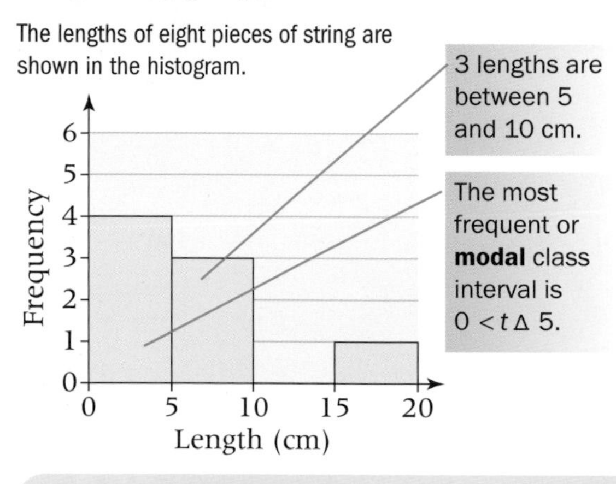

The same eight lengths are shown in a frequency polygon.

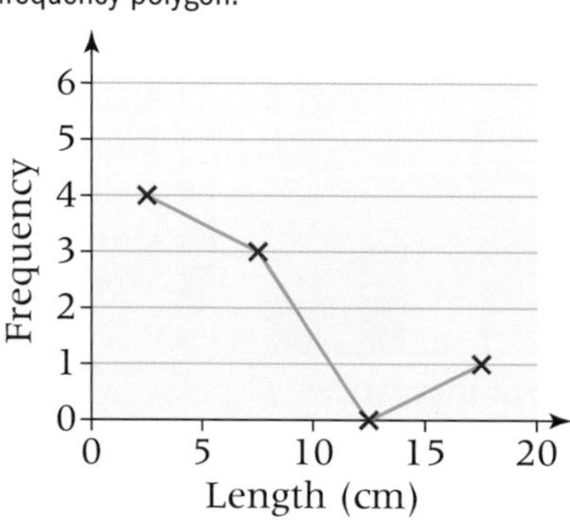

The points are plotted at the midpoints of the class intervals.

- You can interpret two sets of data from a **comparative bar chart**.

The chart shows the attendances at college on Wednesday and Friday.

- Overall attendance is the same on both days.
- Girls' attendance is better than boys'.

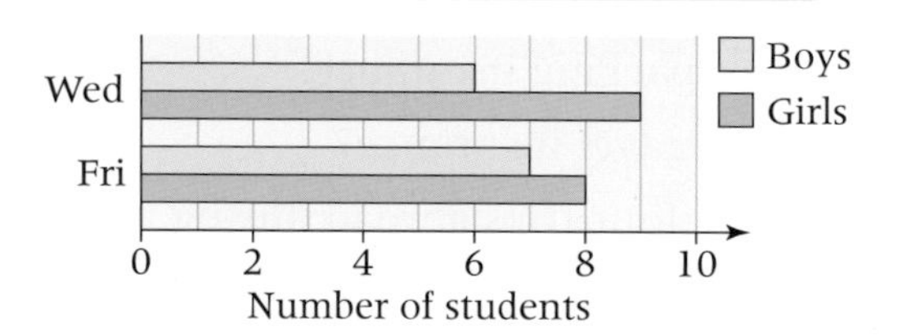

Example

The frequency polygons show the age distribution for the population of two villages.

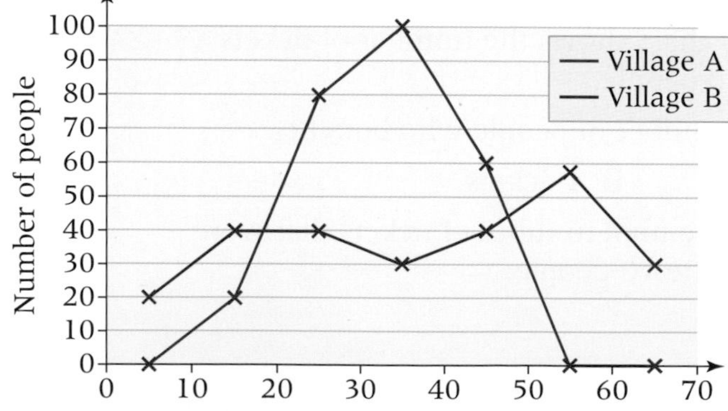

Make two statements to compare the distribution of ages in the two villages.

- The ages in village B are more spread out.
- The modal age is younger in village A than in village B.
  Village A: 35 years Village B: 55 years

Exercise D5.2

1 The bar chart shows the number of minutes per day that men and women spent on household chores in 2000/01. Which household chore do

a men and **b** women spend

i the most time doing

ii the least time doing?

Division of household chores, 2000/01 UK

Source: www.statistics.gov.uk

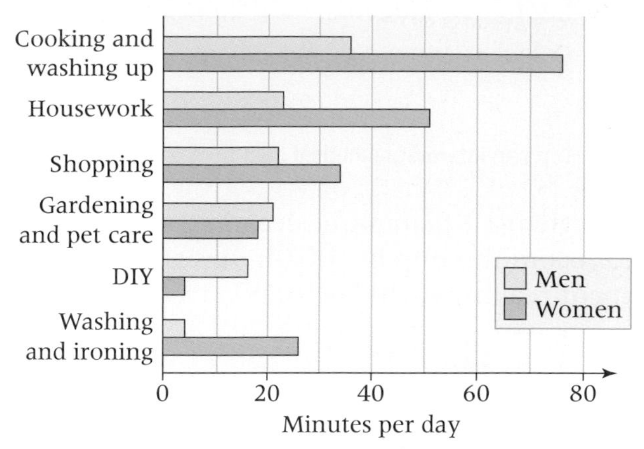

2 The populations of four countries in 1994 and 2004 are shown in the bar chart.

a State the population of

i Bangladesh in 1994

ii India in 2004.

b Which country shows the largest population increase from 1994 to 2004?

c Which country's population is approximately the same in 1994 and 2004?

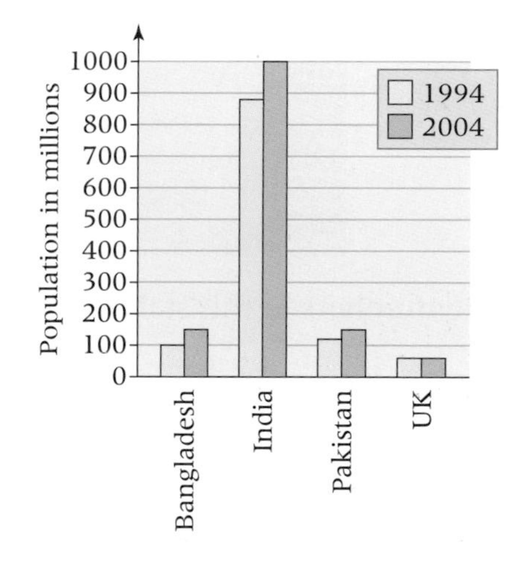

3 The histogram shows the best distances, in metres, that athletes threw a javelin in a competition.

a State the number of athletes who threw the javelin

i between 65 and 70 metres

ii between 80 and 85 metres.

b In which class interval was the winner?

c What is the modal class interval?

d Calculate the total number of athletes who threw a javelin.

4 The lengths of the long jumps for men and women are shown in the frequency polygon. Make two statements to compare the distributions of the length of long jumps for the men and women.

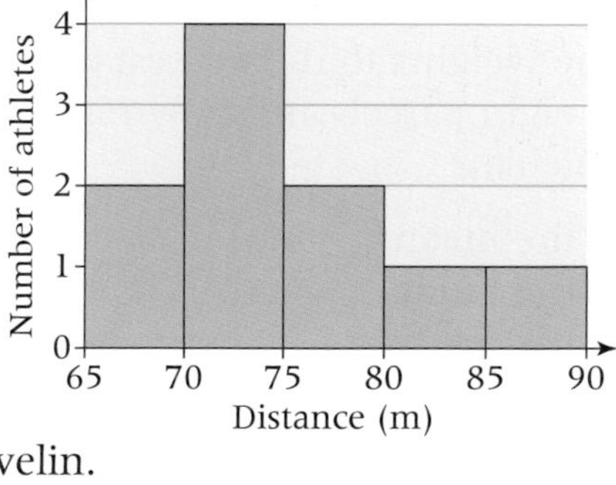

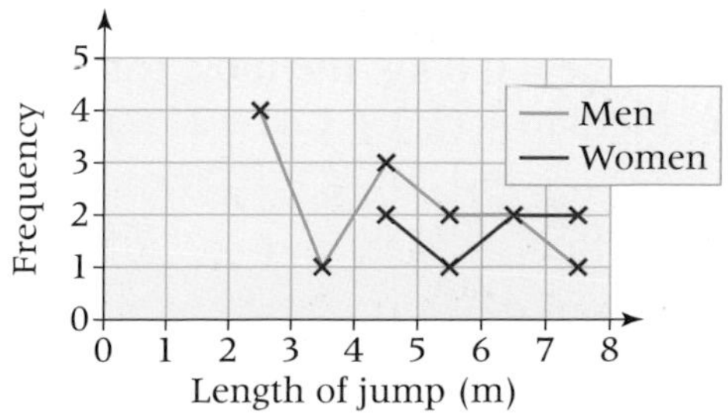

D5.3 Interpreting stem-and-leaf diagrams

This spread will show you how to:

- Interpret a wide range of graphs and diagrams and draw conclusions
- Calculate the mean, median, mode and range of small sets of discrete data.

Keywords
Ordered
Stem-and-leaf diagram

- **You can interpret numerical data from a stem-and-leaf diagram.**

Maggie used a stem-and-leaf diagram to record the number of minutes she spent reading each day for two weeks.

This stem-and-leaf diagram is **ordered** as the data is in numerical order.

stem	leaf
9	7
8	3 4 9
7	0 4 8
6	3 3 6
5	6 9
4	3 8

Key: 8 | 7 means 87

This means 97.

This means 70.

Always give the key.

The numbers in this stem-and-leaf diagram are

43, 48, 56, 59, 63, 63, 66, 70, 74, 78, 83, 84, 89, 97.

- **You can calculate the mean, mode, median and range from a stem-and-leaf diagram.**

Example

The weights to the nearest tenth of a kilogram, of eight parcels are shown in the diagram. Calculate

a the mean
b the mode
c the median
d the range.

3	0
2	5
1	3 4 9
0	7 8 8

Key: 1 | 3 means 1.3 kg

a Mean = (0.7 + 0.8 + 0.8 + 1.3 + 1.4 + 1.9 + 2.5 + 3.0) ÷ 8
= 1.55 kg

b Mode = 0.8 kg, the most common weight

c Median = (1.3 + 1.4) ÷ 2 (2 middle numbers)
= 1.35 kg

d Range = 3.0 − 0.7 (highest value − lowest value)
= 2.3

Exercise D5.3

1 The attendances of nine college classes one Friday afternoon are shown in the stem-and-leaf diagram.

1	0 6
2	2 2 4 6 8 9
3	0

Key:

2	4

means 24 students

a Write out the nine attendances in numerical order, smallest first.

b Calculate

i the mean **ii** the mode **iii** the median **iv** the range.

2 The test marks of 20 students are shown.

31 17 43 19 25 12 7 40 25 21
11 32 37 25 15 9 18 41 23 17

0	
1	
2	
3	
4	

Key:

2	5

means 25 marks

a Copy and complete the stem-and-leaf diagram.

b Redraw the diagram to give an ordered stem-and-leaf diagram.

c Calculate

i the mean **ii** the mode **iii** the median **iv** the range.

3 The times, to the nearest second, for 15 athletes to run 800 m are shown in the stem-and-leaf diagram. Calculate

11	4 6 9
12	0 0 4 5 6 7
13	1 5 6
14	0 2 5

Key:

12	5

means 125 seconds

a the mean

b the mode

c the median

d the range.

4 The speedway scores of a team during one season are shown in the stem-and-leaf diagram.

3	3 4
4	1 1 4 4 5 6 7 9 9 9
5	0 2 7
6	0

Key:

4	5

means 45 points

a Calculate the modal score.

b Calculate the number of scores shown in the diagram.

c If one score is chosen at random, calculate the probability that it is 50 or over.

5 Ten competitors achieve the following distances, measured in centimetres, in the High Jump and the Long Jump.

High Jump

18	1 3 2 5
19	1 6 7 7
20	3 7

Key:

19	6

means 196 cm

Long Jump

73	7 8
74	0 3 6 7 9
75	0 2

Key:

74	3

means 743 cm

a One competitor was injured and could not take part in one of the events. Which event did he miss?

b By calculating the range for each event, make a comparison between the two different events.

D5.4 Interpreting time series graphs

This spread will show you how to:

- Interpret a wide range of graphs and diagrams and draw conclusions

Keywords
Horizontal
Line graph
Time series graph
Trend

- **You can interpret data as it changes with time from a time series graph.**

The percentage of adults who smoke cigarettes is shown by the **line graph**.

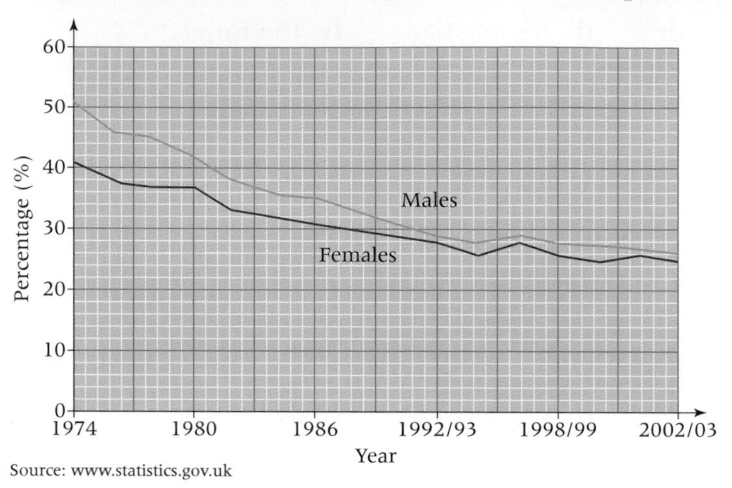

Time is always the **horizontal** axis.

You can see the **trend** from the graphs.

- The percentage of males who smoke is always greater than the percentage of females who smoke.
- The percentages of males and females who smoke are decreasing.

Example

The monthly average temperatures in Madrid and Grand Canaria are shown in the line graphs.

a Which place is usually hotter in April?
b What is the average temperature in Madrid in October?
c In which three months is it colder in Grand Canaria than in Madrid?
d Calculate the range of the temperatures for Grand Canaria.
e Calculate the range of the temperatures for Madrid.

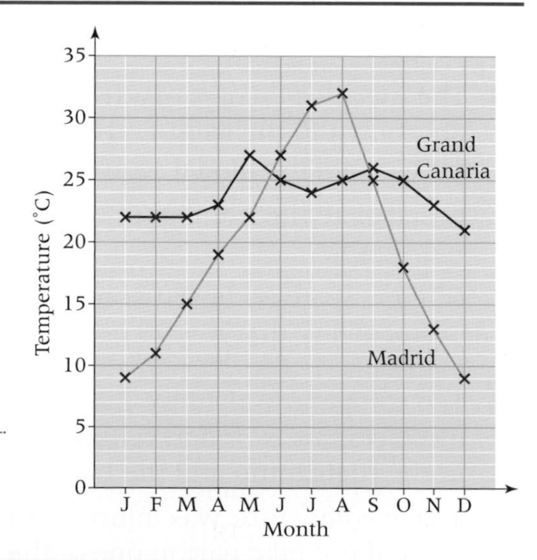

a Grand Canaria
b 18 °C
c June, July, August
d $27 - 21 = 6$ °C
e $32 - 9 = 23$ °C.

The answers to parts **d** and **e** show that the temperature is more variable in Madrid than in Grand Canaria.

Exercise D5.4

1 The line graph shows the temperature, in °C, over 24 hours.

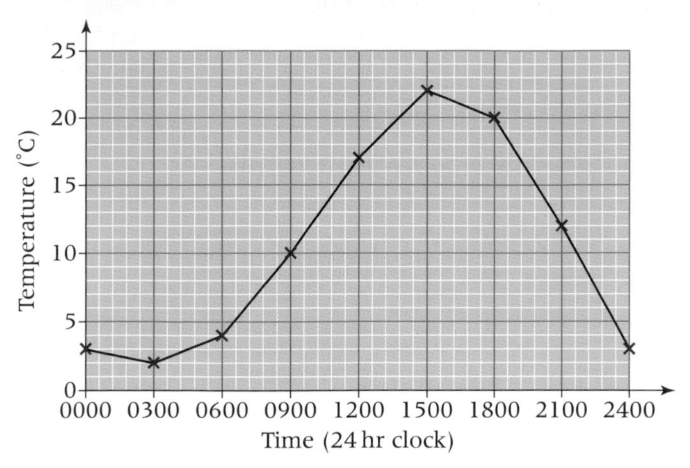

a What time in the morning was the temperature 10 °C recorded?

b What was the maximum temperature recorded? When was the maximum temperature recorded?

c What was the minimum temperature recorded? When was the minimum temperature recorded?

d Calculate the range of the recorded temperatures.

2 The number of passengers, in billions, for local buses and for trains is shown on the line graphs.
Give two observations about the trend shown on the graphs.

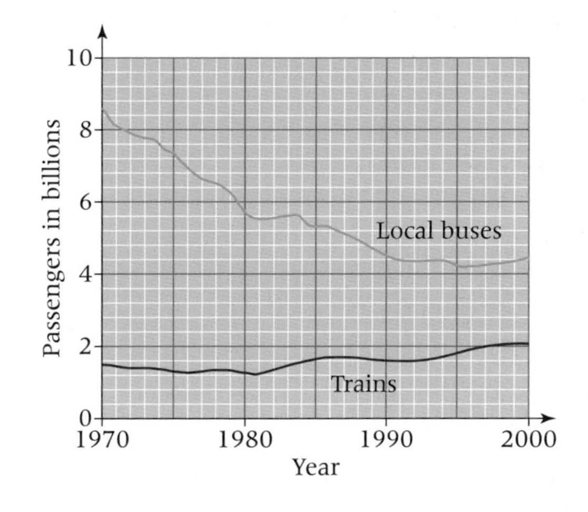

3 The graphs show the expected demand for energy and the supply of fossil fuels for the world.

a Describe the trend for the demand for energy.

b Describe the trend for the supply of fossil fuels.

c What will be needed by 2050 to solve this problem?

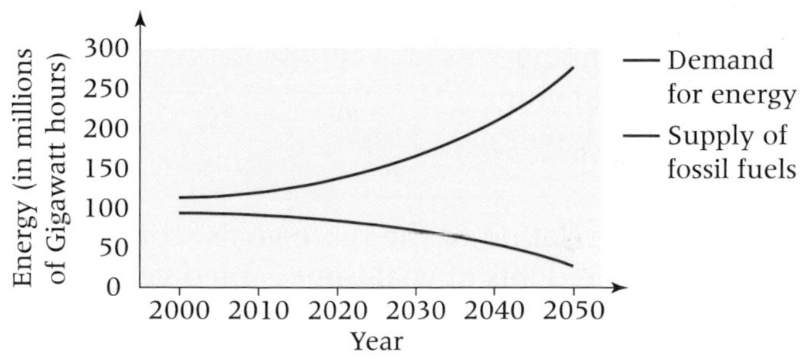

D5.5 Interpreting scatter graphs

This spread will show you how to:

- Have a basic understanding of correlation, including lines of best fit

Keywords
Correlation
Line of best fit
Relationship
Scatter graph
Variable

- **You can interpret two sets of data that have been drawn on a scatter graph.**

If the points are roughly in a straight line, there is a linear **relationship** or **correlation** between the two **variables**.

Positive correlation

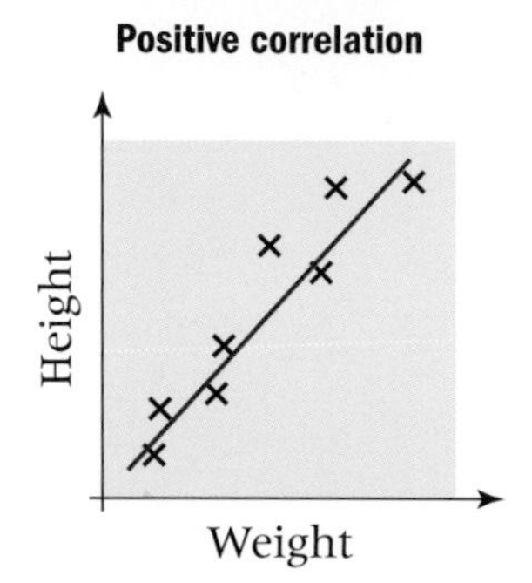

As height increases, weight also increases.

Negative correlation

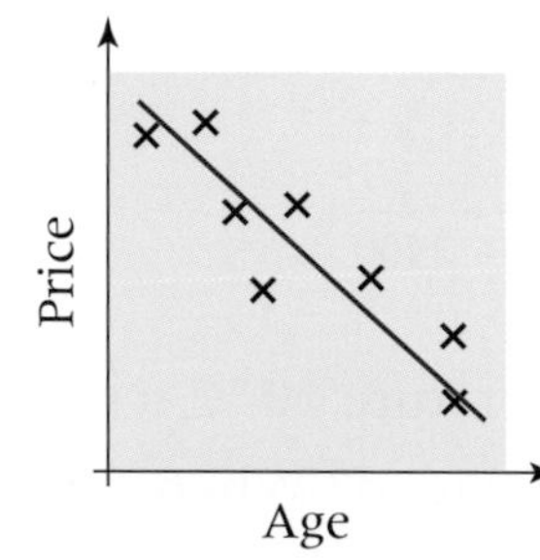

As the age of a car increases, the price decreases.

No correlation

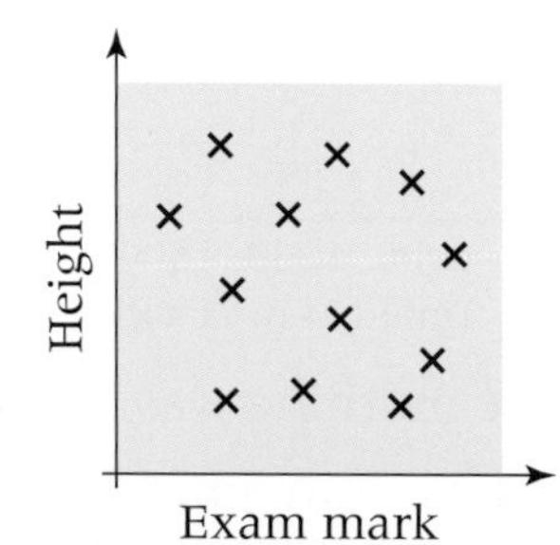

There is no linear relationship between height and exam mark.

The red straight line is the **line of best fit**.

You cannot draw a line of best fit for no correlation.

Example

The scatter graph shows the number of goals scored by 21 football teams in a season plotted against the number of points gained.

a Describe the relationship between the goals scored and the number of points.
b Describe the goals and points for team A.
c If a team scored 45 goals, how many points would you expect it to have?

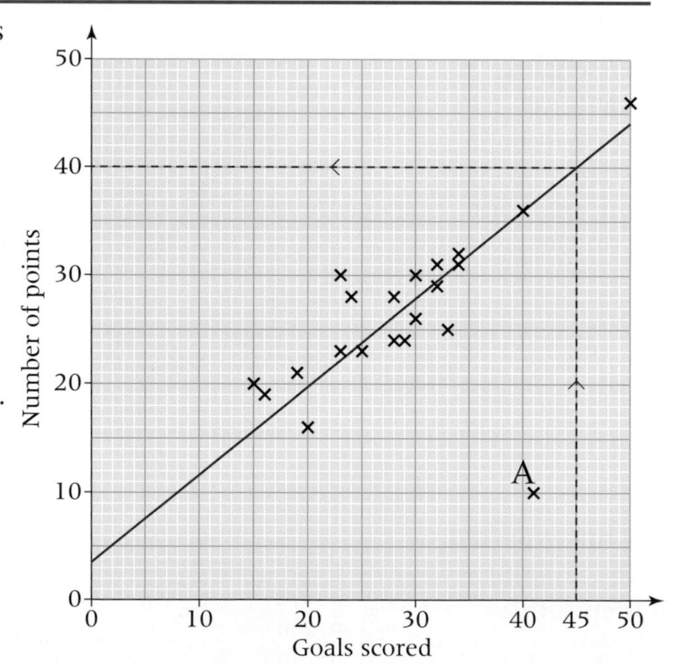

A line of best fit does not have to pass through (0, 0).

a Positive correlation or the more goals scored the more points gained.
b Team A scored lots of goals, but gained very few points.
c See the graph: 45 goals gives 40 points.

Exercise D5.5

1 Describe the type of correlation for each scatter graph.

a

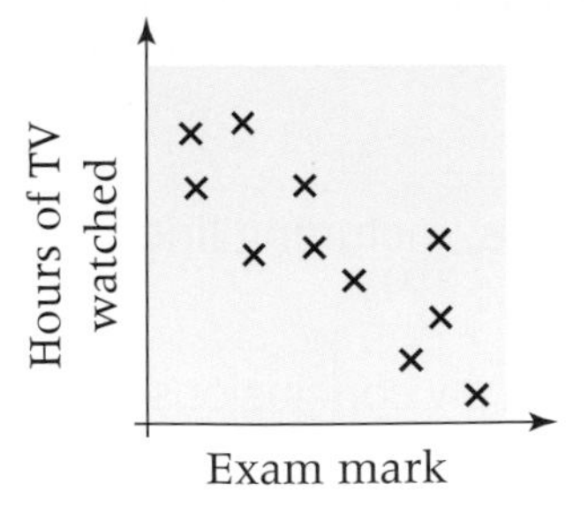

b

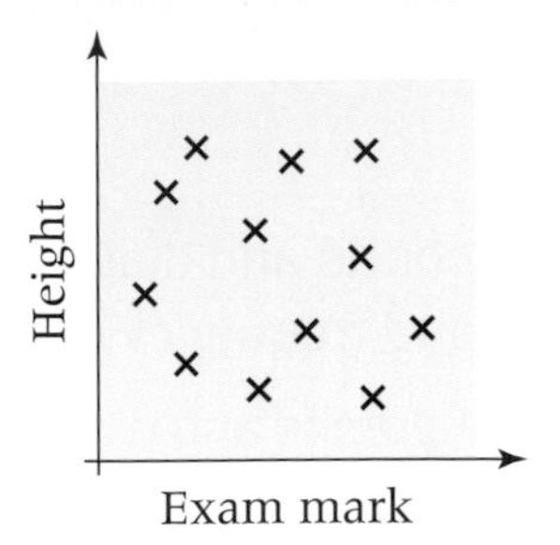

c

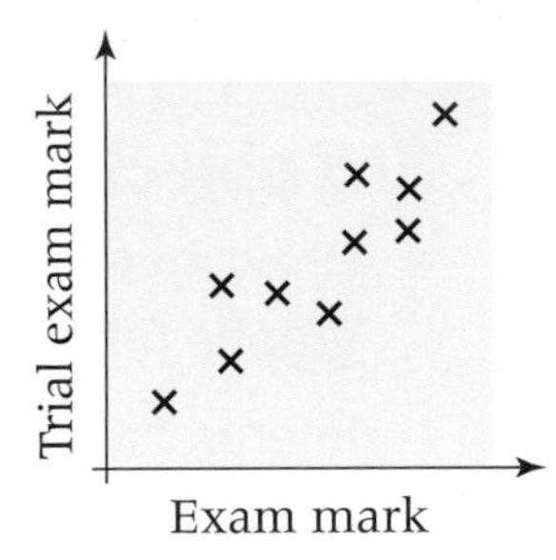

2 Describe the points A, B, C, D and E on each scatter graph.

a

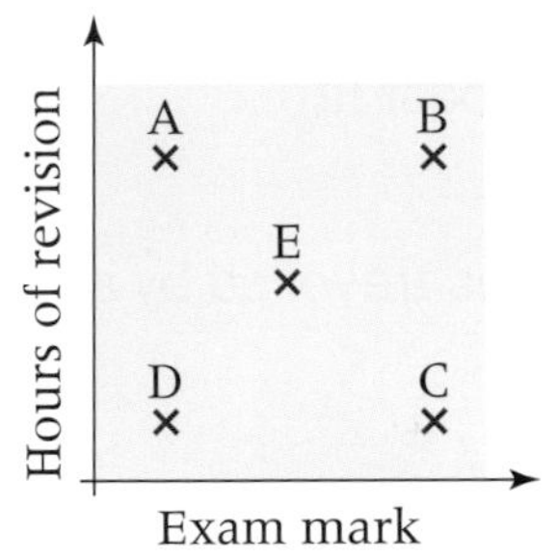

b

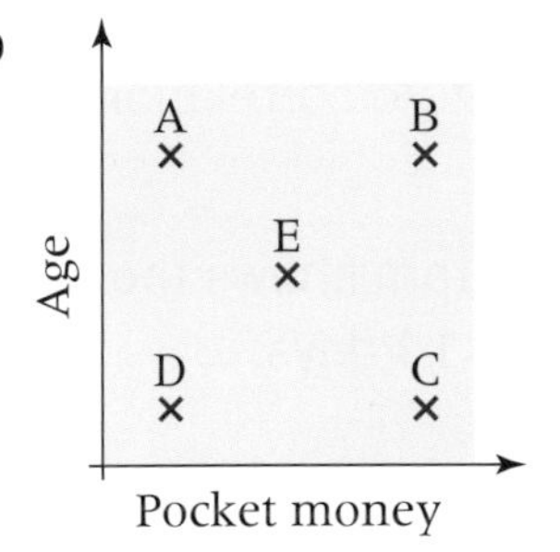

c

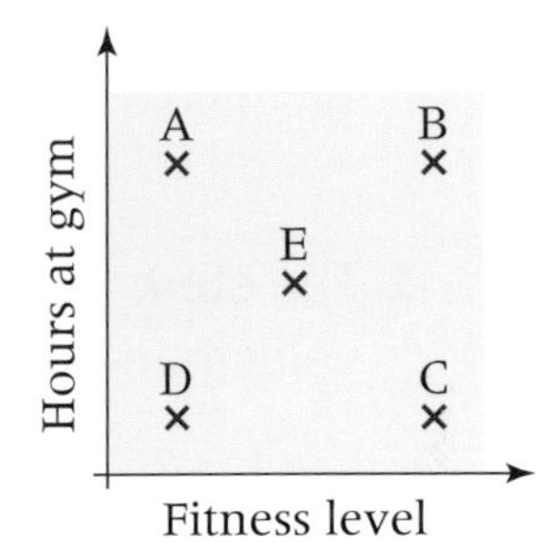

3 The graph shows the marks in two papers achieved by nine students.

Use the line of best fit to estimate

a the Paper 2 mark for a student who scored 12 in Paper 1

b the Paper 1 mark for a student who scored 23 in Paper 2.

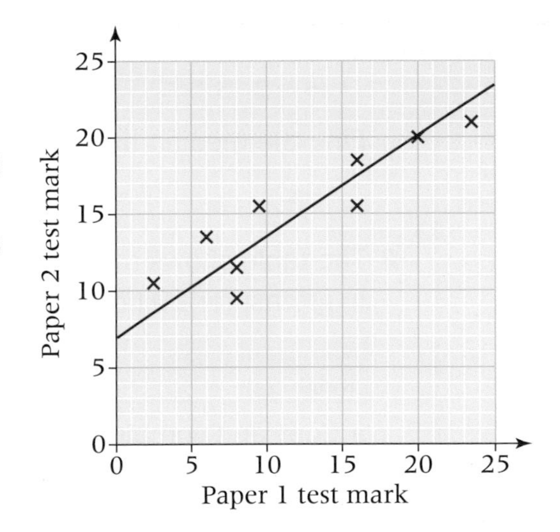

4 The table shows the age and diameter, in centimetres, of trees in a forest.

Age (years)	10	27	6	22	15	25	11	16	21	19
Diameter (cm)	20	78	9	65	38	74	25	44	59	50

a Draw a scatter diagram to show the information.

b State the type of correlation between the age and diameter of the trees.

c Draw a line of best fit.

d If the diameter of a tree is 55 cm, estimate the age of the tree.

Use 2 cm to represent 10 centimetres on the horizontal axis, numbered 0 to 80. Use 2 cm to represent 5 years on the vertical axis, numbered 0 to 30.

D5 Exam review

Key objectives

- Draw and produce pie charts and diagrams for data, including line graphs for time series and frequency diagrams
- Interpret a wide range of graphs and diagrams and draw conclusions
- Calculate the mean, range and median of small discrete data sets
- Appreciate that correlation is a measure of the strength of the association between two variables and distinguish between positive, negative and zero correlation using lines of best fit

1 The stem and leaf diagram shows the number of miles travelled by a salesman each day for 14 days.

1	2 3
2	3 6
3	5 7 7 8
4	1 3 4 8
5	2 5

Key 5 | 2 represents 52 miles

Find the median number of miles travelled per day. (2)

(AQA, 2004)

2 The table shows the number of pages and the weight, in grams, for 10 books:

Number of pages	80	130	100	140	115	90	160	140	105	150
Weight (g)	160	270	180	290	230	180	320	270	210	300

a Copy and complete the scatter graph to show the information in the table.
The first six points have been done for you. (1)

b For these books, describe the relationship between the number of pages and the weight of a book. (1)

c Draw a line of best fit on the scatter diagram. (1)

d Use your line of best fit to estimate
 i the number of pages in a book of weight 280 g
 ii the weight of a book with 120 pages. (2)

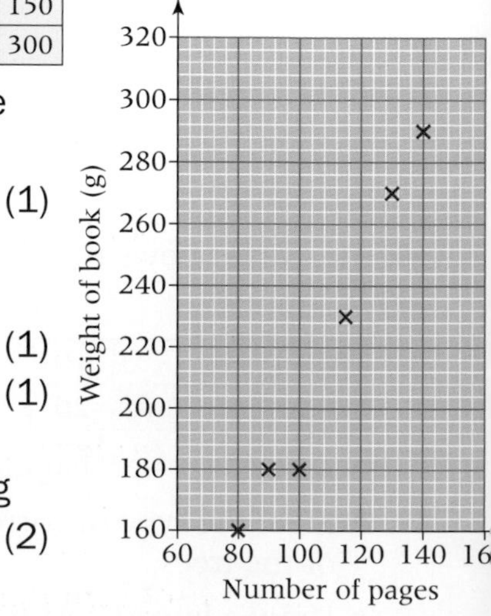

(Edexcel Ltd., 2004)

GCSE formulae – Foundation

In your Edexcel GCSE examination you will be given a formula sheet like the one on this page.

You should use it as an aid to memory. It will be useful to become familiar with the information on this sheet.

Area of a trapezium = $\frac{1}{2}(a+b)h$

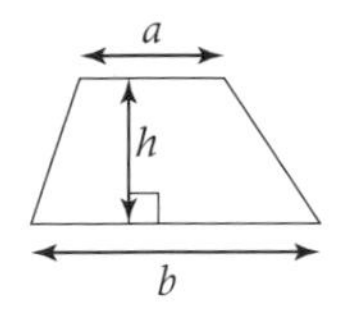

Volume of prism = area of cross section × length

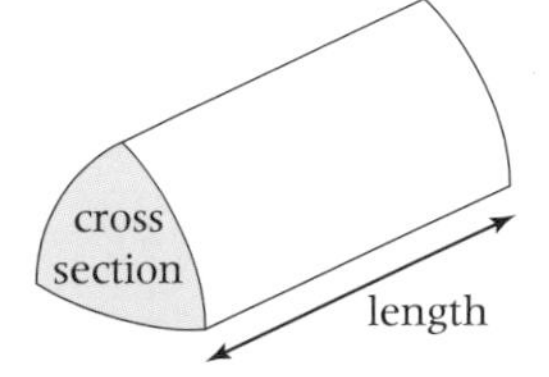

Here are some other formulae that you should learn.

Area of a rectangle = length × width

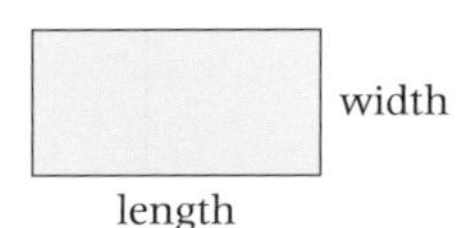

Area of a triangle = $\frac{1}{2}$ × base × height

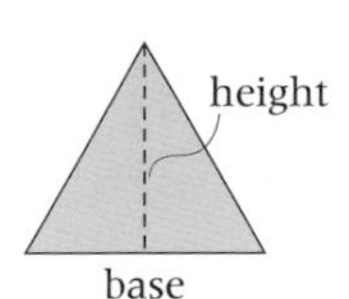

Area of a parallelogram = base × perpendicular height

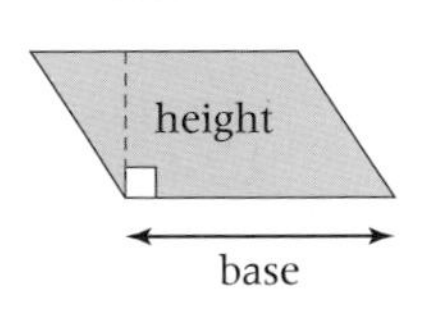

Area of a circle = πr^2

Circumference of a circle = $\pi d = 2\pi r$

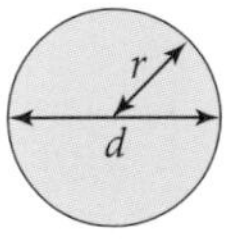

Volume of a cuboid = length × width × height

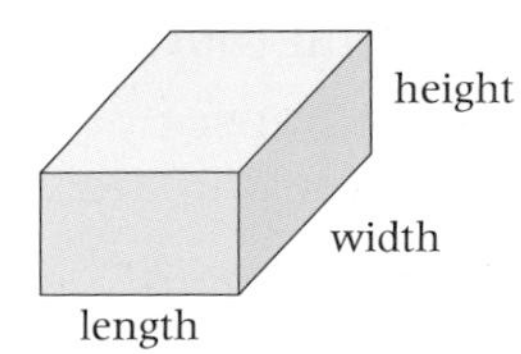

Volume of a cylinder = area of circle × length

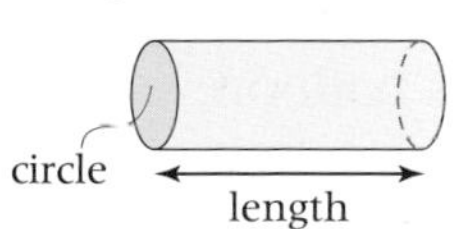

Pythagoras' theorem states,

For any right-angled triangle, $c^2 = a^2 + b^2$ where c is the hypotenuse.

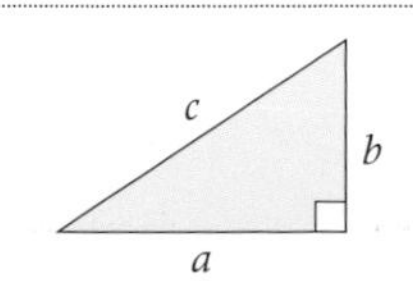

Foundation Practice Exam Paper

This practice exam paper contains 100 marks and should take 100 minutes to complete.

1 **a** Write $\frac{1}{5}$ as a decimal. (1 mark)

b Write 0.4 as a fraction. (1 mark)

c Write 30% as a decimal. (1 mark)

d Write $\frac{1}{5}$, 0.4, 30% in ascending order. (1 mark)

2 **a** Give the coordinates of the point A.

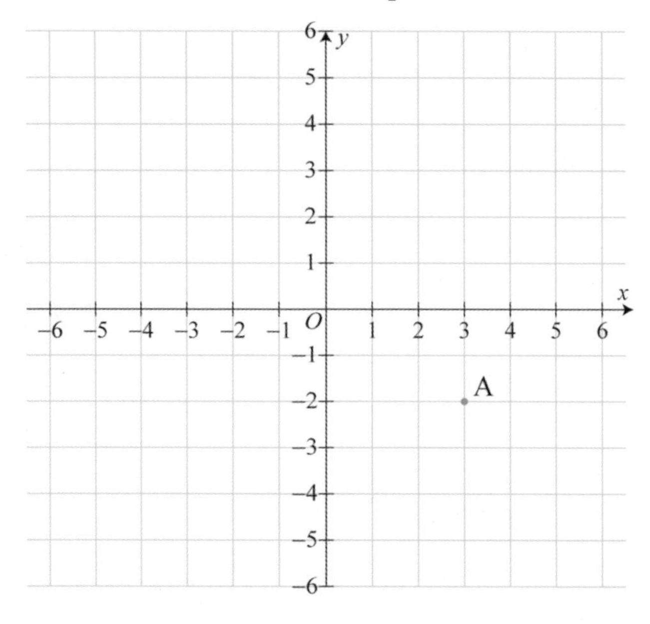

(1 mark)

b On a copy of the grid, plot the point (−3, −5). (1 mark)

3 Give the metric unit that would be used to measure these:

a The height of a house (1 mark)

b The weight of a tomato (1 mark)

c The air in a balloon. (1 mark)

4 Andrew did a survey of the days of the week on which the pupils in his class were born. Here are Andrew's results.

Tuesday	Monday	Sunday	Monday	Friday
Friday	Monday	Tuesday	Saturday	Sunday
Thursday	Friday	Monday	Friday	Friday
Wednesday	Saturday	Thursday	Tuesday	Friday
Thursday	Wednesday	Monday	Friday	Tuesday

a Copy and complete the table to show Andrew's results.

	Tally	Frequency
Monday		
Tuesday		
Wednesday		
Thursday		
Friday		
Saturday		
Sunday		

(3 marks)

b Write the number of children in Andrew's class who were born on a Wednesday. (1 mark)

c Which was the most common day of birth in Andrew's class? (1 mark)

5 Here is a cuboid with length 5 cm, width 2 cm and height 2 cm.

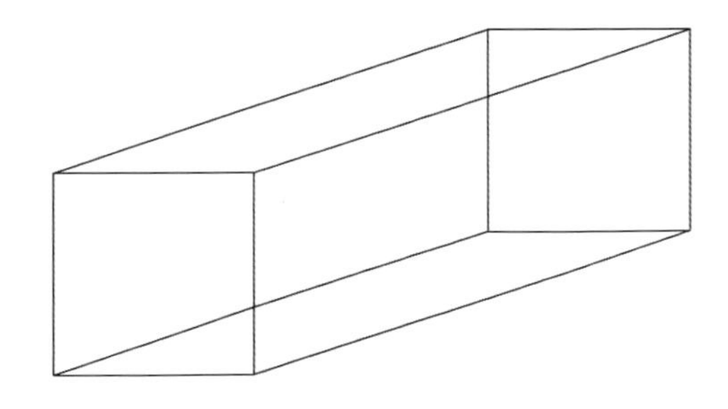

Diagram NOT accurately drawn

a Write down:

i the number of edges of the cuboid. (1 mark)

ii the number of vertices of the cuboid. (1 mark)

iii the number of square faces the cuboid has. (1 mark)

b Draw an accurate net of the cuboid. (3 marks)

6 **a** Estimate how many pieces 28 cm long can be cut from a length of material that is $1\frac{1}{2}$ metres long. (2 marks)

b What change will I get if I buy five stamps at 30 pence each and pay with a £5 note? (2 marks)

7 **a** Estimate $\sqrt{61}$. (1 mark)

b Calculate 20% of 60. (2 marks)

c Calculate $\frac{2}{5}$ of 45. (2 marks)

8 $a = 5$, $b = 2$, $c = \frac{1}{2}$.

Calculate the value of:

a $3a - 2b$ (2 marks)

b $a + b + 4c$ (2 marks)

9 This table gives the temperatures at different times during one day at a weather station.

Time	8 am	10 am	12 noon	2 pm	4 pm	6 pm	8 pm	10 pm
Temp (°C)	–3	–1	0	4	3	2	–1	–4

a By how much did the temperature rise between 10 am and 2 pm? (1 mark)

b What is the difference between the highest and lowest temperatures recorded in the table? (1 mark)

c The temperature falls 3°C between 10 pm and midnight. What is the temperature at midnight? (1 mark)

10 Solve $4x - 3 = 21$. (2 marks)

11 Copy the figure. Shade nine more squares so that the figure has rotational symmetry of order 4.

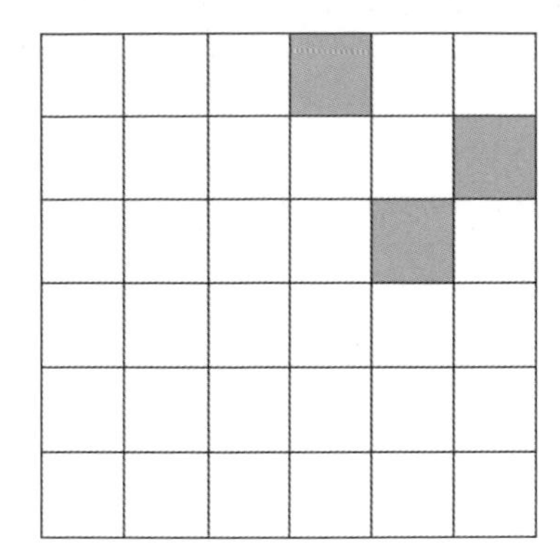

(3 marks)

12 **a** Work out:

i $\frac{3}{4}\times\frac{1}{3}$ (1 mark)

ii $\frac{2}{3}\times\frac{1}{4}$ (2 marks)

b Work out:

i 17.2 – 2.34 (1 mark)

ii 0.3×0.4 (1 mark)

c Work out $\frac{-12}{-2}$ (1 mark)

13 Adele says that the sum of three consecutive integers is always odd.
Roslin says she is wrong and that the sum is always even.
Give examples to show that both Adele and Roslin are wrong. (3 marks)

14 Simplify:

a $3x+x+5x$ (1 mark)

b $5x+2y+y-2x$ (2 marks)

15 **a** i Write down a sum which can be used to estimate the value of
$\frac{59\times 20.3}{29.2}$ (1 mark)

ii Work out the answer to your sum. (1 mark)

b Amber's dog eats $\frac{3}{4}$ of a tin of dog food each day.

Amber is going on holiday for nine days and her mother is going to look after the dog. What is the least number of tins Amber needs to leave with her mother? (3 marks)

16 The table gives information about the nationality of visitors to a National Trust property.

Nationality	**Frequency**
British	30
German	20
American	15
Other	25
Total	**90**

Draw an accurate pie chart to show this information. (4 marks)

17 The picture shows a man standing next to a building. The man and the building are drawn to the same scale.

a Write down an estimate for the height, in metres, of the man. (1 mark)

b Estimate the height, in metres, of this building. (3 marks)

18 Angelie keeps a list of the marks she scores in ten weekly tests.

7, 8, 8, 6, 7, 8, 9, 6, 8, 9.

Calculate her mean mark in the tests. (3 marks)

19 p is a prime number. q is an odd number.

a Is pq

A a prime number **B** not a prime number **C** could be either? (1 mark)

b Is pq

A an odd number **B** an even number **C** could be either? (1 mark)

c Is $\frac{p}{q}$

A an integer **B** not an integer **C** could be either? (1 mark)

20 a Copy the diagram. Enlarge triangle A by scale factor 2, centre (2,1). Label the image as B.

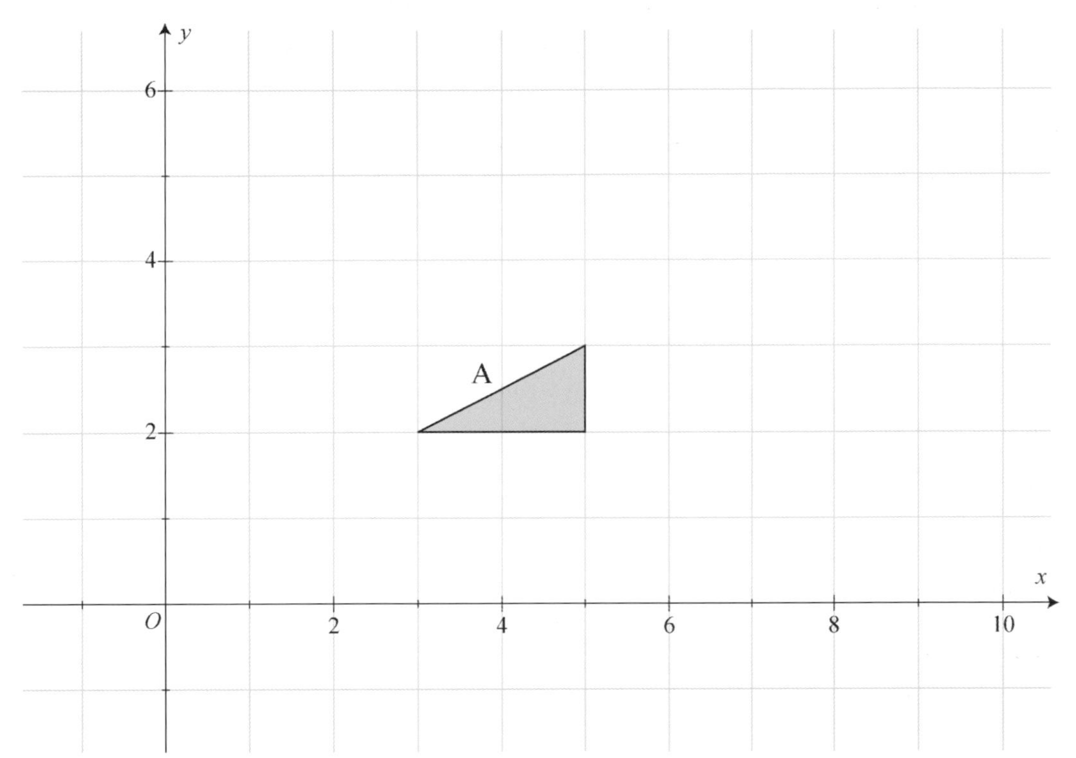

(3 marks)

b Translate triangle A by vector $\begin{pmatrix} 3 \\ -1 \end{pmatrix}$. Label the image C. (2 marks)

21 a Write the first **three** terms of the sequence whose nth term is n^2-1. (2 marks)

b Find an expression for the nth term of the sequence which starts

13 16 19 22 25 (2 marks)

22 Bag A has 3 red, 5 blue and 2 green balls in it. A ball is chosen at random.

a What is the probability the ball is:

i red? (1 mark)

ii not red? (1 mark)

Bag B has 5 red, 6 blue, 2 green and 2 black balls in it.

b i Katie says that you are more likely to get a blue from bag B than from bag A because there are more blue balls in bag B.
Explain why Katie is wrong. (2 marks)

(Question 22 continued on next page.)

ii Dorothy says you are more likely to get a red from bag A because it has only three colours in it, than from bag B which has four colours in it.

Explain why Dorothy is wrong. (2 marks)

23 **a** Simplify $a^3 \times a^2$ (1 mark)

b Simplify $\dfrac{m^3}{m^9}$ (1 mark)

c Simplify $\dfrac{x^6 + 3x^6}{x^2}$ (2 marks)

24 A shape has been shaded on a grid of 1 cm squares.

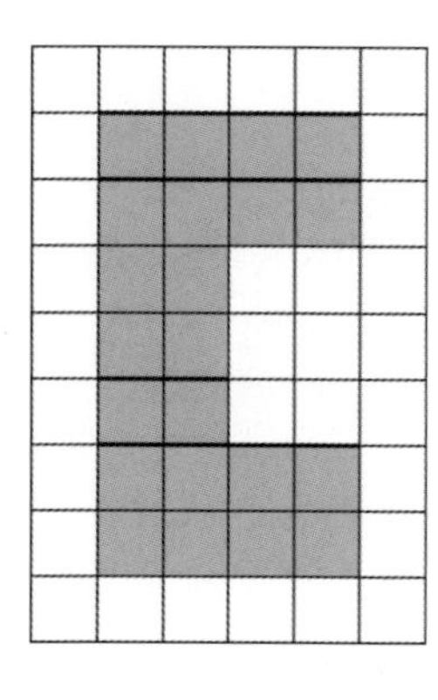

Diagram NOT accurately drawn

a Work out the perimeter of the shape. (2 marks)

b Work out the area of the shape. Give units with your answer. (2 marks)

25 **a** Draw the graph of $y = 2x + 3$. (3 marks)

b Rearrange $y = 2x + 3$ to make x the subject. (2 marks)

GCSE formulae – Higher

In your Edexcel GCSE examination you will be given a formula sheet like the one on this page.

You should use it as an aid to memory. It will be useful to become familiar with the information on this sheet.

Volume of a prism = area of cross section × length

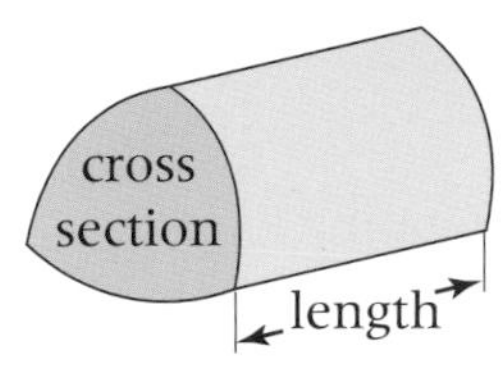

Volume of sphere = $\frac{4}{3}\pi r$

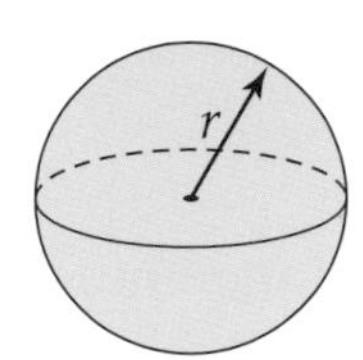

Surface area of sphere = $4\pi r^2$

Volume of cone = $\frac{1}{3}\pi r^2 h$

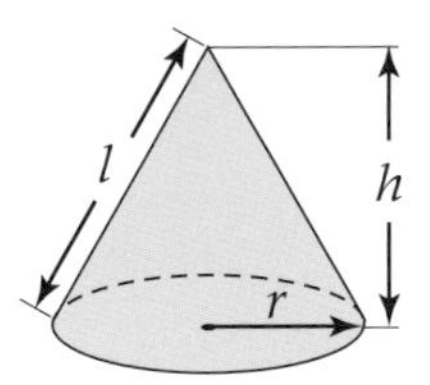

Curved surface area of cone = $\pi r l$

In any triangle ABC

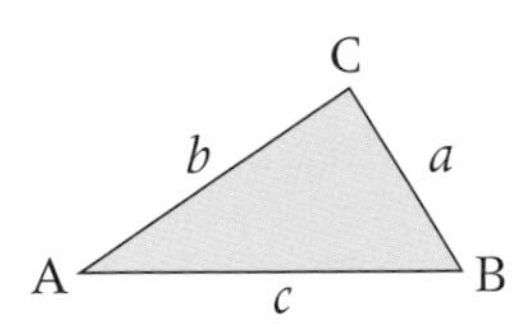

Sine rule $\frac{a}{\sin A} = \frac{b}{\sin B} = \frac{c}{\sin C}$

Cosine rule $a^2 = b^2 + c^2 - 2bc \cos A$

The Quadratic Equation
The solutions of $ax^2 + bx + c = 0$ where $a = 0$, are given by

$$x = \frac{-b \pm \sqrt{(b^2 - 4ac)}}{2a}$$

Higher Practice Exam Paper

This practice exam paper contains 100 marks and should take 100 minutes to complete.

This paper contains topics not covered in this book, as the grade range stretches to A*.

1 Michaela buys cuddly toys at £2.50 each and sells them at a 60% profit.
How much does she sell each cuddly toy for? (3 marks)

2 $y = 3x - 1$ crosses the line $y = 5$ at the point P.
Find the coordinates of the point P. (3 marks)

3 Show why the interior angle of a regular pentagon is 108°. (2 marks)

4 The stem-and-leaf diagram shows information about the scores a batsman made in 19 one-day cricket matches during a season.

Stem	Leaves
0	0 0 0 2 3 7 8 9
1	0 4 8 8 9
2	3
3	5 8
4	
5	2 6
6	
7	
8	7

Key : 2 | 3 means 23

a What was his median score? (1 mark)

b Construct a box plot to show this information. (3 marks)

c The box plot shows the scores for another batsman.

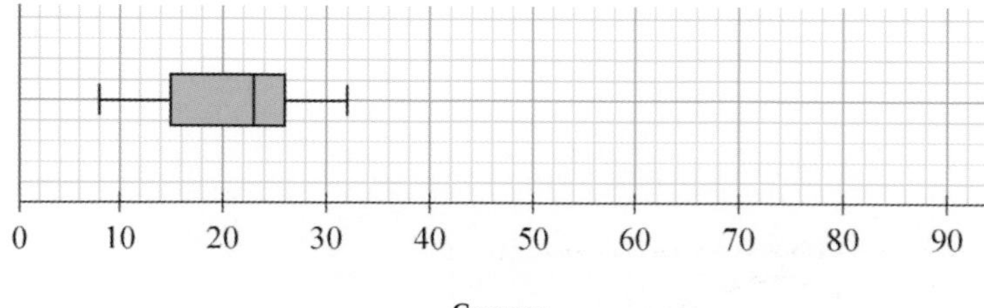

Score

Compare the performances of the two batsmen. (2 marks)

5 **a** Calculate $\dfrac{16.5+8.2}{\sqrt{4.1+3.25}}$. Write all the figures on your calculator. (2 marks)

b Divide £420 in the ratio 4:2:1. (2 marks)

c $4x^3 = -1372$. What is x? (2 marks)

6 The ingredients for a recipe for Irish stew to serve four are listed here.

0.8 kg lamb
2 large onions, sliced
225 g carrots, sliced
450 g potatoes, peeled and sliced
Salt and pepper
Some springs of parsley

a Catriona is going to make Irish stew for six.

i How much lamb does she need? (1 mark)

ii What weight of potatoes does she need? (1 mark)

b The recipe says to cook the stew at 350°F. Catriona's oven is in °C.

Use the formula $C = \frac{5}{9}(F - 32)$ to convert 350°F into °C. (3 marks)

7 The diagram uses a scale of 1 cm = 10 m.
On a copy of the diagram, show the region that is at least 40 m from the river, and not more than 30 m from the pylon.

Pylon

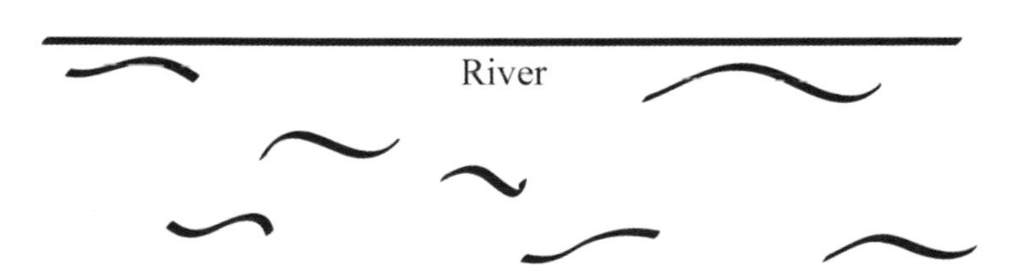

(3 marks)

8 Michael tries to work how much running a car will cost him over a year.
Some of the amounts he knows but others he has to estimate.
The table shows the figures he writes, with (est) meaning it is an estimate.

Tax	165
Insurance	842
Maintenance (est)	700
Petrol (est)	900
Depreciation on car (est)	1000

Construct a pie chart to represent his car expenses. (3 marks)

9 A catalogue shows an MP3 player for £64.99 plus VAT at $17\frac{1}{2}\%$.
The same player can be bought in another shop for £75.
Which is cheaper, and by how much? (3 marks)

10 A circle has diameter 8 cm.

a Calculate the circumference of the circle. (2 marks)

b Calculate the area of the circle. (2 marks)

11 **a** Solve $3x + 7 = 3$ (2 marks)

b Solve $7 + 2n < 5n + 11$ (2 marks)

12 **a** Write 420 as a product of prime factors. (2 marks)

b Given that $495 = 3^2 \times 5 \times 11$, find the least common multiple of 420 and 495. (2 marks)

13 a Find the gradient of the line shown on the grid.

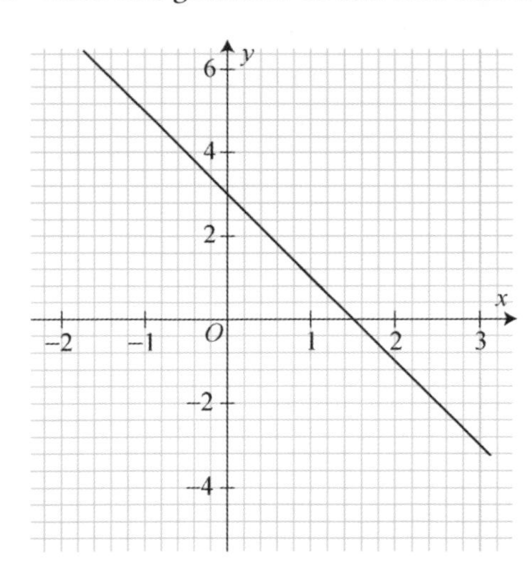

(2 marks)

b Find the equation of the line. (1 mark)

14 A school opens a tuck shop.
They record the takings each day, correct to the nearest £.

M	Tu	W	Th	F	M	Tu	W	Th	F
25	14	17	16	28	27	16	18	19	30

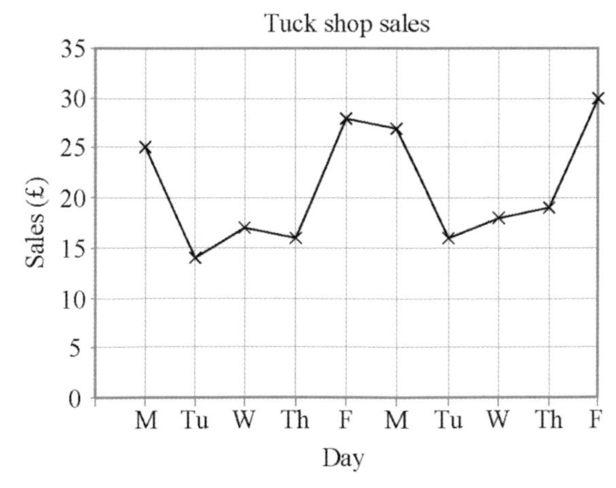

a Work out appropriate moving averages and plot them on a copy of the graph. (3 marks)

b Describe the trend. (1 mark)

c Calculate an estimate of the takings on the next Monday. (2 marks)

15 Where appropriate give your answers correct to 3 significant figures.

a Solve $x^2 - 7x - 10 = 0$ (3 marks)

b Solve $x^2 - 7x + 12 = 0$ (3 marks)

16 From a rectangular sheet 15 cm by 18 cm, a circle of radius 5 cm is cut out.
All measurements are correct to the nearest cm.
Calculate the **least** possible area left. (5 marks)

17 Rearrange $y = \frac{k(1-x)}{1+x}$ to make x the subject. (4 marks)

18 A warden in a game reserve notes the numbers of certain types of animals seen on safari for a week.
He records sightings of 18 elephants, 26 giraffes, 12 white rhinos and 6 lions.

a On a safari the following week, he is asked by a member of the tour what the likelihood is that the next animal they see from these types will be a white, rhino.
Estimate the probability of this outcome. (2 marks)

b During the next month, the warden sees 224 animals of these types altogether.
How many of these would you expect to have been white rhinos? (2 marks)

19 Ronan is planning a tunnel under a hill on his property. He cannot measure the distance directly, so he measures the distance from the two points he wants the tunnel to join to a third point. He also measures the angle between the two lines.
The diagram shows his results.

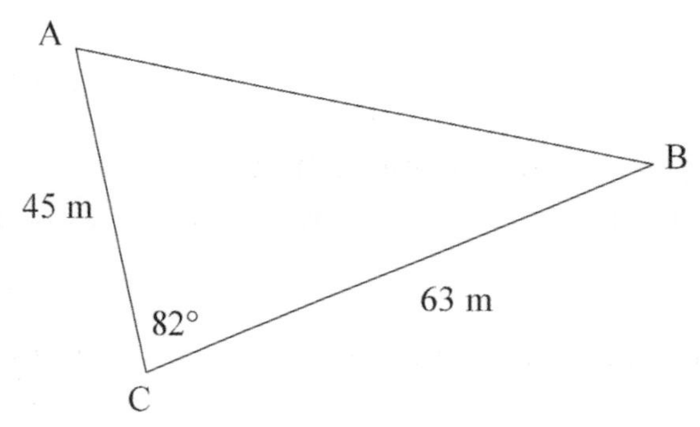

Diagram NOT accurately drawn.

Calculate the distance from A to B. (3 marks)

20 A lift can safely carry 280 kg.

A man weighs 68 kg and has 12 identical boxes to deliver.

The boxes each weigh 17 kg.

All weights are correct to the nearest kg.

Is it possible that the total weight of the man and the boxes will exceed the safety limit? (4 marks)

21 a Complete the table for the values of $y = x^2 - 3x + 3$.

x	-1	0	1	2	3	4
y		3		1		

(2 marks)

b Draw the graph of $y = x^2 - 3x + 3$ on a copy of the grid.

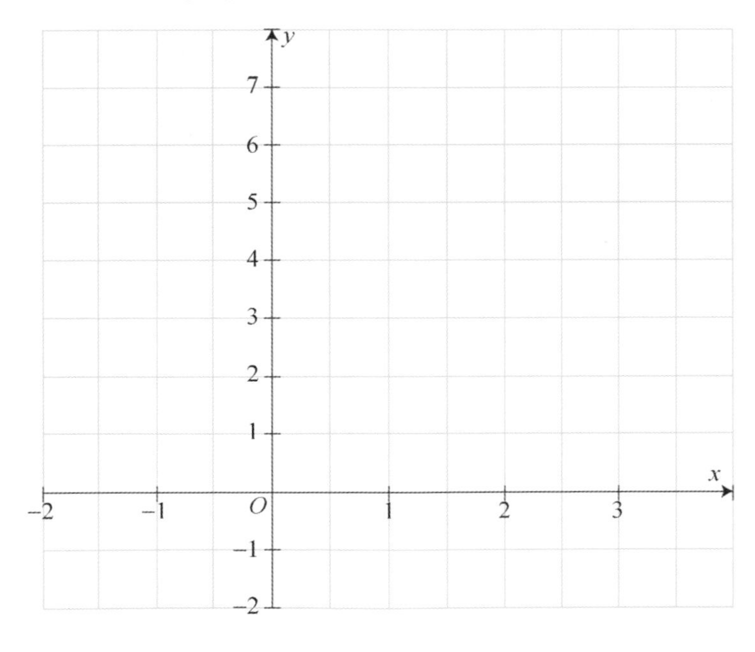

(2 marks)

c By drawing the line $y = 4$ on the diagram, write the x-values where the line $y = 4$ intersects $y = x^2 - 3x + 3$ (2 marks)

d Write down the equation which has these values as solutions. (1 mark)

22 a A sketch of the curve of $y = \sin x$ for x from 0° to 360° is shown.

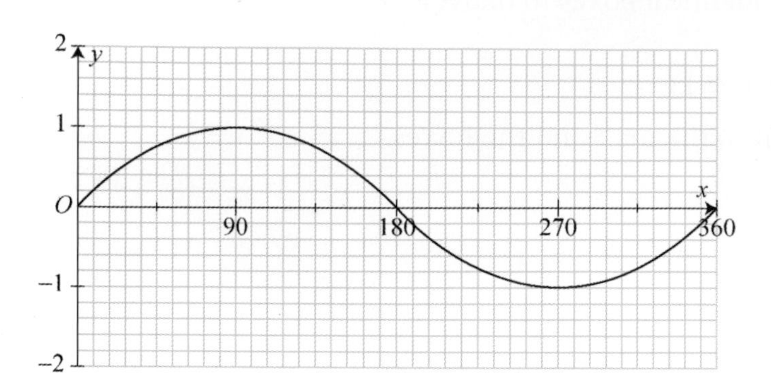

In triangle ABC, AB = 12 cm, BC = 7 cm and angle BAC = 30°.
Use the sine rule and the sketch of $\sin x$ to calculate the size of the two possible angles ACB. Give your answers correct to 1 decimal place (1 dp). (4 marks)

b i The graph of $y = \tan x$ is shown.
Copy the graph and show $y = \tan 3x$ on the same graph.

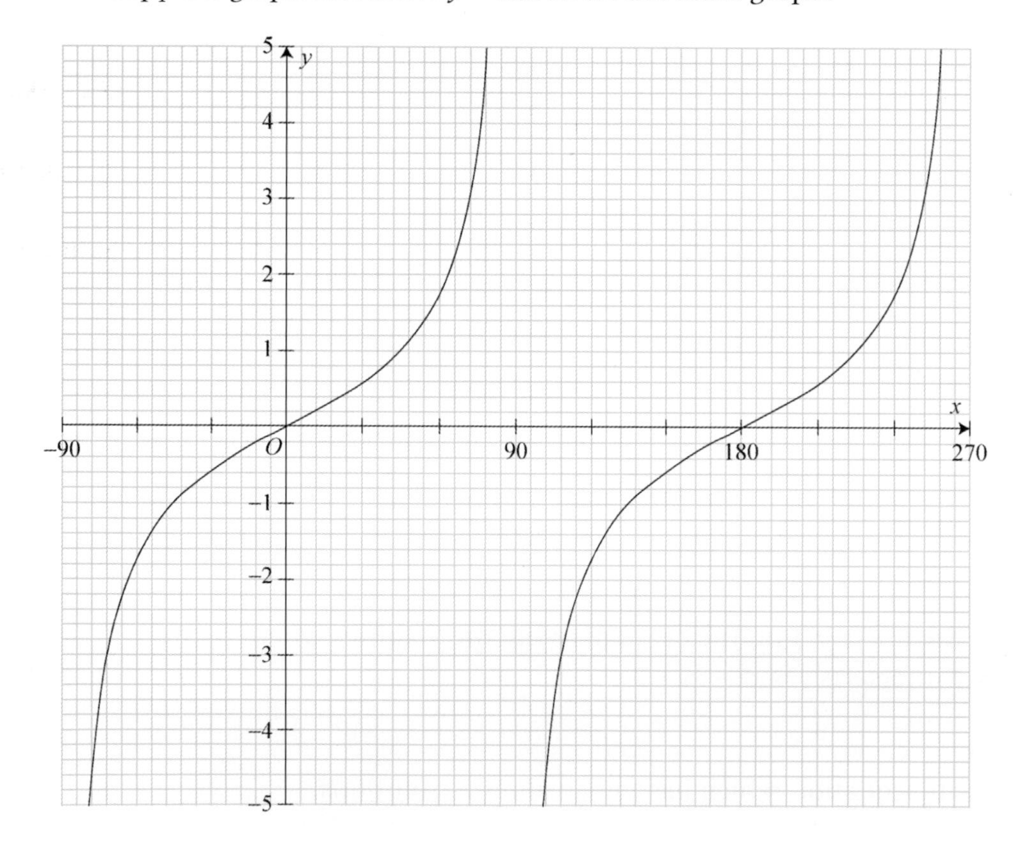

(2 marks)

ii The graph of $y = \cos x$ is shown.
Copy the graph and show $y = 1 + 2\cos x$ on the same graph.

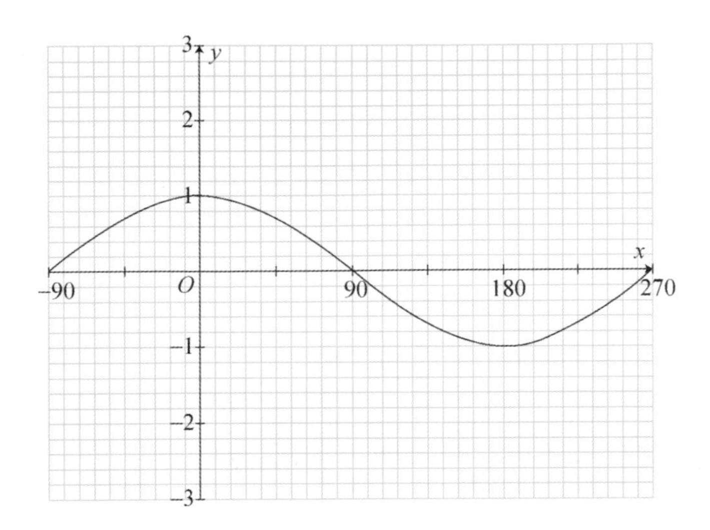

(2 marks)

23 y is inversely proportional to x.
$y = 12$ when $x = 3$.

a Find the relationship between y and x. (2 marks)

b Find y when $x = 9$. (1 mark)

c Find x when $y = 6$. (1 mark)

Answers

N1 Before you start ...

1 0.312, 0.35, 0.37, 0.4
2 2.34
3 −8, −5, −3, −1, 2, 4
4 a 7 b −10 c −12 d −3
5 1, 2, 3, 4, 6, 8, 12, 16, 24, 48

N1.1

1 a Four hundred and fifty-six
b Thirteen thousand, two hundred
c One hundred and fifteen thousand and twenty
d Four hundred and sixty thousand, three hundred and forty
e Four million, three hundred and twenty-five thousand, four hundred
f Fifty-five million, six hundred and seventy thousand, three hundred and forty-five
g Forty-five point eight
h Three hundred and sixty-seven point zero three
i Four thousand, five hundred and three point three four
j Two thousand seven hundred point zero two
2 a 538 b 15 603 c 417.3 d 537.403
3 a 25.5 b 1.85 c 50 d 4.95
e 1.375 f 0.705
4 a 5.007, 5.099, 5.103, 5.12, 5.2
b 0.5, 0.509, 0.525, 0.545, 0.55
c 7.058, 7.302, 7.35, 7.387, 7.403
d 0.4, 0.42, 2.4, 4.2, 42
e 26.9, 26.97, 27.06, 27.1, 27.6
f 13.19, 13.3, 13.43, 14.03, 14.15
5 a 320 b 4 c 1.52
d 0.146 e 23.7 f 2430
g 0.0123 h 4.59 i 3400
j 135.6 k 0.0236 l 17.45
m 0.392 n 0.728 o 0.0124
p 8.14
6 a 96.6 b 937.3 c 22.23 d 24 140
7 a 182.4 b 1.824

N1.2

1 a 3.8 b 4.25 c 540 d 4.8
2 a 6.5 cm b 4.75 kg
c 1154 °C d 3.24 tonnes
3 a About 7.3 cm b About 68.3 ml
c About 46.7 mph d About 8.37 °C
e About 54 g f About 3.3 cm
g About 3.73 ml h About 2800 °C
4 a 13 : 59 b 15 : 21
c 2 h 7 min d 1 h 20 min

N1.3

1 a −13, −12, −6, 0, 15, 17
b −8, −7, −6, −5, −3, 0
c −5, −2, 1, 2, 3, 4
d −8, −3, −1.5, 2, 3, 9
e −5, −4.5, −3, −2, 2, 3
f −9, −1, 2, 3, 6, 8
g −4.5, −3, −2.5, −1, 0, 5.5
h −6, −5.8, −5.7, −5.4, −5.1, −5
2 a 16 b −7 c 4
d 37 e 17 f −7
g 9 h −11 i −8
j 21 k −8 l −2
m −18 n 4 o −5
p 2 q −20 r 10
s −8 t −28 u −18
v −3 w −9
3 a 31 b −11 c 4
d −7 e 5.5 f 2.5
g 7.5 h −0.5
4 a 8 b −2 c 9
d 0 e −13 f −14
g −14 h 20 i 12
j 3 k −5 l −6
m 4 n −25 o −1
p 1 q −9 r −23
s −9 t −23
5 a 1 b −4 c −9
d 5 e 2 f 9
6 a 14°C b −10°C c 4°C

N1.4

1 a −2 b 2 c −3 d 3
e −4 f 4
2 a 6 b 30 c −21 d 8
e −20 f 24 g −24 h 42
i 16 j 50 k 5 l −8
m 5 n 5 o −9 p −63
q −49 r 72 s −9 t −6
u −4 v −10 w 9 x −56
y −13
3 a + × − = −, −20 c − × − = +, 30
d − × − = +, 28 f + × + = +, 40
g + × − = −, −35 h − ÷ − = +, 8
j − × − = +, 70
4 a −120 b 132 c −225 d −147
e −117 f −133 g −414 h −40
i −13 j −67.2
5 a −1 b −15 c −1 d −14
e −9 f 5 g −16

N1.5

1 a 2, 4, 5, 10, 20 b 2, 3, 4, 6, 8, 12, 16
c 5, 10, 15, 20 d 2, 3, 5, 17, 19

2 a 1×24, 2×12, 3×8, 4×6
b 1×45, 3×15, 5×9
c 1×66, 2×33, 3×22, 6×11
d 1×100, 2×50, 4×25, 5×20, 10×10
e 1×120, 2×60, 3×40, 4×30, 5×24, 6×20, 8×15, 10×12
f 1×132, 2×66, 3×44, 4×33, 6×22, 11×12
g 1×160, 2×80, 4×40, 5×32, 8×20, 10×16
h 1×180, 2×90, 3×60, 4×45, 5×36, 6×30, 9×20, 10×18, 12×15
i 1×360, 2×180, 3×120, 4×90, 5×72, 6×60, 8×45, 9×40, 10×36, 12×30, 15×24, 18×20
j 1×324, 2×162, 3×108, 4×81, 6×54, 9×36, 12×27, 18×18
k 1×224, 2×112, 4×56, 7×32, 8×28, 14×16
l 1×264, 2×132, 3×88, 4×66, 6×44, 8×33, 11×24, 12×22
m 1×312, 2×156, 3×104, 4×78, 6×52, 8×39, 12×26, 13×24
n 1×325, 5×65, 13×25
o 1×432, 2×216, 3×144, 4×108, 6×72, 8×54, 9×48, 12×36, 16×27, 18×24

3 a 17, 34, 51 **b** 29, 58, 87
c 42, 84, 126 **d** 25, 50, 75
e 47, 94, 141 **f** 35, 70, 105
g 90, 180, 270 **h** 120, 240, 360
i 95, 190, 285 **j** 208, 416, 624

4 324

5 a 2 **b** 5 **c** 6 **d** 8
e 15 **f** 18 **g** 25 **h** 12
i 15

6 a 12 **b** 40 **c** 36 **d** 75
e 42 **f** 150

7 a 120 seconds **b** 18 cm × 18 cm

N1.6

1 a 75 **b** 40 **c** 63 **d** 180 **e** 441

2 a 2×3^2 **b** $2^3 \times 3$ **c** $2^3 \times 5$
d 3×13 **e** $2^4 \times 3$ **f** 2×41
g $2^2 \times 5^2$ **h** $2^4 \times 3^2$ **i** $2^2 \times 3^2 \times 5$
j $3^2 \times 5 \times 7$ **k** $2^2 \times 3 \times 37$ **l** $2 \times 3^3 \times 5^2$

3 a 7 missing in the answer; $126 = 2 \times 3^2 \times 7$
b Divided 105 by 5 but recorded it as 3; $210 = 2 \times 3 \times 5 \times 7$
c 221 is not prime, so he should not have stopped; $221 = 13 \times 17$

4 a 3 **b** 5 **c** 6 **d** 48
e 3 **f** 17

5 a 72 **b** 120 **c** 72 **d** 60
e 180 **f** 432

6 a $\frac{3}{4}$ **b** $\frac{2}{3}$ **c** $\frac{5}{8}$ **d** $\frac{2}{3}$
e $\frac{9}{13}$ **f** $\frac{2}{5}$

7 a $x = 4$, $y = 3$ **b** $a = 2$, $b = 3$, $c = 5$

N1 Exam review

1 a $3^2 \times 5$ **b** 9 **c** 180

2 ai 7 °C **aii** −10 °C
bi 6 °C **bii** 8 °C
c −7 °C

N2 Before you start ...

1 47

2 a 1.3 **b** 0.3

3 a 42 **b** 35

4 a 48 **b** 3.8

5 $7 \times 12 \div 4$

N2.1

1 a 26 **b** 37 **c** 52 **d** 10
e 33 **f** 5 **g** 180 **h** 9

2 a 28 **b** 72 **c** 5 **d** 16
e 2 **f** 75

3 a $5 \times (2 + 1) = 15$ **b** $5 \times (3 - 1) \times 4 = 40$
c $20 + 8 \div 2 - 7 = 17$ **d** $2 + 3^2 \times (4 + 3) = 65$
e $2 \times (6^2 \div 3) + 9 = 33$ **f** $(4 \times 5 + 5) \times 6 = 150$

4 a Pete, because the contents of the brackets are $2 \times 9 - 4 = 18 - 4 = 14$.
b No; $(5 \times 4)^2 = 20^2 = 400$, whereas $5 \times 4^2 = 5 \times 16 = 80$.
ci 55.7685 **cii** 55.8

5 a 1 **b** 2 **c** 2 **d** 14 **e** 40
f 7 **g** 11 **h** 3

6 a 170 **b** 0.58 **c** 1.78

7 3

N2.2

1 ai 3490 **aii** 3500 **aiii** 3000
bi 3390 **bii** 3400 **biii** 3000
ci 14 850 m **cii** 14 900 m **ciii** 15 000 m
di £57 790 **dii** £57 800 **diii** £58 000
ei 92 640 kg **eii** 92 600 kg **eiii** 93 000 kg
fi £86 190 **fii** £86 200 **fiii** £86 000
gi 3440 **gii** 3400 **giii** 3000
hi 74 900 **hii** 74 900 **hiii** 75 000

2 a 4 **b** 29 **c** 469 **d** 369
e 20 **f** 27 **g** 101 **h** 0

3 ai 3.447 **aii** 3.45 **aiii** 3.4
bi 8.948 **bii** 8.95 **biii** 8.9
ci 0.128 **cii** 0.13 **ciii** 0.1
di 28.387 **dii** 28.39 **diii** 28.4
ei 17.999 **eii** 18.00 **eiii** 18.0
fi 10.000 **fii** 10.00 **fiii** 10.0
gi 0.004 **gii** 0.00 **giii** 0.0
hi 2785.556 **hii** 2785.56 **hiii** 2785.6
ii 158.852 **iii** 158.85 **iiii** 158.9

4 ai 8.37 **aii** 8.4 **aiii** 8
bi 18.8 **bii** 19 **biii** 20

ci 35.8 cii 36 ciii 40
di 279 dii 280 diii 300
ei 1.39 eii 1.4 eiii 1
fi 3890 fii 3900 fiii 4000
gi 0.008 37 gii 0.0084 giii 0.008
hi 2400 hii 2400 hiii 2000
ii 8.99 iii 9.0 iiii 9
ji 14.0 jii 14 jiii 10
ki 1400 kii 1400 kiii 1000
li 140 000 lii 140 000 liii 100 000
mi 3280 mii 3300 miii 3000

5 a $5 \times 6 = 30$ b $18 + 22 = 40$ c $\frac{6 \times 3}{9} = 2$
d $35 - 10 = 25$ e $\frac{33 \times 5}{3} = 55$
f $(10^2 + 9)^2 \approx 100^2 = 10\,000$

6 ai 66.5 cm aii 67.5 cm
bi 34.65 litres bii 34.75 litres
ci 8.355 kg cii 8.365 kg
di 0.3865 mm dii 0.3875 mm

7 a 2.65 m b 2.55 m

N2.3

1 ai 2000 aii 1500 aiii 1550
bi 6000 bii 5800 biii 5790
ci 18 000 cii 17 800 ciii 17 790
di 35 000 dii 35 100 diii 35 130
ei 237 000 eii 236 900 eiii 236 870

2 ai 4.356 aii 4.36 aiii 4.4 aiv 4
bi 9.857 bii 9.86 biii 9.9 biv 10
ci 0.937 cii 0.94 ciii 0.9 civ 1
di 19.496 dii 19.50 diii 19.5 div 19
ei 26.808 eii 26.81 eiii 26.8 eiv 27
fi 20.000 fii 20.00 fiii 20.0 fiv 20
gi 0.005 gii 0.00 giii 0.0 giv 0
hi 3896.657 hii 3896.66
hiii 3896.6 hiv 3897

3 a 0.3 b 150 c 0.08 d 280
e 38 f 0.04 g 92.3 h 4460

4 a 10^2 b 0.01 c 0.01 d 1000
e 0.1 f 0.01

5 a 2.4 b 0.56 c 50 d 20
e 0.48 f 400

6 a $4 \times 4 = 16$ b $20 \times 20 = 400$
c $\frac{5 \times 8}{20} = 2$ d $54 \div 9 = 6$

7 a $\frac{30 \times 40}{3 \times 4} = 100$ b $\frac{16 \times 0.5}{0.2 \times 32} = 1.25$
c $(25 + 4)^2 \approx 30^2 = 900$ d $\frac{64 \times 4}{4} = 64$
e $\sqrt{2 \div 0.04} = \sqrt{50} \approx 7$ f $\sqrt{30 \div 0.6} = \sqrt{50} \approx 7$

8 Sean Used $\frac{30 \times 10}{5}$

N2.4

1 a 63 b 12.1 c 5.4 d 24.2
e 360 f 4.3 g 236 h 0.0078

2 a 15.4 b 189 c 58.3 d 114.8
e 133 f 63.6 g 49 h 69
i 35.2 j 173.6 k 134.1 l 784.3
m 112 n 34 o 18 p 19.8

3 a 11.9 b 65.1 c 111.2 d 93.6
e 100.8 f 211.2 g 43 h 73

4 a 11.6 b 35.6 c 148.6 d 153.9
e 140.4 f 100.3

5 a 49.3 b 726 c 67.2 d 66.7
e 26.6 f 67.5 g 12 h 40

6 a £116.91 b 30.38 m^2 c £35.76
ai 101 430 aii 1 014 300 aiii 1014.3
aiv 10.143 av 1.0143 avi 1014.3
avii 1.0143 aviii 0.10143
bi 65.49 bii 654.9 biii 3.7 biv 17.7

7 a £116.91 b £1.247

N2.5

1 a 26.08 b 32.71 c 2.81
d 31.99 e 26.47 f 13.49

2 a 51 b 100.8 c 109.2
d 174.8 e 117.6 f 475.3

3 a 3.9 b 6.1 c 8.8 d 12.3
e 13.3 f 14.9 g 45 h 55
i 22

4 a 161.98 kg b 19.22 g c 1.102 kg

5 a 4.002 b 4.4 c 75.31 d 14.7
e 41.673 f 181.64

6 a £3.33 b £11.31 c 75 trees d £79.25
e £86.55

N2.6

1 a $2.4 \times (4.3 + 3.7) = 19.2$
b $6.8 \times (3.75 - 2.64) = 7.548$
c $(3.7 + 2.9) \div 1.2 = 5.5$
d $(2.3 + 3.4^2) \times 2.7 = 37.422$
e $5.3 + 3.9 \times (3.2 + 1.6) = 24.02$
f $3.2 + 6.4 \times (4.3 + 2.5) = 46.72$

2 a 178.412 383 5
b 0.196 708 95
c 3.210 178 253
d 3.350 190 476
e 1.157 007 415
f 0.135 604 5

3 ai 15.3 m^2 aii £103.40 b £66.67

4 a £21.13 b £16.37
c Yes, the new bill is cheaper than the old bill.

N2 Exam review

1 a 2000 b 4320 c 3

2 a 1.962 631 579 b 1.96 or 2.0

N3 Before you start ...

1 $\frac{1}{3}$

2 a $x = 10$ b $y = 4$

3 3

4 0.1, $\frac{1}{4}$, $\frac{1}{2} = 0.5$, $23\% = \frac{23}{100}$

5 0.7, 0.75, 0.8, 0.875

N3.1

1 **ai** $\frac{8}{12}$ **aii** $\frac{2}{3}$
bi $\frac{14}{16}$ **bii** $\frac{7}{8}$
ci $\frac{12}{20}$ **cii** $\frac{3}{5}$
di $\frac{10}{15}$ **dii** $\frac{2}{3}$

2 **a** $\frac{1}{3}$ **b** $\frac{3}{4}$ **c** $\frac{3}{5}$ **d** $\frac{4}{9}$
e $\frac{5}{8}$ **f** $\frac{1}{3}$ **g** $\frac{4}{9}$ **h** $\frac{23}{93}$

3 **a** $\frac{3}{2}$ **b** $\frac{11}{3}$ **c** $\frac{35}{8}$ **d** $\frac{20}{9}$
e $\frac{41}{7}$ **f** $\frac{39}{5}$ **g** $\frac{96}{11}$ **h** $\frac{88}{7}$
i $\frac{163}{13}$

4 **a** $1\frac{1}{4}$ **b** $1\frac{3}{5}$ **c** $1\frac{4}{7}$ **d** $2\frac{1}{4}$
e $2\frac{1}{5}$ **f** $2\frac{6}{7}$ **g** $4\frac{3}{5}$ **h** $3\frac{1}{9}$
i $8\frac{3}{8}$

5 **a** 8 **b** 27 **c** 56 **d** 56
e 2 **f** 90 **g** 85 **h** 7

6 **a** $\frac{2}{5} > \frac{1}{3}$ **b** $\frac{1}{3}, \frac{7}{18}, \frac{4}{9}$

7 **a** $\frac{2}{5}$ **b** $\frac{2}{3}$ **c** $\frac{4}{7}$ **d** $\frac{5}{6}$
e $\frac{4}{7}$ **f** $\frac{10}{7}$

8 **a** $\frac{3}{15}, \frac{1}{3}, \frac{2}{5}$ **b** $\frac{1}{2}, \frac{15}{28}, \frac{4}{7}$ **c** $\frac{4}{7}, \frac{5}{8}, \frac{9}{14}$

N3.2

1 **a** $\frac{2}{3}$ **b** $\frac{5}{8}$ **c** $\frac{5}{11}$ **d** $\frac{13}{17}$
e $\frac{3}{23}$ **f** $\frac{13}{27}$

2 **a** 1 **b** $\frac{2}{3}$ **c** $1\frac{2}{11}$ **d** $\frac{7}{13}$
e $1\frac{2}{3}$ **f** $\frac{2}{3}$ **g** $2\frac{1}{3}$ **h** $3\frac{4}{7}$

3 **a** $\frac{5}{6}$ **b** $\frac{17}{20}$ **c** $\frac{4}{15}$ **d** $\frac{18}{35}$
e $\frac{23}{24}$ **f** $\frac{38}{45}$ **g** $\frac{59}{99}$ **h** $\frac{94}{105}$

4 **a** $\frac{1}{3}$ **b** $\frac{1}{6}$ **c** $\frac{3}{4}$ **d** $\frac{1}{3}$

5 **a** $1\frac{7}{15}$ **b** $2\frac{1}{10}$ **c** $2\frac{7}{12}$ **d** $1\frac{31}{35}$
e $2\frac{1}{15}$ **f** $1\frac{7}{8}$ **g** $1\frac{7}{12}$ **h** $\frac{43}{63}$

6 **a** $5\frac{11}{12}$ miles **b** $1\frac{13}{16}$ lb **c** $1\frac{79}{80}$ kg **d** $\frac{2}{15}$
ei $29\frac{9}{28}$ feet **eii** $62\frac{4}{45}$ feet **eiii** $16\frac{1}{9}$ m

N3.3

1 **a** $1\frac{1}{2}$ **b** 2 **c** $3\frac{1}{3}$ **d** $2\frac{1}{7}$
e $2\frac{1}{2}$ **f** $4\frac{1}{3}$

2 **a** 2 **b** 4 **c** $3\frac{1}{3}$ **d** $\frac{7}{12}$
e $\frac{3}{5}$ **f** $1\frac{3}{5}$ **g** 6 **h** $23\frac{3}{8}$

3 **a** 8 **b** 10 **c** 14 **d** 20
e 48 **f** 220

4 **a** $2\frac{4}{5}$ kg **b** $2\frac{1}{7}$ m

5 **a** 6 **b** $17\frac{1}{2}$ **c** $2\frac{2}{5}$ **d** 14
e 48 **f** $6\frac{3}{7}$ **g** 2 **h** $2\frac{1}{7}$

6 **a** $\frac{3}{10}$ **b** $\frac{9}{20}$ **c** $\frac{15}{28}$ **d** $\frac{12}{35}$
e $\frac{2}{3}$ **f** $\frac{7}{24}$ **g** $\frac{2}{3}$ **h** $2\frac{1}{4}$
i $\frac{9}{49}$ **j** $\frac{1}{2}$ **k** $2\frac{1}{3}$ **l** $1\frac{37}{40}$

7 **a** 10 **b** $\frac{5}{6}$ **c** $1\frac{1}{15}$ **d** $\frac{6}{7}$
e $\frac{27}{28}$ **f** $1\frac{1}{5}$ **g** $\frac{1}{4}$ **h** $\frac{4}{35}$
i $\frac{4}{55}$ **j** $2\frac{5}{8}$ **k** $1\frac{1}{6}$ **l** $1\frac{2}{25}$
m 2 **n** $3\frac{3}{8}$ **o** $1\frac{13}{15}$

8 **a** $3\frac{17}{25}$ **b** $4\frac{4}{55}$

N3.4

1 **a** $\frac{3}{10}$ **b** $\frac{3}{5}$ **c** $\frac{16}{25}$ **d** $\frac{9}{20}$
e $\frac{3}{8}$ **f** $1\frac{2}{25}$ **g** $3\frac{23}{100}$

2 **a** 0.3 **b** 0.44
c 1.04 **d** 0.62 **e** 0.45 **f** 0.52
g 0.28 **h** 3.35

3 **a** 0.44 **b** 0.67 **c** 1.35 **d** 0.73
e 1.14 **f** 1.4 **g** 2.17 **h** 0.85

4 **a** $\frac{2}{5}$ **b** $\frac{9}{10}$ **c** $\frac{7}{20}$ **d** $\frac{13}{20}$
e $\frac{1}{100}$ **f** $3\frac{31}{50}$ **g** $\frac{61}{400}$ **h** $\frac{17}{800}$

5 **a** 54% **b** 40% **c** 85% **d** 52%
e 66.7% **f** 24% **g** 120% **h** 44%

6 **a** 0.6 **b** 0.375 **c** 0.4375 **d** $0.\dot{4}$
e $0.\dot{7}1428\dot{5}$ **f** $0.58\dot{3}$
g 0.652173913 **h** 0.228571428

7 **a** 0.37 **b** 0.07 **c** 1.89
d 0.45 **e** 1.45 **f** 0.008
g 2.5 **h** 1.232

8 **a** 72% **b** 20% **c** 125%
d 3% **e** 102% **f** 3.25%
g 33.3% **h** 137.2%

9 **a** 68.6% **b** 64% **c** 89.5%
d 191.7% **e** 26.3%

10 Recurring decimals are:

$\frac{1}{3}, \frac{1}{6}, \frac{1}{7}, \frac{1}{9}, \frac{1}{11}, \frac{1}{12}, \frac{1}{13}, \frac{1}{14}, \frac{1}{15}, \frac{1}{18}, \frac{1}{21}, \frac{1}{22}, \frac{1}{24}, \frac{1}{27}, \frac{1}{30}$

N3.5

1

Fraction	Decimal	Percentage
$\frac{3}{8}$	0.375	37.5%
$\frac{7}{25}$	0.28	28%
$\frac{3}{20}$	0.15	15%
$\frac{3}{8}$	0.375	37.5%
$\frac{4}{5}$	0.8	80%
$\frac{7}{40}$	0.175	17.5%

2 a $\frac{5}{8}$ b $\frac{4}{5}$ c $\frac{5}{7}$ d $\frac{3}{8}$
e $\frac{16}{11}$ f $\frac{14}{9}$ g $1\frac{7}{23}$ h $2\frac{8}{11}$
3 a $<$ b $>$ c $>$ d $>$
4 a 47%, $\frac{12}{25}$, 0.49 b 78%, $\frac{4}{5}$, 0.81
c $\frac{7}{12}$, $\frac{4}{5}$, 66% d 29%, 0.3, $\frac{5}{16}$, $\frac{7}{22}$
5 a $\frac{19}{28} = 67.9\%$ b $\frac{11}{16} = 68.8\%$
c $\frac{19}{24} = 79.2\%$ d $\frac{1}{3} = 33.3\%$
6 a German, as $\frac{37}{54} = 68.5\%$
b $\frac{7}{31} = 23\%$, so Sarah's class is in accord with the rest of the college.

N3 Exam review

1 a $\frac{1}{10}$ b $\frac{3}{20}$
2 a $\frac{1}{6}, \frac{3}{8}, \frac{1}{2}, \frac{2}{3}, \frac{3}{4}$ b $\frac{3}{5}$, 65%, $\frac{2}{3}$, 0.72, $\frac{3}{4}$

N4 Before you start ...

1 $\frac{3}{8}$ 2 £28.00 3 1.5
4 4 5 £68 6 50 mph

N4.1

1 ai $\frac{3}{10}$ aii 30% bi $\frac{5}{8}$ bii 62.5%
ci $\frac{4}{9}$ cii 44.4% di $\frac{5}{12}$ dii 41.7%
ei $\frac{7}{25}$ eii 28%
2 a $\frac{3}{5}$, 61%, 0.63 b $\frac{17}{25}$, 69%, $\frac{7}{10}$, 0.71
c 34%, $\frac{7}{20}$, 0.36, $\frac{3}{8}$, $\frac{2}{5}$ d $\frac{2}{5}$, 42%, $\frac{3}{7}$
e 0.14, 15%, $\frac{3}{19}$, $\frac{1}{5}$ f 81%, $\frac{8}{9}$, 0.9, 0.93, $\frac{19}{20}$
3 ai $\frac{3}{5}$ aii 60% bi $\frac{13}{25}$ bii 52%
ci $\frac{3}{20}$ cii 15% di $\frac{7}{10}$ dii 70%
ei $\frac{4}{5}$ eii 80%
4 a $\frac{28}{45}, \frac{17}{25}, \frac{29}{40}, \frac{15}{20}, \frac{23}{30}$, 79%
b Maths

N4.2

1 a 5 b 8 c 7
d 4.5 e 1.5 f 3.3
g 3.1 h 7.2 i 6.7
j 8.6 k 1.8 l 1.4
2 a $\frac{1}{3}$ b $\frac{1}{2}$ c $\frac{3}{10}$ d $\frac{1}{4}$ e $\frac{3}{8}$
f $\frac{2}{5}$ g $\frac{1}{6}$ h $1\frac{1}{2}$ i 2
3 a Column B is 4 times column A.
b Not in direct proportion.
c Column F is 1.8 times column E.
d Column H is 3.1 times column G.
4 £50, £150, £1500, £1900, £2400
5 a £18.75 b £5.95 c £11.13
d £6.24 e £9.48 f 720 Mb
g £1.87 h 12 litres i £55.65 j 6300g

N4.3

1 a 1 : 2 b 8 : 5 c 8 : 5 d 3 : 2
e 19 : 9 f 1 : 6 g 2 : 4 : 3 h 4 : 5 : 8
2 a 2 : 5 b 11 : 16 c 5 : 2 d 5 : 3
e 5 : 3 f 8 : 5
3 a 1 : 3 b 7 : 3 c 15 : 2 d 9 : 100
4 a 1 : 3 b 3 : 2 c 2 : 1 d 3 : 4
5 a 1 : 2.5 b 1 : 3.85 c 1 : 4.17 d 1 : 41.67
e 1 : 23.68 f 1 : 150 g 1 : 12.5
6 a 1 : 50 b 1 : 20 000 c 1 : 36
7 a 35 kg b 28 kg

N4.4

1 b 3 : 1; height of 144 cm = 3 × width of 48 cm; width of 48 cm = $\frac{1}{3}$× height of 144 cm
c 16 : 7; limousine length of 6.4 m = $\frac{16}{7}$ × car length of 2.8 m; car length of 2.8 m = $\frac{7}{16}$ × limousine length of 6.4 m
d 3 : 4; can containing 330 ml = $\frac{3}{4}$ × can containing 0.44 litre; can containing 0.44 litre = $\frac{4}{3}$ × can containing 330 ml
2 a 15 women b 88 kg
c 60 purple flowers d 504 students
3 a 1.2 cm b 44 lecturers c 39 cm
4 a 325 m b 0.6 cm
5 a £27 : £63 b 287 kg : 82 kg
c 64.5 tonnes : 38.7 tonnes
d 19.5 litres : 15.6 litres e £6 : £12 : £18

N4.5

1 Leonard £7, Pavel £6.75, Andy £11.55
2 a £7.42 per hour b 170 bricks per hour
c 55 km/h
3 £1 = 5 litas, £1 = 11.55 dollars, £1 = 6.3 riyals
4 €345.60, €1036.80, €9072
5 a £820 b £94.30

N4.6

1 9 mph
2 £2.10 per metre
3 32 mph
4 89.25 mph
5

Speed (km/h)	Distance (km)	Time
105	525	5 hours
48	106	2 hours 12.5 minutes
$37\frac{1}{3}$	84	2 hours 15 minutes
86	215	2 hours 30 minutes
37.1	65	1 hour 45 minutes

6 a 5 g/cm^3 b 87.88 g

7 a 7.59 g/cm^3 to 3 sf b 105
8 a 9.46 kg to 3 sf b 6.15 litres to 3 sf
9 1358 m

N4 Exam review

1 a £770 b £3500
2 a 100 hamburgers b USA by £1.99

N5 Before you start ...

1 a 144 b 9
2 a 125 b 2
3 a 16 b 1 000 000
4 1, 2, 3, 4, 6, 8, 12, 24
5 2, 3, 5, 7, 11, 13
6 a 45 b 100

N5.1

1 a 25 b 121 c 225 d 289
2 a 16, 36 b 49 c 121, 144 d 225
3 a 256 b 13.69 c 2500 d 44.89
e 316.84 f 17.64 g 3.61 h 0.01
i 15.21 j 4.41 k 0.49 l 175.56
4 a ±23 b ±12.53 c ±6.40 d ±0.4
e ±2.6 f ±28.28 g ±36.67 h ±6.21
i ±84.22 j ±15.32
5 a 7 b 9 c 5 d 6
e 11 f 12 g 3 h 2
6 a $(2.645\,751)^2 = 6.999\,998$; 2.645 751 is only accurate to 6 d.p. so its square is not exactly 7.
b 56 and 57
7 a 4.5
bi 6.3 bii 7.7 biii 9.7

N5.2

1 a 343 b 1000 c 2197 d 6859
2 a Square: 4, 16; Cube: 27
b Square: 64, 144; Cube: 64
c Square: 196, 256; Cube: 216
d Square: 900; Cube: 1000
3 a 512 b 13.82 c 8000 d 59.32
e 1601.61 f −21.95 g 704.97 h 0.125
i −157.46 j 970.30 k −0.001 l 4784.09
4 a 9 b 4.64 c 4 d 4.41
e 1.97 f 1.39 g 1.1 h 3.83
i 23 j −6 k −4.12 l 0.25
5 ai 2.7 aii 3.7 aiii 4.3
aiv 5.3 av 6.7 avi 7.9
avii 9.7 aviii 11.4

N5.3

1 a 16 b 32 c 125 d 2401 e 729
2 a 3375 b 729 c 1024 d 217 678.23
e 2197
3 a 25 b 40 c 259 947
d 8000 e $\frac{5}{256}$ or 0.01953
4 a 3 b 4 c 4 d 7
e 23
5 a 340 b 76 600 c 0.085 d 23 000
e 312 000 f 56 200 g 2.96
6 a 3^4 b 7^5 c 2^{12} d 10^3
e 3^4 f 4^7 g 10^{10} h 7^6
i 2^6 j 10^5 k 4^0
7 a 0.1 b 0.125 c 0.001 d 0.333 ...
e 0.142857 ... f 0.076923 ...
8 a 1 b $\frac{1}{5} = 0.2$ c 1
9 a 5.4 b 3 c 10^4
d 1 730 000
10 a y^5 b 4^{10} c w^5 d 4^{y-2}
e g^4 f 576

N5.4

1 a 10^2 b 10^1
c 10^5 d 10^0
2 a 2×10^2 b 8×10^2
c 9×10^3 d 6.5×10^2
e 6.5×10^3 f 9.52×10^2
g 2.358×10 h 2.5585×10^2
3 a 500 b 3000
c 100 000 d 250
e 4900 f 3 800 000
g 750 000 000 000
h 8 100 000 000 000 000 000
4 a 6×10^2 b 4.5×10^4
c 6.5×10^0 d 5×10^6
5 a 4×10^5 b 9×10^7
c 2.5×10^8 d 2.4×10^{13}
6 a 2×10^2 b 2×10^4
c 5×10 d 7.5×10^2
7 a 9.75×10^9 b 1.37×10^4
c 4.01×10^{11} d 2.06×10^8
8

Planet	Mean distance from Sun (m)	Light travel time
Mercury	5.79×10^{10}	3 minutes 13 seconds
Earth	1.50×10^{11}	8 minutes 20 seconds
Mars	2.28×10^{11}	12 minutes 40 seconds
Jupiter	7.78×10^{11}	43 minutes 13 seconds
Pluto	5.90×10^{12}	5 hours 27 minutes 47 seconds

N5.5

1 a 3×10^{-1} b 4.7×10^{-3}
c 7.8×10^{-5} d 4.485×10^{-1}
2 a 2.8×10^{-1} b 4×10^{-2}
c 1.35×10^{-3} d 1.2×10^{-7}
3 a 1×10^{-2} km b 2×10^{-3} g
c 5×10^{-6} m d 1.1×10^{-2} litre
4 a 5×10^{-1} b 9.2×10^{-8}
c 2×10^{-2} d 4.2×10^{-8}
5 a 1.15×10^{-7} b 1.83×10^{-1}
c 4.85×10^{-6} d 5.01×10^{-1}
6 a 5.2×10^{-1} b 4.6×10^{-2}
c 2.09×10^{-3} d 1.3×10^{-2}
7 See Q6
8 9.11×10^{-8} m^3
9 5.3×10^{-3} kg
10 About 385 atoms

N5 Exam review

1 11.6

2 a $2^2 \times 3^3$ b 12

N6 Before you start ...

1 £33 2 $43 3 57 kg
4 £5.40 5 £367.50

N6.1

1 a $2\frac{1}{2}$ b 2 c $2\frac{2}{3}$ d $1\frac{6}{7}$ e 2 f $1\frac{1}{3}$

2 a 4 b $3\frac{3}{4}$ c 4 d $4\frac{2}{3}$ e $2\frac{1}{4}$
f $22\frac{2}{5}$ g $13\frac{1}{3}$ h $8\frac{5}{9}$ i 10

3 a $\frac{4}{5}$ kg b $3\frac{1}{2}$ kg c $7\frac{1}{5}$ litres d $14\frac{2}{5}$ kg

4 a €12 b £28 c $37\frac{1}{2}$ m d $36\frac{4}{7}$ km
e £375 f $58\frac{1}{3}$ mm g 1375 m h $18\frac{6}{13}$ g

5 a 264 kg b $4500 c 4.44 kg d 952 cups
e 21.67 tonnes f 96° g 139.35°
h 0.87 hours or 52 minutes i £260.67
j 20 hours

6 a $\frac{2}{3}$ b $\frac{3}{5}$ c $\frac{1}{5}$ d $\frac{11}{60}$

7 a £24 b £7.50 c 73 days

N6.2

1 a £150 b 2 kg c £40
d 18.5 kg e £0.30 or 30p f 34.28 m

2 a £9 b 82 kg c $5
d £0.75 or 75p e £31.50 f 0.19 m

3 a £51 b 72 Mb c £45
d £40 e 136.5 m f £22
g 1099 mm h 6.3 kg i 31.5 mm

4 a Find 10% by dividing by 10; find 5% by halving 10%; add the two answers together.
b Find 10% and then halve it.
c Find 10% and times by 3; find 5% by halving 10%; add the two answers together.
d Find 10%; halve 10% to find 5%; halve 5% to find 2.5%; add the three answers together.
e Find 10% and halve it to find 5%; subtract 5% from 100% (the original amount).

5 a 11.2 Mb b 13.2 tonnes c 90.85 km
d £98.28

6 a £2.04 b 13.92 km c £3.04
d €108.80 e 11.05 m f 33.58 cm
g 125.8 m h £1.53 i £2.13

N6.3

1 a 4.5 kg b 10.2 m c 54°
d 0.74 cm e 331.5 ml f 63°
g 18.2 kg h 85.87 kg i 5.544 kg
j 3.96 m^2

2 a £385 b 70.3 kg c £550.20
d 491.4 km e 1128 kg

3 a £397.80 b 524.9 kg c £1758.96
d 599.56 km e $3423.55

4 New wage: £364, £296.92, £428.74, £217.64, £206.59

5 a 492.8 ml b £166.50 c £250 185
d 1081 students

N6.4

1 Selling price: £10.20, £25.96, £7.96, £28.12, £176.49

2 a £116.33 b £2.10 c £36.74 d £21.89

3 aii 742.4 ml aiii 740 ml
bii 4.914 m biii 4.9 m
cii 2.7356 tonnes ciii 2.7 tonnes
dii 519.6 g diii 520 g

4 Payment by installments costs £189.36, so cash payment is cheaper by 36p.

N6.5

1 a £612.50 b £2335.80 c £26 684.80

2 a £1358.20 b £8828.30 c £4658.63
d £2634.91

3 a £790 b £1109.25 c £54.60
d £132.43

4 a £477.30 b £14 105 c £149 582

5 a £8820 b £15 099.37 c £4324.91

N6 Exam review

1 a £88.80 b £46.75

2 £9720

A1 Before you start ...

1 ai 3 × 2 aii 6 × 5 aiii 4 × 10 aiv 3 × 7
bi 6 bii 30 biii 40 biv 21

2 ai 9 aii 9 aiii 9
b Order does not matter when adding (commutative).

3 ai 30 aii 30 aiii 30
b Order does not matter when multiplying (commutative).

4 a 5 b 9 c 4 d 10

5 ai 1, 2, 3, 6, 9, 18 aii 1, 2, 3, 4, 6, 12
aiii 1, 2, 3, 4, 6, 8, 12, 24
b 1, 2, 3, 6 c 6

A1.1

1 a $4b$ b $2y$ c $3a$
d $9p$ e $3x$ f $6z$

2 a $5p + 6q$ b $9x + 7y$ c $2m + 8n$
d $x + 5y$ e $8r - 6s$ f $2g - 4f$
g $2a + 6b + 5c$ h $5u - 2v + 3w$ i $3x - 4y + 5z$
j $6r + 5s + 2t$

3 a $6t$ b $3n$ c $4x$

4 a $4x$ b $4x + 8$

5 a $2m$ b $3m$ c $12m$

6 a $6c$ b $10d$ c $6c + 10d$

7 a $50f + 30g$ b $80j + 40k$
c $50x + 60y + 30z$ d $60r + 80s + 40t$

A1.2

1 a y^4 b m^6 c x^3 d p^2

2 a $3t^2$ b $4pq^2$ c $6v^2w^3$ d $2r^4s$

3 a $6m^2n$ **b** $8y^3z^2$ **c** $12gh^3$ **d** $10xy^4$
4 a $6m^2$ **b** $12p^3$ **c** $6xy^2$ **d** $10r^2s^2$
5 a n^5 **b** s^7 **c** p^4 **d** t^4
6 a x^7 **b** x^8 **c** x^9 **d** x^7
7 a r^2 **b** r **c** r^5 **d** r^3
8 a m^4 **b** x **c** t^2 **d** y^3
e 1
9 a x **b** m^2 **c** s^3 **d** v^2
e q^3 **f** t^4 **g** 1 **h** y^2
10 $2n^3 = 2 \times n^3$, $2 \times n \times n = 2n^2$, $n^2 = \frac{n^4}{n^2}$, $5 \times n = 5n$

A1.3

1 a $3m + 6$ **b** $4p + 24$ **c** $2x + 8$
d $5q + 5$ **e** $12 + 2n$ **f** $6 + 3t$
g $12 + 4s$ **h** $8 + 2v$
2 a $6q + 3$ **b** $8m + 4$ **c** $12x + 9$
d $6k + 2$ **e** $10 + 10n$ **f** $12 + 6p$
g $4 + 12y$ **h** $10 + 8z$
3 a $n + 5$ **b** $3(n + 5)$
4 a $5p + 9$ **b** $7m + 8$ **c** $2x + 4$
d $10 + 5k$ **e** $9t + 10$ **f** $4r + 7$
5 a $s + 6$ **b** $2(s + 6)$ **c** $2s + 19$
6 a $5n + 12$ **b** $6p + 10$ **c** $10x + 10$ **d** $18n + 7$
7 a $12y$ **b** $y + 2$ **c** $20(y + 2)$ **d** $32y + 40$
8 a $4x^2 + x$ **b** $m^3 + 2m$ **c** $2t^3 + 8t$ **d** $3p^3 + 3p$
9 a $p^2 + 3p + 2$ **b** $15w^2 + 48w + 9$
c $-2m^2 + 5m + 3$ **d** $y^2 + 2y + 1$

A1.4

1 a $9r$ **b** $4m^2$ **c** $4x$ **d** $8tv$
e $10mn$ **f** $6xy^2$ **g** $2x^2 + x$ **h** $9w - 8$
i $z^3 + 3z + 1$
2 a $8y + 12$ **b** $6x - 4$ **c** $6k - 6$ **d** $4 - 4n$
3 a $k^2 + k$ **b** $m^2 + 7m$ **c** $10t + 5$
4 a $2m^2 - 6m$ **b** $8p^2 - 4p$ **c** $r^3 + 3r$ **d** $2s^3 - 8s$
5 a $5r + 4$ **b** $2s$ **c** $4j + 5$ **d** $15t - 9$
6 a $12m^2 - 9m - 3$ **b** $6p^2 - 14p + 4$
c $-10q^2 + 15q + 45$ **d** $8v^2 - 20v + 12$
7 a $n + 4$, $3(n + 4)$, $4n$, $4n - 2$, $2(4n - 2)$
b $5n - 16$

A1.5

1 a 1, 2 **b** 1, 2, 4 **c** 1, 2, 5, 10
d 1, 2, 3, 6 **e** 1, 3 **f** 1, 2
g 1, 2 **h** 1, 2, 4, 8
2 a 3 **b** 2 **c** 2 **d** 4
3 a y **b** s **c** m **d** $2y$
4 a $2(x + 5)$ **b** $3(y + 5)$ **c** $4(2p - 1)$
d $3(2 + m)$ **e** $5(n + 1)$ **f** $6(2 - t)$
g $2(7 + 2k)$ **h** $3(3z - 1)$
5 a $w(w + 1)$ **b** $z(1 - z)$ **c** $y(4 + y)$
d $m(2m - 3)$ **e** $p(4p + 5)$ **f** $k(7 - 2k)$
g $n(3n^2 - 2)$ **h** $r(5 + 3r)$
6 $4(x + 3) = 4x + 12$, $4x^2 - 3x = x(4x - 3)$, $3(x - 4) = 3x - 12$, $4x + 3x^2 = x(4 + 3x)$
7 a $4(y - 3)$ **b** $x(2x + 3)$ **c** $y(3y - 1)$
d $5(3 + t^2)$ **e** $3m(1 + 3m)$ **f** $2r(r - 1)$
g $v(4v^2 + 1)$ **h** $3w(w + 1)$
8 a Kate
b Debbie should have an 8 in front of the x^2-term inside the bracket. If she had included this, she would have been able to take out more factors. Bryn should have $8x$ rather than $8x^2$ as the first term in the bracket. He can take out a further factor of 2.

A1 Exam review

1 a $2y$ **b** $3p^2$ **c** $x(x - 3)$
2 ai $7x - 2$ **aii** $n^2 + 6n + 9$
bi $a(2a + 1)$ **bii** $4xy^2(2x^2 - y)$

A2 Before you start ...

1 a 8 **b** 11 **c** 7 **d** 6
2 a 7 **b** 6 **c** 15 **d** 2
3 16 cm
4 a x **b** m **c** $3n$ **d** $2p$

A2.1

1 a $a = 5$ **b** $b = 18$ **c** $x = 5$ **d** $x = 5$
e $x = 8$ **f** $x = 6$ **g** $x = 5.5$ **h** $x = 50$
i $y = 18$ **j** $z = 40$ **k** $y = 3$ **l** $p = 2$
2 a $m = 7$ **b** $p = -7$ **c** $n = 2$ **d** $q = 5$
e $r = -11$ **f** $k = 6$ **g** $m = 5$ **h** $a = 8$
3 $16p - 15 = 33$ ($p = 3$, $m = n = 8$)
4 a $6n + 6 = 36$, $n = 5$
b $5m - 6 = 54$, $m = 12$
5 a $2y + 16 = 36$ **b** $y = 10$
6 a 5 cm **b** 6 cm **c** 8 cm
In **c** all sides are equal.

A2.2

1 a $4x + 12$ **b** $2y - 8$ **c** $15 - 5a$ **d** $6 - 3b$
2 a $x = 4$ **b** $s = 6$ **c** $t = 3$ **d** $v = 3$
3 a $-2c - 8$ **b** $-3d + 9$ **c** $8m - 2$ **d** $12 - 8n$
4 a $a = -6$ **b** $b = -1$ **c** $c = 2$ **d** $d = -2$
5 a $e = 1.5$ **b** $f = \frac{2}{3}$ **c** $g = 0.75$ **d** $h = -0.5$
6 a $x = -8.5$ **b** $y = 2.5$ **c** $z = 2.5$
x is the odd one out.
7 a $3(x - 2)$ **b** $x = 6$
8 a $4(2y + 5)$ **b** $y = 1$
9 $4(z - 6)$, $z = 8$

A2.3

1 a $m = 5$ **b** $p = 3$ **c** $t = 3$
d $n = 7$ **e** $q = 8$ **f** $s = 8$
2 a $s = -3$ **b** $t = -4$ **c** $u = -2$ **d** $v = -1$
3 a $a = 5$ **b** $b = -4$ **c** $c = 2.5$ **d** $d = -3$
4 a $x = 7.5$ **b** $x = -3$ **c** $x = 0.25$ **d** $x = -1.5$
5 a $21 = 2n + 5$ **b** $4n - 11 = 21$
c $2n + 5 = 4n - 11$ **d** $n = 8$
6 a $n = 10$
b $5n - 8 = 2n + 10$, $n = 6$
c $3n + 4 = 5n + 12$, $n = -4$

A2.4

1 a $r=4$ b $s=-3$ c $t=2$ d $v=-5$
2 a $a=3$ b $b=-2$ c $c=5$ d $d=-4$
3 a $x=6$ b $y=-6$ c $z=5.5$ d $m=2$
4 a $e=3.5$ b $f=0.5$ c $g=-1.5$ d $h=-1$
5 a–d

	$3x-2$	$4x+1$	$8x-3$
$2(x+3)$	$x=8$	$x=2.5$	$x=1.5$
$4(2x-1)$	$x=0.4$	$x=1.25$	No solution
$3(4x+1)$	$x=-\frac{5}{9}$	$x=-0.25$	$x=-1.5$

6 a $3x+9$ b $4x+4$
c $4x+4=3x+9$ d $x=5$
7 a $4(x+2)$ b $8x=4(x+2)$ c $x=2$
8 a $4m$ b $3(m+2)$
c $4m=3(m+2)$, $m=6$ d 8

A2.5

1 a $x=15$ b $m=-8$ c $n=-18$ d $m=20$
2 a $s=9$ b $t=6$ c $u=-10$ d $v=12$
3 a $x=6$ b $y=6$ c $z=15$ d $q=-8$
4 a $x=1$ b $x=11$ c $x=-17$ d $x=6$
5 a $x=12$ b $x=6$ c $x=7$ d $x=5$
6 a $\frac{n}{4}+6$ b $\frac{n}{4}+6=10$ c $n=16$
7 a $\frac{n}{3}-4=7$, $n=33$ b $\frac{n}{2}+8=3$, $n=-10$
8 a $\frac{2x+6}{3}$ b $x=9$
9 $\frac{4+x}{4}=10$, $x=36$

A2.6

1 a

x	x^2	Too big or too small
5.5	30.25	Too big
5.4	29.16	Too big

d $x=5.4$
2 $x=4.1$
3 $x=5.2$
4 a 0, 6, 24 b x is between 2 and 3.
e $x=2.2$
5 $x=4.1$

A2 Exam review

1 a $x=20$ bi $x+50$ pence bii 95 pence
2 a $x=8$ b $y=6.5$ c $p=\frac{5}{8}$

A3 Before you start ...

1 a 3 b 6 c 5 d 6
2 a 3 b 4 c 5 d 4
3 a 4, 8, 12, 16, 20, 24
b 3, 6, 9, 12, 15, 18
c 5, 10, 15, 20, 25, 30
d 6, 12, 18, 24, 30, 36
4 a 14 b 9 c 3 d 7
5 a 4 b 16 c 25 d 9
e 36 f 81

A3.1

1 6, 10, 14, 18, 22; 26, 21, 16, 11, 6; 10, 7, 4, 1, −2; −6, −4, −2, 0, 2; −23, −16, −9, −2, 5
2 a 2, 4, 6, 8, 10 b 17, 19, 21, 23, 25
c 4, 8, 12, 16, 20 d 24, 30, 36, 42, 48
e 7, 12, 17, 22, 27 f 1, 4, 9, 16, 25
g 2, 5, 10, 17, 26 h 2, 4, 8, 16, 32
3 a 4, 7, 10, 13, 16 b 25, 19, 13, 7, 1
c 4, 8, 12, 16, 20 d 30, 22, 14, 6, −2
e −16, −10, −4, 2, 8
4 ai 31, 38
aii First term 3, increase by 7 each time
bi 25, 33
bii First term −7, increase by 8 each time
ci −7, −4
cii First term −19, increase by 3 each time
di 11, 27
dii First term 7, increase by 4 each time
ei 11, 19
eii First term 3, increase by 8 each time
fi 1, −4
fii First term 16, decrease by 5 each time
gi −11, 1
gii First term −11, increase by 3 each time
hi 13, −1
hii First term 13, decrease by 7 each time
5 ai 2, 3, 5, 8, 12
aii First term 2, increase by 1, 2, 3, 4
bi 18, 22, 25, 27, 28
bii First term 18, increase by 4, 3, 2, 1
ci −6, −5, −3, 0, 4
cii First term −6, increase by 1, 2, 3, 4
di 6, 8, 11, 15, 20
dii First term 6, increase by 2, 3, 4, 5
ei −24, −23, −21, −17, −9
eii First term −24, increase by 1, 2, 4, 8 (powers of 2)
fi −15, −7, −1, 3, 5
fii First term −15, increase by 8, 6, 4, 2
6 a 1, 3, 5, 7, 9, 11, 13 and 4, 7, 10 (13 could go in the second sequence instead)
b 2, 4, 8, 16, 32 and 7, 11, 15, 19, 23

A3.2

1 a 6, 11, 16; 51 b 11, 14, 17; 38
c 4, 12, 20; 76 d −2, 4, 10; 52
e 22, 20, 18; 4 f 10, 5, 0; −35
g −13, −6, 1; 50 h −2, 2, 6; 34
2 a 14, 17, 23, 38 b −3, 3, 15, 45
3 a 5, 8, 13, 20, 29 b −1, 2, 7, 14, 23
c 2, 8, 18, 32, 50 d 11, 8, 3, −4, −13
4 a 12, 33, 108 b −2, 19, 94
c 15, 57, 207
5 a 5, 9, 13
b Each term is one more than a multiple of 4.
c No, 222 is not one more than a multiple of 4.

6 a 3, 9, 19 b $2 \times 5^2 + 1 = 51$ not 101

7 a 1, 4, 9, 16, 25 b +3, +5, +7, +9

c

4th square number	16	1+3+5+7
5th square number	25	1+3+5+7+9
6th square number	36	1+3+5+7+9+11

d 10th square number = 100

A3.3

1 a 8, 6, 4, 2, 0 b 1, 0, −1, −2, −3

c 14, 10, 6, 2, −2 d 13, 6, −1, −8, −15

e 13, 10, 7, 4, 1 f 1, −4, −9, −14, −19

g −7, −9, −11, −13, −15

h 25, 20, 15, 10, 5 i 5, 2, −1, −4, −7

j 4, −6, −16, −26, −36 k 5, 2, −3, −10, −19

l −8, −2, 8, 22, 40

2 a +4

b The nth term contains the term $4n$.

c

Sequence	5	9	13	17	21
$4n$	4	8	12	16	20

d $4n + 1$

3 a $6n + 5$ b $9n - 8$ c $7n + 8$ d $4n - 14$

e $23 - 3n$ f $19 - 4n$ g $24 - 8n$ h $39 - 8n$

4 a $4n + 3$ b $4n - 10$ c $41 - 9n$ d $21 - 6n$

5 a Term 3 = 13, Term $n = 3n + 4$

b Term 4 = 14, Term 10 = 50, Term $n = 6n - 10$

A3.4

1 a

b 8, 11, 14, 17, 20 c 35

2 a

× × × × × × × × × ×
×   ×
×   ×
× × × × × × × × × ×

b 10, 14, 18, 22, 26

c Four crosses are added each time – two to the top and two to the bottom

d +4 e $4n + 6$

3 a 13, 17 (add 4) b $4n - 3$ c 197

4 a 18, 22 (add 4); $4n + 2$; 202

b 9, 11 (add 2); $2n + 1$; 101

A3.5

1 a 4, 7, 10, 13, 16

b $3n + 1$

c The nth pattern has three branches of n dots and one extra dot in the centre.

2 a 6, 11, 16, 21, 26 b $5n + 1$

c The nth pattern has five branches of n beads and one extra bead in the centre.

3 a 3 b 5, 8, 11, 14, 17 c $3n + 2$

d Start with two cards leaning against each other, then add three more cards each time (another pair of leaning cards and one bridging card).

e No, 52 is not two more than a multiple of 3.

4 a Add one vertical and two horizontal posts each time.

b 16

c $3n + 1$

d Start with one vertical post then add three posts each time.

e 79

5 a

Type of car	Number of days						
	1	2	3	4	5	6	7
Small	70	90	110	130	150	170	190
Medium	85	110	135	160	185	210	235
Large	100	130	160	190	220	250	280

bi $10 \times 30 + 70 = £370$

bii $14 \times 25 + 60 = £410$

biii $20n + 50$

A3 Exam review

1 a $3n - 1$ b $3n - 1$ can never be divisible by 3, 99 is

2 $m = 6n$

A4 Before you start ...

1 a 6 b 11 c 9 d 6

2 a $3x + 3$ b $2x - 2$ c $8x + 12$ d $12x - 6$

3 a $x = 4$ b $y = 6$

4 a $4(x + 2)$ b $2(3x + 1)$ c $3(y - 3)$

5 2, 3, 5, 7, 11, 13, 17, 19

6 a < b > c > d <

A4.1

1 a 24 b 6 c 18 d 8

e 27 f 5 g −2 h 16

i 104 j 4

2 a −12 b 15 c 14 d 9

e 4 f −3.5

3 ai 19 aii −5 bi 2 bii −10

ci 12 cii 9 di 23 dii −13

ei 16 eii 4 fi 36 fii 12

4 a $x = 6$ b $y = 11$ c $x = -2$

d $y = 1$ e $b = 3$ f $c = -4$

g $f = 2$ h $g = -4$

5 b $3x - 6$ c $6 + 2x$ d $10 - 5x$

6 a $4(2m + 1)$ b $3(4n - 3)$ c $5(3p + 11)$

d $q(q + 2)$ e $4(4r - 7)$ f $2q(2p - 5)$

7 a Identity b Function c Formula

d Equation e Expression f Formula

g Formula h Function

A4.2

1 a 60 cm^2 b 100 cm^2 c 5 m^2 d 11.5 m^2

2 ai £4 aii £7.25 aiii £3.10 b 10p

3 a £115 b £290

4 a 13 b 48

5 144

6 a 47.5 b 84

7 a 2 b 2

8 4

9 a 3 b 2

10 a 2 b 2

11 Paul; when $x = 6$, $2x^2 = 2 \times 36 = 72$

A4.3

1 a $C = 35 + 20h$ b £95

2 a $C = 2 + 0.6m$ bi £5 bii £11

3 a $C = 75 + 35d$ b €320 c 12 days

4 a $P = 2b + 2$ b 12 pencils c 18 pencils d 23

5 14 children

6 5 days

7 a $C = 3n + 1.5m$ b $m = 4 \times n$ c $C = 9n$ d 6 adults

A4.4

1 a 8 b −2

2 a 4 b 14

3 a 11 b 6.5

4 a 60 b 144

5 a $\frac{1}{2}$ b 4

6 a $x = \frac{y - c}{m}$ b $t = \frac{v - u}{a}$ c $x = 2(y - d)$
d $y = \frac{4 - x}{3}$

7 a $t = \frac{s + 6}{3}$ b $t = \frac{2x - 9}{5}$ c $t = \frac{3x - 12}{2}$

8 a $y = \frac{2 - 4x}{6}$ b $y = \frac{2x + 6}{3}$ c $y = \frac{3x - z}{5}$

9 a $n = \frac{p - 1}{4}$
bi 9 bii 13 biii 57

10 a 200 b 150 c 570

A4.5

1 $A = \text{length} \times \text{width} = m(m + 2) = m^2 + 2m$

2 a $x(x + 2)$ b $x(x - 2)$
c $A = x(x + 2) - x(x - 2) = x^2 + 2x - x^2 + 2x = 4x$

3 a odd, even; even + even = even; odd + odd = even; odd + even = odd
b For any two consecutive numbers, one is odd and one is even. odd + even = odd.
c odd + even + odd + even = (odd + even) + (odd + even) = odd + odd = even

4 a odd
b $2n(2n + 1)$
c $4n^2 + 2n$
d $4n^2 \div 2 = 2n^2$, multiple of 2
e $4n^2$ and $2n$ are even, so $4n^2 + 2n$ is even + even = even.

5 a e.g. if $x = 2$, $3x = 6$ which is even
b e.g. if $x = 2$, $x^2 = 4$ which is even
c e.g. if $x = 2$ and $y = 3$, $x^2 + y^2 = 13$ which is odd

A4.6

1 a 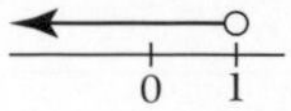 0 1 b 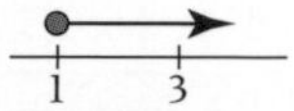 1 3
c 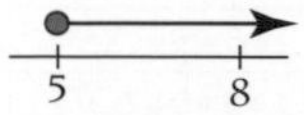 5 8 d 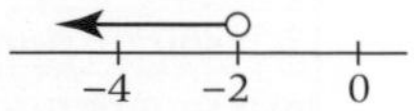 −4 −2 0
e 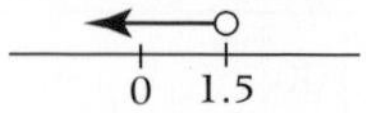 0 1.5 f 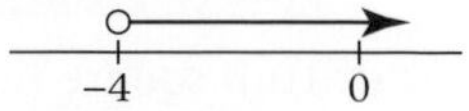 −4 0
g 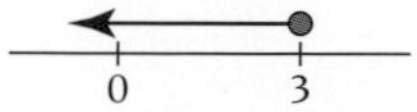 0 3 h −3 −1.5 0

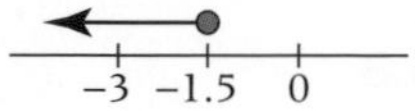

2 ai $2x > 10$ aii $4x > 20$
bi $3y \leqslant 18$ bii $5y \leqslant 30$
c $5x \geqslant -20$ d $6m < -18$

3 a $x \leqslant 2$ b $x < 5$ c $x > -2$ d $x \geqslant -4$

4 a $3x > 9$, $x > 3$ b $7x > 35$, $x > 5$
c $2x \leqslant 2$, $x \leqslant 1$ d $5x \geqslant 15$, $x \geqslant 3$

5 a $x \leqslant 5$ b $x \geqslant 6$ c $x \geqslant 2$
d $x < 4$ e $x \geqslant -1$ f $x \leqslant 3$
g $x > 3$ h $x \leqslant 2.5$

6 a $x \geqslant 3$ b $x > 2$ c $x \geqslant 0$
d $x > -1$ e $x \geqslant -2$ f $x \leqslant 4$
g $x \geqslant -2.5$ h $x < -3.75$

7 a ii b v c vi
d i e iv f iii

8 $x \geqslant 3$

A4.7

1 a $x < 5$ b $y \geqslant 6$ c $y \leqslant 4$ d $r > 9$
e $w \leqslant 15$ f $s > 2$ g $u < -12$ h $v \geqslant -4$

2 a $x > 1$, $x < 5$ b $x > -5$, $x < -1$
c $x > -2$, $x < 4$ d $x \geqslant -6$, $x \leqslant -1$
e $x > 2$, $x < 7$ f $x \geqslant -1$, $x \leqslant 2$

3 a $-10 \leqslant x < 5$ b $5 < y < 12$
c $-2 < z \leqslant 6$ d $-4 \leqslant t \leqslant 2$

4 a 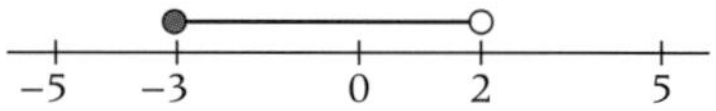 −5 −3 0 2 5
b 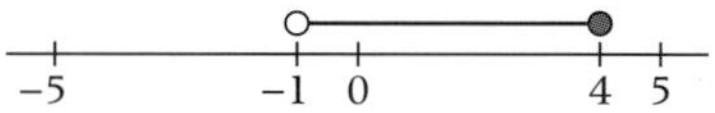 −5 −1 0 4 5
c 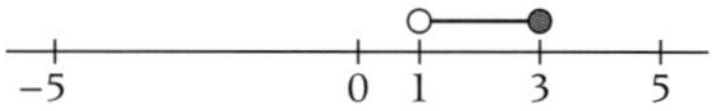 −5 0 1 3 5
d −5 −4 0 5

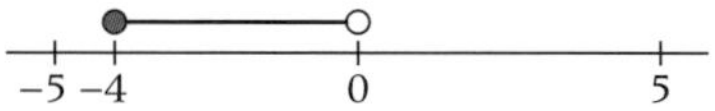

5 a −3, −2, −1, 0, 1 b 0, 1, 2, 3, 4
c 2, 3 d −4, −3, −2, −1

6 **a**, **b**, **c** and **e**

7 a 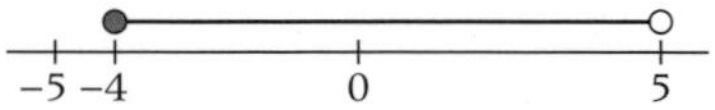 −5 −4 0 5
b −4, −3, −2, −1, 0, 1, 2, 3, 4

8 1, 2, 3, 4

9 a −2, −1, 0, 1, 2
b −1, 0, 1, 2, 3, 4
c 2, 3, 4, 5

A4 Exam review

1 19

2 a Bryani; $4x^2 = 4 \times 3^2 = 4 \times 9 = 36$ b 64

A5 Before you start ...

1

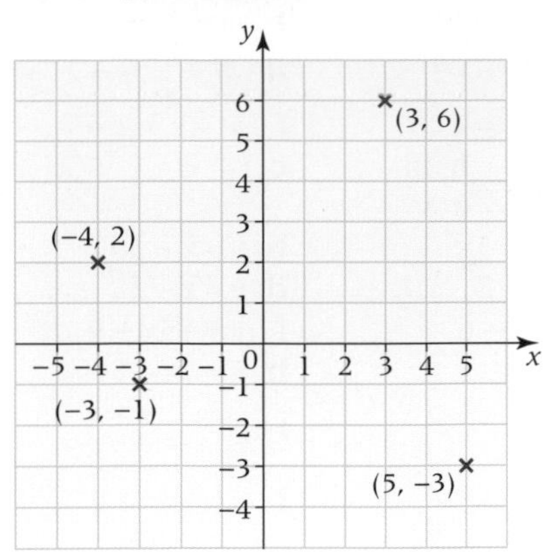

2 $y = 1, 3, 5, 7, 9$

3 a, b, e, f

4 c

5 **a** $y = 2x - 5$

b $y = 4 - 2x$

A5.1

1 **a** $y = -1, 3, 5, 7, 11$

b

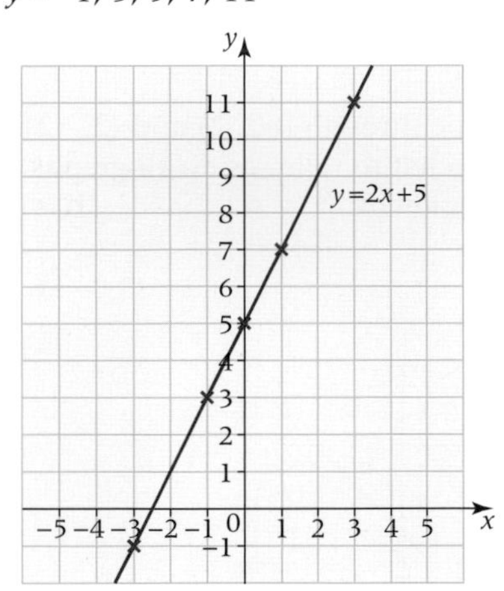

2 a–d

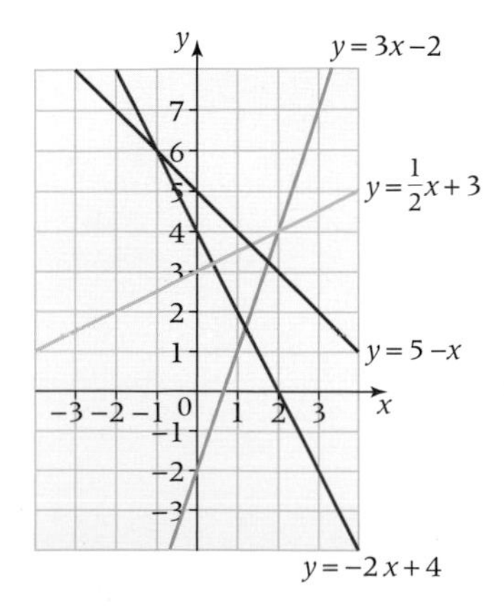

3 a–d

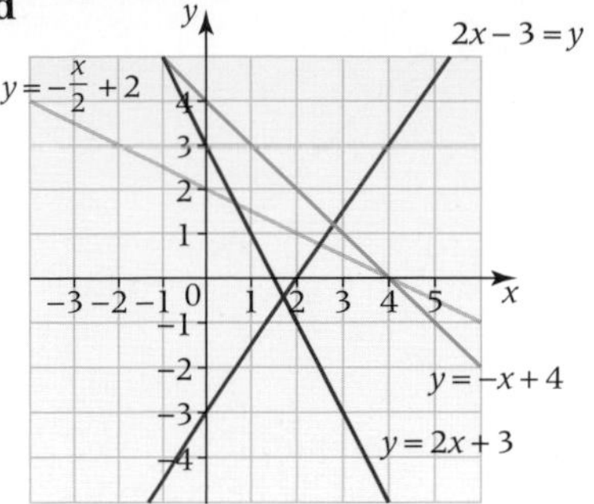

4 a–d

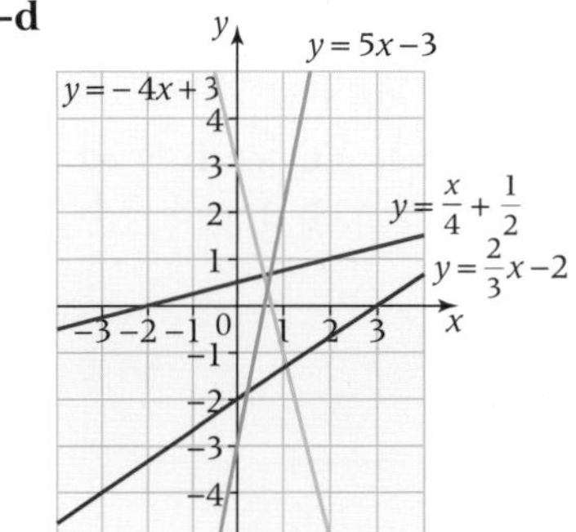

5 $4x + 2y = 1$ is the odd one out, as it gives $y = \frac{1}{2} - 2x$ not $y = 2x + \frac{1}{2}$

6 **a**

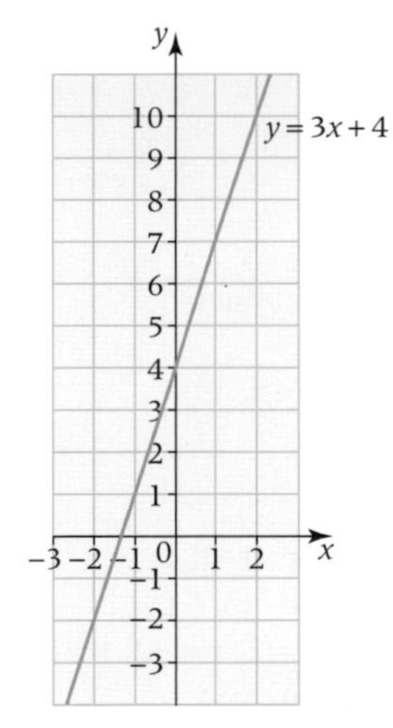

bi 5.5 **bii** −1.5

7 **a–d**

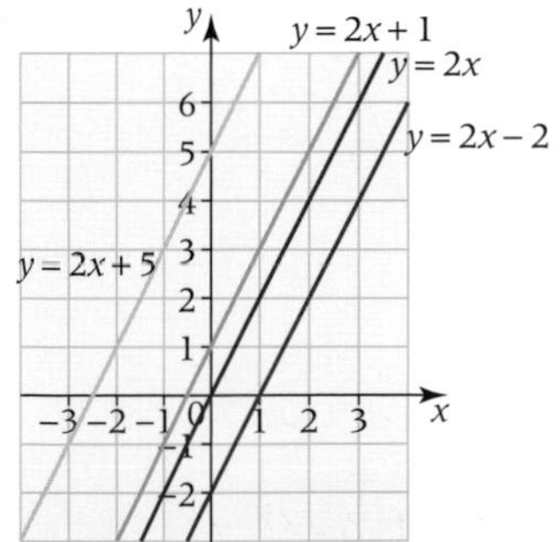

The lines are all parallel.

8 a–c

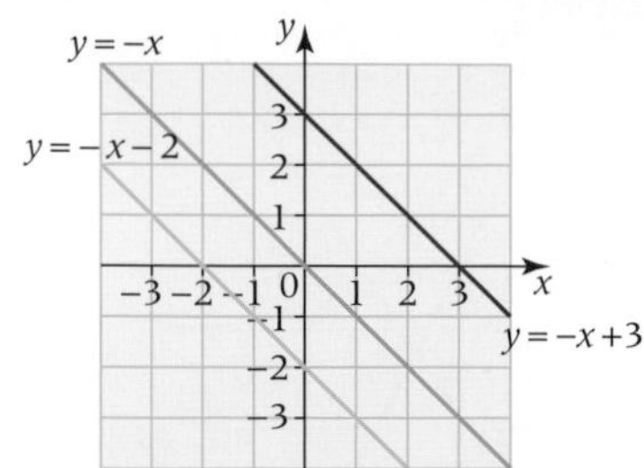

The lines are all parallel.
$y = -x + 1$ would lie between $y = -x$ and $y = -x + 3$, with the same slope and passing through (0, 1).

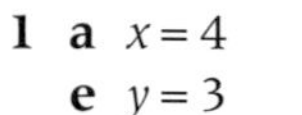

A5.2

1 a $x = 4$ **b** $x = -2$ **c** $x = 1$ **d** $y = 5$
e $y = 3$ **f** $y = -1$ **g** $y = -3$ **h** $y = -5$

2 a–e

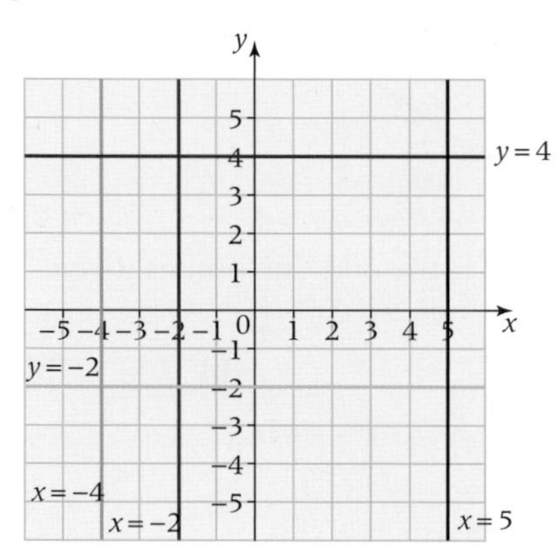

3

Horizontal lines	Vertical lines
$y = 4$	$x = 5$
$y = -10$	$x = 15$
$y = -2$	$x = -3$
$y = -6$	$x = 4$

4 a (−2, 4) **b** (5, −2) **c** (−4, 4)

5 a (2, −3) **b** (1, 6) **c** (3, −1)

6 For example, the lines $x = 2$, $x = -2$, $y = 2$ and $y = -2$ make a square.

7 b x-axis **c** y-axis

8 a–c

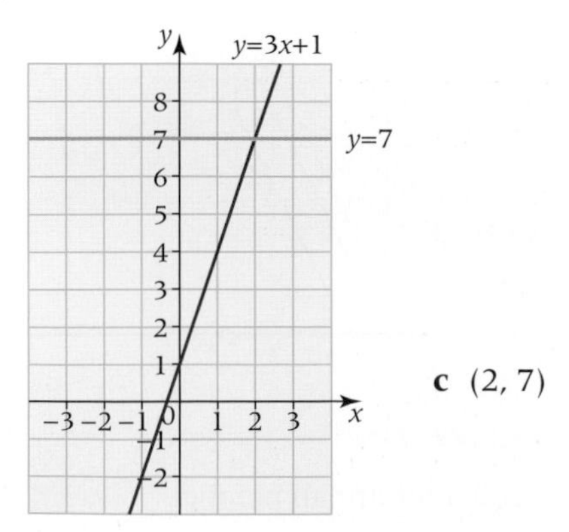

c (2, 7)

A5.3

1 a $m = 3, c = -1$ **b** $m = 2, c = 5$
c $m = 4, c = -3$ **d** $m = \frac{1}{2}, c = 2$
e $m = 5, c = 1$ **f** $m = -3, c = 7$

2 a ii **b** iii **c** i
d iv **e** v

3 a $y = -x + 5$ **b** $y = x + 3$
c $y = -x - 2$ **d** $y = x - 3$
e $y = -2x + 6$ **f** $y = -5x + 9$
g $y = -3x - 2$ **h** $y = 2x + 5$
i $y = -\frac{1}{2}x + 2$ **j** $y = \frac{1}{2}x + 4$
k $y = -\frac{1}{2}x + 2$ **l** $y = -3x + 4$

4 a $y = x + 3$ and $y = x - 3$ (**b** and **d**)
b $y = -x + 5$ and $y = -x - 2$ (**a** and **c**)
c $y = -3x - 2$ and $y = -3x + 4$ (**g** and **l**)
d $y = -\frac{1}{2}x + 2$ and $y = -\frac{1}{2}x + 4$ (**i** and **k**)

5 a, **b** and **e**

6 $y = \frac{1}{2}x - 9$, $y = x + 11$, $y = 2x - 2$, $y = 3x + 1$, $y = 4x + 3$

7 $y = 3x + c$ for any c

8 $y = -4x + c$ for any c

9 $y = \frac{1}{2}x + 4$

10 Straight line through (−2, 2) and (2, −2); slopes down from left to right, as do all graphs with negative gradient

A5.4

1 a

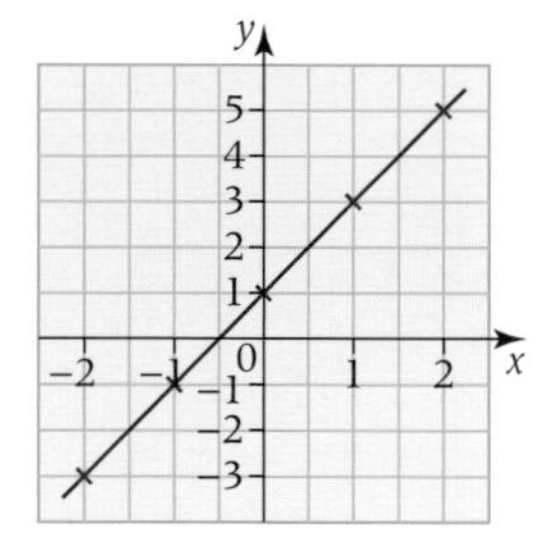

b $y = 2x + 1$

2 a $y = x + 1$
b $y = 2x - 2$
c $y = -2x + 2$
d $y = -x + 3$

3 A: $x = 1$
B: $y = x$
C: $y -2$
D: $y = -\frac{1}{2}x - 1$
E: $y = 2x + 4$

4 a

x	0	1	2	3	4
y	30	70	110	150	190

b

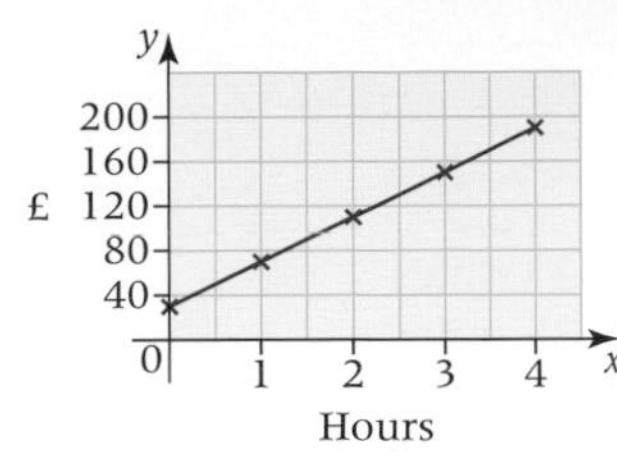

c $y = 40x + 30$
d £130

A5.5

1 **a** $y = 2$ **b** $y = 8$ **c** $x = 2$ **d** $x = -3$

2 **a**

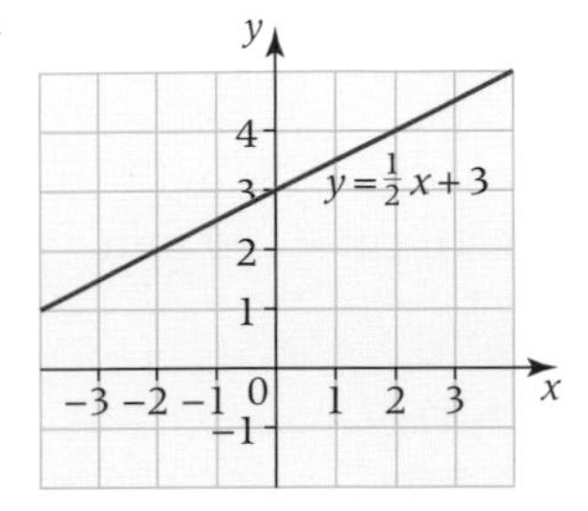

b $x = 8$
c $(2, 4)$ and $(-3, 3.5)$

3 **a** $y = 4x - 7$ **b** $x = -2$

5 **ai** $x = 1.5$ **aii** $x = 5$ **aiii** $x = -3$
b $3x + 2 = 11$, $x = 3$

6 **a** $x = 2$
bi $x = 4$ **bii** $x = -4.5$ **biii** $x = 1.5$

A5.6

1 **a** $(2, 3)$ **b** $(-1, -4)$ **c** $(3, 7)$ **d** $(-2, -4)$
e $(7, -2)$ **f** $(4, 1)$

2 **a** Yes **b** Yes **c** No **d** Yes

3 **b** $y = 2x + 1$ and $y = 4x + 2$ cross at $\left(-\frac{1}{2}, 0\right)$, $y = 3$ and $y = x + 1$ cross at $(2, 3)$, $y = x$ and $y = -x$ cross at $(0, 0)$

4 **b** $(1, 1)$
c No, because the lines do not cross at $(1, 2)$.

5 **b** $(3, 2)$

6 **b** $x = 4$, $y = 2$

7 **a** $5x + 2y = 9$
b–c

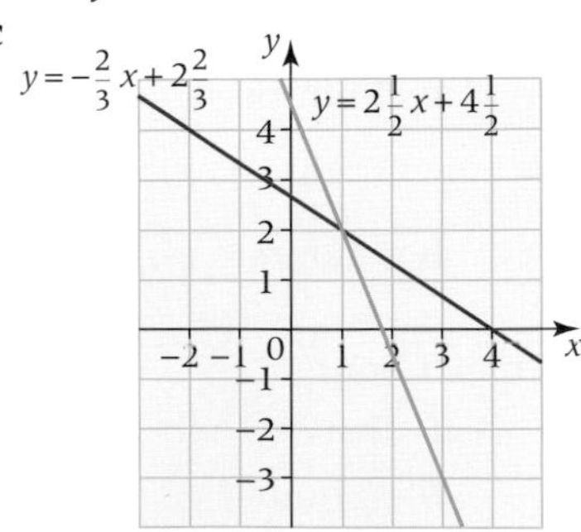

d $(1, 2)$
e £1
f £2

A5.7

1 **a** $y = 9, 1, 0, 1, 9$
b–d

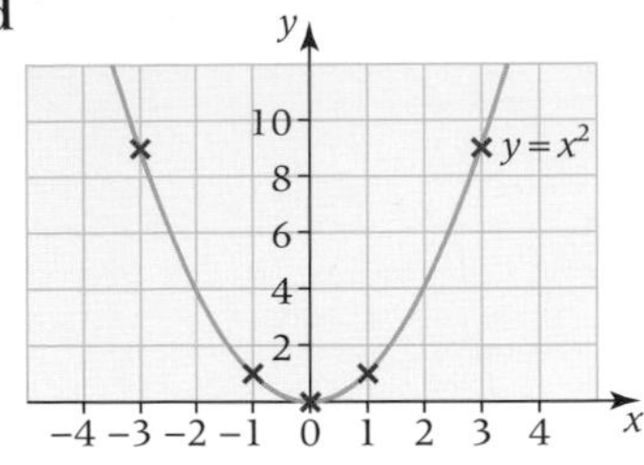

e $y = 0$

2

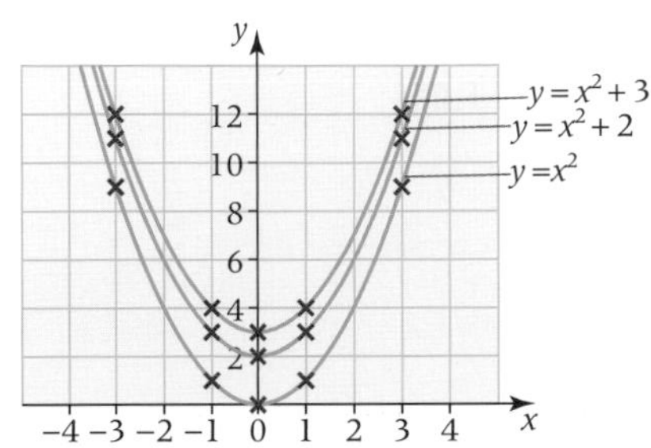

$y = x^2 + 2$ is the same shape as $y = x^2$ but shifted up 2; $y = x^2 + 3$ is the same shape as $y = x^2$ but shifted up 3.

3

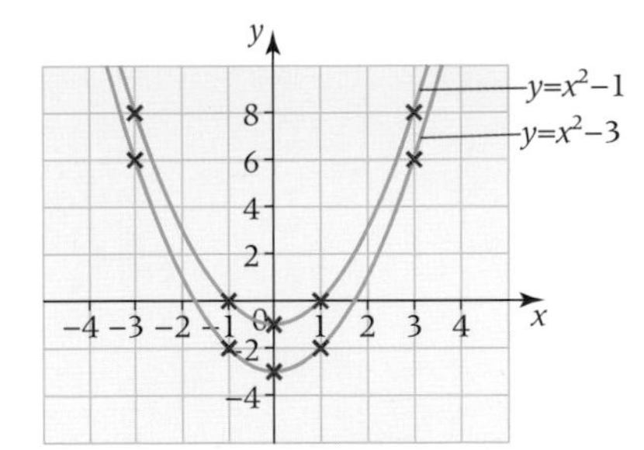

$y = x^2 - 1$ is the same shape as $y = x^2$ but shifted down 1; $y = x^2 - 3$ is the same shape as $y = x^2$ but shifted down 3.

4 **a** iv **b** i **c** ii **d** iii

5 **a**

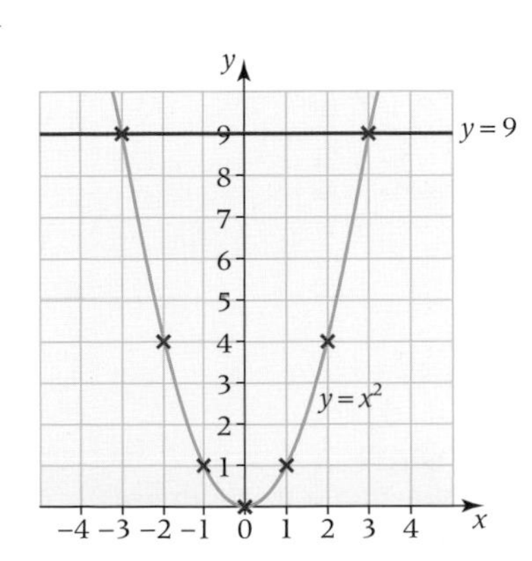

b $(3, 9)$ and $(-3, 9)$ **c** $x = -3, 3$

6 a

x	−2	−1	0	1	2	3	4
y	9	4	1	0	1	4	9

b,c

$y = x^2 - 2x + 1$

d $x \approx 3.25$ or -1.25

A5 Exam review

1 a

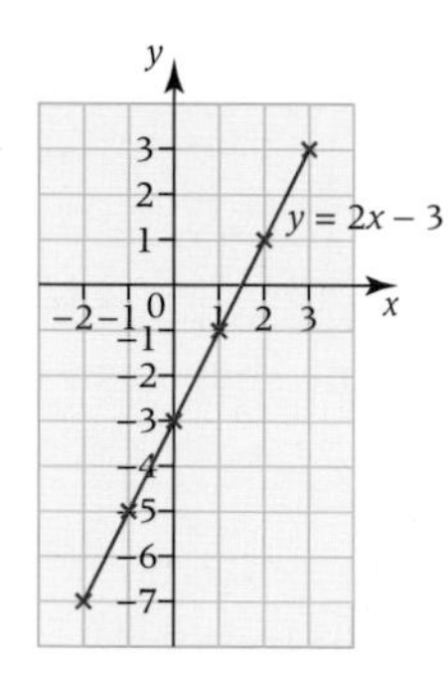

b (2.5, 2)

2 a 5, −1

b

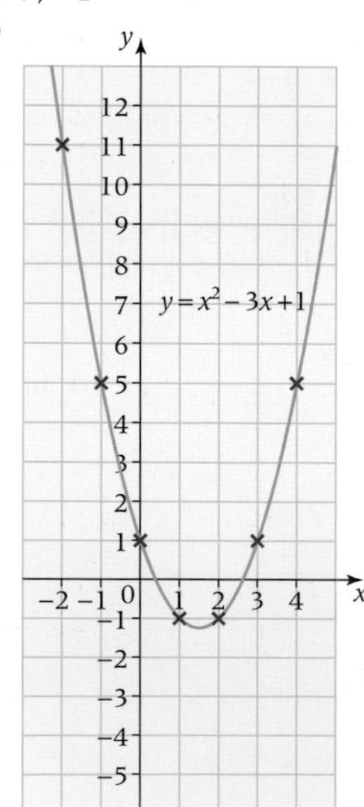

c $-1\frac{1}{4}$

A6 Before you start ...

1 10 miles = 16 km, 20 miles = 32 km
2 10 m = 1000 cm, 20 m = 2000 cm
3 7 km
4 3.8 litres
5 4.5 kg

A6.1

1 a 4.8 km b 6.25 miles
c 7.2 km d 1.25 miles
1 mile is longer than 1 km.

2 a 2.7 kg b 4.5 kg c 8.8 lb
d 4 lb e 3.2 kg f 1 lb
g 5.5 kg h 11 lb
1 kg is heavier than 1 lb, as 1 kg ≈ 2.2 lb

3 ai 68 °F aii 14 °C
aiii 86 °F aiv −7 °C
b 32 °F
c 28 °F, 36 °F, 43 °F, 54 °F, 75 °F

A6.2

1 a e.g. 0 cm = 0 mm, 10 cm = 100 mm
b 100 mm

2 a 0 lb, 22 lb, 11 lb
ei 4.5 kg eii 2.3 kg
eiii 6.6 lb eiv 5.5 lb

3 a e.g. £0 = $0, £10 = $17, £20 = $34
b $51
di $8.50 dii £17.60

4 b The NZ cap is cheapest.

A6.3

1 a 1 pm b 60 miles
c $1\frac{1}{2}$ hours d 25 miles
e

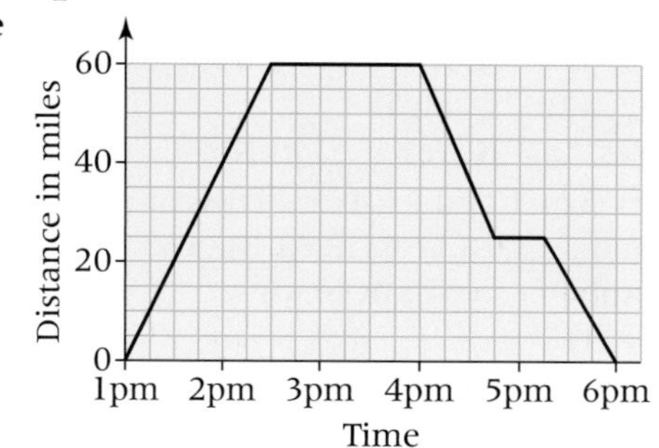

2 a F b C c D d A
e G f B g E

3 a 8 hours b 260 miles c 2 hours
d $\frac{1}{2}$ hour e 2 hours

A6.4

1 a 80 km b 5 hours
c 16 km/h
d $\frac{50}{2} = 25$ km per hour
e 15 km/h f first g faster

2 a 5 miles b $\frac{1}{2}$ hour
c 2:30 pm to 3 pm, as the graph is steeper here.
d 5 mph e 10 miles f $\frac{1}{2}$ hour
g 20 mph h 15 miles

i

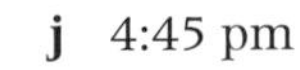

j 4:45 pm

3 **a** 2 km **b** 4 km/h

c

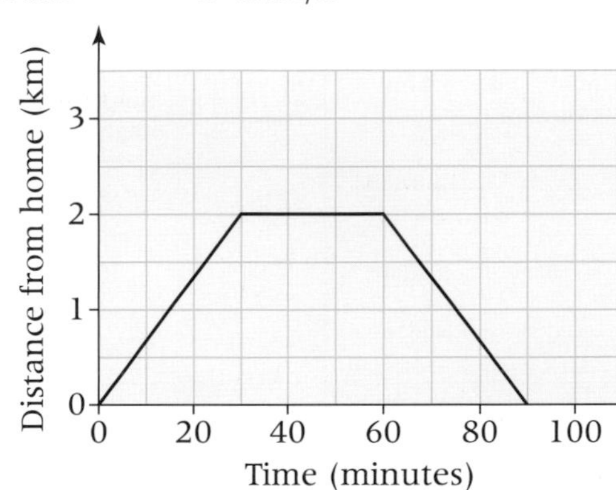

A6.5

1 **a** 200 **b** 225 **c** 25

d

e Sales rose sharply – due to Christmas season.

f Increase in sales

2 **a** 0.7 m **b** 0.5 m **c** 1.9 m **d** 0.1 m

e No, as growth has slowed down.

3 **a** iii **b** ii **c** i

A6 Exam review

1 **a** 40 km/h

b

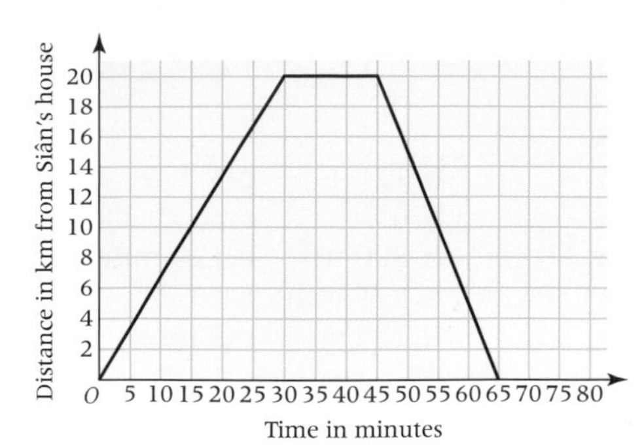

2 **a** 09:05 **b** 7 km **c** 10 minutes

d 21 km/h

S1 Before you start ...

1 **a** 600 **b** 71 000 **c** 48 **d** 2630

e 4500 **f** 6 **g** 75 **h** 6.5

i 3.2 **j** 0.24

2 **a** 27 **b** $22\frac{1}{2}$ **c** $10\frac{1}{2}$ **d** 40

e $9\frac{3}{5}$

3 **a** 9 **b** 15.5 **c** 7.8

4 **a** 48 mm **b** 4.8 cm

5 Perimeter = 26 cm, Area = 40 cm^2

6 5 cubes

S1.1

1 **a** cm or m **b** ml or cl **c** kg **d** cm

e kg **f** km **g** ml or cl **h** litre

i tonne **j** g

2 **a** 2 cm **b** 4 m **c** 4.5 m **d** 4 km

e 5 mm **f** 4500 g **g** 6 kg **h** 6.5 kg

i 2.5 t **j** 3000 ml

3 **a** 10 miles **b** 25 miles **c** 55 miles

d 52.5 miles

4 **a** 2.5 cm **b** 12.5 cm **c** 15 cm

d 30 cm **e** 90 cm

5 **a** 4.4 lb **b** 88 lb **c** 110 lb

d 1.1 lb **e** 5.5 lb

6 **a** 180 g butter, 360 g caster sugar

b 240 g rice, 120 g raisins, 90 g sugar, 120 g currants

c 180 g self-raising flour, 60 g corn flour, 60 g cornflakes, 30 g drinking chocolate, 180 g margarine, 90 g sugar

7 112 km/h

S1.2

1 **a** 12 m, 8 m^2 **b** 19 cm, 12 cm^2

c 39 mm, 81 mm^2 **d** 26.8 cm, 43.2 cm^2

e 30.4 m, 38.4 m^2

2 **a** 12 cm^2 **b** 30 m^2 **c** 14 cm^2

d 72 mm^2 **e** 13.5 cm^2

3 a 5 cm b 9 cm c 12 m

4 a 32 cm, 44 cm^2
  b 42 cm, 74 cm^2
  c 56 cm, 188 cm^2

S1.3

1 a 6 b 12 c 3 d 6

2 a 6 b 6 c 8 d 12

3 a 80 cm^2 b 800 m^2 c 120 mm^2 d 384 cm^2

4 a 50 cm^2 b 375 mm^2 c 28 m^2 d 160 cm^2

5 a 6 m b 14 cm c 8 mm d 8 cm

6 a $\frac{1}{2}\times 10\times(10+20)=150\,cm^2$
  b $25 + 100 + 25 = 150\ cm^2$

S1.4

1 a 31.4 cm b 25.12 m
  c 37.68 cm d 62.8 m
  e 12.56 m f 50.24 cm
  g 9.42 m h 21.98 cm

2 a 6 cm b 5 m
  c 9 cm d 15 m
  e 100 cm

3 a 153.86 cm^2 b 78.5 m^2
  c 50.24 cm^2 d 28.26 m^2
  e 314 m^2 f 200.96 cm^2
  g 113.04 mm^2 h 254.34 cm^2

4 a 3 m b 9.4 m
  c 7.1 m^2

S1 Exam review

1 3.6 cm

2 116.8 cm^2

S2 Before you start ...

1 a 73 b 142 c 177
  d 143 e 163

2 a 90 b 120 c 90

3 a 40° b 120°

S2.1

1 a 80° b 60° c 70°
  d 155° e 27° f 120°

2 a 54° b 60° c 118°
  d $d = 28°$, $e = 104°$ e $f = 37°$, $g = 71°$

3 a 23°, isosceles b 60°, equilateral
  c 90°, right-angled

4 a 71° b 66° c 60°

S2.2

1 a 50° b 50° c $c = 70°$, $d = 110°$
  d 144° e 128°

2 a 90°, rectangle b 115°, kite
  c 106°, parallelogram d 108°, rhombus
  e 67°, isosceles trapezium

3 a 70° b 50° c 100°

S2.3

1 a 60° b 60°, 120°, 60°, 120° c 360°

2

Number of sides	Number of triangles	Sum of the interior angles
4	2	360°
5	3	540°
6	4	720°
7	5	900°
8	6	1080°
9	7	1260°
10	8	1440°

3 a 1080° b 135°
  c

Number of sides	Name	Number of triangles	Sum of the interior angles	One interior angle
3	Equilateral triangle	1	180°	60°
4	Square	2	360°	90°
5	Regular pentagon	3	540°	108°
6	Regular hexagon	4	720°	120°
7	Regular heptagon	5	900°	128.6°
8	Regular octagon	6	1080°	135°
9	Regular nonagon	7	1260°	140°
10	Regular decagon	8	1440°	144°

S2.4

1 a 360 b 45°
  c

Number of sides	Name	Sum of exterior angles	One exterior angle
3	Equilateral triangle	360°	120°
4	Square	360°	90°
5	Regular pentagon	360°	72°
6	Regular hexagon	360°	60°
7	Regular heptagon	360°	51.4°
8	Regular octagon	360°	45°
9	Regular nonagon	360°	40°
10	Regular decagon	360°	36°

2 a 18° b 360° c 20 d 20 sides

3 a 24° b 156°

4 a 146° b 115°

5 a 45° b 135° c Octagon

S2.5

1 a 47° (vertically opposite angles)
  b 117° (vertically opposite angles)
  c $c = 35°$ (vertically opposite angles), $d = 145°$ (angles on a straight line add to 180°)

d $d = 103°$ (vertically opposite angles), $e = 77°$, (angles on a straight line add to 180°), $f = 77°$ (vertically opposite angles)

e 60° (angles at a point add to 360°)

2 a 110° (corresponding angles)

b 47° (alternate angles)

c 115° (alternate angles)

d 63° (corresponding angles)

e 130° (corresponding angles)

f $f = 68°$ (corresponding angles), $g = 112°$ (angles on a straight line add to 180°)

g $h = 50°$ (angles on a straight line add to 180°), $i = 50°$ (corresponding angles)

h $j = 63°$ (alternate angles), $k = 63°$ (vertically opposite angles)

i $l = 118°$ (alternate angles), $m = 118°$ (vertically opposite angles)

3 a $a = 36°$ (alternate angles), $b = 63°$ (alternate angles), $c = 81°$ (angles on a straight line/in a triangle add to 180°)

b $a = 61°$ (corresponding angles), $b = 49°$ (corresponding angles), $c = 70°$ (angles in a triangle add to 180°)

c $a = 113°$ (alternate angles), $b = 67°$ (angles on a straight line add to 180°), $c = 113°$ (corresponding angles), $d = 67°$ (angles on a straight line add to 180°), $e = 113°$ (corresponding/alternate angles)

S2.6

1 a 47° (alternate angles)

b 63° (corresponding angles)

c $c = 56°$ (alternate angles), $d = 43°$ (corresponding angles)

d $e = 52°$ (alternate angles), $f = 48°$ (corresponding angles), $g = 80°$ (angles in a triangle/on a straight line add to 180°)

e 125°

2 a $a = b = c = 61°$

b $d = e = f = 70°$, $g = 40°$

c $h = 36°$, $i = j = 108°$, $k = 36°$

d $l = m = n = o = p = 60°$

e $q = r = s = t = 25°$

3 a 108° **b** 72° **c** 112° **d** 68° **e** 40°

4 a $a = 55°$ **b** $b = 42°$ $c = 105°$ $d = 35°$ $e = 35°$

S2 Exam review

1 36°

2 a 54°

b i 72°

b ii The triangle is isosceles, so the other angle at Q is 54°. Angles in a triangle add to 180°.

S3 Before you start ...

1 a (1, 3) **b** (1, −2) **c** (−3, −2) **d** (−1, 3)

2 a Clockwise

b Anticlockwise

3 a $x = 2$ **b** $y = 2$

c $y = -x$ **d** $y = x$

4 180°

S3.1

1 a

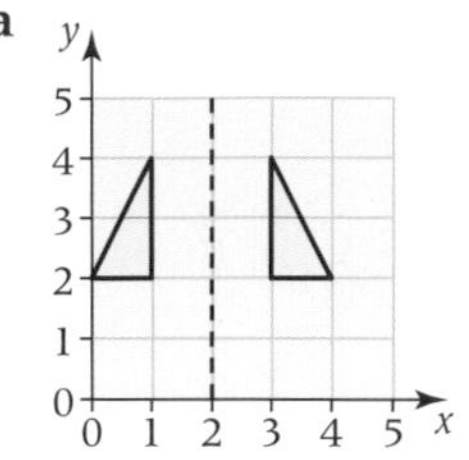

b

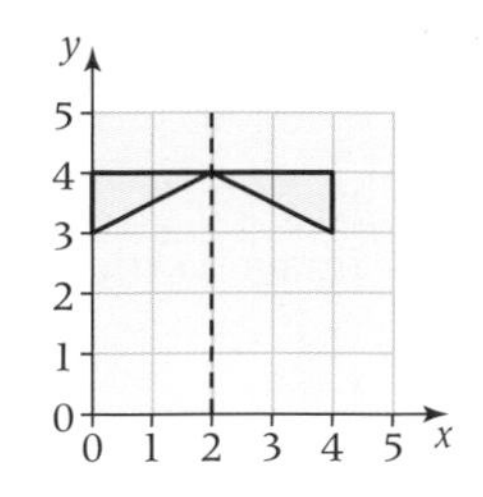

c

2 a $x = 3$ **b** $y = 1$ **c** $x = -1$

3 a–b

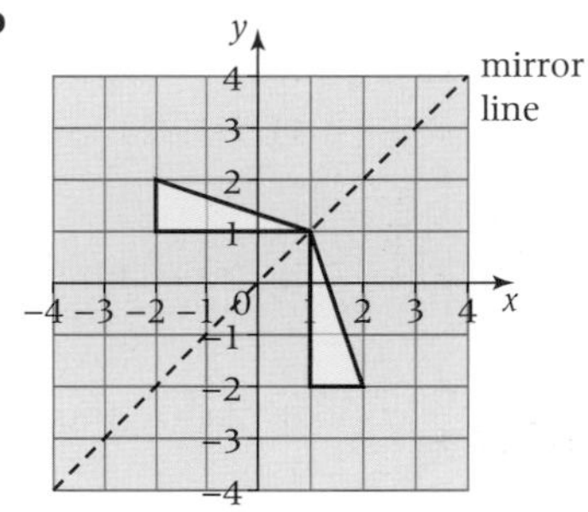

3 c $y = x$

4 a–c

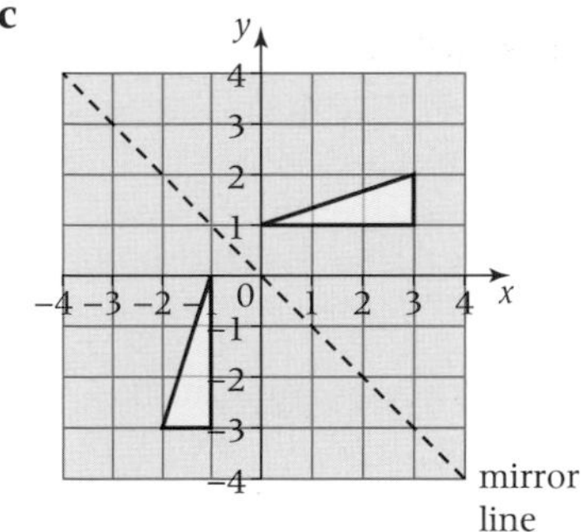

d (−1, 0), (−1, −3), (−2, −3)

e $y = -x$

S3.2

1 a 90° anticlockwise **b** 90° clockwise

c 180° clockwise **d** 90° anticlockwise

e 90° clockwise **f** 90° clockwise

g 90° anticlockwise **h** 90° anticlockwise

i 180° clockwise **j** 90° clockwise

k 90° clockwise **l** 180° anticlockwise

m 90° clockwise **n** 180° clockwise

o 90° anticlockwise

2

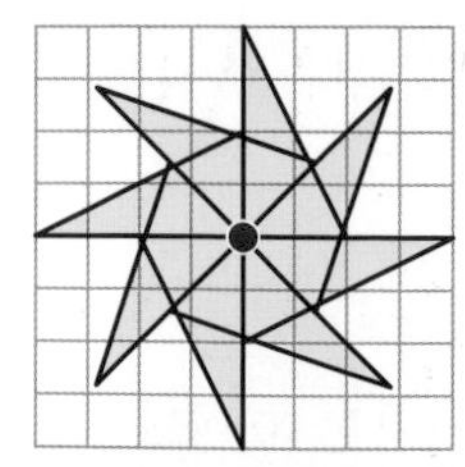

3 **b** Kite
c Image vertices: (0, 0), (−1, 2), (0, 3), (1, 2)
d (−1, 2)
e Yes

4 **c** (−1, −1), (2, −1), (2, −3)

S3.3

1 D, F, H, J

2 **A** Translation $\begin{pmatrix} 4 \\ 3 \end{pmatrix}$ **B** Translation $\begin{pmatrix} 4 \\ 0 \end{pmatrix}$

C Translation $\begin{pmatrix} 0 \\ 3 \end{pmatrix}$ **D** Translation $\begin{pmatrix} 0 \\ -3 \end{pmatrix}$

E Translation $\begin{pmatrix} 4 \\ -3 \end{pmatrix}$ **F** Translation $\begin{pmatrix} 0 \\ -4 \end{pmatrix}$

G Translation $\begin{pmatrix} -4 \\ 3 \end{pmatrix}$ **H** Translation $\begin{pmatrix} -4 \\ -3 \end{pmatrix}$

3 **a** Translation $\begin{pmatrix} 5 \\ 0 \end{pmatrix}$ **b** Translation $\begin{pmatrix} 0 \\ 4 \end{pmatrix}$

c Translation $\begin{pmatrix} 0 \\ -4 \end{pmatrix}$ **d** Translation $\begin{pmatrix} 2 \\ -6 \end{pmatrix}$

e Translation $\begin{pmatrix} 3 \\ 2 \end{pmatrix}$ **f** Translation $\begin{pmatrix} -5 \\ 4 \end{pmatrix}$

g Translation $\begin{pmatrix} -2 \\ 6 \end{pmatrix}$ **h** Translation $\begin{pmatrix} -5 \\ 0 \end{pmatrix}$

i Translation $\begin{pmatrix} -3 \\ -6 \end{pmatrix}$ **j** Translation $\begin{pmatrix} 5 \\ -4 \end{pmatrix}$

4 **a** (−4, 2) **b** Isosceles trapezium
c Image vertices: (1, 0), (4, 0), (2, 2), (3, 2)
d Congruent **e** (1, 0)

S3.4

1 **a** 2 **b** 3 **c** 2 **d** 2
e 2 **f** 3 **g** 5 **h** 2

2 **a** A: Yes, 2; D: Yes, 3; B, C, E: No
b A, D

3 **a** A: Yes, 2; C: Yes, 3; B, D, E: No
b A, C

S3.5

1 **a** 2 **b** 3
c 2 **d** 2

2 **a**

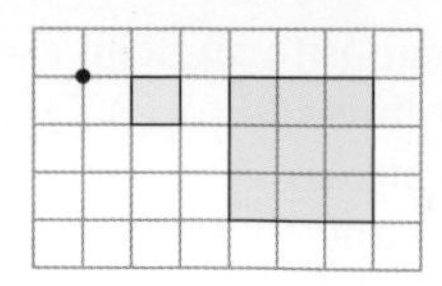

b

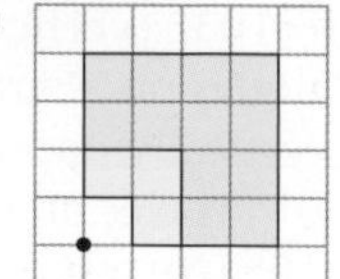

c

d

e

f

3

S3.6

1 Depends on handwriting, for example:

A B C D E F G H I
J K L M N O P Q R
S T U V W X Y Z

2 **a** 3 **b** 2 **c** 2
d 3 **e** 2 **f** 3
g 2 **h** 2 **i** 2 **j** 2

3 a 3 lines, 4 lines, 5 lines, 6 lines, 8 lines
  b 3, 4, 5, 6, 8

4 a

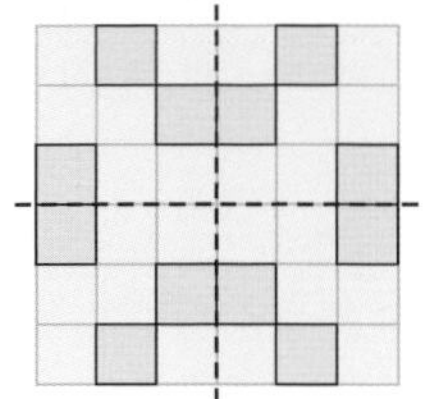

  b

5 a 3
  b 4
  c 2

S3 Exam review

1 a Reflection in $x = 3$

  b

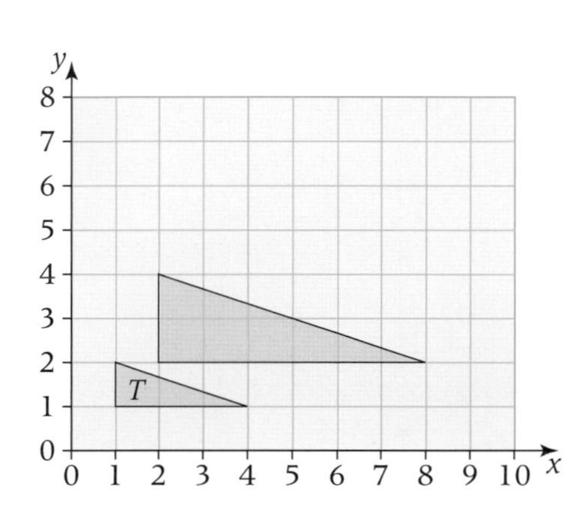

2 a–b

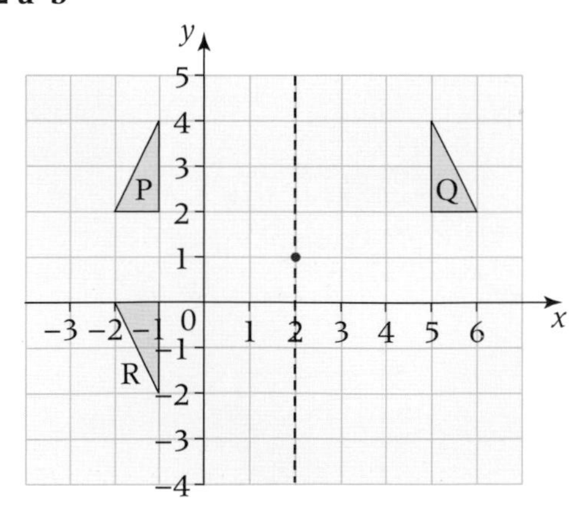

S4 Before you start ...

1 a ii   b iii   c i

2 4

3 48 cm^2

S4.1

1 a e.g. (1, 1)   b e.g. (3, 0)
  c e.g. (3, 2)   d e.g. (−2, 0)
  e (1, 1.5)   f e.g. (3, 0)

2

Shape	Equal in length	Bisect each other	Perpen-dicular
Rectangle	✓	✓	✗
Kite	✗	✗	✓
Isosceles trapezium	✓	✗	✗
Square	✓	✓	✓
Parallelogram	✗	✓	✗
Rhombus	✗	✓	✓
Ordinary trapezium	✗	✗	✗

3 a (−1, 2)   c 6 square units

S4.2

1 a Triangular prism
  b Square-based pyramid
  c Sphere
  d Cuboid
  e Pentagon-based pyramid
  f Cone
  g Cylinder
  h Cube
  i Tetrahedron
  j Pentagonal prism

2 a

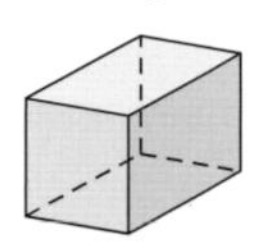

  b

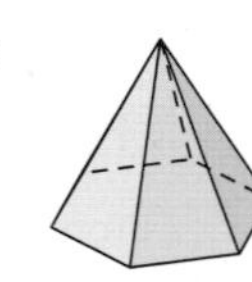

  c

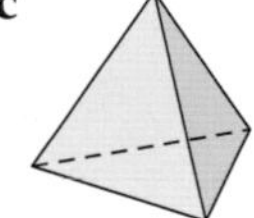

3 a 8   b 12   c 6

4 a

Name of solid	No. of faces (f)	No. of edges (e)	No. of vertices (v)
Triangular prism	5	9	6
Square-based pyramid	5	8	5
Tetrahedron	4	6	4
Pentagonal prism	7	15	10
Square-based prism	6	12	8
Cube	6	12	8
Hexagonal pyramid	7	12	7
Octagonal prism	10	24	16
Pentagonal pyramid	6	10	6

  b $e + 2 = f + v$

5 For example,

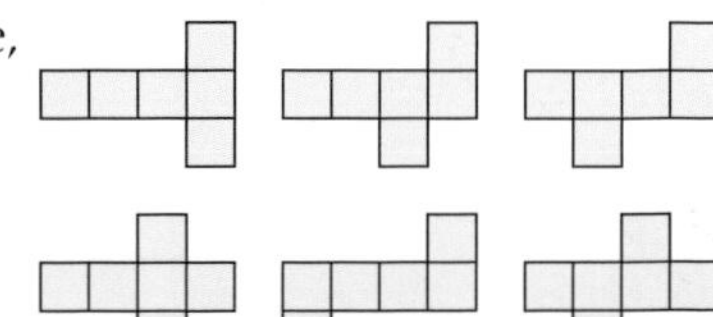

S4.3

1 a

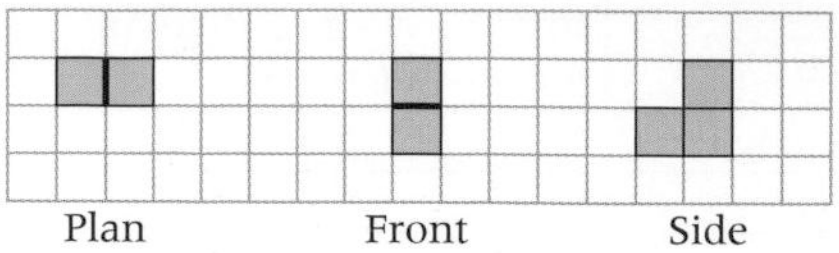
Plan
Front
Side

b

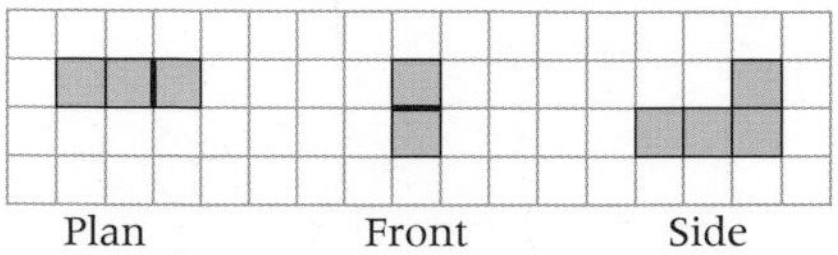
Plan
Front
Side

c

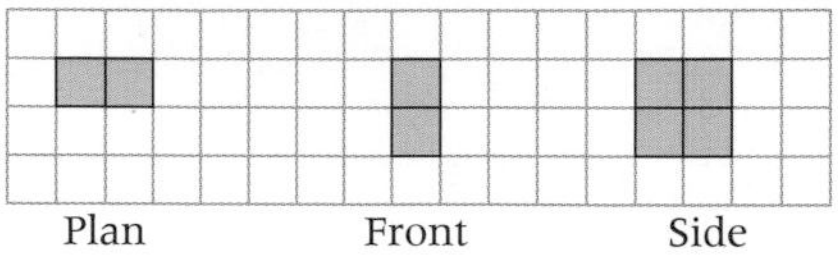
Plan
Front
Side

d

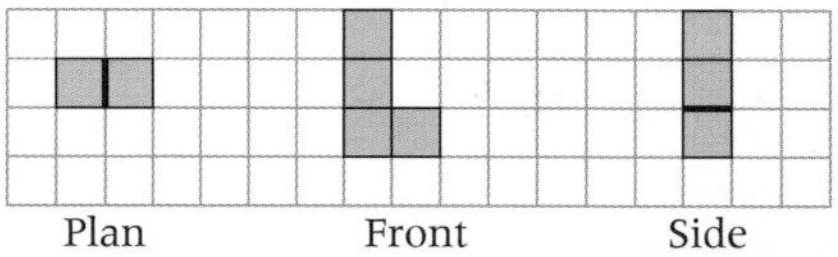
Plan
Front
Side

e

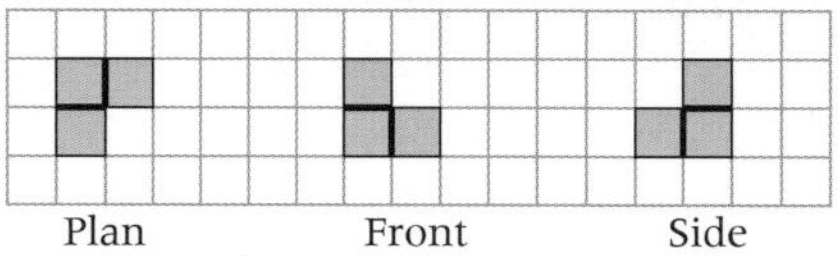
Plan
Front
Side

f

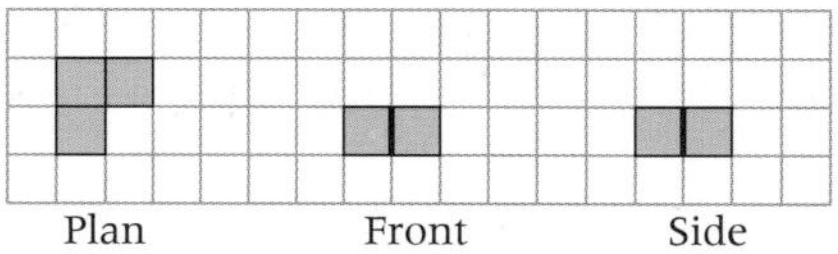
Plan
Front
Side

2 a

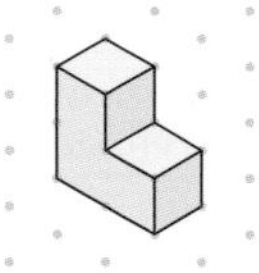

3 cubes

b

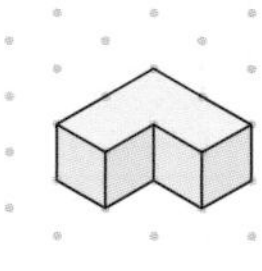

3 cubes

c

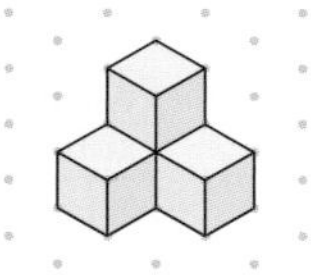

4 cubes

3 a

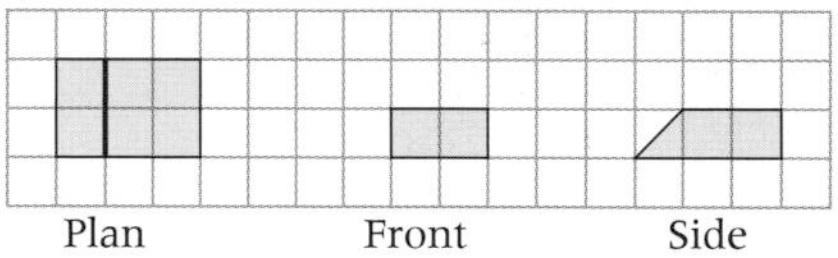
Plan
Front
Side

b

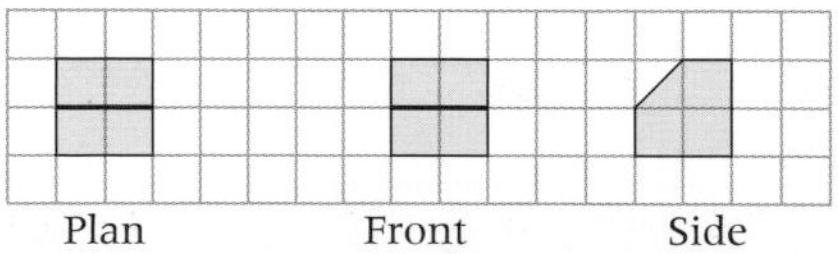
Plan
Front
Side

c

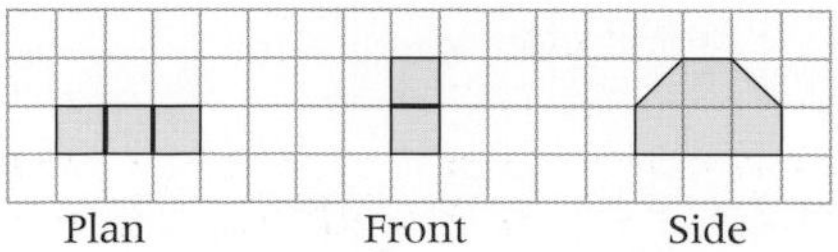
Plan
Front
Side

S4.4

1 a (2, 0, 3)
b (4, 5, 0)
c (3, 5, 4)
d (4, 5, 4)

2 a

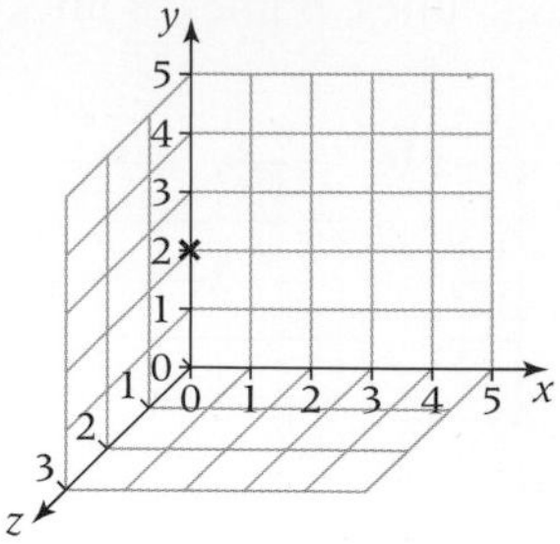
y
x
z
0 1 2 3 4 5

b

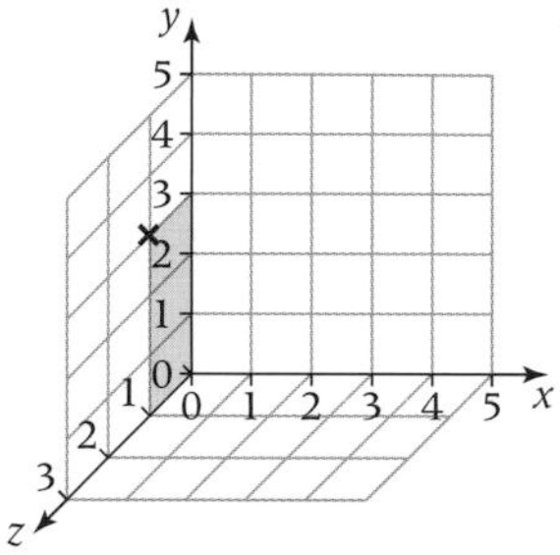
y
x
z
0 1 2 3 4 5

c

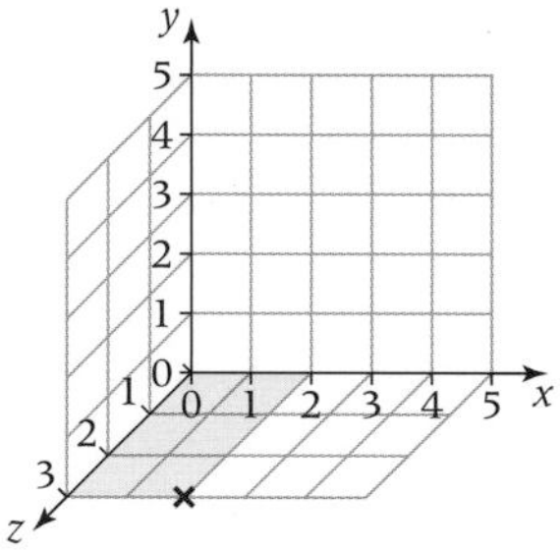
y
x
z
0 1 2 3 4 5

d

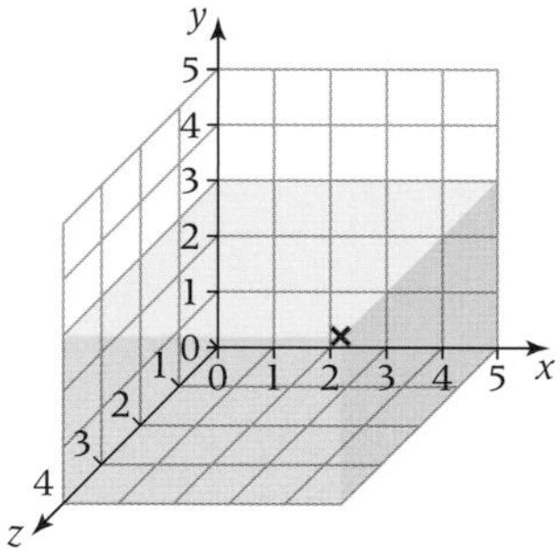
y
x
z
0 1 2 3 4 5

e

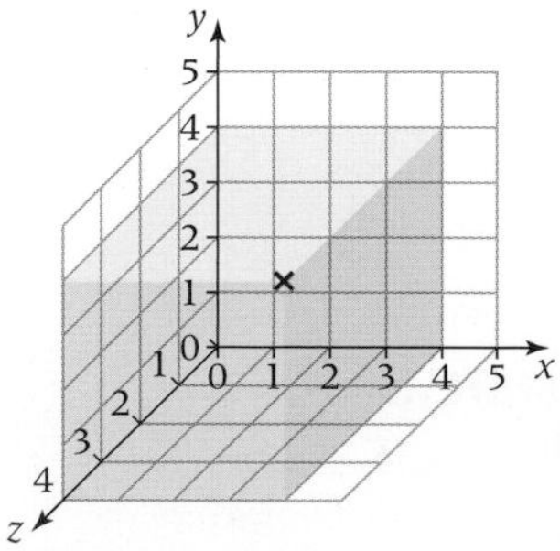
y
x
z
0 1 2 3 4 5

3 a (0, 3, 0) **b** (3, 3, 0) **c** (3, 3, 3) **d** (0, 3, 3)
e (0, 0, 0) **f** (3, 0, 0) **g** (3, 0, 3) **h** (0, 0, 3)

4 a $p = 4$, $q = 5$, $r = 2$
b A(0, 5, 0), B(4, 5, 0), C(4, 5, 2), D(0, 5, 2), E(0, 0, 0), F(4, 0, 0), G(4, 0, 2), H(0, 0, 2)
c 40 cubic units

S4.5

1 a 48 cm^2 b 32 cm^2 c 24 cm^2 d 208 cm^2

2 a 136 cm^2 b 47.12 m c 160 cm^2 d 118 cm^2
 e 50.27 cm

3 a 24 cm^2 b 40 cm^2
 c 32 cm^2 d 6 cm^2
 e 108 cm^2

4 a i 52 cm^2 ii length = 4 cm, width = 3 cm, height = 2 cm
 b i 94 cm^2 ii length = 5 cm, width = 4 cm, height = 3 cm

S4.6

1 a 160 cm^3 b 108 m^3 c 134.4 m^3

2 a 203.125 m^3 b 9.3 cm^3 c 3.125 m^3

3 a 24 cm^2, 120 cm^3
 b 4 m^2, 20 m^3 c 32 cm^2, 256 cm^3

4 a 402 cm^3 b 170 cm^3
 c 503 cm^3 d 157 cm^3

5 a i 39.27 cm^2 ii 3926.99 cm^3
 b i 45 cm^2 ii 4500 cm^3

S4.7

1 a 1800 mm b 4.5 cm c 3.5 m
 d 2 km e 3.5 km f 4.5 m
 g 0.85 m h 250 cm i 2.5 m
 j 0.8 km

2 a 8 m^2 b $80\,000\text{ cm}^2$

3 a 24 m^2 b $240\,000\text{ cm}^2$

4 a 400 mm^2 b 730 mm^2 c 1090 mm^2
 d 250 mm^2 e $40\,000\text{ mm}^2$

5 a 6 cm^2 b 12 cm^2 c 8.5 cm^2
 d 65 cm^2 e 100 cm^2

6 a 4 m^2 b 8.5 m^2 c 100 m^2
 d 12.5 m^2 e 0.5 m^2

7 a $50\,000\text{ cm}^2$ b $100\,000\text{ cm}^2$ c $65\,000\text{ cm}^2$
 d $77\,500\text{ cm}^2$ e 6000 cm^2

8 a 4 km^2 b 18 km^2 c 0.5 km^2
 d 1.5 km^2

9 a 1000 litres b 6000 litres c 7500 litres

S4.8

1 a Area b Volume
 c Length d None of these
 e Volume f None of these
 g Length h Volume
 i Area j None of these
 k Length l Length
 m None of these n None of these
 o None of these p Area
 q Area r None of these
 s Volume t Volume

2 a Area b Length c Volume
 d Area e Area f Length
 g Length h Area

3 c, as it has dimensions L^3

4 c, as it has dimensions L^2

5 b, as it has dimensions L^3

S4 Exam review

1 a

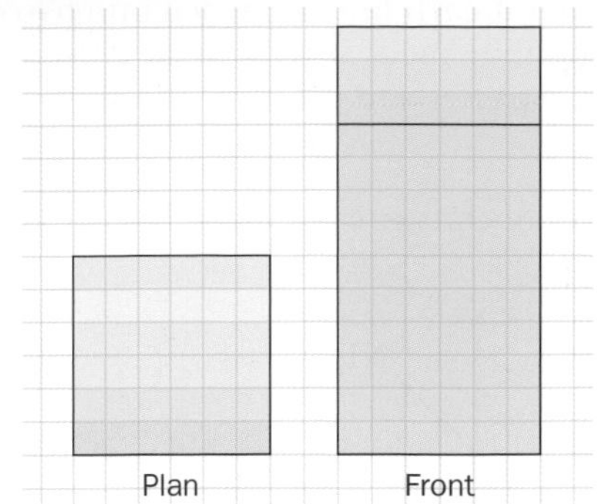

 b $55\,603\text{ cm}^3$ to nearest cm

2 a Cylinder, cone
 b Cuboid, square-based pyramid

S5 Before you start ...

1 a 42° b 123° c 230°

2 360°

3 Circle with 4.6 cm diameter

4 10 cm^2

5 a 67 mm b 6.7 cm

S5.1

1 a Rectangle, 27.3 cm b Rhombus, 22.4 cm
 c Rectangle, 27.9 cm

2 a 057° b 168° c 237° d 276° e 348°

3 a

North
Truro
170°
7 cm
80°
Falmouth
St. Mawes

Scale: 1cm represents 2 km

 b 2.4 km c 13.8 km

S5.2

1 a 8.2 cm b 5.3 cm c 11.6 cm

2 a 5.7 cm, 5.7 cm b 7.6 cm, 5.0 cm
 c 5.8 cm, 3.4 cm

3 a 60° b 76° c 90°

4 a 5.7 cm, 55° b 6.0 cm, 45° c 6 cm, 67°

5 a 9.6 cm b 2.4 cm c two

S5.3

For examples of constructing perpendicular bisectors, see page 222.

1 d 4 cm

4 c 90°

5 The diagonals are perpendicular bisectors of each other if each line is divided into two equal parts and they meet at right angles. Use your ruler, compasses and protractor to check.

S5.4

For examples of constructing angle bisectors, see page 224.

6 c

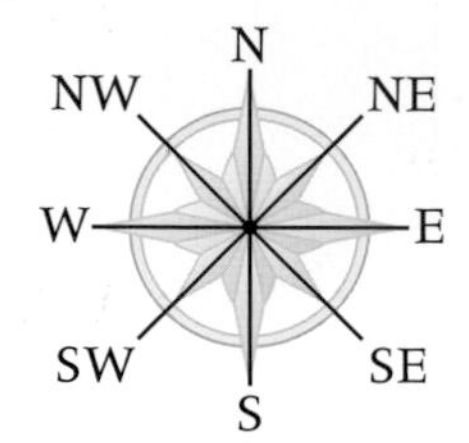

7 d perpendicular

e The obtuse angles are equal (vertically opposite angles) and the acute angles are equal (vertically opposite angles). One acute angle and one obtuse angle sum to 180° (angles on a straight line). The angle between the bisectors is half an acute angle and half an obtuse angle, so it is half 180°, which is 90°.

8 e 2.2 cm

S5.5

1 b Angle bisector

2 Circle of radius 3 cm

3 A parallel line centred between the existing parallel lines

4 b Perpendicular bisector

c The region to the left of the perpendicular bisector

5

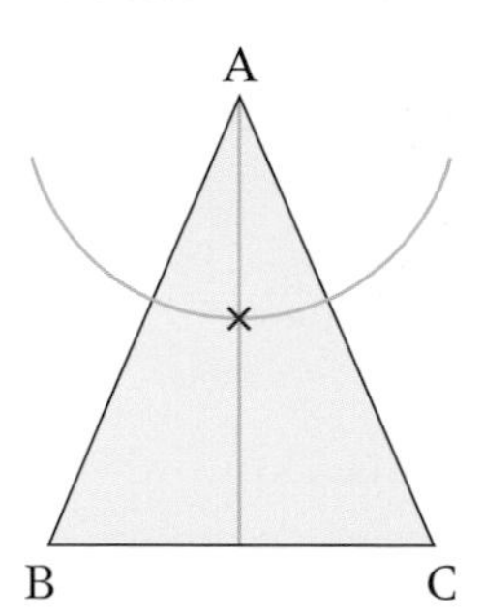

5

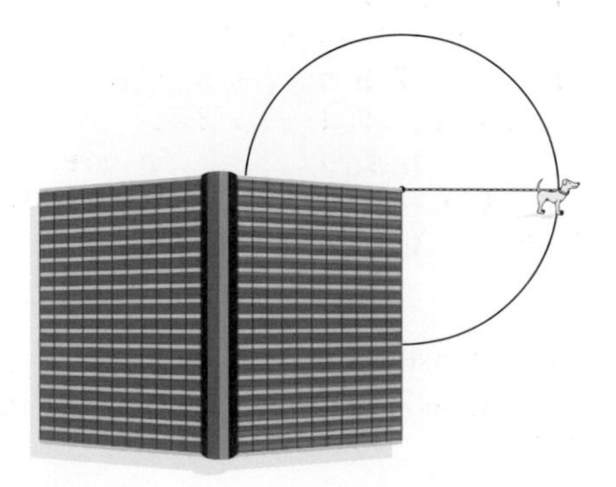

S5 Exam review

2

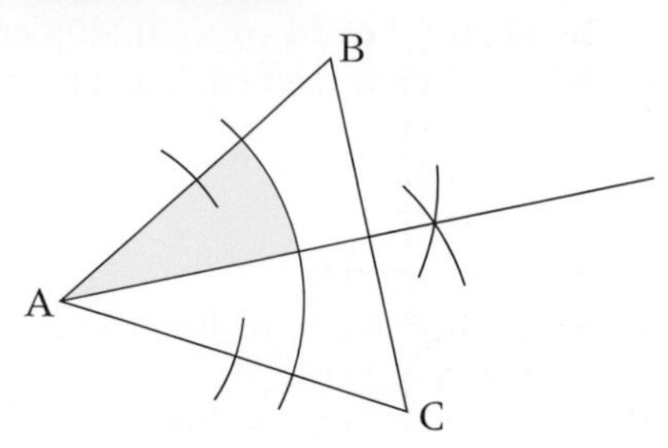

S6 Before you start ...

1 a 47°

b 123°

3 a

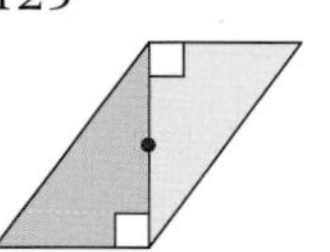

b

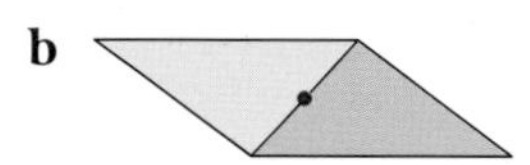

4 200 cm²

S6.1

1 3

2 B

3 a–b

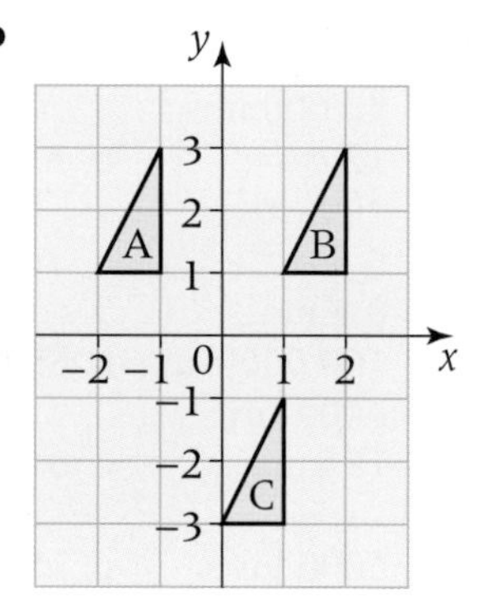

c Translation $\begin{pmatrix} 1 \\ 4 \end{pmatrix}$

4 a–b

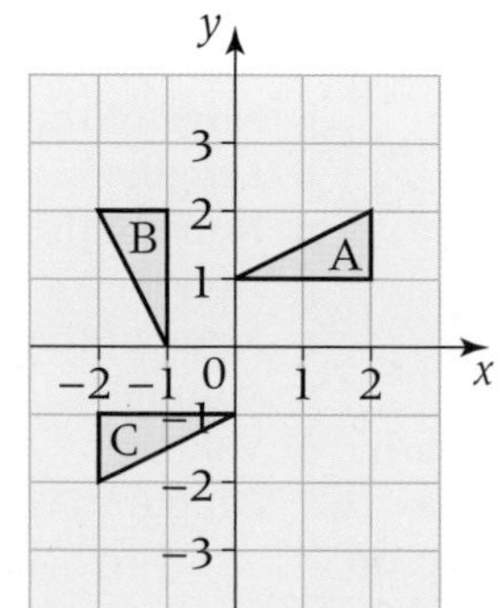

c Rotation through 90° clockwise

5 a–b

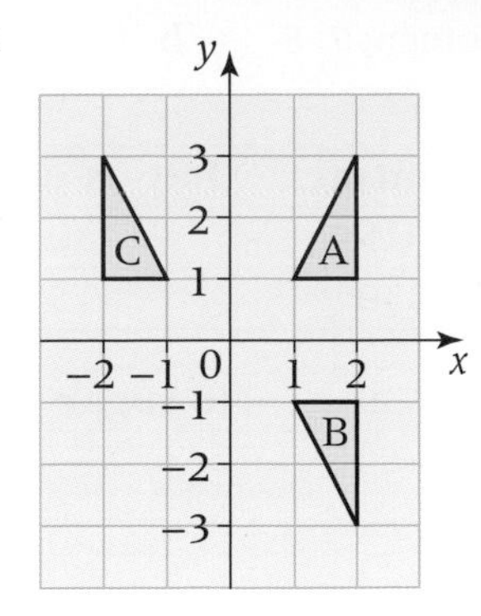

c Rotation through 180°

S6.2

1 a $a = 40°$, $b = 50°$
b $c = 20°$, $d = 40°$
c $h = 75°$, $i = 75°$, $j = 30°$, $k = 75°$, $l = 75°$
d All 60°

2 a Yes, 2 **b** Yes, 5 **c** No **d** No
e Yes, 4 **f** No

3 a 2, 8 cm **b** 3, 12 cm

4 a 6 cm **b** 12 cm

S6.3

1 a 16 cm **b** 12 cm^2
c Length = 30 cm, width = 10 cm
d 80 cm **e** 300 cm^2 **f** 5, 25

2 a 32 cm^3 **b** 6 cm, 12 cm, 12 cm
c 864 cm^3 **d** 27

3 a 30 cm^2 **b** 120 cm^2

4 a 250 cm^3 **b** 2000 cm^3

5

Scale factor	Multiplier for length	Multiplier for area	Multiplier for volume
4	4	16	64
5	5	25	125
6	6	36	216
7	7	49	343

6 56 cm, 160 cm^2

S6.4

1 a 40 cm **b** 100 cm **c** 5 cm
d 65 cm **e** 125 cm

2 a 110 cm **b** 80 cm

3 75 cm by 200 cm

4 a 6 m, 9 m, 18 m **b** 162 m^2

5 a 1000 m **b** 4000 m **c** 5000 m
d 250 m **e** 7250 m

S6.5

1 a 64 cm^2 **b** 100 m^2 **c** 3.24 m^2
d 1296 mm^2 **e** 20.25 m^2

2 a 9 m **b** 2 cm **c** 14 cm
d 2.7 m **e** 1 mm

3 a 20 cm^2 **b** 8 cm^2 **c** 100 mm^2

4 a 10 cm **b** 25 m **c** 17 cm
d 26 mm **e** 30 cm **f** 65 mm

S6.6

1 a 4 cm^2 **b** 15 cm^2

2 a 9 cm **b** 12 mm **c** 24 m
d 16 cm **e** 2.5 m

3 a 5.2 cm **b** 7.1 m **c** 8.9 cm
d 53 m **e** 9 cm **f** 39 cm

4 a 6 m **b** 15 m **c** 7.5 m^2

S6.7

1 a

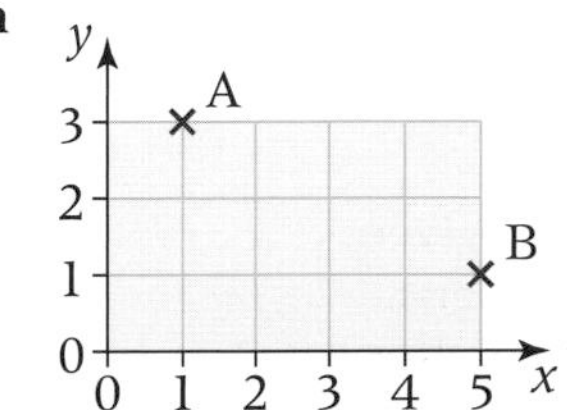

b (3, 2)

2 a (3, 3) **b** (2, 4) **c** (2, 1)
d (5, 4) **e** (2, 3)

3 $x = 5$, $y = 2$

4 a 5 units **b** 5 units **c** 3.2 units
d 5.7 units **e** 6.7 units

5 a

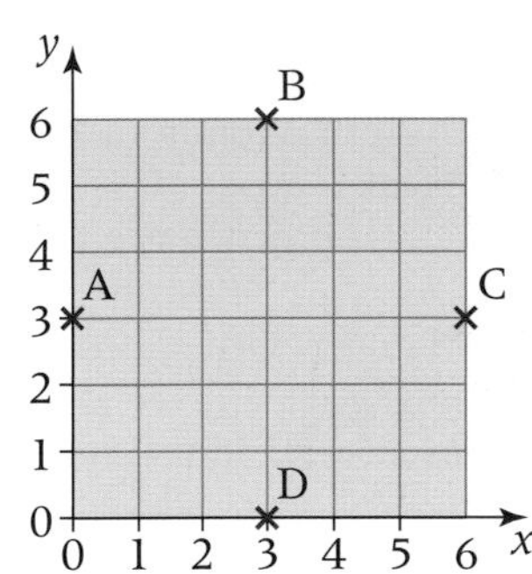

b Square **c** 4.2 units
d 4.2 units **e** 18 square units

S6 Exam review

1 5.2 m

2 a 12.5 cm **b** 7.2 cm

D1 Before you start ...

1 a 44, 47, 48, 55, 56, 59, 61, 65
b 0.5, 0.8, 1.5, 1.7, 2.1, 2.5
c 13.1, 13.2, 21.3, 23.1, 31.2, 32.1

2 a 81 **b** 179 **c** 105 **d** 205 **e** 441
f 74 **g** 135 **h** 177 **i** 219 **j** 41

3 a 07.48 **b** 18.17

D1.1

1 a 6, 7, 8, 5, 4 **b** i **c** u **d** 30

2 a 5, 4, 7, 4, 2, 2, 1, 2, 3 **b** 5 **c** 92 mm

3 a 9, 17, 11, 3 **b** 808

4 a 28 **b** 67

D1.2

1 a Controlled experiment b Observation
c Controlled experiment d Data logging
e Controlled experiment f Observation
g Observation h Data logging
i Observation j Data logging

2 a Observation b 1, 21; 2, 11; 3, 3; 4, 4; 5, 1
c 40 d 73

3 a Controlled experiment
b–c 1, 13; 2, 14; 3, 4; 4, 4; 5, 5; 6, 5
d 2 e 45
f Yes, the dice seems to be biased in favour of 1 and 2.
g By increasing the number of rolls.

D1.3

1 a Choices should be given.
b 'Recently' is too vague.

2 a Which is your favourite fruit? b 100

3 a 30 is in two options and over 40s are missing.
b The options are too vague.
c Using more than one shop would improve the reliability of the data.
d Examples:
How old are you?
Under 20 20–29 30–39 40 or over
How often do you go shopping?
At most once a month 3 times a month
At least once a week

4 a Which channel do you watch the most?
b 30
c There are far too many 'Other' possibilities.

5 a 'Regularly' is too vague.
b How many times a week do you buy a newspaper?
0 1 2 3 4 5 6 7
c There is no option for less than 1 year. If someone last bought a book a number of years ago, they are unlikely to remember how many years exactly. Better options are:
Less than a week ago 1–4 weeks ago
1–6 months ago More than 6 months ago

D1.4

1 a discrete b discrete c continuous
d discrete e continuous f discrete
g continuous h continuous i discrete
j discrete

2 a 4, 9, 16, 11, 10 b 50

3 a 8, 6, 7, 10, 3, 6 b 40

4 a 0, 4, 7, 6, 8 b 25

5 a 8, 13, 10, 9 b $1.0 < m \leqslant 2.0$ c 40

D1.5

1 a i 15 a ii 12
b i 39 b ii 22
c 58

2 Men: 12, 5; Women: 7, 8 b 12

3 a–b Example:

	Contract	Pay as you go
Black	25	30
Silver	30	15

4

	Sugar	No sugar
Tea		
Coffee		

5

	Part-exchange	Cash
Saloons		
Hatchbacks		

6 a

	Broken	Not broken
PC		
Laptop		

b 26 broken computers

D1 Exam review

1 a 59
b Lots of people will buy flowers for their mothers on Monther's Day

2 a There are no 'Bad' options.
b i The options are too vague.
b ii How much money do you normally spend in the canteen?
£0–£1 £1.01–£1.50 £1.51–£2.00
More than £2

D2 Before you start ...

1 a 90° b 130°

2 a 120 b 45 c 60 d 72
e 6 f 20 g 30 h 18
i 10

3 a ii b i

4 a 661, 665, 714, 741, 746, 751, 756
b 0.5, 0.9, 1.5, 1.8, 2.1, 2.2
c 44.6, 45.6, 45.9, 46.4, 49.5

5 a (3, 5) b (5, 1)

D2.1

1 a 1550
b 3 guitars, 2.5 guitars, 1.5 guitars

2 a 13 b 10°
c Win = 150°, Draw = 80°, Lose = 130°
d Pie chart with angles given in part **c**

3 a 6°
b Sunny = 90°, Cloudy = 108°, Rainy = 84°, Snowy = 18°, Windy = 60°
c Pie chart with angles given in part **b**

4 a 0.75°

b Pie chart with: Cod 90°, Plaice 75°, Haddock 72°, Sardines 87°, Mackerel 36°

D2.2

1 a 8, 8, 13, 6, 5

b–c

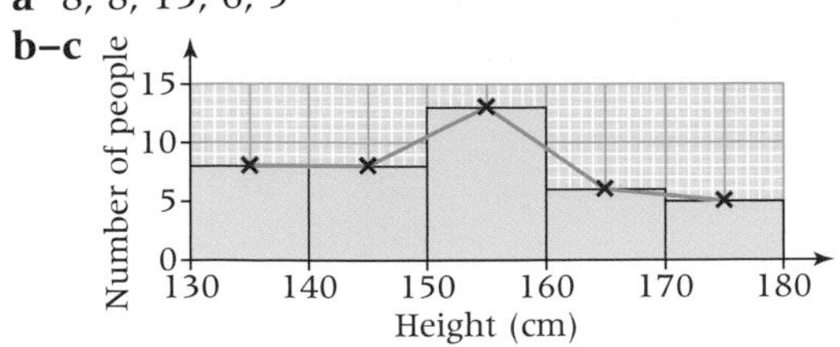

2

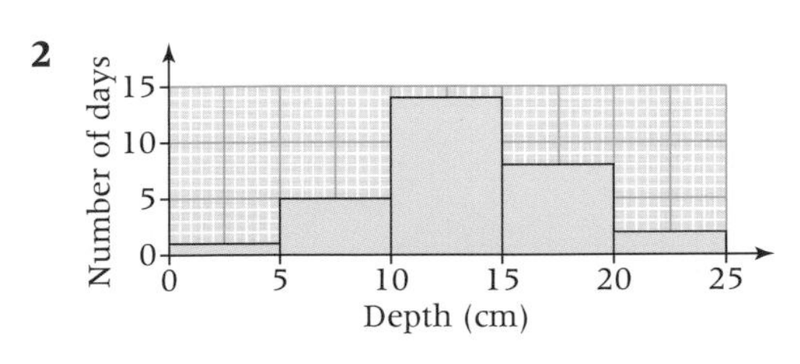

3

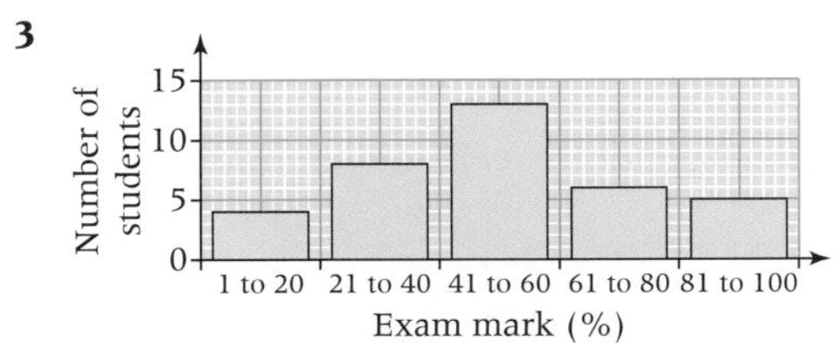

4 b–c

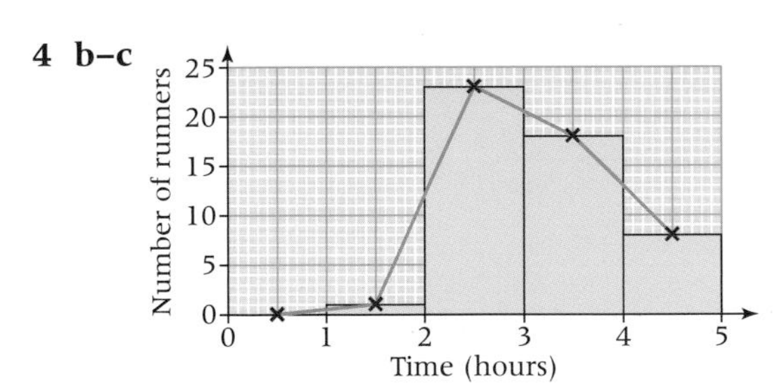

D2.3

1

10	8 8 8
11	0 0 1 2 6 7 7 9 9
12	0 0 1 1 5 5 6 6
13	0 1 4 5 6 7 8
14	0 2 4

Key: 12 | 5 means 125 seconds

2

43	0 2 3 8 8 9
44	0 1 3 4 5 7
45	0 0 1 2 6 9
46	0 1 3 5 5 9 9

Key: 44 | 7 means 44.7 seconds

3 a

1	0 4 4 4 7 9 9 9
2	0 0 0 1 2 2 3 3 3 4 5 6 6 6 6 7 8 9 9
3	1 2 4

Key: 1 | 7 means 17°C

b

5	0 7 7 7
6	3 6 6 6 8 8 8
7	0 2 2 3 3 3 5 7 9 9 9 9
8	1 2 4 4 8
9	0 3

Key: 5 | 7 means 57°F

D2.4

1

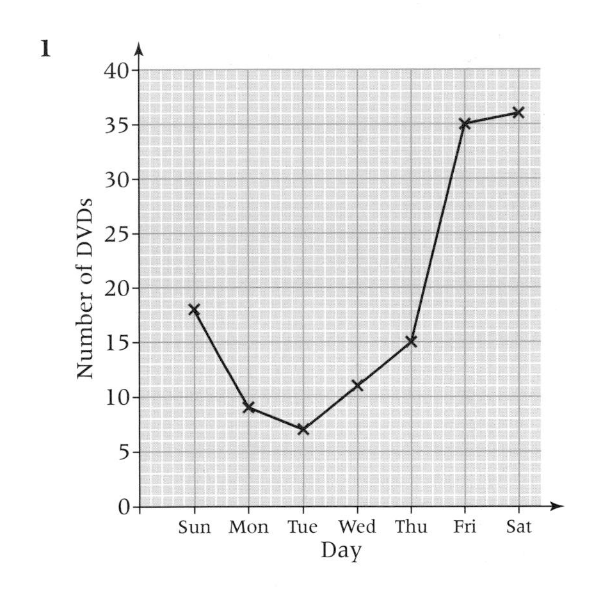

2

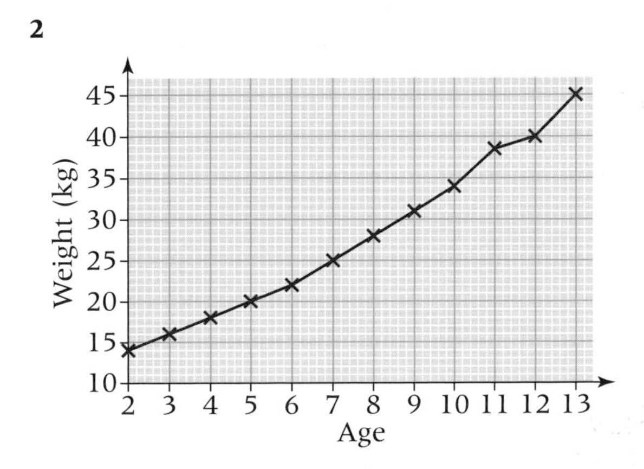

3

4

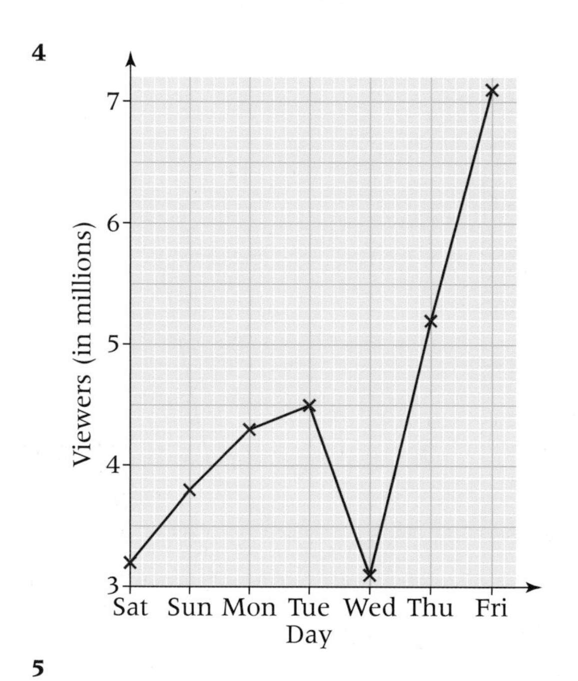

5

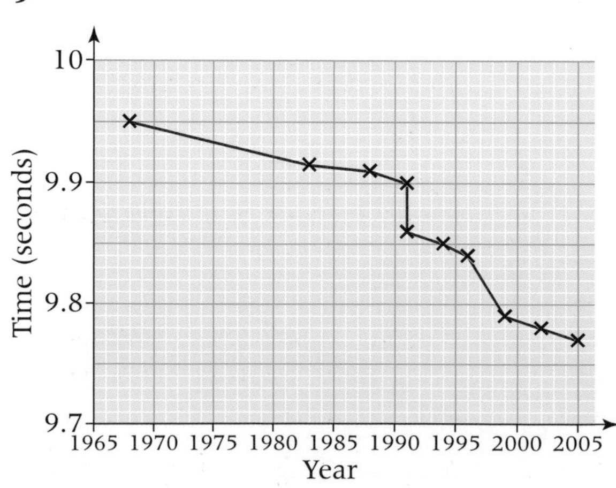

D2.5

1 **a** No correlation **b** Positive correlation
c Negative correlation

2 **a**

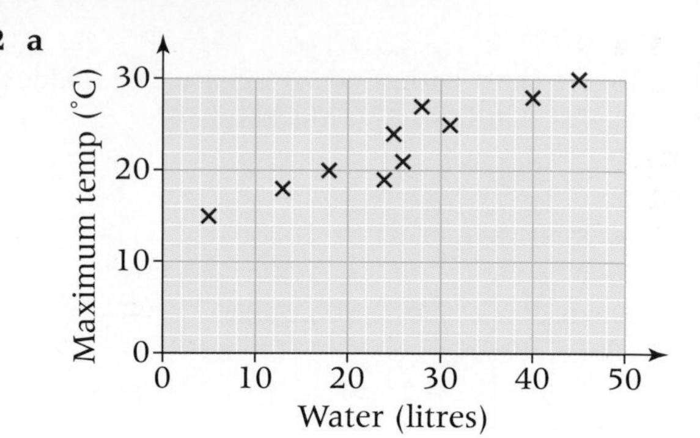

b Positive correlation
c i Increases **c ii** Decreases

3 **a**

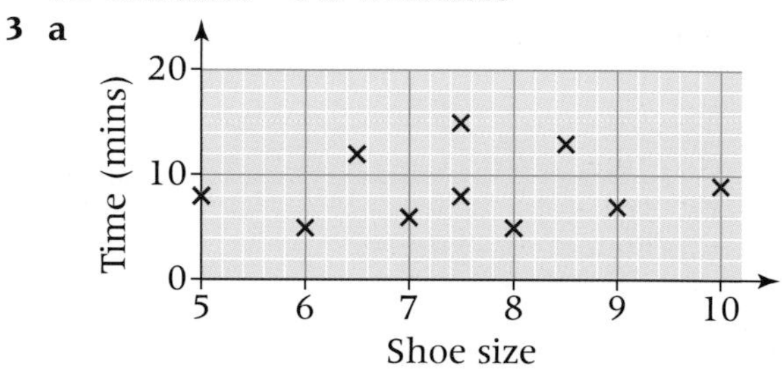

b No correlation **c** No relationship

D2 Exam review

1 **a**

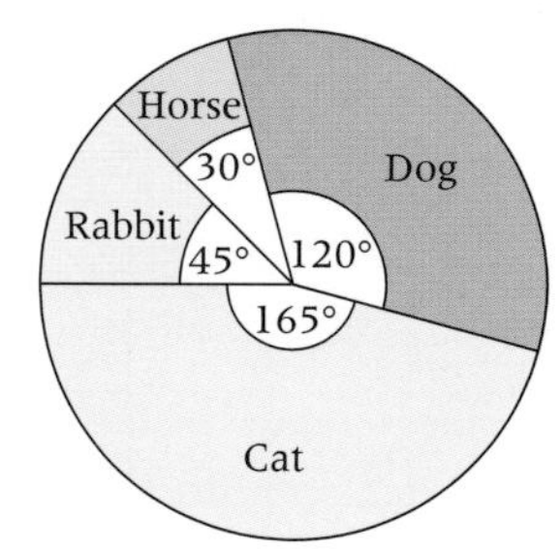

b Iqbal only asked school students, which is not a random sample of the town population.

2

0	5 7 8 8
1	0 0 0 0 2 5 5 5 6
2	0 0 0 4 5
3	3 5

Key: | 1 | 5 | means 15 minutes

D3 Before you start ...

1 **a** 2, 3, 5, 7 **b** 1, 4, 9 **c** 1, 3, 6, 10
d 3, 6, 9 **e** 1, 2, 4, 8

2 **a** $\frac{2}{3}$ **b** $\frac{1}{2}$ **c** $\frac{3}{4}$ **d** $\frac{1}{4}$ **e** 1

3 **a** 1 **b** 1 **c** $\frac{3}{10}$ **d** $\frac{2}{5}$

4 **a** 0.9 **b** 0.4 **c** 0.85

5 **a** 7 **b** 14 **c** 16 **d** 20 **e** 18
6 **a** 0.01 **b** 0.25 **c** 0.35 **d** 0.05 **e** 0.36

D3.1

1 Impossible: **b**, **c**, **e**; Certain: **a**, **d**
2 **a** head, tail **b** 1, 2, 3, 4
c c, h, a, n, g, e **d** yellow, green, blue
e Mon, Tue, Wed, Thu, Fri, Sat, Sun
3 **Aa** blue, blue, blue, blue
Ab 1
Ac 0
Ad blue
Ae red
Ba blue, blue, blue, red
Bb $\frac{3}{4}$
Bc $\frac{1}{4}$
Bd blue
Be red
Ca blue, blue, red, red
Cb $\frac{1}{2}$
Cc $\frac{1}{2}$
Cd–e equal chance of each colour
Da blue, red, red, red
Db $\frac{1}{4}$
Dc $\frac{3}{4}$
Dd red
De blue
Ea red, red, red, red
Eb 0
Ec 1
Ed red
Ee blue
4 **a** $\frac{1}{250} = 0.004$ **b** $\frac{10}{250} = \frac{1}{25} = 0.04$
5 **a** $\frac{48}{120} = \frac{2}{5}$ **b** $= \frac{3}{5}$

D3.2

1 **a** 0.0 **b** 0.5 **c** 1.0
2 **a i** $\frac{1}{10}$ **a ii** $\frac{3}{10}$ **a iii** $\frac{6}{10} = \frac{3}{5}$
b Yellow 0.1, Green 0.3, Red 0.6
c Yellow **d** Red **e** 1
3 **a i** $\frac{6}{8} = \frac{3}{4}$ **a ii** $\frac{2}{8} = \frac{1}{4}$
b Green 0.25, Pink 0.75
c Green **d** Pink **e** 1
4 0.99
5 0.6

D3.3

1 **a** $\frac{20}{50} = \frac{2}{5}$ **b** $\frac{15}{50} = \frac{3}{10}$ **c** $\frac{10}{50} = \frac{1}{5}$ **d** $\frac{5}{50} = \frac{1}{10}$
2 **a i** 24 **a ii** 8 **a iii** 16
a iv 6 **a v** 18
b i $\frac{2}{24} = \frac{1}{12}$ **b ii** $\frac{12}{24} = \frac{1}{2}$ **b iii** $\frac{8}{24} = \frac{1}{3}$
b iv $\frac{16}{24} = \frac{2}{3}$ **b v** $\frac{6}{24} = \frac{1}{4}$ **b vi** $\frac{18}{24} = \frac{3}{4}$
3 **a** 50 Tiles
b i $\frac{2}{5}$ **ii** $\frac{8}{25}$ **c** 0 **d** $\frac{2}{10} = \frac{1}{5}$ **e** $\frac{8}{10} = \frac{4}{5}$
c

	Cross(X)	Circle(O)
Green	$\frac{2}{5}$	$\frac{1}{5}$
Red	$\frac{2}{25}$	$\frac{5}{25}$

d $\frac{25}{25} = 1$ The probabilities of mutually exclusive outcomes add up to 1.

D3.4

1 75
2 80
3 9
4 **a** 25 **b** 20 **c** 15
5 **a** 0.15
b i 10, 15, 30, 20, 10, 15
b ii 50, 75, 150, 100, 50, 75
b iii 100, 150, 300, 200, 100, 150
6 £1

D3.5

1 **a** 11, 16, 14, 9 **b** Red
c i $\frac{11}{50} = 0.22$ **c ii** $\frac{16}{50} = 0.32$ **c iii** $\frac{40}{50} = 0.28$
2 **a** 9, 10, 6, 9, 6 **b** 2 **c** 40
d i $\frac{9}{40} = 0.225$ **d ii** $\frac{10}{40} = \frac{1}{4} = 0.25$
d iii $\frac{6}{40} = \frac{3}{20} = 0.15$ **d iv** $\frac{9}{40} = 0.225$
d v $\frac{6}{40} = \frac{3}{20} = 0.15$
e 15
3 **a** 50
b i $\frac{9}{50} = 0.18$ **b ii** $\frac{14}{50} = \frac{7}{25} = 0.28$
b iii $\frac{27}{50} = 0.54$
c 2 red, 3 green, 5 blue
d By increasing the number of times a ball is taken out.

D3.6

1 **a** Yes **b** Yes **c** No **d** Yes
e No **f** Yes **g** No **h** Yes
2 **a** $\frac{1}{4}$ **b** $\frac{1}{4}$ **c** $\frac{1}{2}$
3 **a** $\frac{2}{5}$ **b** $\frac{1}{2}$ **c** $\frac{1}{10}$ **d** $\frac{1}{2}$ **e** $\frac{3}{5}$
f 1
4 **a** $\frac{1}{5}$ **b** $\frac{1}{5}$ **c** $\frac{2}{5}$ **d** $\frac{2}{5}$ **e** $\frac{4}{5}$
5 **a** 30, 5, 5, 10
b i $\frac{3}{5}$ **b ii** $\frac{1}{10}$ **b iii** $\frac{3}{10}$

D3.7

1 a RA, RB, RC, RD, RE, YA, YB, YC, YD, YE, GA, GB, GC, GD, GE, PA, PB, PC, PD, PE

b

	R	Y	G	P
A	(R, A)	(Y, A)	(G, A)	(P, A)
B	(R, B)	(Y, B)	(G, B)	(P, B)
C	(R, C)	(Y, C)	(G, C)	(P, C)
D	(R, D)	(Y, D)	(G, D)	(P, D)
E	(R, E)	(Y, E)	(G, E)	(P, E)

c $\frac{1}{20}$

2 a

	1	2	3	4
Heads	(1, H)	(2, H)	(3, H)	(4, H)
Tails	(1, T)	(2, T)	(3, T)	(4, T)

bi $\frac{1}{8}$ bii $\frac{1}{4}$

3 a

	Club (C)	Diamond (D)	Spade (S)	Heart (H)
Club (C)	(C,C)	(D,C)	(S,C)	(H,C)
Diamond (D)	(C,D)	(D,D)	(S,D)	(H,D)
Spade (S)	(C,S)	(D,S)	(S,S)	(H,S)
Heart (H)	(C,H)	(D,H)	(S,H)	(H,H)

b $\frac{1}{4}$

4 a

	1	2	3	4	5	6
1	(1, 1)	(2, 1)	(3, 1)	(4, 1)	(5, 1)	(6, 1)
2	(1, 2)	(2, 2)	(3, 2)	(4, 2)	(5, 2)	(6, 2)
3	(1, 3)	(2, 3)	(3, 3)	(4, 3)	(5, 3)	(6, 3)
4	(1, 4)	(2, 4)	(3, 4)	(4, 4)	(5, 4)	(6, 4)
5	(1, 5)	(2, 5)	(3, 5)	(4, 5)	(5, 5)	(6, 5)
6	(1, 6)	(2, 6)	(3, 6)	(4, 6)	(5, 6)	(6, 6)

P(Double six) = $\frac{1}{36}$

D3 Exam review

1 a 0.28
b 1600

2 40

D4 Before you start ...

1 a 26 b 18 c 5.6

2 a 4, 5.5, 6.5, 7, 8
b 2.3, 2.4, 3.2, 3.4, 4.2, 4.3
c 7.5, 7.9, 8.3, 8.6, 9.1

3 a 6, 6, 7, 8, 8, 9, 9, 9, 9, 9
b 100, 100, 101, 102, 102, 104, 104, 104, 104, 104

4 a 3 b 8 c 42 d 47 e 52

D4.1

1 a Discrete b Discrete c Continuous
d Continuous e Continuous f Discrete
g Continuous h Continuous i Discrete
j Continuous k Discrete l Discrete

2 a 5 b 4 c 9 d 1 e 7
f 5 g 17 h 7 i 2 j £3.85

3 a 2 b 3 c 30

4 41 kg

5 a 31 °F b 34 °F

6 23 or 42

D4.2

1 a 4

b–d

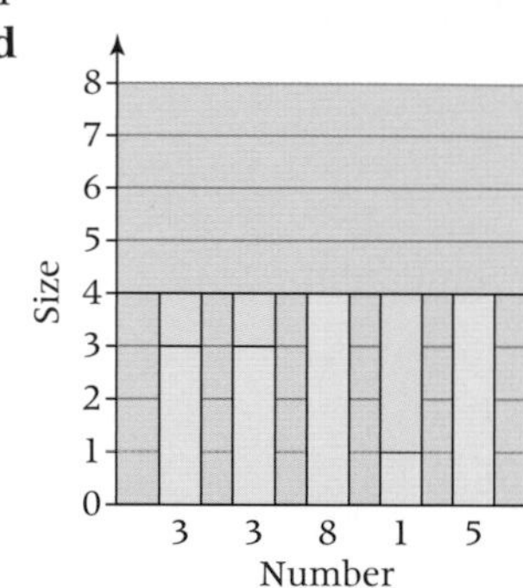

2 a 11, 4, 25 b Car

3 a 38p, 85p, £1.12, £2.15, £2.47 b £1.12

4 a 80.4 °F b 80 °F c 80 °F

5 a 5.0 b 5.8

D4.3

1 a 3, 3, 3, 3, 4, 4, 5, 5
b Range = 2, Mode = 3, Median = 3.5, Mean = 3.75

2 a 1, 1, 1, 1, 2, 2, 2, 2, 2, 2, 3, 3, 4, 4, 4, 4, 4, 5, 5, 5, 5, 5, 5, 5, 5
b 8
c Range = 4, Mode = 5, Median = 4, Mean = 3.28

3 a 2.4 b 1 c 2 d 3

4 a 2.7 b 4 c 3 d 3

5 a 3, 4, 0, 3, 3, 3, 3, 0
b Mean = 5.85, Mode = 4, Median = 6, Range = 7

D4.4

1 a 2, 4
b Number 47's data is more spread out than Number 45's.

2 a 1, 3, 4, 1, 1 b 1.8, 2, 2
c On average, houses in Ullswater Drive have more cars than those in Ambleside Close.

3 a Ireland: 22, 23, 24, 24, 24, 25, 26, 26, 26, 27, 27, 28
Spain: 2, 5, 6, 7, 10, 10, 10, 11, 11, 11, 11, 12
b Ireland 25.5, Spain 10
c On average, Ireland has more days of rain than Spain.
d Ireland 6, Spain 10
e There is a bigger spread or larger variation in the number of rainy days in Spain.

D4.5

1 a 1, 6, 8, 3, 3, 3
  b 70 to 74
  c 70 to 74
2 a 3
  b 23, 28, 33, 38
  c 23, 168, 66, 38; Mean speed = 29.5 mph
3 a 30 students
  b $160 < h \leqslant 170$
  c $170 < h \leqslant 180$
  d 163.3 cm

D4 Exam review

1 57kg
2 a 1 b 3 c 2.1

D5 Before you start ...

1 a 360° b 270° c 95°
2 a 30 b 20 c 15
  d 10 e 3
3 a 14, 24, 10, 20 b 6, 2, 5, 10 c 7, 3, 5, 13
4 a (1, 3) b (4, 1)

D5.1

1 ai 24 aii 12 aiii 36
  aiv 4 av 76
  b Cotton
2 ai £20 aii £10
  b Children
  c 4 × World Single Trip = £20, which is cheaper than World Annual insurance.
3 ai 1 person aii 3 people
  b 28 tickets
4 a 120° b 3°
  ci 30 cii 50 ciii 40

D5.2

1 ai Cooking and washing up
  aii Washing and ironing
  bi Cooking and washing up
  bii DIY
2 ai 100 million
  aii 1000 million
  b India
  c UK
3 ai 2 aii 1
  b 85–90 m
  c 70–75 m
  d 10 athletes
4 The women jumped further, on average. The men had a greater spread of long jumps than the women.

D5.3

1 a 10, 16, 22, 22, 24, 26, 28, 29, 30
  bi 23 bii 22
  biii 24 biv 20
2 a

0	7 9
1	7 9 2 1 5 8 7
2	5 5 1 5 3
3	1 2 7
4	3 0 1

  b

0	7 9
1	1 2 5 7 7 8 9
2	1 3 5 5 5
3	1 2 7
4	0 1 3

  c i 23.4 c ii 25
  c iii 22 c iv 36
3 a 128 sec b 120 sec
  c 126 sec d 31 sec
4 a 49 b 16 c $\frac{1}{4}$
5 a Long jump
  b High jump 26 cm, long jump 15 cm; there is a greater spread of distances for the high jump than there is for the long jump.

D5.4

1 a 0900 b 22 °C, 1500
  c 2 °C, 0300 d 20 °C
2 There were always more bus passengers than train passengers from 1970 to 2000. The number of bus passengers was decreasing during this period, whereas the number of train passengers was increasing.
3 a Increasing rapidly
  b Decreasing slowly
  c An alternative form of energy supply

D5.5

1 a Negative correlation
  b No correlation
  c Positive correlation
2 a A: Poor exam mark, lots of revision; B: Very good exam mark, lots of revision; C: Very good exam mark, little revision; D: Poor exam mark, little revision; E: Average exam mark, average amount of revision
  b A: Not much pocket money, equal eldest; B: Lots of pocket money, equal eldest; C: Lots of pocket money, equal youngest; D: Not much pocket money, equal youngest; E: Average pocket money, middle age
  c A: Low fitness level, lots of hours in gym, B: Good fitness level, lots of hours in gym, C: Good fitness level, few hours in gym, D: Low fitness level, few hours in gym, E: Medium fitness level, medium hours in gym
3 a 15 b 24

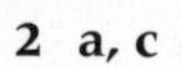

4 a, c

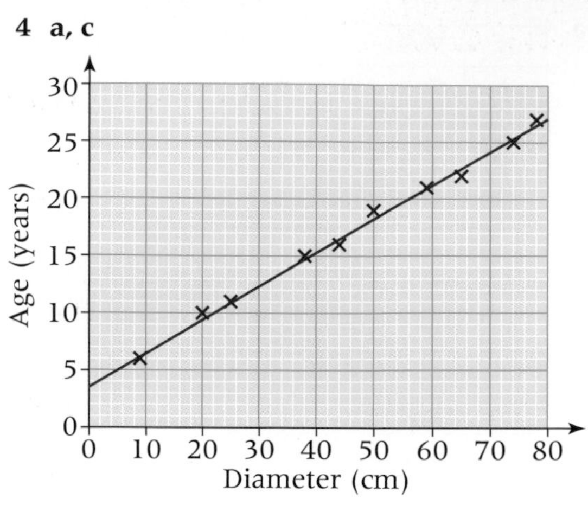

b Positive correlation

d 20 years

D5 Exam review

1 $37\frac{1}{2}$ miles

2 a, c

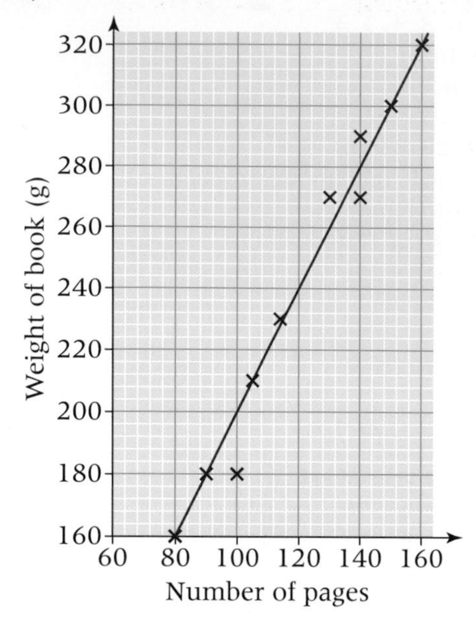

b The greater the number of pages, the heavier the book (positive correlation).

d i 140 pages **d ii** 240 g

Foundation Practice Paper Answers

TOTAL 100 marks

1 **a** 0.2 (1 mark)

b $\frac{2}{5}$ (1 mark)

c 0.3 (1 mark)

d $\frac{1}{5}$, 30%, 0.4 (1 mark)

2 **a** (3, –2) (1 mark)

b (1 mark)

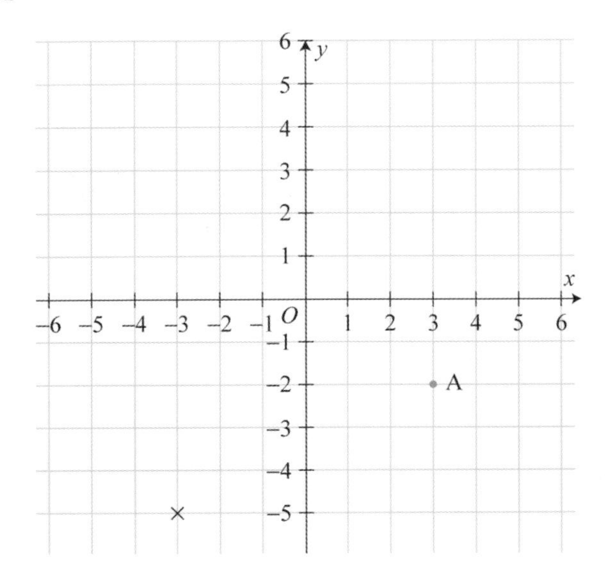

3 **a** metres (1 mark)

b grams (1 mark)

c cm^3 *or* litres (1 mark)

4 **a** (3 marks)

	Tally	Frequency
Monday	𝍸	5
Tuesday	IIII	4
Wednesday	II	2
Thursday	III	3
Friday	𝍸 II	7
Saturday	II	2
Sunday	II	2

b 2 (1 mark)

c Friday (1 mark)

5 **a** **i** 12 (1 mark)

ii 8 (1 mark)

iii 2 (1 mark)

b (3 marks)

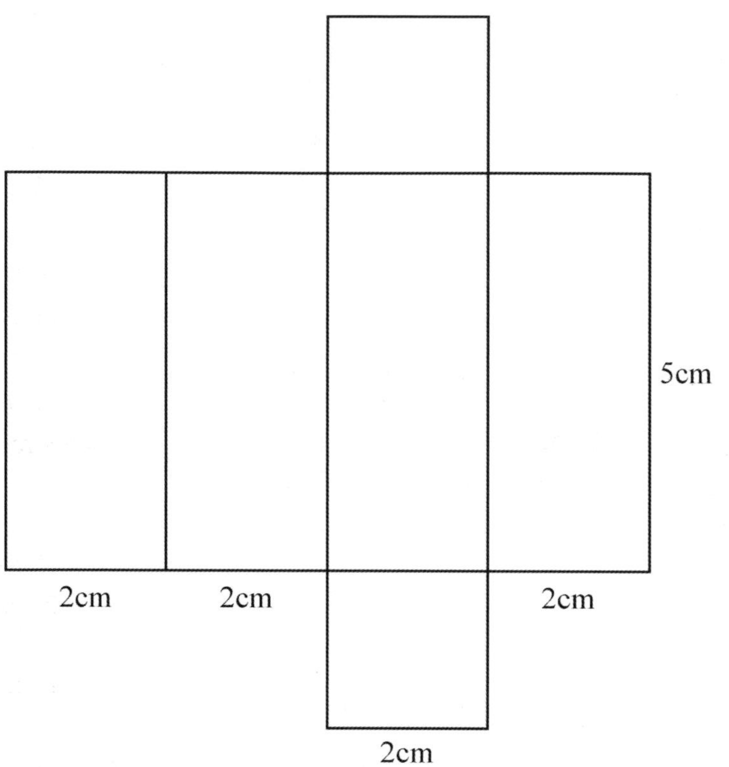

Other nets are possible

6 **a** 5 (2 marks)

b £3.50 (2 marks)

7 **a** 8 (1 mark)

b 12 (2 marks)

c 18 (2 marks)

8 **a** 11 (2 marks)

b 9 (2 marks)

9 **a** 5°C (1 mark)

b 8°C (1 mark)

c –7°C (1 mark)

10 $x = 6$ (2 marks)

11 (3 marks)

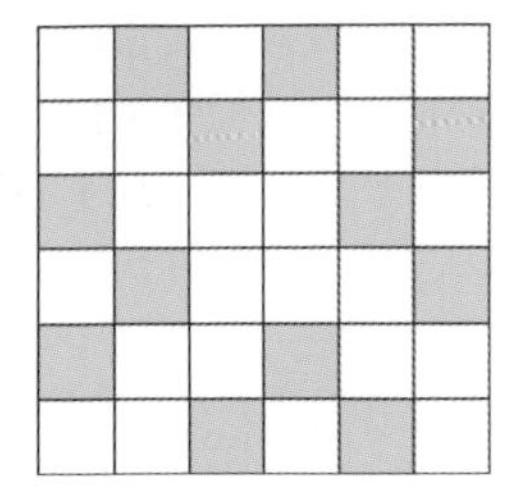

12 a i $\frac{1}{4}$ (1 mark)

ii $\frac{11}{12}$ (2 marks)

b i 14.86 (1 mark)

ii 0.12 (1 mark)

c 6 (1 mark)

13 $2 + 3 + 4 = 9$ (odd) whereas $3 + 4 + 5 = 12$(even) (3 marks)
(If you start with an even, you will get an odd answer, but if you start with an odd, you will get an even answer.)

14 a $9x$ (1 mark)

b $3x + 3y$ (2 marks)

15 a i $\frac{60 \times 20}{30}$ (1 mark)

ii 40 (1 mark)

b 7 (3 marks)

16 (4 marks)

Nationalities of visitors to a National Trust property

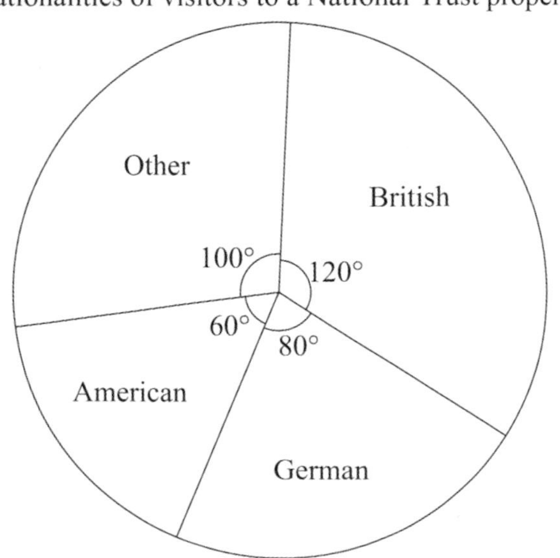

17 a 1.8 m (allow 1.6 m – 2 m) (1 mark)

b 6m (allow 5 to 7 m) (3 marks)

18 7.6 (3 marks)

19 a Not prime (1 mark)

b Could be either (1 mark)

c Could be either (1 mark)

20 a, b (3 marks), (2 marks)

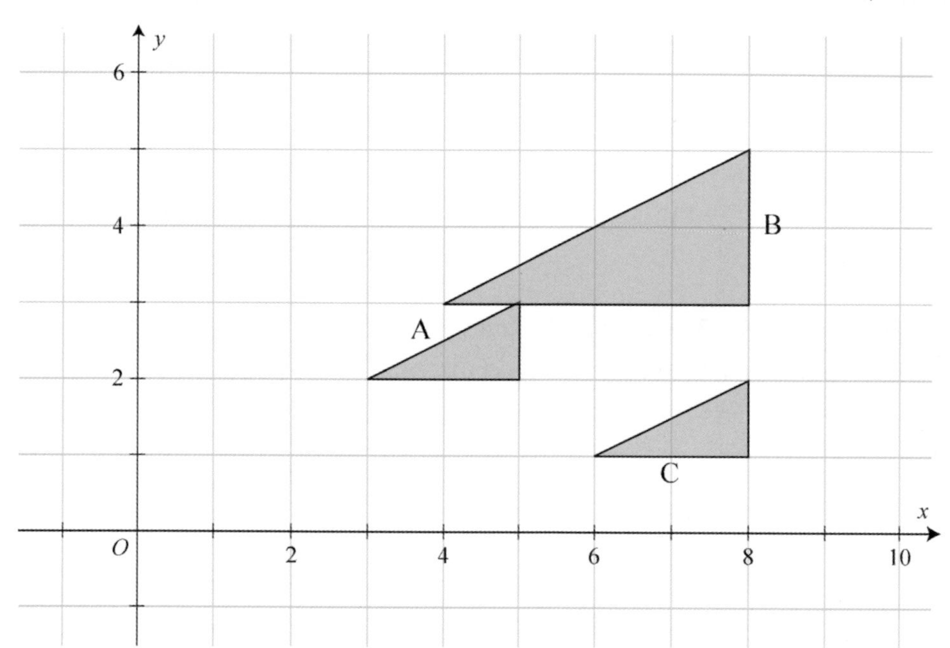

21 a 0, 3, 8 (2 marks)

b $3n+10$ (2 marks)

22 a i $\frac{3}{10}$ (1 mark)

ii $\frac{7}{10}$ (1 mark)

b i Probability of a blue from bag $A = \frac{5}{10} = \frac{15}{30}$ (2 marks)

Probability of a blue from bag $B = \frac{6}{15} = \frac{12}{30}$

Therefore more likely to get a blue from bag A.

ii Probability of a red from bag $A = \frac{3}{10} = \frac{9}{30}$ (2 marks)

Probability of a red from bag $B = \frac{5}{15} = \frac{10}{30}$

Therefore more likely to get a red from bag B.

23 a a^5 (1 mark)

b m^{-6} or $\frac{1}{m^6}$ (1 mark)

c $4x^4$ (2 marks)

24 a 26 cm (2 marks)

b 22 cm^2 (2 marks)

25 a (3 marks)

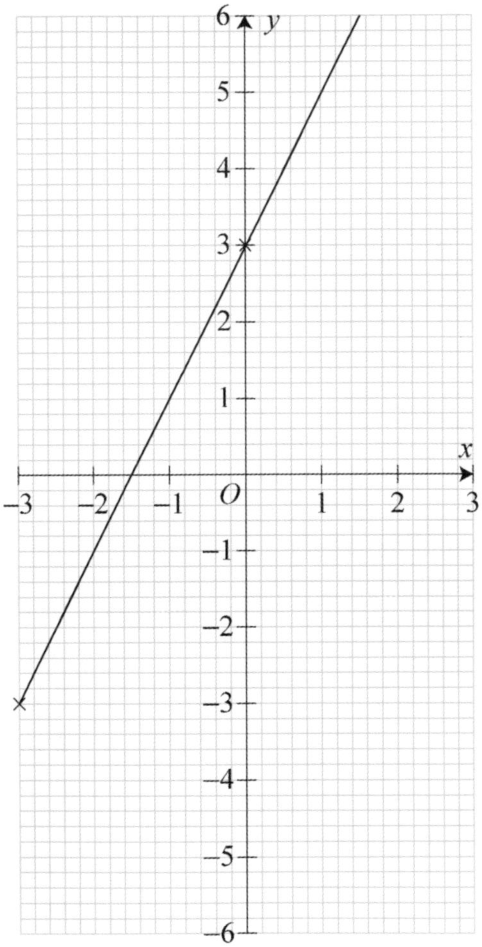

b $x = \frac{y-3}{2}$ (2 marks)

Higher Practice Paper Answers

TOTAL 100 marks

1 £4 (3 marks)

2 (2, 5) (3 marks)

3 Exterior angles add to 360° (2 marks)

$\therefore$ each exterior angle $= \dfrac{360°}{5}$

$= 72°$

$\therefore$ each interior angle $= 180° - 72°$

$= 108°$

4 **a** 14 (1 mark)

b (3 marks)

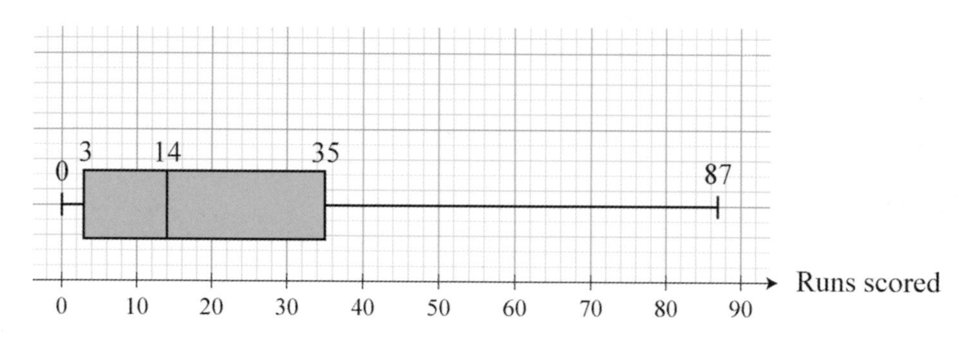

c The second batsman has a higher median score and shows much less variability in his scores. However his highest score is 32, so he is much less likely to produce a very high score. The first batsman, when he scores, scores well, but he is far more likely to do badly. (2 marks)

5 **a** 9.110732253 (2 marks)

b £240:£120:£60 (2 marks)

c –7 (2 marks)

6 **a** **i** 1.2 kg (1 mark)

ii 675 g (1 mark)

b 177° C (3 marks)

7 (3 marks)

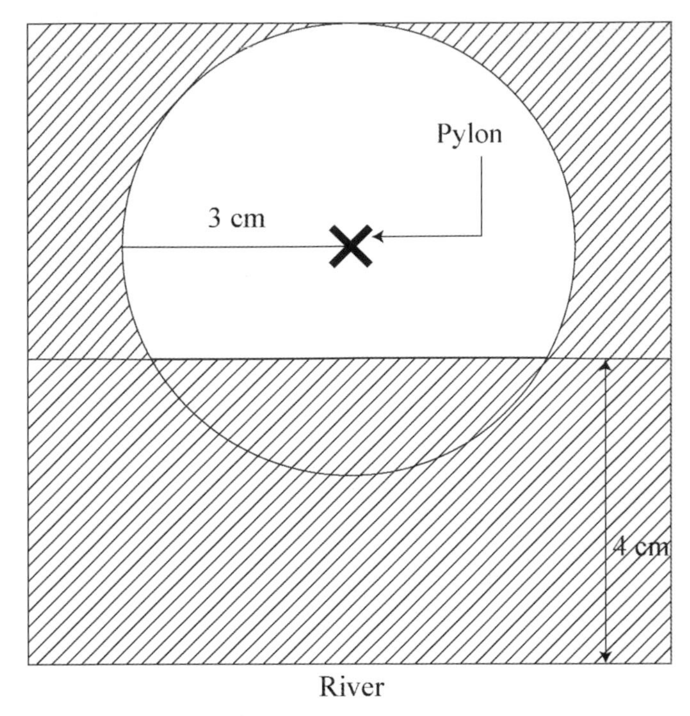

8 (3 marks)

Tax	£165	16°
Insurance	£842	84°
Maintenance (est)	£700	70°
Petrol (est)	£900	90°
Depreciation on car (est)	£1000	100°

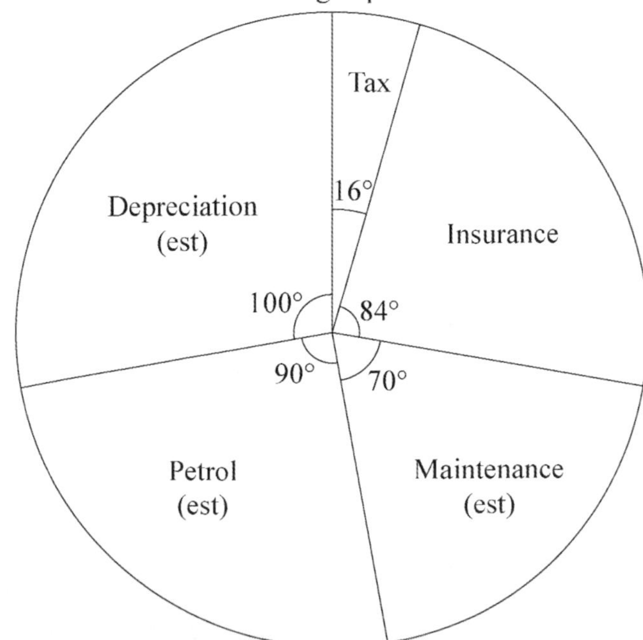

Check for :
1. a title
2. correct angles
3. labelled sectors

9 £1.36, cheaper in shop (3 marks)

10 a 25.1 cm (2 marks)

b 50.3 cm^2 (2 marks)

11 a $x = -1\frac{1}{3}$ (2 marks)

b $n > -1\frac{1}{3}$ (2 marks)

12 a $2^2 \times 3 \times 5 \times 7$ (2 marks)

b 13 860 (2 marks)

13 a –2 (2 marks)

b $y = -2x + 3$ (1 mark)

14

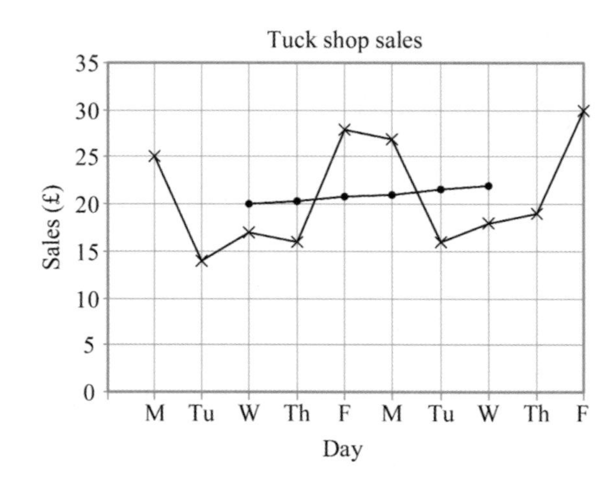

a Weekly Moving Averages: 20, 20.4, 20.8, 21, 21.6, 22 (3 marks)

b A very slight upward trend. (1 mark)

c Moving averages are increasing at a rate of 0.4 each time. (2 marks)

Therefore estimate next moving average = 22.4

$$\therefore \frac{16+18+19+30+x}{5} = 22.4$$

$\therefore x = £29$

15 a –1.22 or 8.22 (3 marks)

b 3 or 4 (3 marks)

16 158.7 cm^2 (5 marks)

17 $x = \frac{k-y}{k+y}$ (4 marks)

18 a $\frac{6}{31}$ (2 marks)

b 43 (2 marks)

19 72.1 m (3 marks)

20 No

Max possible weight = $68.5 + 12 \times 17.5\text{kg}$ (4 marks)

= 278.5 kg which is within the safe limit (i. e. < 279.5 kg)

21 a (2 marks)

x	−1	0	1	2	3	4
y	**7**	3	**1**	1	**3**	**7**

b (2 marks)

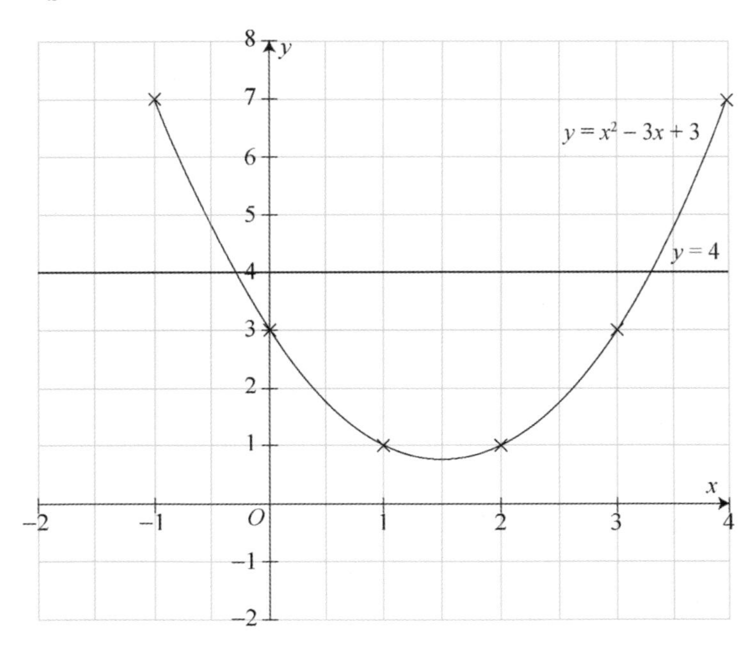

c − 0.3 or 3.3 (2 marks)

d $x^2 - 3x + 3 = 4$ (1 mark)

or $x^2 - 3x - 1 = 0$

22 a 59.0° or 121.0° (4 marks)

b **i** (2 marks)

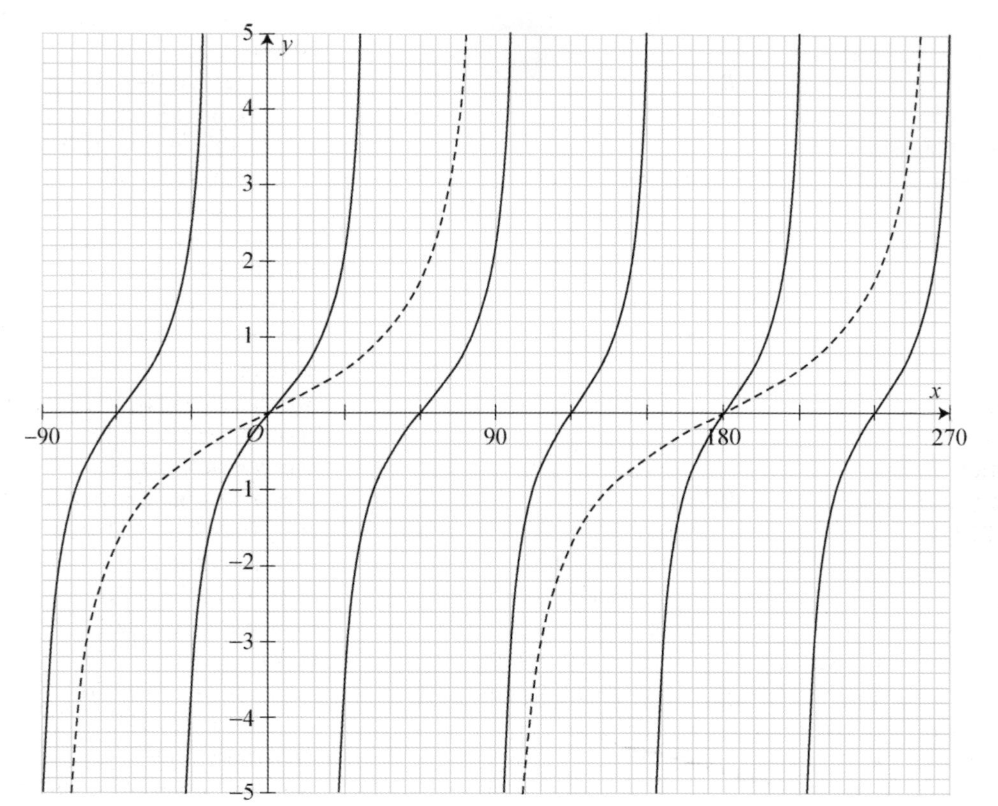

ii (2 marks)

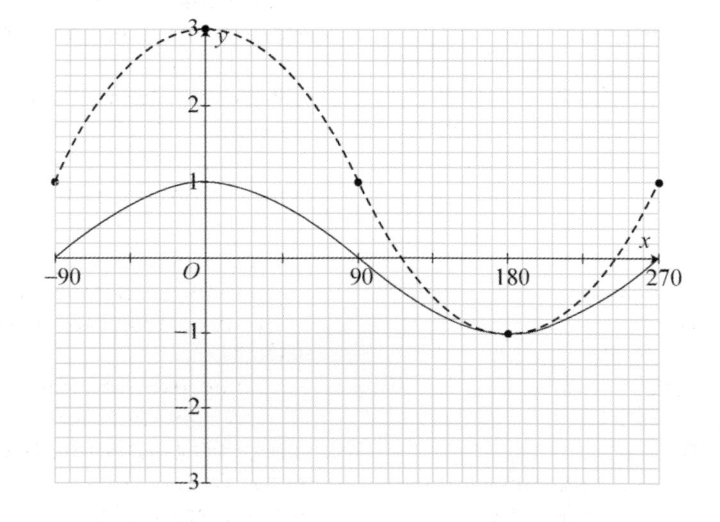

23 **a** $y = \frac{36}{x}$ (2 marks)

b $y = 4$ (1 mark)

c $x = 6$ (1 mark)